Wireless Sensing and Networking for the Internet of Things

Wireless Sensing and Networking for the Internet of Things

Editors

Zihuai Lin
Wei Xiang

MDPI • Basel • Beijing • Wuhan • Barcelona • Belgrade • Manchester • Tokyo • Cluj • Tianjin

Editors
Zihuai Lin
The University of Sydney,
Camperdown, Australia

Wei Xiang
La Trobe University,
Melbourne, Australia

Editorial Office
MDPI
St. Alban-Anlage 66
4052 Basel, Switzerland

This is a reprint of articles from the Special Issue published online in the open access journal *Sensors* (ISSN 1424-8220) (available at: https://www.mdpi.com/journal/sensors/special_issues/sensing_iot).

For citation purposes, cite each article independently as indicated on the article page online and as indicated below:

LastName, A.A.; LastName, B.B.; LastName, C.C. Article Title. *Journal Name* **Year**, *Volume Number*, Page Range.

ISBN 978-3-0365-7448-6 (Hbk)
ISBN 978-3-0365-7449-3 (PDF)

© 2023 by the authors. Articles in this book are Open Access and distributed under the Creative Commons Attribution (CC BY) license, which allows users to download, copy and build upon published articles, as long as the author and publisher are properly credited, which ensures maximum dissemination and a wider impact of our publications.

The book as a whole is distributed by MDPI under the terms and conditions of the Creative Commons license CC BY-NC-ND.

Contents

About the Editors vii

Zihuai Lin and Wei Xiang
Wireless Sensing and Networking for the Internet of Things
Reprinted from: *Sensors* **2023**, *23*, 1461, doi:10.3390/s23031461 1

Xiaosa Xu, Wen-Kang Jia, Yi Wu and Xufang Wang
On the Optimal Lawful Intercept Access Points Placement Problem in Hybrid Software-Defined Networks
Reprinted from: *Sensors* **2021**, *21*, 428, doi:10.3390/s21020428 7

Abebe Diro, Naveen Chilamkurti, Van-Doan Nguyen and Will Heyne
A Comprehensive Study of Anomaly Detection Schemes in IoT Networks Using Machine Learning Algorithms
Reprinted from: *Sensors* **2021**, *21*, 8320, doi:10.3390/s21248320 29

Ethan Chen, John Kan, Bo-Yuan Yang, Jimmy Zhu and Vanessa Chen
Intelligent Electromagnetic Sensors for Non-Invasive Trojan Detection
Reprinted from: *Sensors* **2021**, *21*, 8288, doi:10.3390/s21248288 43

Khalid Haseeb, Amjad Rehman, Tanzila Saba, Saeed Ali Bahaj and Jaime Lloret
Device-to-Device (D2D) Multi-Criteria Learning Algorithm Using Secured Sensors
Reprinted from: *Sensors* **2022**, *22*, 2115, doi:10.3390/s22062115 63

Kenneth E. Schackart III and Jeong-Yeol Yoon
Machine Learning Enhances the Performance of Bioreceptor-Free Biosensors
Reprinted from: *Sensors* **2021**, *21*, 5519, doi:10.3390/s21165519 81

Muhammad Zubair Islam, Shahzad, Rashid Ali, Amir Hadier and Hyung Seok Kim
IoTactileSim: A Virtual Testbed for Tactile Industrial Internet of Things Services
Reprinted from: *Sensors* **2021**, *21*, 8363, doi:10.3390/s21248363 109

Shuangshuang Li and Wenming Cao
SEMPANet: A Modified Path Aggregation Network with Squeeze-Excitation for Scene Text Detection
Reprinted from: *Sensors* **2021**, *21*, 2657, doi:10.3390/s21082657 129

Paula Fraga-Lamas, Sérgio Ivan Lopes and Tiago M. Fernández-Caramés
Green IoT and Edge AI as Key Technological Enablers for a Sustainable Digital Transition towards a Smart Circular Economy: An Industry 5.0 Use Case
Reprinted from: *Sensors* **2021**, *21*, 5745, doi:10.3390/s21175745 145

Janis Eidaks, Romans Kusnins, Ruslans Babajans, Darja Cirjulina, Janis Semenjako and Anna Litvinenko
Fast and Accurate Approach to RF-DC Conversion Efficiency Estimation for Multi-Tone Signals
Reprinted from: *Sensors* **2022**, *22*, 787, doi:10.3390/s22030787 181

Vyacheslav Begishev, Dmitri Moltchanov, Anna Gaidamaka, and Konstantin Samouylov
Closed-Form UAV LoS Blockage Probability in Mixed Ground- and Rooftop-Mounted Urban mmWave NR Deployments
Reprinted from: *Sensors* **2022**, *22*, 977, doi:10.3390/s22030977 205

Georgios Fevgas, Thomas Lagkas, Vasileios Argyriou and Panagiotis Sarigiannidis
Coverage Path Planning Methods Focusing on Energy Efficient and Cooperative Strategies for Unmanned Aerial Vehicles
Reprinted from: *Sensors* **2022**, *22*, 1235, doi:10.3390/s22031235 . . . . **221**

Md Abdulla Al Mamun, David Vera Anaya, Fan Wu and Mehmet Rasit Yuce
Landmark-Assisted Compensation of User's Body Shadowing on RSSI for Improved Indoor Localisation with Chest-Mounted Wearable Device
Reprinted from: *Sensors* **2021**, *21*, 5405, doi:10.3390/s21165405 . . . . **241**

Rui Wang, Yue Wang, Yanping Li, Wenming Cao and Yi Yan
Geometric Algebra-Based ESPRIT Algorithm for DOA Estimation
Reprinted from: *Sensors* **2021**, *21*, 5933, doi:10.3390/s21175933 . . . . **271**

Zaheer Allam, Simon Elias Bibri, David S. Jones, Didier Chabaud and Carlos Moreno
Unpacking the '15-Minute City' via 6G, IoT, and Digital Twins: Towards a New Narrative for Increasing Urban Efficiency, Resilience, and Sustainability
Reprinted from: *Sensors* **2022**, *22*, 1369, doi:10.3390/s22041369 . . . . **287**

Dinesh Tamang, Alessandro Pozzebon, Lorenzo Parri, Ada Fort and Andrea Abrardo
Designing a Reliable and Low-Latency LoRaWAN Solution for Environmental Monitoring in Factories at Major Accident Risk
Reprinted from: *Sensors* **2022**, *22*, 2372, doi:10.3390/s22062372 . . . . **305**

Pisana Placidi, Renato Morbidelli, Diego Fortunati, Nicola Papini, Francesco Gobbi and Andrea Scorzoni
Monitoring Soil and Ambient Parameters in the IoT Precision Agriculture Scenario: An Original Modeling Approach Dedicated to Low-Cost Soil Water Content Sensors
Reprinted from: *Sensors* **2021**, *21*, 5110, doi:10.3390/s21155110 . . . . **323**

Yonghui Tu, Haoye Tang and Wenyou Hu
An Application of a LPWAN for Upgrading Proximal Soil Sensing Systems
Reprinted from: *Sensors* **2022**, *22*, 4333, doi:10.3390/s22124333 . . . . **351**

Xue Han, Zihuai Lin, Cameron Clark, Branka Vucetic and Sabrina Lomax
AI Based Digital Twin Model for Cattle Caring
Reprinted from: *Sensors* **2022**, *22*, 7118, doi:10.3390/s22197118 . . . . **371**

About the Editors

Zihuai Lin

Zihuai Lin received a Ph.D. degree in Electrical Engineering from the Chalmers University of Technology, Sweden, in 2006. Prior to this, he worked at Ericsson Research, Stockholm, Sweden. Following his Ph.D. graduation, he worked as an Associate Professor at Aalborg University, Denmark. He is currently an Associate Professor at the School of Electrical and Information Engineering at the University of Sydney, Australia. His research interests include IoT wireless sensing and networking, 5G/6G cellular systems, IoT in healthcare, TeraHertz communications, see-through wall radar imaging, Ghost Imaging, wireless artificial intelligence (AI), AI-based ECG/EEG signal analysis, information theory, communication theory, source/channel/network coding, coded modulation, MIMO, OFDMA, SC-FDMA, radio resource management, cooperative communications, small-cell networks and others.

Wei Xiang

Professor Wei Xiang is Cisco Research Chair of AI and IoT, and the Director of the Cisco La Trobe Centre for AI and IoT at La Trobe University. Previously, he was the Foundation Chair and Head of the Discipline of IoT Engineering at James Cook University, Cairns, Australia. Due to his instrumental leadership in establishing Australia's first accredited Internet of Things Engineering degree program, he was inducted into Pearcy Foundation's Hall of Fame in October 2018. He is a TEDx speaker and an elected Fellow of the IET in UK and Engineers Australia. He received the TNQ Innovation Award in 2016, the Pearcey Entrepreneurship Award in 2017, and Engineers Australia Cairns Engineer of the Year in 2017. He was a co-recipient of four Best Paper Awards at WiSATS'2019, WCSP'2015, IEEE WCNC'2011, and ICWMC'2009. He has been awarded several prestigious fellowship titles. He was the Vice Chair of the IEEE Northern Australia Section from 2016 to 2020. He is currently an Associate Editor for IEEE Communications Surveys and Tutorials, IEEE Transactions on Vehicular Technology, IEEE Internet of Things Journal, IEEE Access and the Nature journal of Scientific Reports. He has published over 300 peer-reviewed papers, including three books and 220 journal articles. He has severed in a large number of international conferences in the capacity of General Co-Chair, TPC Co-Chair, Symposium Chair, etc. His research interests include the Internet of Things, wireless communications, machine learning for IoT data analytics and computer vision.

Editorial

Wireless Sensing and Networking for the Internet of Things

Zihuai Lin [1,*] and Wei Xiang [2,*]

[1] School of Electrical & Information Engineering, University of Sydney, Camperdown, NSW 2006, Australia
[2] School of Engineering and Mathematical Sciences, La Trobe University, Melbourne, VIC 3086, Australia
* Correspondence: zihuai.lin@sydney.edu.au (Z.L.); W.Xiang@latrobe.edu.au (W.X.)

Citation: Lin, Z.; Xiang, W. Wireless Sensing and Networking for the Internet of Things. *Sensors* **2023**, *23*, 1461. https://doi.org/10.3390/s23031461

Received: 7 December 2022
Accepted: 9 December 2022
Published: 28 January 2023

Copyright: © 2023 by the authors. Licensee MDPI, Basel, Switzerland. This article is an open access article distributed under the terms and conditions of the Creative Commons Attribution (CC BY) license (https://creativecommons.org/licenses/by/4.0/).

In recent years, we have witnessed the exponential proliferation of the Internet of Things (IoT)-based networks of physical devices, vehicles, and appliances, as well as other items embedded with electronics, software, sensors, actuators, and connectivity, which enable these objects to connect and exchange data. Facilitating the introduction of highly efficient IoT, wireless sensing, and network technologies will reduce the need for traditional processes that must currently be manually carried out, thus freeing up the precious resources of a dwindling workforce, and informing more meaningful and necessarily human-centered work.

This Special Issue aims to collate innovative developments in areas relating to IoT, wireless sensing, and networking. The eighteen papers published in this Special Issue cover software-defined network (SDN)-based IoT networks, artificial intelligence (AI) for IoT, industrial IoT, smart sensors, energy efficiency optimization for IoT and wireless sensor networks, IoT applications for agriculture, smart cities, healthcare, localization, and environment monitoring.

In [1], an IoT network with intercept access points (IAPs), SDN nodes, and non-SDN nodes was developed for the purpose of lawful interception. Different from traditional networks with centralized management, this paper optimized the deployment of IAPs in hybrid software-defined networks containing both SDN and non-SDN nodes. This work presented an enhanced equal-cost multi-path shortest-path algorithm for IAP deployment and three SDN interception models in accordance. In addition, the authors proposed the use of a restriction minimal vertex cover algorithm (RMVCA) in hybrid SDN nodes to consider the geographic importance of all intercepted targets and the global cost of operator operations and maintenance. By applying a variety of SDN interception algorithms based on the RMVCA to actual network topologies, the authors were able to significantly optimize the deployment efficiency of IAPs and improve the intercept link coverage in hybrid SDN nodes, as well as reasonably deploy the best intercept access point and intercept the whole hybrid SDN with the fewest SDN nodes, thereby aiding in the introduction of lawful interception.

The second paper [2] developed anomaly detection methods by utilizing machine learning to safeguard an IoT system. The authors provided a thorough analysis of prior work in creating machine-learning-based anomaly detection methods for safeguarding IoT systems. Additionally, they claimed that blockchain-based systems used for anomaly detection are capable of jointly building efficient machine learning models for anomaly detection.

The authors of [3] outlined a comprehensive self-testing method that used energy-efficient learning modules and nanoscale electromagnetic (EM) sensing devices to identify security concerns and malicious attacks at the front-end sensors. The development of a built-in threat detection method employing intelligent EM sensors dispersed on the power lines was proven to facilitate the efficient use of energy while detecting unusual data activity without compromising performance. Energy-constrained wireless devices may also be able to have an on-chip detection system to quickly foresee hostile attacks on the front lines due to the minimal energy and space usage.

Ref. [4] introduced a D2D multi-criteria learning technique for secured IoT networks to enhance data exchange without adding extra fees or data diversions for mobile sensors. Additionally, machine learning was shown to lower the risks of compromise in the presence of anonymous devices and increase the reliability of the IoT-enabled communication system. Broad simulation-based experiments were also used to evaluate and assess the proposed work, showing significantly better performance for realistic network topologies in terms of packet delivery ratio, packet disruptions, data delays, energy consumption, and computing complexity.

The authors of [5] demonstrated how machine learning can improve the functionality of biosensors without biological receptors. The performance of these biosensors was enhanced by machine learning, which effectively substitutes modeling for the bioreceptor to increase specificity. Since their introduction, simple regression models have been commonly used in biosensor-related fields to determine analyte compositions based on the biosensor's signal strength. Traditionally, bioreceptors offer good sensitivity and specificity to the biosensor. However, a growing number of biosensors without bioreceptors have been created for a variety of purposes. The usage of ML for imaging, E-nose and E-tongue, and surface-enhanced Raman spectroscopy (SERS) biosensors was discussed in this study. It is also particularly noteworthy that several artificial neural network (ANN) methods paired with principal component analysis (PCA), support vector machine (SVM), and other algorithms performed remarkably in a variety of tasks.

The authors of [6] stressed the exigency of using a virtual testbed dubbed IoTactileSim to implement, investigate, and manage QoS provisioning in tactile industrial IoT (IIoT) services. The study demonstrated that tactile IIoT enables the real-time control and manipulation of remote industrial environments via a human operator. The authors also showed that a communication network with ultra-low latency, ultra-high reliability, availability, and security is required by TIoT application cases. Furthermore, it has become more difficult to research and enhance the quality of services (QoSs) for tactile IIoT applications due to the absence of the tactile IIoT testbed. IoTactileSim uses the robotic simulator CoppeliaSim and network emulator Mininet to carry out real-time haptic teleoperations in both virtual and actual surroundings. This allows the real-time monitoring of network impairments, operators, and teleoperator data flow, as well as various implemented technology parametric values.

In [7], a novel feature fusion-based approach to scene text detection was created. Rather than solely relying on feature extraction from SENet, this technique incorporated MPANet's features to make up the difference. By using the suggested fusion technique, the text detection model could achieve better detection performance than the baseline network. In addition, the model was post-processed with a progressive expansion technique to provide rapid and precise text detection. This method was shown to be important for in studying natural scene text detection technology that is oriented toward actual application scenarios because it aims to improve experimental results without introducing end-to-end networks with too many parameters, and it ultimately achieves high accuracy and fast text detection.

The energy-efficient design of IoT is a very challenging topic. As mentioned in [8], although IoT technologies and paradigms such as edge computing have enormous potential for the digital transition towards sustainability, they do not yet contribute to the IoT industry's sustainable development. Due to its use of scarce raw materials and its energy consumption in manufacturing, operation, and recycling processes, this industry has a substantial carbon footprint. To address these challenges, the green IoT (G-IoT) paradigm was developed as a study field to lower this carbon footprint; nevertheless, its sustainable vision directly clashes with the arrival of edge artificial intelligence (edge AI), which mandates the use of additional energy. The authors of [8] addressed this issue by investigating various factors that influence the design and development of edge AI G-IoT systems. In addition, their study provided an Industry 5.0 use case that highlights the various principles that were discussed. In particular, the proposed scenario involved an

Industry 5.0 smart workshop that aims to improve operator safety and operation tracking, employing a mist computing architecture built of IoT nodes with AI capabilities.

For the energy harvesting of IoT in paper [9], a fast and accurate numerical method was given to determine the RF–DC power conversion efficiency (PCE) of energy harvesting circuits in the case of power-carrying signals with multiple tones and periodic envelopes. In recent years, extensive research has been conducted on this kind of signal. For low-to-medium input power levels, their use was shown to produce a potentially higher PCE than the usual sine wave signal. Because of this, the authors wanted to devise a fast and accurate two-frequency harmonic balance method (2F-HB) because a fast PCE calculation could speed up the process of optimizing the converter circuit by a lot. A comparison study was conducted to show how well the 2F-HB works when it comes to computing. The results of [9] show that the 2F-HB performs much better than widely used methods such as the transient analysis (TA) method, the harmonic balance method (HB), and the multidimensional harmonic balance method (MHB). This method also proved to be more effective than Keysight ADS, a commercial non-linear circuit simulator that uses both HB and MHB. The proposed method could also be easily added to commercially available non-linear circuit simulation software, such as Keysight ADS and Ansys HFSS, as used by many people.

Unmanned aerial vehicles (UAVs) represent one of the new types of devices that use 5G and 6G networks. One possible way of supporting advanced services for UAVs, such as video monitoring, is to use the recently standardized millimeter-wave (mmWave) frequency band for new radio (NR) technology. However, buildings may cause frequent outages if they block the paths between NR base stations (BSs) and UAVs. In [10], the authors used the tools of integral geometry to describe the connectivity properties of UAVs in terrestrial urban deployments of mm-wave NR systems. The main metric of interest is the likelihood of UAV line-of-sight (LoS) blockage. Unlike other studies, the proposed approach made it possible to obtain a close approximation of the likelihood of line-of-sight blockage as a function of city and network deployment parameters.

In another review [11], early-stage coverage path planning (CPP) methods were presented in the robotics field. The objective of CPP algorithms is to reduce the overall coverage path and execution time. Significant research has been conducted in the field of robotics, particularly in the areas of multi-unmanned unmanned aerial vehicle (UAV) collaboration and energy efficiency in CPP challenges. In addition, this paper also addressed multi-UAV CPP techniques and focused on CPP algorithms that conserve energy.

In [12], the authors investigated a method used to mitigate the user's body shadowing effect on the RSSI to improve localization accuracy. They also examined the effect of the user's body on the RSSI. The idea of a landmark was then used to develop an angle estimate method. An inertial measurement unit (IMU)-aided decision tree-based motion mode classifier was used to accurately identify different landmarks. A compensation strategy was then proposed to fix the RSSI. The closest neighbor method was used to estimate the unknown location. The results show that the suggested system can greatly increase localization accuracy. After adjusting for the body effect, a median localization accuracy of 1.46 m was attained, compared to 2.68 m before the compensation using the traditional K-nearest neighbor approach. Additionally, when comparing the suggested system's performance to that of the two other relevant works, it clearly surpassed the competition. By using a weighted K-nearest neighbor approach, the median accuracy was further increased to 0.74 m.

Direction-of-arrival (DOA) estimation is integral in array signal processing, and the estimating signal parameter via rotational invariance techniques (ESPRIT) algorithm is one of the typical super-resolution algorithms used for finding directions in an electromagnetic vector sensor (EMVS) array. However, existing ESPRIT algorithms treat the output of the EMVS array as a "long vector", which leads to a loss of signal orthogonality. Ref. [13] proposed a geometric algebra-based ESPRIT algorithm (GA-ESPRIT) to estimate 2D-DOA with double parallel uniform linear arrays. The approach integrated GA with ESPRIT to describe multidimensional signals holistically. Direction angles were determined by different GA matrix operations to retain correlations among EMVS components. Experimental results show that GA-ESPRIT is robust to model mistakes and requires less time and memory.

The '15-min city' concept offers new perspectives on livability and urban health in post-pandemic cities. Smart city network technologies can offer personalized pathways to respond to contextualized difficulties through data mining and processing to better enhance urban decision-making processes. The authors of [14] argued that digital twins, IoT, and 6G can benefit from the '15-min city' concept. The data collected by these devices and analyzed by machine learning reveal urban fabric patterns. Unpacking these dimensions to support the '15-min city' notion can illuminate new ways of redefining agendas to better respond to economic and societal requirements and line with environmental commitments, including UN Sustainable Development Goal 11 and the New Urban Agenda. This study argued that these new connectivities should be examined so that relevant protocols can be created and urban agendas can be recalibrated to prepare for impending technology breakthroughs, offering new avenues for urban regeneration and resilience crafting.

Environment monitoring is one of the commonly used IoT applications. Ref. [15] proposed a low-latency LoRaWAN system for environmental monitoring in factories at major accident risk (FMARs). Low-power wearable devices for sensing dangerous inflammable gases in industrial plants are meant to reduce hazards and accidents. Detected data must be provided immediately and reliably to a remote server to trigger preventive steps and then optimize the functioning of a machine. In these scenarios, the LoRaWAN system is the best connectivity technology due to off-the-shelf hardware and software. The authors examined LoRaWAN's latency and reliability restrictions and proposed a strategy to overcome them. The suggested solution also used downlink control packets to synchronize ED transmissions (DCPs). These experiments validated the proposed technique for the FMAR scenario.

For low-cost IoT precision agriculture applications such as greenhouse sensing and actuation, the authors of [16] created a LoRaWAN-based wireless sensor network with low power consumption. All of the research's subsystems were entirely constructed using only commercially available components and freely available or open-source software components and libraries. This entire system was established to demonstrate the possibility of creating a modular system using low-cost commercially available components for sensing purposes. The data generated by the experiments were compiled and kept in a database maintained by a cloud-based virtual computer. Using a graphical user interface, the user had the ability to observe the data in real time. In a series of experiments conducted with two types of natural soil, loamy sand and silty loam, the overall system's dependability was demonstrated. The system's performance in terms of soil characteristics was then compared to that of a Sentek reference sensor. Temperature readings indicate good agreement within the rated accuracy of the implemented sensors, whereas readings from the inexpensive volumetric water content (VWC) sensor revealed variable sensitivity. The authors made several conclusions using a unique approach to maximize the parameters of the non-linear fitting equation connecting the inexpensive VWC sensor's analog voltage output with the standard VWC.

The authors of [17] integrated LPWAN technology to an existing proximate soil sensor device by building an attachment hardware system (AHS) and accomplishing technical upgrades for low-cost, low-power, wide-coverage, and real-time soil monitoring in fields. The testing results demonstrate that, after upgrading, the sensor device can run for several years with only a battery power supply, and that the effective wireless communication coverage is nearly 1 km in a typical suburban farming context. As a result, the gadget not only keeps the sensor device's original mature sensing technology, but also displays ultra-low power consumption and long-distance transmission. The proposed method also serves as a model for extending LPWAN technology to a broader spectrum of inventoried sensor devices for technical advancements.

The final paper [18] of this Special Issue focused on digital twins for cattle care. The authors established cutting-edge artificial-intelligence-powered digital twins of cattle status in this research (AI). The project was based on an IoT farm system that can record and monitor the health of livestock from a distance. The sensor data obtained from the farm IoT system was used to create a digital twin model of cattle based on deep learning (DL). It was shown that the real-time monitoring of the physiological cycle of cattle is possible, and by applying this model, the next physiological cycle of cattle can be predicted. An enormous amount of data to confirm the accuracy of the digital twins model acted as the foundation of this effort. The loss error of training for this digital twin model, predicting the future behavioral state of cattle, was approximately 0.580, and the loss error of doing so after optimization was approximately 5.197. This work's digital twins model could be used to predict the cattle's future time budget.

Conflicts of Interest: The authors declare no conflict of interest.

References

1. Xu, X.; Jia, W.; Wu, Y.; Wang, X. On the Optimal Lawful Intercept Access Points Placement Problem in Hybrid Software-Defined Networks. *Sensors* **2021**, *21*, 428. [CrossRef] [PubMed]
2. Diro, A.; Chilamkurti, N.; Nguyen, V.; Heyne, W. A Comprehensive Study of Anomaly Detection Schemes in IoT Networks Using Machine Learning Algorithms. *Sensors* **2021**, *21*, 8320. [CrossRef] [PubMed]
3. Chen, E.; Kan, J.; Yang, B.; Zhu, J.; Chen, V. Intelligent Electromagnetic Sensors for Non-Invasive Trojan Detection. *Sensors* **2021**, *21*, 8288. [CrossRef] [PubMed]
4. Haseeb, K.; Rehman, A.; Saba, T.; Bahaj, S.; Lloret, J. Device-to-Device (D2D) Multi-Criteria Learning Algorithm Using Secured Sensors. *Sensors* **2022**, *22*, 2115. [CrossRef] [PubMed]
5. Schackart, K.; Yoon, J. Machine Learning Enhances the Performance of Bioreceptor-Free Biosensors. *Sensors* **2021**, *21*, 5519. [CrossRef] [PubMed]
6. Zubair Islam, M.; Shahzad, Ali, R.; Haider, A.; Kim, H. IoTactileSim: A Virtual Testbed for Tactile Industrial Internet of Things Services. *Sensors* **2021**, *21*, 8363. [CrossRef] [PubMed]
7. Li, S.; Cao, W. SEMPANet: A Modified Path Aggregation Network with Squeeze-Excitation for Scene Text Detection. *Sensors* **2021**, *21*, 2657. [CrossRef] [PubMed]
8. Fraga-Lamas, P.; Lopes, S.; Fernández-Caramés, T. Green IoT and Edge AI as Key Technological Enablers for a Sustainable Digital Transition towards a Smart Circular Economy: An Industry 5.0 Use Case. *Sensors* **2021**, *21*, 5745. [CrossRef] [PubMed]
9. Eidaks, J.; Kusnins, R.; Babajans, R.; Cirjulina, D.; Semenjako, J.; Litvinenko, A. Fast and Accurate Approach to RF-DC Conversion Efficiency Estimation for Multi-Tone Signals. *Sensors* **2022**, *22*, 787. [CrossRef] [PubMed]
10. Begishev, V.; Moltchanov, D.; Gaidamaka, A.; Samouylov, K. Closed-Form UAV LoS Blockage Probability in Mixed Ground- and Rooftop-Mounted Urban mmWave NR Deployments. *Sensors* **2022**, *22*, 977. [CrossRef] [PubMed]
11. Fevgas, G.; Lagkas, T.; Argyriou, V.; Sarigiannidis, P. Coverage Path Planning Methods Focusing on Energy Efficient and Cooperative Strategies for Unmanned Aerial Vehicles. *Sensors* **2022**, *22*, 1235. [CrossRef] [PubMed]
12. Mamun, M.; Anaya, D.; Wu, F.; Yuce, M. Landmark-Assisted Compensation of User's Body Shadowing on RSSI for Improved Indoor Localisation with Chest-Mounted Wearable Device. *Sensors* **2021**, *21*, 5405. [CrossRef] [PubMed]
13. Wang, R.; Wang, Y.; Li, Y.; Cao, W.; Yan, Y. Geometric Algebra-Based ESPRIT Algorithm for DOA Estimation. *Sensors* **2021**, *21*, 5933. [CrossRef] [PubMed]
14. Allam, Z.; Bibri, S.; Jones, D.; Chabaud, D.; Moreno, C. Unpacking the '15-Minute City' via 6G, IoT, and Digital Twins: Towards a New Narrative for Increasing Urban Efficiency, Resilience, and Sustainability. *Sensors* **2022**, *22*, 1369. [CrossRef]
15. Tamang, D.; Pozzebon, A.; Parri, L.; Fort, A.; Abrardo, A. Designing a Reliable and Low-Latency LoRaWAN Solution for Environmental Monitoring in Factories at Major Accident Risk. *Sensors* **2022**, *22*, 2372. [CrossRef]
16. Placidi, P.; Morbidelli, R.; Fortunati, D.; Papini, N.; Gobbi, F.; Scorzoni, A. Monitoring Soil and Ambient Parameters in the IoT Precision Agriculture Scenario: An Original Modeling Approach Dedicated to Low-Cost Soil Water Content Sensors. *Sensors* **2021**, *21*, 5110. [CrossRef] [PubMed]
17. Tu, Y.; Tang, H.; Hu, W. An Application of an LPWAN for Upgrading Proximal Soil Sensing Systems. *Sensors* **2022**, *22*, 4333. [CrossRef]
18. Han, X.; Lin, Z.; Clark, C.; Vucetic, B.; Lomax, S. AI-Based Digital Twin Model for Cattle Caring. *Sensors* **2022**, *22*, 7118. [CrossRef] [PubMed]

Disclaimer/Publisher's Note: The statements, opinions and data contained in all publications are solely those of the individual author(s) and contributor(s) and not of MDPI and/or the editor(s). MDPI and/or the editor(s) disclaim responsibility for any injury to people or property resulting from any ideas, methods, instructions or products referred to in the content.

Article

On the Optimal Lawful Intercept Access Points Placement Problem in Hybrid Software-Defined Networks

Xiaosa Xu, Wen-Kang Jia *, Yi Wu and Xufang Wang

Fujian Provincial Engineering Technology Research Center of Photoelectric Sensing Application, Key Laboratory of OptoElectronic Science and Technology for Medicine of Ministry of Education, College of Photonic and Electronic Engineering, Fujian Normal University, Fuzhou 350007, China; xiaosaxu521@163.com (X.X.); wuyi@fjnu.edu.cn (Y.W.); fzwxf@fjnu.edu.cn (X.W.)
* Correspondence: wkjia@fjnu.edu.cn

Citation: Xu, X.; Jia, W.-K.; Wu, Y.; Wang, X. On the Optimal Lawful Intercept Access Points Placement Problem in Hybrid Software-Defined Networks. *Sensors* **2021**, *21*, 428. https://doi.org/10.3390/s21020428

Received: 16 December 2020
Accepted: 4 January 2021
Published: 9 January 2021

Publisher's Note: MDPI stays neutral with regard to jurisdictional claims in published maps and institutional affiliations.

Copyright: © 2021 by the authors. Licensee MDPI, Basel, Switzerland. This article is an open access article distributed under the terms and conditions of the Creative Commons Attribution (CC BY) license (https://creativecommons.org/licenses/by/4.0/).

Abstract: For the law enforcement agencies, lawful interception is still one of the main means to intercept a suspect or address most illegal actions. Due to its centralized management, however, it is easy to implement in traditional networks, but the cost is high. In view of this restriction, this paper aims to exploit software-defined network (SDN) technology to contribute to the next generation of intelligent lawful interception technology, i.e., to optimize the deployment of intercept access points (IAPs) in hybrid software-defined networks where both SDN nodes and non-SDN nodes exist simultaneously. In order to deploy IAPs, this paper puts forward an improved equal-cost multi-path shortest path algorithm and accordingly proposes three SDN interception models: T interception model, ECMP-T interception model and Fermat-point interception model. Considering the location relevance of all intercepted targets and the operation and maintenance cost of operators from the global perspective, by the way, we further propose a restrictive minimum vertex cover algorithm (RMVCA) in hybrid SDN. Implementing different SDN interception algorithms based RMVCA in real-world topologies, we can reasonably deploy the best intercept access point and intercept the whole hybrid SDN with the least SDN nodes, as well as significantly optimize the deployment efficiency of IAPs and improve the intercept link coverage in hybrid SDN, contributing to the implementation of lawful interception.

Keywords: lawful interception; hybrid SDN; intercept access point; minimum vertex cover

1. Introduction

National security and social stability, in today's world, have been shaken by some security threats such as terrorist attacks, cybercrime and information warfare. For the law enforcement agencies (LEAs; L), therefore, lawful interception (LI) is still one of the main means to intercept a suspect or address these illegal actions at present. As we all know, lawful interception is a kind of data acquisition of communication network based on lawful authorization for the purpose of analysis or evidence collection. Thus, it allows the law enforcement agencies with court orders or other legitimate authorities to selectively eavesdrop on individual users. Most countries require those licensed telecom operators to provide legitimate interception gateways and nodes on their networks for communication interception. To deploy the gateways and nodes in legacy networking where traditional gateways or nodes rely on dedicated devices and backhaul links to intercept network traffic, however, leads to unimaginable cost. On the contrary, software-defined networking (SDN) [1], different from the traditional networking, can simplify the traditional network' architecture [2] and thus enable efficient management and centralized control [3] for intercepting network traffic at an extremely low cost because of its property of software definition with OpenFlow protocol [4]. The deployment of SDNs, however, is not a one-step process, but a long process, namely, in the wake of the increasing deployment of SDNs [5], a situation where both SDN nodes and non-SDN (N-SDN) nodes exist simultaneously is formed gradually. Therefore, it is of great significance to study how to design a brand-new network information lawful interception system architecture based on the software-defined network (SDN) technology and to discuss its challenges such as the deployment of intercept access point (IAPs), route selection of intercept, the minimum cost of intercept, the minimum number of intercept access points etc. in a hybrid SDN.

In this paper, we propose the deployment and optimization strategy of intercept access points, which includes single intercept access point selection, the shortest route optimization algorithm between three points, the minimum intercept traffic cost algorithm, and the restrictive minimum vertex cover algorithm.

The problem of single intercept access point selection is the shortest path problem that is to solve the shortest path between two given vertices in a weighted graph. At this time, the shortest path not only refers to the shortest path in the sense of pure distance, but also in the sense of economic distance, time and network. In this paper, the cost of shortest path between two points can refer to hop-count, traffic, transmission delay, transmission bandwidth, energy consumption etc. As is known to all, Dijkstra Algorithm [6] is the most typical single source shortest path algorithm, which is used to calculate the shortest path from one node to all nodes, and yet not all equal-cost multi-path shortest path. Meanwhile, Li [7] proposed an improved Dijkstra Algorithm that can find most of the shortest paths using the initial shortest path set through applying for concept of precursor node but cannot find all shortest paths. Moreover, a lot of related work with respect to the shortest path have been done by [8–14] in various fields.

In view of this, we develop an improved equal-cost multi-path shortest path algorithm (i.e., ECMP-Dijkstra) which can find all shortest paths between the source (S) and the destination (D), and accordingly put forward three SDN interception models based on ECMP-Dijkstra Algorithm in hybrid SDN. The three SDN interception models can be viewed as a cost-effective three-point shortest path algorithm with low time and space complexity, and thus can be used to deploy the best intercept access point reasonably in hybrid SDN.

The optimization of traffic engineering in hybrid SDN, like [15–17], is also one of our focuses. This study mainly concerns with the best transmission quality of intercepted data, the minimum cost of returning intercepted data to the interception center (i.e., LEA; L), the total traffic in global network, the transmission quality of traffic normally accepted by users when deploying intercept access points.

In reality, the deployment of intercept access points in the Internet does not simply corresponds to the micro perspective of a single data flow between three points. There is a very dynamic and complex traffic matrix [18] relationship and interactive influence among hundreds of millions of nodes in the large-scale Internet. A certain intercept access point (IAP; I) can meet the demand of traffic between S-D (from S to D) path, but there are also tens of millions of other traffic between intercept target node pairs, which may also flow through I node at the same time. Therefore, it is very important to select the deployment location of intercept access point, which must occupy the hub position, and greatly covers all intercepted traffic and must go through the critical path. For this reason, the location relevance of all intercepted targets and the operation and maintenance cost of operators must be taken into consideration from the global perspective, and thus the deployment problem of intercept access points is viewed as the minimum vertex cover problem (MVCP) that is NP-complete [19] to find its solution.

A lot of investigations have been done on MVCP in theory and applications for the last several decades [20–22]. Some parameterized algorithms about MVCP have been applied in biochemistry [23,24]. Moreover, the optimal approximation algorithm for MVCP have been proposed in [25–30]. Authors in [25–30] proposed the approximate optimization algorithm for MVCP by using the concept of degree.

Referring to their proposed algorithm, we develop a restrictive minimum vertex cover algorithm (RMVCA) in hybrid SDN networks to optimize the deployment efficiency of IAPs and to improve the link coverage of the whole interception system.

The ultimate aim of this paper is to contribute to the theory of lawful interception technology, the development of Internet and national security. In summary, the main contributions of this paper are as follows:

- To solve the problem of single intercept access point selection and routing between three points, we develop an improved equal-cost multi-path shortest path algorithm (i.e., ECMP-Dijkstra) and accordingly put forward three SDN interception models (e.g., T model, ECMP-T model and Fermat-point model) to deploy the best intercept access point reasonably in hybrid SDN, realizing the effective deployment of intercept access point in lawful interception system.
- Considering the location relevance of all intercepted targets and the operation and maintenance cost of operators of the whole interception system, we proposed a restrictive minimum vertex cover algorithm (RMVCA) to intercept the whole interception system with the least SDN nodes, optimize the deployment efficiency and improve the intercept link coverage for the whole interception system when deploying IAPs.
- Based RMVCA, we put forward three approaches PA, RA, and HA for experiments, and study and analyze the impact of different approaches on the efficiency of deploying intercept access points and on the intercept link coverage in hybrid SDN, to seek out the best RMVCA approach.
- We study and analyze the impact of different SDN interception models on various performance metrics of lawful interception system by using three real-world topologies, to seek out the best interception model.

In this paper, we first analyze various SDN interception models in hybrid software-defined networks and propose their algorithms, and then develop a restricted minimum vertex coverage algorithm from a global perspective. Extensive simulation results based on real-world network topology show that RMVCA can significantly improve network interception link coverage and deployment efficiency of IAPs of whole interception system, and that the performance metrics of the interception system are the best when Fermat-point interception model is adopted.

The remainder of this article is structured as follows. Section 2 surveys relevant work and Section 3 presents ECMP-Dijkstra Algorithm and SDN interception models. We propose the RMVCA in Section 4, followed by the performance evaluation of RMVCA and SDN interception models in Section 5. Then, Section 6 concludes the paper.

2. Related Works

Table 1 presents comparisons between our proposed and the related works according to different parameters.

Table 1. Comparisons of related works.

Lawful Interception	[31]	[32]	Our Proposed
SDN based	No	Yes	Yes
Cost	Very High	Low	Low
The shortest path algorithm	[6]	[7–14]	Our proposed
Time-Space complexity	Low	Medium	Medium-Low
The number of ECMP	One	The most	All
Minimum Vertex Cover	[33]	[20–30]	Our proposed
Time-Space complexity	Low	Medium	Medium-Low
Results	Very bad	Near-optimal	Near-optimal

2.1. Lawful Interception (LI) and Hybrid Software-Defined Networks (H-SDNs)

With the dramatic development of the Internet, an increasing number of people commit crimes on the Internet, and criminal activities are extremely rampant, which seriously affect people's security and national stability. Thus, lawful interception (LI) is still one of the momentous means for the law enforcement agencies (LEAs) to maintain national security, crack down on crime and prevent cybercrime. For interception system, an intercept device is installed to intercept network traffic, and copies it back to LEAs, and then carries out identifying and analyzing by manual or machine. With the development of new network technology and the continuous increase of network traffic, it is a more and more common and difficult task to carry out lawful interception on the Internet [31] for helping tracking culprits and to understand the nature and behavior of current Internet traffic.

With the development of SDN technology, legacy Ether-net switches are gradually migrating to SDN, and this process is harmless [32]. Although the emerging SDN networks that provides programmability to networks can have an improvement in implementing traffic engineering (TE), management departments still hesitate to deploy SDN fully because of various reasons such as budget constraints, risk considerations as well as service level agreement (SLA) guarantees. This results in developing SDN network incrementally, i.e., to deploy the SDN network only through migrating fewer SDN switches in legacy network, thus, to form the hybrid SDN networks (H-SDNs). H-SDN network provides a coexistence and cooperation environment for N-SDN nodes and SDN nodes, which brings many benefits to traditional IP networks. For the near-optimal performance of traffic engineering, therefore, it is crucial to maximize the benefits of SDN with minimal SDN deployment. Therefore, it is imperative to deploy SDN intercept access point in a hybrid SDN (H-SDN) network where SDN nodes (routers) and legacy nodes coexist and operate in perfect harmony, realizing lawful interception. In H-SDN, the links between SDN nodes and between SDN nodes and N-SDN nodes can be intercepted (i.e., SDN links), and the links between N-SDN nodes cannot be intercepted (i.e., N-SDN links) due to the lack of special equipment and dedicated return link in hybrid SDN. In other words, in the interception system based SDN, the law enforcement agencies (LEAs) do not have to set up special equipment and a dedicated line in traditional IP networks, but can intercept traffic of links through SDN intercept access point to respond to requests from the interception center, which can greatly reduce the cost of traditional special equipment and leased lines. The interception system based SDN will be no longer restricted by the bandwidth of the intercepting dedicated equipment and link. By deploying intercept access point, the interception system will have a lot of redundant links or paths to be employed to return data flow, thus, to reduce or avoid the risk of single point failure or to further guarantee the multi-path routing method.

Therefore, the deployment of SDN intercept access point in interception system is helpful to perfect the route of intercepting traffic, to make full use of Internet bandwidth resources, to improve user's quality of service, and to further optimize the performance of the whole interception system.

2.2. Dijkstra Algorithm

The most classic single source shortest path algorithm is Dijkstra Algorithm [6], which was proposed in 1956 and became well-known three years later. Dijkstra Algorithm can calculate the shortest path from one node to all nodes, yet not all equal-cost multi-path shortest path.

Many modified algorithms based on Dijkstra Algorithm are proposed in [7–14]. An improved algorithm of Dijkstra Algorithm was proposed by Li [7]. Under the concept of precursor node, Li exploited the initial shortest path set calculated by Dijkstra Algorithm, to calculate most but not all of the shortest paths. The authors of [8] improved Dijkstra Algorithm for solving three issues, such as the ineffective mechanism to digraph. In addition, the work [9] proposed some modifications on Dijkstra Algorithm and made the number of iterations less than the number of the nodes. Work [10] proposed an optimized algorithm based on Dijkstra Algorithm to optimize logistics route for the supply chain. On the other hand, the study [11] modified Dijkstra Algorithm and the modified algorithm is very of efficiency for public transport route planning. Work [12] used Dijkstra Algorithm towards shortest path computation in navigation systems for making sensible decision and time saving decisions. By the way, the study [13] improved Dijkstra Algorithm to find the maximum load path. Work [14] introduced an improved Dijkstra Algorithm for analyzing the property of 2D grid map and increased significantly the speed of Dijkstra Algorithm.

Referring to their proposed algorithms, we also improve Dijkstra Algorithm and propose an improved equal-cost multi-path shortest paths algorithm (ECMP-Dijkstra), which can calculate all equal-cost shortest paths from one node to all nodes, thus developing a cost-effective shortest path optimization algorithm between three points (i.e., S, D and L) with low time and space in hybrid SDN.

2.3. The Minimum Vertex Cover Problem (MVCP)

The traditional algorithm to solve the minimum vertex cover algorithm (MVCP) is 2-approximation [33]. This algorithm can find the set of vertex cover which is no more than twice of the optimal vertex cover, and the time complexity of the algorithm is O (E+V). More importantly, the results obtained by this algorithm are different each time, and thus may be inaccurate and not approximate solution. However, this algorithm has its advantages: every time a vertex is selected, and all the edges connected by the vertex are deleted.

The authors in [20–22] made much contribution to MVCP in theory and applications. The authors in [23,24] proposed parameterized algorithms for MVCP, and applied them in biochemistry. Work [25] proposed an improved greedy algorithm for minimum vertex cover problem, and the algorithm used the concept of degree (i.e., the number of links connected by a node) to carry out an order of degree and to select the node with the largest degree to add to the minimum vertex cover set until the degree of all nodes is 0 (i.e., the vertexes in the minimum vertex cover set has covered all the edges). Thus, the result is a very excellent approximate solution. However, the process of judging the degree of the algorithm is too complicated. Authors in [26] presented a greedy heuristic algorithm for MVCP to offer better results on dense graphs. The study [27] presented a breadth first search approach, which can get the exact result of MVCP for grid graphs. Work [28] proposed a near-optimal algorithm named MAMA to optimize the unweighted MVCP, and MAMA can return near optimal result in quick-time. Authors in [29] proposed a NHGA for MVCP to yield near-optimal solutions. In [30], authors studied an ameliorated genetic algorithm for the partial VCP to skip the local optimum by powerful vertex and adaptive mutation. All of their algorithm are based on the concept of degree.

Combining with the advantages of the above algorithms, we proposed an ameliorated restrictive minimum vertex cover algorithm (RMVCA) in hybrid SDN using the concept of degree to significantly simplify the process of degree judgment and to yield near-optimal result, thus, in the whole interception system, realizing the optimization of the deployment efficiency of IAPs and the improvement of intercept link coverage.

3. ECMP-Dijkstra Algorithm and SDN Interception Models
3.1. ECMP-Dijkstra Algorithm

When deploying the best intercept access point in hybrid SDN, we have to calculate all equivalent shortest paths between two points and then select out the best route from all equal-cost shortest paths to choose the best node as IAP. The most typical single source shortest path algorithm is Dijkstra Algorithm [6]. Accordingly, an improved algorithm of Dijkstra Algorithm was proposed by Li [7]. Under the concept of precursor node, Li exploited the initial shortest path set calculated by Dijkstra Algorithm, to calculate most but not all of the shortest paths. In view of this, on the basis of Dijkstra Algorithm and Li's Algorithm, we propose an improved equal-cost multi-path shortest paths algorithm (ECMP-Dijkstra), which can calculate all equal-cost shortest paths from one node to all nodes. The notations used in the algorithms and in the following equations are listed in Table 2.

Table 2. Notations.

Notation	Meaning
N_{SDN}	the SDN nodes selected randomly from all nodes in H-SDN
S, D, L	the source and the destination and the interception center or the LEA
I	the set of the best intercept access points
$sp_{S,D}$ or sp_{S-D}	the shortest path from node S to node D
$N_{S,D}$	the set of nodes in the shortest path $sp_{S,D}$
SN	the set of SDN nodes
i	the SDN node or SDN devices
j	the index of the j-th element of a vector
$h(i)$	the set of hop-count, $h(i)$ denotes the hop-count or cost of node i
$numh(j)$	the set of costs, $numh(j)$ denotes the cost of the j-th element
$inh(j)$	$inh(j)$ denotes the node with the index of j
$minhops$	the minimum cost or hop-count
β	the maximum index
N	the number of nodes
$hops(i,j)$ or $hops_{i-j}$	the minimum hop-count or cost from node i to node j

The pseudo code of Dijkstra Algorithm is given in Algorithm 1. We input the source node s and an undirected graph $G(V,E)$ where V denotes the set of all nodes and E denotes the set of all edges. We explain Algorithm 1 that inf denotes an infinity and sps,i denotes the shortest path from the source node s to node i. In lines 11–14, we get the minimum hop-count value minhops and the corresponding node key. In lines 15–19, we remove node key from U and then add node key to S and add node key to the shortest path sps,i to get the shortest path sps,key. Finally, we obtain the shortest path set SP from the source node s to all nodes in V.

Algorithm 1 Dijkstra Algorithm

Input: $s; G(V,E)$
Output: SP
1: $S(s) = 0; U(i) = inf, i \in V, i \neq s; SP = \emptyset$
2: $SP \leftarrow SP \cup sp_{s,s}$
3: **while** $U \neq \emptyset$ **do**
4:     $tsp = \emptyset; minhops= inf; key = None$
5:     **for** edge $e_{i,j}$ in E **do** // node $i, j \in V, i \neq j$
6:         **if** $hops(e_{i,j}) + S(j) \leq U(i)$ **then**
7:             $U(i) \leftarrow hops(e_{i,j}) + S(j)$
8:             $tsp(i) = j$
9:         **end if**
10:     **end for**
11:     $numu(k), inu(k) \leftarrow sort(U(i))$
12:     $\beta \leftarrow Num(numu(k))$
13:     $minhops= numu(\beta)$
14:     $U \leftarrow U - key$
15:     $S(key) = minhops$
16:     $S(key) = minhops$
17:     **for** shortest path $sp0_{s,i}$ in SP **do**
18:         **if** $i == tsp(key)$ **then**
19:             $sp_{s,key} \leftarrow Merge(sp_{s,i}, key)$
20:             $SP \leftarrow SP \cup sp_{s,key}$
21:         **end if**
22:     **end for**
23: **end while**
24: **return** SP

Based on Dijkstra Algorithm, we propose an improved equal-cost multi-path shortest paths algorithm (ECMP-Dijkstra) so as to calculate all equal-cost shortest paths from the source node *s* to all nodes. Detailed pseudo code of ECMP-Dijkstra Algorithm is summarized in Algorithm 2. At beginning, we input the source node *s* and the shortest path set *SP* calculated by Dijkstra Algorithm, which contains only one shortest path from *s* to all nodes. In line 1, we use the shortest path set *SP* to calculate the minimum hop-count or cost set *S* from *s* to all nodes by the function *hops()* and *S(i)* denotes the minimum cost from node *s* to node *i*. In line 5, *rsp(i)* denotes all equal-cost shortest paths from the source node *s* to the destination node *i*. We loop through the edge-set *E(ei,j)* and judge whether the hop-count or cost from node *s* to node *i* (i.e., *S(i)*) plus the hop-count of *edgei,j* equals the hop-count from node *s* to node *j* (i.e., *S(j)*). If it does, then we add node *j* to all equal-cost shortest paths from node *s* to node *i* in lines 11–12, thus obtaining multiple shortest paths from node *s* to node *j* and adding them to the shortest path set *SP* in line 13. In lines 2–13, we exploit the precursor node and the initial shortest path set *SP* repeatedly, to add equal-cost shortest paths to *SP* and thus update *SP* constantly. In line 18, we delete the duplicate shortest path from *SP* using the function *DeleteDup()*. Thus, we update the shortest path set *SP* repeatedly until the number of shortest paths in SP does not increases.

Algorithm 2 ECMP-Dijkstra Algorithm

Input: s; $G(V,E)$; SP
Output: SP
1: $S \leftarrow hops(SP)$
2: **repeat**
3:     $nSP \leftarrow Num(SP)$
4:     **for** shortest path $sp_{s,i}$ in SP **do**
5:         $rsp(i) = sp_{s,i}$
// $sp_{s,i}$ may contain more than one shortest path.
6:     **end for**
7:     **for** edge $e_{i,j}$ in E **do**
// node $i, j \in V, i \neq j$
8:         $sp0_{i,j} \leftarrow Sp(e_{i,j})$
// Convert $e_{i,j}$ to shortest path $sp0_{i,j}$.
9:         **if** shortest path $sp0_{i,j} \notin SP$ **then**
10:           **if** $S(i) + hops(sp0_{i,j}) == S(j)$ **then**
11:             **for** shortest path $sp'_{s,i}$ in $rsp(i)$ **do**
12:                 $sp'_{s,i} \leftarrow Merge(sp'_{s,i}, j)$
13:                 $SP \leftarrow SP \cup sp'_{s,i}$
14:             **end for**
15:           **end if**
16:         **end if**
17:     **end for**
18:     $SP \leftarrow DeleteDup(SP)$
19:     $nSP' \leftarrow Num(SP)$
20: **until** $nSP == nSP'$

We use three real-world topologies CRN, COST 239, NSFNet for simulation experiments, where China's 156 major railway nodes network (China Railway Network; CRN) [34] has 156 nodes and 226 links, Pan-European fiber-optic network (COST 239) [35] has 28 nodes and 41 links and T1 NSFNet network topology [36] has 14 nodes and 21 links. Under the three topologies, we compared ECMP-Dijkstra Algorithm with Dijkstra Algorithm and Li's Algorithm and the experimental results are shown in Figure 1 where TSP denotes the total number of shortest paths from one node to all nodes. Moreover, the higher the TSP, the better the intercept access points deployment may be. From the figures, we know that TSP of ECMP-Dijkstra Algorithm is higher than Dijkstra Algorithm and Li's Algorithm, thus, to deploy intercept access point reasonably.

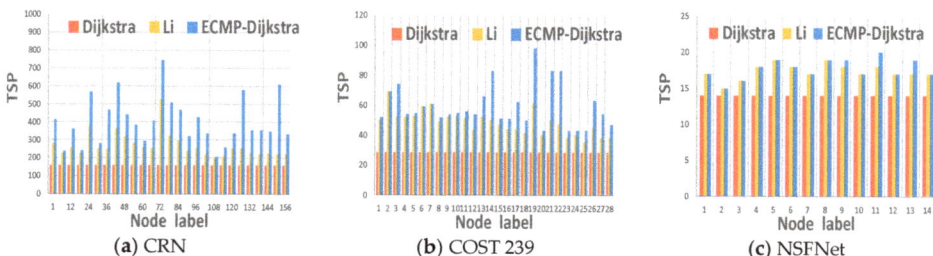

Figure 1. The impact of three shortest path algorithm on TSP in three topologies. TSP. The higher, the better.

3.2. SDN Interception Models

For lawful interception in hybrid SDN, we first need to analyze how to intercept, that is, how to deploy intercept access point between the source (S), the destination (D) and the interception center (L). In this section, we will analyze various network interception models (i.e., the deployment strategies of IAP) in hybrid SDN. The deployment of intercept access point includes the single "IAP selection problem" in the shortest path S-D (i.e., the shortest path from S to D) and its derived "the shortest path algorithm problem between three points (i.e., S, D and L)". The above two problems can be viewed as the same problem. Once the location of the intercept access point is determined, then the fourth point (IAP; I) can meet the service traffic between S, D and L. Under the condition that S-I, D-I, and L-I path are the shortest at the same time, the shortest path between three points can be solved to meet the needs of interception system.

We aim to solve the problem of selecting single intercept access point and routing between three-points, namely to deploy the best intercept access point in the shortest paths between S, D, and L. Analyzing interception models in hybrid SDN, we divide them into two interception models by the deployment location of intercept access point: legacy interception models and SDN interception models as shown in Figures 2 and 3.

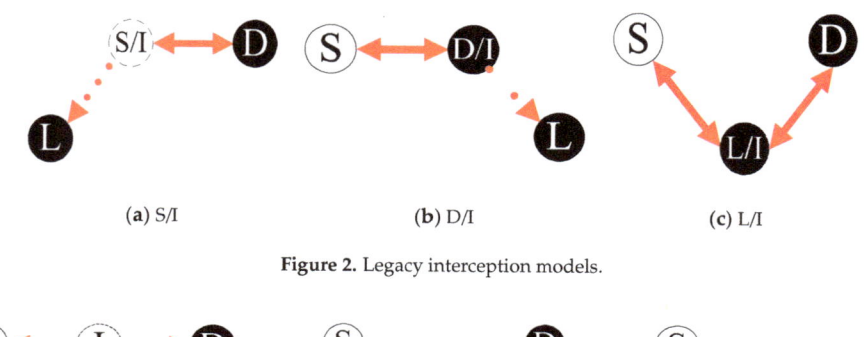

Figure 2. Legacy interception models.

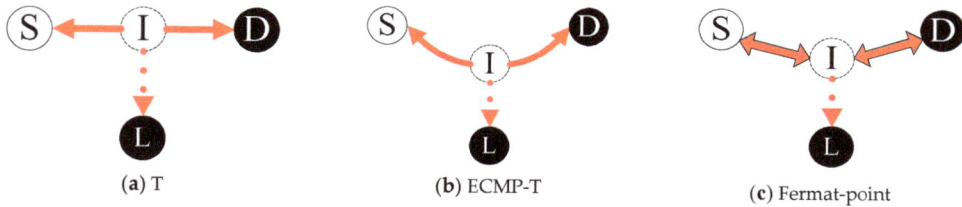

Figure 3. SDN interception models.

The legacy interception models include: S/I model, D/I model, and L/I model as shown in Figure 2a–c. As we all know, the interception service in legacy networks is limited by the deployment location of intercept access point due to the unimaginable cost of setting up special equipment and dedicated return link, to intercept network traffic. Thus, S or D or L is usually adopted as intercept access point I used to respond to the requirements of the interception center and to perform the traffic interception action in legacy network.

In this paper, we mainly study and analyze the SDN interception models, which includes T model, ECMP-T model and Fermat-point model as shown in Figure 3a–c. In view of the performance metrics of lawful interception system, the three SDN interception models are used to thoroughly study to find the optimal algorithm of deploying intercept access point.

Figure 3a shows T model: its name comes from the topology similar to the T-word. Under the concept of SDN networking in an undirected and weighted network $G(V,E)$, any SDN node on the shortest path S-D can be selected as intercept access point (I) under the premise of not affecting the existing shortest path arrangement of S-D (i.e., maintaining the existing end-to-end transmission quality). While only the node with the minimum hop-count (or cost) to the interception center (L) should be adopted as the best I-point to run the function to capture traffic transferred to the interception center.

Figure 3b shows ECMP-T model: based on the operation mode of T model, the path I-L must be the shortest path, but this shortest path S-I-D does not necessarily meet the optimal path. In fact, there may be more than one shortest path S-D, namely, the shortest path S-D is equal-cost multi-path (ECMP). Hence there may be a southward equal-cost shortest path in the T-word path theoretically, in which there is another intercept access point (I) and the hop-count (or cost) of I-L path is lower than the current one, so this interception model is called ECMP-T model that the nearest I-point from the interception center (L) is selected as the best intercept access point I among all the equivalent shortest paths between S and D.

Detailed pseudo code of T or ECMP-T model is presented in Algorithm 3. At the beginning of the algorithm, the set I used to store the best intercept access point is set to be empty in line 1. In lines 2–4, we calculate the shortest path $sp_{S,D}$ from S to D using Dijkstra or ECMP-Dijkstra Algorithm and then obtain the node-set $N_{S,D}$ in the shortest path $sp_{S,D}$, and next select out SDN nodes from the node-set $N_{S,D}$ to get the SDN node-set SN. If the SDN node-set SN is not empty, we traverse SDN nodes in SN and implement lines 6–16; otherwise, we fail to deploy intercept access point (IAP) between S, D and L, and thus save the wrong node combination of S, D and L in line 18. Line 7 calculates the lowest hop count (or cost) from SDN nodes to L and get the cost vector $h(i)$. In line 9, we sort the cost vector $h(i)$ by size of hop count in descending order and then get the sorted vector $numh(j)$ and the corresponding label vector $inh(j)$, where j denotes the subscript of j-th element of a vector. Line 11 takes the minimum hop-count value $minhops$ from the sorted vector $numh(j)$. Finally, in lines 13–14, we select the node with the minimum cost $minhops$ as the best intercept access point and then add the selected IAP $inh(j)$ to the set I.

Algorithm 3 T or ECMP-T Model

Input: N_{SDN}; S; D; L
Output: I
1: $I = \varnothing$
2: $sp_{S,D} \leftarrow Dijkstra(S,D)$ or $ECMP\text{-}Dijkstra(S,D)$
3: $N_{S,D} \leftarrow Onodes(sp_{S,D})$
4: $SN \leftarrow Select(N_{S,D}, N_{SDN})$
5: **if** the SDN node-set $SN \neq \varnothing$ **then**
6:   **for** node i in SN **do**
7:     $c(i) \leftarrow hops(Dijkstra(i,L))$ or
    $hops(ECMP\text{-}Dijkstra(i,L))$
8:   **end for**

9: $numh(j), inh(j) \leftarrow sort(h(i))$
10: $\beta \leftarrow Num(SN)$
11: $minhops \leftarrow numh(\beta)$
12: **for** key j in $numh$ **do**
13:    **if** $numh(j) = minhops$ **then**
14:      $I \leftarrow I \cup inh(j)$
15:    **end if**
16: **end for**
17: **else**
18: $SaveFail(S,D,L)$
19: **end if**
20: **return** I

The only difference of pseudo code of T model and ECMP-T model is whether to use Dijkstra Algorithm or ECMP-Dijkstra Algorithm to calculate the shortest path.

Figure 3c shows Fermat-point model: In geometry, Fermat-point refers to the point with the smallest sum of the distances from the three vertices of the triangle. Accordingly, we extend it to the node with the smallest sum of the distances from the three nodes of S, D and L in SDN network, and at the same time with meeting the constraints of the shortest path of S-D, S-L and D-L between the three points. Theoretically, Fermat-point model is optimal.

Details of pseudo code of Fermat-point model are summarized in Algorithm 4. In lines 2–4, we calculate all equal-cost shortest paths of S-D, S-L, D-L using ECMP-Dijkstra Algorithm, and then obtain all node sets in the equal-cost shortest paths in lines 5–7, and next combine these node sets to get the node-set $N_{S,D,L}$ in line 8, and further select out SDN nodes from the node-set $N_{S,D,L}$ to get the SDN node-set SN. If the SDN node-set SN is not empty, we traverse SDN nodes in SN and implement lines 11–24; otherwise, we fail to deploy intercept access point (IAP) between S, D and L, thus to save the wrong node combination of S, D and L. Lines 12–14 calculate the lowest hop count (or cost) of i-S, i-D, i-L, and then add the results to the sum, to get the cost vector $h(i)$ in line 15. In lines 17–24, we sort the cost vector $h(i)$ by size of cost value in descending order and then take the minimum cost value $minhops$, and next select the node $inh(j)$ with the minimum cost $minhops$ as the best intercept access point and finally add the selected IAP $inh(j)$ to the set I.

Algorithm 4 Fermat-point Model

Input: N_{SDN}; S; D; L
Output: I
1: $I = \varnothing$
2: $sp_{S,D}, sp_{S,L}, sp_{D,L} \leftarrow ECMP\text{-}Dijkstra((S,D),(S,L),(D,L))$
3: $N_{S,D}, N_{S,L}, N_{D,L} \leftarrow Onodes((sp_{S,D}, sp_{S,L}, sp_{D,L}))$
4: $N_{S,D,L} \leftarrow N_{S,D} \cup N_{S,L} \cup N_{D,L}$
5: $SN \leftarrow Select(N_{S,D,L}, N_{SDN})$
6: **if** the SDN node-set $SN \neq \varnothing$ **then**
7:   **for** node i in SN **do**
8:     $hs(i) \leftarrow hops(ECMP\text{-}Dijkstra(i,S))$
9:     $hd(i) \leftarrow hops(ECMP\text{-}Dijkstra(i,D))$
10:    $hl(i) \leftarrow hops(ECMP\text{-}Dijkstra(i,L))$
11:    $h(i) \leftarrow hs(i) + hd(i) + hl(i)$
12:   **end for**
13: $numh(j), inh(j) \leftarrow sort(h(i))$
14:   $\beta \leftarrow Num(SN)$
15:   $minhops \leftarrow numh(\beta)$
16:   **for** key j in $numh$ **do**
17:     **if** $numh(j) = minhops$ **then**
18:       $I \leftarrow I \cup inh(j)$

19:     **end if**
20:   **end for**
21: **else**
22:     *SaveFail(S,D,L)*
23: **end if**
24: **return** *I*

We use $sp_{i\text{-}j}$ to denote the shortest path from node *i* to node j, and $hops_{i\text{-}j}$ denotes the lowest hop-count or cost from node *i* to node *j*. We use '→' to denote that the next-node is N-SDN node and use '⇒' to denote that the next-node is SDN node. Examples of three interception models are illustrated in Figure 4, where we select node 154, node 9, node 105 all marked by red as S, D and L respectively and select 30 nodes randomly in Figure 4 as SDN nodes which includes node i∈{4, 8, 11, 19, 23, 25, 31, 38, 49, 50, 58, 60, 65, 67, 77, 82, 89, 92, 100, 103, 117, 120, 121, 125, 128, 134, 140, 150, 152, 156}, to construct a hybrid SDN.

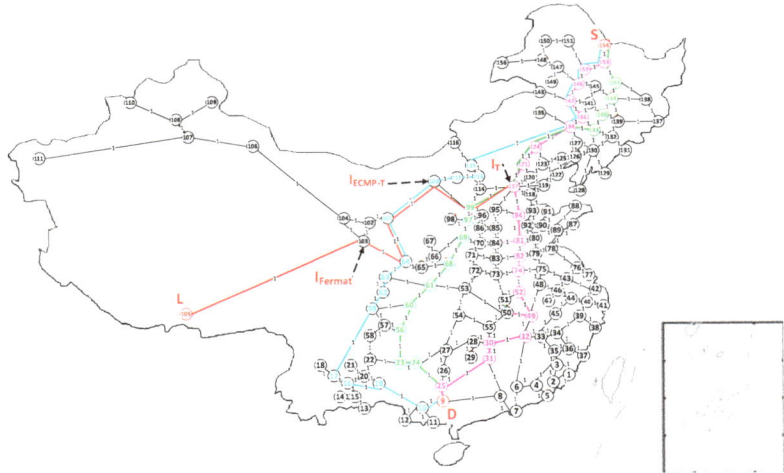

Figure 4. China's 156 major railway nodes network (China Railway Network; CRN).

We run T interception model: One shortest path from node 154 to node 9 is $sp_{154\text{-}9}$ marked by pink in Figure 4 that is 154 → 153 ⇒ 152 → 146 → 142 → 136 ⇒ 134 → 124 ⇒ 121 ⇒ 117 → 94 → 81 ⇒ 82 → 74 → 52 ⇒ 49 → 32 → 30 ⇒ 31 ⇒ 25→9, and $hops_{154\text{-}9}$ = 20. Among all nodes in $sp_{154\text{-}9}$, node 117 that is an SDN node has the lowest hop count to node 105 due to $hops_{117\text{-}105}$ = 6, and thus node 117 can be used as the best intercept access point I in T interception mode.

We run ECMP-T interception model: There are 22 equivalent shortest paths from node 154 to node 9, but we only show three shortest paths (i.e., $sp_{154\text{-}9}$ contains $sp1_{154\text{-}9}$, $sp2_{154\text{-}9}$, and $sp3_{154\text{-}9}$) from node 154 to node 9 marked by pink, bright green, turquoise respectively in Figure 4. $sp1_{154\text{-}9}$ is 154 → 153 ⇒ 152 → 146 → 142 → 136 ⇒ 134 → 124⇒ 121 ⇒ 117 → 94 → 81 ⇒ 82 → 74 → 52 ⇒ 49 → 32 → 30 ⇒ 31 ⇒ 25 → 9, and $sp2_{154\text{-}9}$ is 154 → 153 → 155 → 144 ⇒ 140 → 133 ⇒ 134 → 124 ⇒ 121 ⇒ 117 → 99 → 97 → 69 → 68 → 61 ⇒ 60 → 56 ⇒ 23 → 24 ⇒ 25→9, and $sp3_{154\text{-}9}$ is 154 → 153 → 155 → 144 ⇒ 140 → 133 ⇒ 134 → 115 → 113 → 112 ⇒ 100 → 101 → 64 → 63 → 62 → 59 → 17 → 16 ⇒ 19 → 10 → 9, and $hops_{154\text{-}9}$ = 20. Among all nodes in $sp_{154\text{-}9}$, node 100 that is SDN node in $sp3_{154\text{-}9}$ has the lowest hop count to node 105 due to $hops_{100\text{-}105}$ = 4, and thus node 100 can be used as the best intercept access point (I) in ECMP-T interception mode. Apparently, $hops_{100\text{-}105}$ < $hops_{117\text{-}105}$, namely, this I-point outperforms the one in the T model.

We run Fermat-point interception model: the node-set $N_{154\text{-}9\text{-}105}$ with no repeat is obtained by all $sp_{154\text{-}9}$, $sp_{154\text{-}105}$, $sp_{9\text{-}105}$ (i.e., $sp_{S\text{-}D}$, $sp_{S\text{-}L}$, $sp_{D\text{-}L}$). Namely, $N_{154\text{-}9\text{-}105}$ contains all nodes of all the shortest paths from node 154 to node 9, from node 154 to node 105, and from node 9 to node 105. And then, the sum of hop-count from node 103 in $N_{154\text{-}9\text{-}105}$ to node 154, node 9, node 105 (i.e., $hops_{103\text{-}154,9,105}$) is the smallest and $hops_{103\text{-}154,9,105} = 23$, This means that node 103 that is SDN node in $N_{154\text{-}9\text{-}105}$ can be used as the best intercept access point I in Fermat-point interception mode.

We have solved the problem of single intercept access point deployment above, and then expand to deploy intercept access points in hybrid SDN.

Running different network interception models, we will study and analyze the influence on the best transmission quality of intercepted data (the minimum cost from intercept access point (I) to interception center (L); MILC), the total cost of running intercept operation in global network (TOC), and the quality of service of normal user's data stream (UQoS) with different proportion of SDN node. According to the proposed three models in Figure 3, MILC, TOC, UQoS are calculated in respectively in (1), (2) and (3), where N denotes the maximum node label or index, and any node can be selected as S, D and L in hybrid SDN topology, i.e., there are N^3 possibilities for node-combination of S, D and L. After the node-combination selection (S, D, L), the best intercept access point (I) can be got by the SDN interception models, then the hop count or cost of the shortest path S-I, D-I and L-I can be calculated by the function $hops(i,j)$, thus calculating MILC, TOC and UQoS.

4. Restricted Minimum Vertex Cover Algorithm

There is no exception that most network optimization deployment problems can be viewed as the minimum vertex cover problem (MVCP) in graph theory. In the process of migration of SDN technology for large-scale Internet, it may be faced with the situation of hybrid deployment of SDN nodes and non-SDN nodes (N-SDN). In this hybrid SDN, not all nodes have software-defined functions to play the role of intercept access point. Only some nodes with the function of software definition can respond to the requirements of the interception center and to run interception operation. Therefore, it is very critical to select the best deployment location of intercept access point. And IPA must occupy the position of the hub, greatly covering all traffic through the critical path, and under a certain proportion of threshold, it may not achieve 100% intercept link coverage. Therefore, the minimum vertex cover problem must be transformed into the restricted minimum vertex cover problem question (RMVCP).

$$\text{MILC} = \sum_{S=1}^{N} \sum_{D=1}^{N} \sum_{L=1}^{N} hops(L, I) \tag{1}$$

$$\text{TOC} = \sum_{S=1}^{N} \sum_{D=1}^{N} \sum_{L=1}^{N} hops(S, I) + hops(D, I) + hops(L, I) \tag{2}$$

$$\text{UQoS} = \sum_{S=1}^{N} \sum_{D=1}^{N} \sum_{L=1}^{N} hops(S, I) + hops(I, L) \tag{3}$$

Considering overall situation (e.g., the location relevance of all intercepted targets, the operation and maintenance cost of operators) from the whole interception system, we intend to develop a restricted minimum vertex cover algorithm (RMVCA) to achieve the best intercept link coverage of the whole network with the minimum number of intercept access points as well as optimize the efficiency of deployment when deploying intercept access points in the hybrid SDN.

RMVCP: given a network graph $G(V,E)$, where V denotes the set of all nodes, and E denotes the set of all links in the network. There exists non-SDN nodes and SDN nodes at the same time in the network where $V = S \cup N$, and S denotes the set of SDN nodes, and N denotes the set of non-SDN nodes. To find a P set ($P \subseteq S \subseteq V$), so that every link in the network is covered (intercepted) by at least an SDN node in the P set.

Figure 5 shows an example of solving RMVCP. In this hybrid SDN, SDN nodes (i.e., solid circle) set $S = \{1, 3, 8, 9, 11, 12, 13, 17, 19, 20, 21, 22, 25, 26, 27, 28\}$ and non-SDN nodes (i.e., light circle) set $N= \{2, 4, 5, 6, 7, 10, 14, 15, 16, 18, 23, 24\}$. Using RMVCA, the SDN nodes set $P = \{1, 8, 9, 11, 13, 20, 22, 25, 26, 27\}$ is recommended to be selected as the intercept access points set, but 7 links (marked as dotted lines) in the example failed to be covered due to the hybrid deployment of SDN and N-SDN nodes, and thus only about 80% of the links (marked as solid lines) are completely covered by 10 intercept access points.

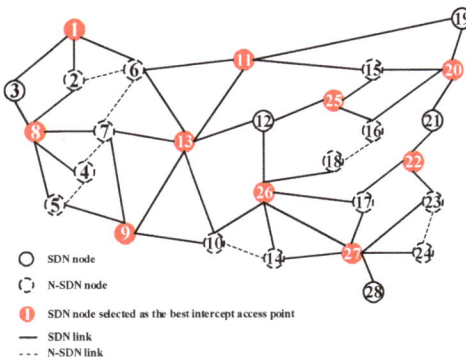

Figure 5. Hybrid SDN covered by minimum vertexes.

RMVCA ensure the result a near-optimal solution or one of the approximate solutions, so as to meet the optimal solution of the deployment problem of intercept access points.

Based on the concept of degree, we, at a time, use greedy algorithm to select one approximate or equivalent optimal intercept access point to reduce the scale of the problem recursively, so as to obtain the minimum vertex approximation set covering all SDN links and achieve the best intercept link coverage with the minimum number of intercept access point.

Details of pseudo code of RMVCA are summarized in Algorithm 5. We input an undirected and weighted network cover set P is set to empty originally. In line 2–3, we get the set N of N-SDN nodes and the accordingly edge-set $E_{N\text{-}SDN}$ of N-SDN nodes by the set N. Line 4 removes $E_{N\text{-}SDN}$ from the edge-set E, to get the edge set E_{SDN} of SDN nodes. Lines 6–7 traverse each SDN node and calculate its degree $d(i)$. In line 9, we sort the degree vector $d(i)$ in the ascending order and get the sorted degree vector $numd(j)$ and the accordingly label vector $ind(j)$ where j denotes the index or subscript of j-th element. Line 11 selects the maximum degree $numd(\beta)$ from $numd(j)$ where β denotes the number of SDN nodes in the set S. In lines 12–18, we judge the degree of node and implement accordingly measures. If the maximum degree of node is not equal to zero in line 12, we first add the node $ind(\beta)$ where β denotes the subscript of β-th element to the minimum vertex cover set P in line 13 and then calculate the adjacent edge-set $\theta(ind(\beta))$ of node $ind(\beta)$ in line 14, and next remove the adjacent edge-set $\theta(ind(\beta))$ from the edge-set E in line 15, which leads to the degree reduction of each SDN node. Finally, we return line 5 to judge whether E_{SDN} is empty and then calculate the degree of each SDN node again. Otherwise, if the maximum degree of node is equal to zero, we break the loop and end the algorithm.

Algorithm 5 RMVCA (Restricted Minimum Vertex Cover Algo-rithm)

Input: $G(V,E)$; S: the SDN node-set
Output: P
1: $P = \varnothing$
2: $N \leftarrow V - S$
3: $E_{N\text{-}SDN} \leftarrow Edge(N(i))$
4: $E_{SDN} \leftarrow E - E_{N\text{-}SDN}$
5: **while** $E_{SDN} \neq \varnothing$ **do**
6:    **for** node i in S **do**
7:       $d(i) \leftarrow i$
8:    **end for**
9:    $numd(j), ind(j) \leftarrow sort(d(i))$
10:   $\beta \leftarrow Num(S)$
11:   $maxnumd \leftarrow numd(\beta)$
12:   **if** $maxnumd > 0$ **then**
13:      $P \leftarrow P \cup ind(\beta)$
14:      $\theta(ind(\beta)) \leftarrow ind(\beta)$
15:      $E_{SDN} \leftarrow E_{SDN} - \theta(ind(\beta))$
16:   **else**
17:      break
18:   **end if**
19: **end while**
20: **return** P

Using RMVCA proposed above, we will study and analyze the influence of different SDN node proportion on the maximum intercept link coverage of the whole network (i.e., max-ILC) and the accordingly needed minimum number of SDN nodes in the P set for realizing the maximum intercept link coverage (i.e., numP), as well as the influence of RMVCA on the intercept link coverage (i.e., ILC) and the efficiency of deploying intercept access points in whole hybrid SDN.

5. Simulation and Results

5.1. Simulation Environment and Performance Metrics of Lawful Interception System

In our simulation, we choose three real-world backbone topologies CRN, COST 239, NSFNet to evaluate the performance of three SDN interception models. Under the three network topologies, we randomly select different number of nodes as SDN nodes to construct the hybrid SDN network and the weight of each link is set to 1 by default, and the source node (S), the destination node (D) and the interception center (L) are selected randomly and thus there are 3,796,416 (156^3), 21,952, 2,744 node combinations of S, D and L.

Under different proportion of SDN nodes, we will study and analyze the influence of different SDN interception models on the best transmission quality of intercepted data (the minimum cost from intercept access point (I) to the interception center (L); MILC), the total cost of running intercept operation in global network (TOC), and the quality of service of normal user's data stream (UQoS), the deployment efficiency of IAP (the total number of times to calculate the shortest path during the process of deploying IAP; TTC)), and the total number of failures to deploy IEP (i.e., NFD).

According to the proposed three SDN interception models, MILC, MRLC, TOC, and UQoS are calculated respectively in (1), (2) and (3). Based RMVCA, we run different SDN interception models and calculate and count up MILC, TOC, UQoS, TTC and NFD of each node combination of S, D and L and then compare and analyze the results to evaluate the performance of three SDN interception models.

5.2. Benchmark Approach

In order to analyze the influence of RMVCA on the intercept link coverage of whole hybrid SDN and the efficiency of deploying intercept access points, we propose three approaches, proactive approach (PA), reactive approach (RA), hybrid approach (HA), and then compare them by running three SDN interception models in real-world topology CRN. To show the effectiveness of HA, we compare it with the following baselines: PA and RA.

Experimental initialization: We randomly select some nodes as SDN nodes (i.e., given a hybrid SDN network topology), and then use RMVCA to calculate the minimum vertex cover set P required to achieve the maximum intercept link coverage in theory and the accordingly number N of SDN nodes in the P set. Additionally, the calculation amount of this initialization process is negligible compared with the one of the whole H-SDN.

Nodes selection: we traverse any node as S, D and L in topology CRN (i.e., there are 3,796,416 (156^3) possibilities for node-combination of S, D, L) and then the node combination of S, D and L is given for experiments.

Proactive approach (PA): when running SDN interception models to deploy intercept access point, we select the best intercept access point from the minimum vertex cover set P calculated by RMVCA. Details of pseudo code of PA in T or ECMP-T model are summarized in Algorithm A1 of Appendix A. The only difference of pseudo code of PA in T model and ECMP-T model is whether to use Dijkstra Algorithm or ECMP-Dijkstra Algorithm to calculate the shortest path.

Reactive approach (RA): according to the selected node combination of S, D and L, we run three interception models without exploiting RMVCA to deploy intercept access points.

Hybrid Approach (HA): running three SDN interception models to deploy intercept access point, we get the node-set $N_{S,D,L}$ where all nodes are selected from the shortest paths between S, D and L, and then obtain the node-set SP whose nodes also exist in the node-set P calculated by RMVCA. If the node-set SP is not empty, we preferentially select node from the SP set to deploy the best intercept access point; otherwise, we implement RA. Details of pseudo code of HA in T or ECMP-T model are summarized in Algorithm A2 of Appendix A.

When implementing PA or RA or HA, we count and calculate the frequency of the nodes selected as the best intercept access point, and then sort the nodes from largest to smallest based their frequency, and next select the first N nodes and calculate their intercept link coverage for studying and analyzing the impact of different approaches on the intercept link coverage (i.e., ILC) of the whole hybrid SDN. Additionally, we count the total times of calculating the shortest path (i.e., TTC) during the process of deploying intercept access points for studying and analyzing the impact of different approaches on the efficiency of deploying IAPs.

5.3. Results and Discussion

5.3.1. ILC

Using RMVCA, we study and analyze the influence of different numbers of N-SDN nodes on the maximum intercept link coverage (i.e., max-ILC) and the accordingly needed minimum number of SDN nodes in the P set for realizing the maximum intercept link coverage (i.e., numP). Moreover, we take the operator's operation and maintenance cost (i.e., the minimum number of SDN nodes) and network intercept link coverage into account comprehensively, so as to find the best proportion of SDN nodes from the experimental results.

Randomly selecting the number of N-SDN nodes (node i ∈ (0,156)) in CRN topology, we conducted 10,000 experiments in the same proportion of N-SDN nodes. Due to the different network topologies under the same SDN node proportion, the results of each experiment are different. The statistical results of 10,000 experiments are shown in Figures 6 and 7.

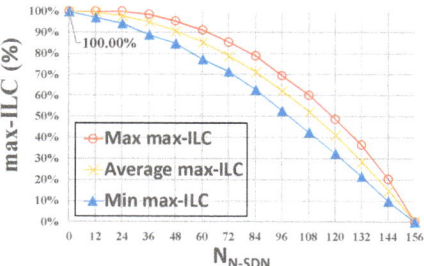

Figure 6. The influence of different numbers of N-SDN nodes on max-ILC.

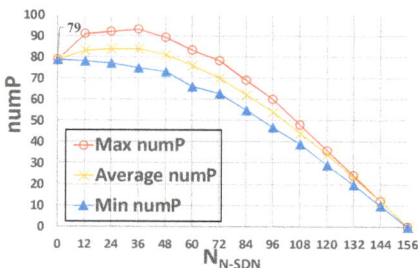

Figure 7. The influence of the number of N-SDN nodes on numP.

Figure 6 shows the influence of different numbers of N-SDN nodes on max-ILC. From the figure, we can see that the number of SDN links in hybrid SDN decreases gradually with the increase of the number of N-SDN nodes (the decrease of the number of SDN nodes), resulting in the gradual decline of the network intercept link coverage. And max-ILC = 0.00% denotes that all links in the whole network are N-SDN links that cannot be intercepted, namely, all nodes in the network are N-SDN nodes. Additionally, we can see that the intercept link coverage of the whole hybrid SDN can reach 80.53~100% when the number of N-SDN nodes is between 0 and 57 (i.e., the number of SDN nodes is between 99 and 156), namely, only when the number of SDN nodes in hybrid SDN is more than 99 can SDN nodes intercept more than 80% of the links of the whole network.

Figure 7 shows the influence of different numbers of N-SDN nodes on numP. From the figure, we can see that when the number of N-SDN is 0 (i.e., the number of SDN nodes is 156), 79 SDN nodes are required to achieve the maximum intercept link coverage; the number of SDN nodes required to intercept the whole network gradually increases first and then decreases gradually. This is because that when the number of N-SDN nodes is between 0 and 37 (i.e., the number of SDN nodes is between 119 and 156), though the increase of N-SDN links results in the decrease of the degree of some SDN nodes, the total number of SDN links does not decrease significantly. Thus, more SDN nodes are needed to intercept the same number of links. Accordingly, the number of SDN nodes required to intercept the whole network increases. While when the number of N-SDN is between 38 and 156 (i.e., the number of SDN is between 0 and 118), the number of SDN links greatly decreases with the increasing number of N-SDN nodes, so the number of SDN nodes needed to achieve maximum intercept link coverage also decreases gradually. Moreover, when all nodes in the network are N-SDN nodes, all links are N-SDN links, and thus the minimum vertex cover set P is empty (i.e., numP = 0). To sum up, according to Figures 4 and 5, we only need 69~95 SDN nodes to achieve 80.53~100% intercept link coverage of the whole interception system when the number of SDN nodes in the whole hybrid SDN is between 99~156.

Next, we will study and analyze the influence of three different approaches and three SDN interception models on intercept link coverage (ILC) as shown in Figure 8. From the figure, we can see that ILC of PA and HA with RMVCA is higher than that of RA without RMVCA in general, and ILC of PA and HA are relatively close, whether using T model, ECMP-T model or Fermat-point interception model. Additionally, compared with RA, PA and HA can significantly improve the intercept link coverage when the number of N-SDN nodes is between 0 and 60 (i.e., the number of SDN nodes is between 96 and 156). And this improvement decreases with the decrease of SDN nodes.

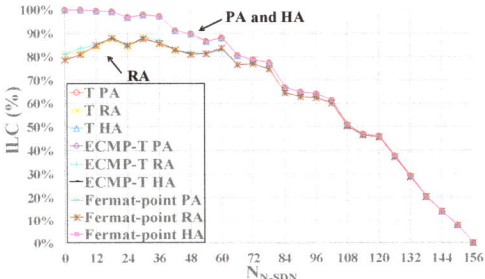

Figure 8. The impact of three approaches in three SDN interception models on ILC under CRN topologies. ILC. The higher, the better.

Meanwhile, another conclusion we can make is that the three SDN interception models have nearly the same intercept link coverage. In other words, the intercept link coverage (ILC) has no relationship with SDN interception models and the SDN interception models have little impact on ILC.

5.3.2. TTC

Using RMVCA, we will analyze the impact of RMVCA on the efficiency of deploying intercept access points in whole hybrid SDN. In many experiments, we run three SDN interception models to deploy IAPs in three approaches during which the shortest paths need to be calculated, and thus the total times of calculating the shortest path (TTC) is different. In order to evaluate the performance of RMVCA, we employ TTC as its most important performance metric. We predict that RMVCA can improve the efficiency of deploying IAPs (i.e., reduce the total deployment time). The experimental results are shown in Figure 9.

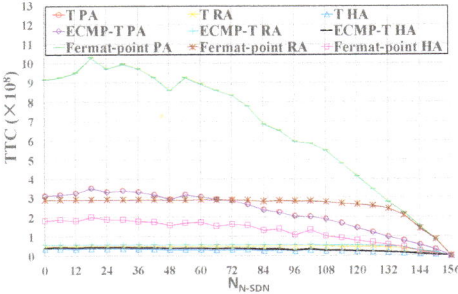

Figure 9. The impact of three approaches in three SDN interception models on TTC under CRN topologies. TTC. The lower, the better.

From Figure 9, we can see that TTC of Fermat-point interception model is the highest whether in PA, RA or HA, namely, running Fermat-point model may take the longest time to deploy IAPs. In addition, TTC of T model and ECMP-T model is similar and is far lower than that of Fermat-point model. Therefore, in terms of the efficiency of deploying IAPs, T model and ECMP-T model are better than Fermat-point model.

Also, Figure 9 show the impact of three approaches in three SDN interception models on TTC under CRN topologies. From the figure, we can see that compared TTC in PA and RA, TTC in HA is the lowest, whether running T model, ECMP-T model or Fermat-point model in hybrid SDN. Namely, HA is the best approach in terms of the efficiency of deploying IAPs based on thorough analysis and comparison.

Meanwhile, we also can see that TTC in PA is the highest and thus PA is the most undesirable approach. Considering that TTC is the most important performance metric of RMVCA, we can abandon PA. According to Figure 9, we can conclude by calculating that with the increasing number of N-SDN nodes (i.e., with the decreasing number of SDN nodes) in hybrid SDN, HA can significantly improve the deployment efficiency of intercept access points for the reason that compared with RA, HA can decrease TTC on average by 41.14%, 44.07%, 53.32% respectively in T model, ECMP-T model and Fermat-point model. In conclusion, PA is the most undesirable approach that should be abandoned. While HA is the best approach in terms of the deployment efficiency of IAPs.

5.3.3. MILC, TOC and UQoS

After deploying the best intercept access point (IAP; I), the interception center (the law enforcement agencies; L) hopes to receive the data intercepted by the intercept access point with the minimum cost (i.e., the minimum cost or hop-count from the intercept access point (I) to the interception center (L); MILC). Therefore, MILC is one of the most important performance metrics of lawful interception system. In addition, the network operators are most concerned about the total cost of running intercept operation in global network (i.e., TOC) which is the prominent performance metrics of lawful interception system. Meanwhile, running different SDN interception models to deploy intercept access point may lead to the different selection of the best intercept access point (namely the placement location of IAP differs) and the different amount of calculation, thus affecting the quality of service of normal user's data stream (UQoS). Thus, UQoS is also one of the important performance metrics of lawful interception system. In a word, MILC, TOC and UQoS are of great significance for the Law Enforcement Agencies, the network operators and the users, respectively. Focused on three hybrid SDN topologies CRN, NSFNet and COST 239, we study and analyze the impact of running three different SDN interception models to deploy the best IAPs on MILC, TOC, and UQoS of whole lawful interception system under different number of SDN nodes. The experimental results of the three topologies are shown in Figure 10a–c.

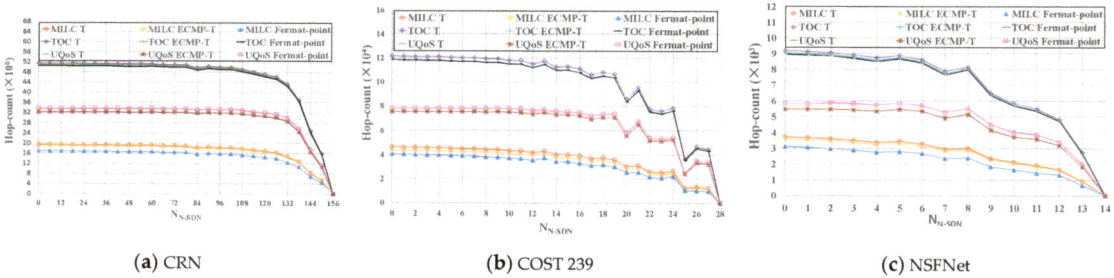

(a) CRN (b) COST 239 (c) NSFNet

Figure 10. The impact of three SDN interception models on MILC, TOC, UQoS under three topologies. Hop-count. The lower, the better.

From the figures, we can see that MILC, TOC in T model are relatively close to the ones in ECMP-T model. And MILC and TOC consumed by ECMP-T model are lower than that of T model, so ECMP-T model is better than T model. More importantly, compared with MILC and TOC in T model and ECMP-T model, MILC and TOC in Fermat-point model are the lowest in all number of SDN nodes. In other words, Fermat-point model can decrease MILC and TOC compared with T model and ECMP-T model. More specifically, compared with T model and ECMP-T model, Fermat-point model can decrease MILC on average by 13.41%, 11.11% in CRN topology, 14.91%, 8.73% in COST 239, and 19.72%, 16.04% in NSFNet, and TOC on average by 1.91%, 0.99% in CRN topology, 2.82%, 0.46% in COST 239, and 2.65%, 1.03% in NSFNet. These simulation results verify that the performance of Fermat-point model outperforms T model and ECMP-T model and thus Fermat-point model is the best SDN interception model in terms of MILC and TOC.

Meanwhile, from the figures, we can see that no matter in CRN, COST 239 or NSFNet, ECMP-T model and T model have the same UQoS. In other words, T model and ECMP-T model have little impact on the transmission quality of traffic normally accepted by users and on deployment efficiency of IAP. According to the principle of T model and ECMP-T model, we know the simulation results in three hybrid SDN topologies are consistent with the theory, so these results are true and reliable. In addition, we can also clearly observe from the figures that UQoS in Fermat-point model is higher than the one in T model and ECMP-T model, which means that Fermat-point model slightly affect UQoS. Thus, Fermat-point model has poor performance in terms of UQoS.

5.3.4. NFD

Due to the hybrid SDN topologies where N-SDN nodes cannot be selected as IAP, not every combination of S, D and L can successfully deploy intercept access point. We count the total number of failures to deploy IEP (i.e., NFD), to evaluate the performance of SDN interception models. The statistical results are shown in Figure 11a–c. We can clearly observe from the figures that in the three hybrid SDN topologies, the total number of failures to deploy IAPs (NFD) in Fermat-point model is the least compared with NFD in T model and ECMP-T model, which means that Fermat-point model has a high success rate to deploy intercept access point. More specifically, compared with T model and ECMP-T model, Fermat-point model decreases NFD on average by 88.21%, 86.87% in CRN topology, 76.9%, 74.68% in COST 239, and 67.53%, 66.26% in NSFNet. To sum up, the performance of Fermat-point model outperforms T model and ECMP-T model and thus Fermat-point model is the best interception model in terms of NFD.

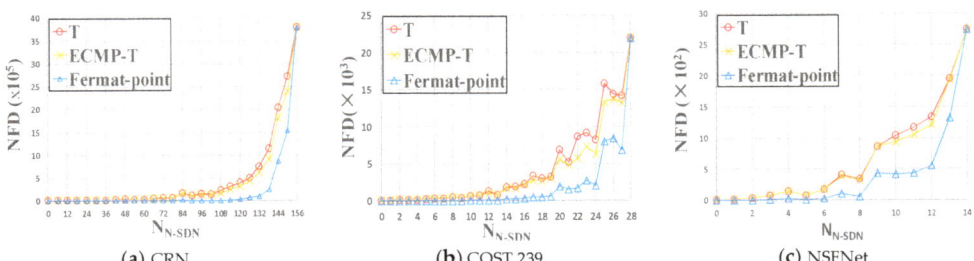

Figure 11. The impact of SDN interception models on NFD under three topologies. NFD. The lower, the better.

6. Conclusions

In this paper, we proposed an improved equal-cost multi-path shortest path algorithm (ECMP-Dijkstra) and accordingly proposed three SDN interception models T model, ECMP-T model, and Fermat-point model, to deploy the best intercept access point reasonably in three real-world hybrid SDN topologies. Subsequently, we proposed a restrictive minimum vertex coverage algorithm (RMVCA) to intercept the whole interception system with the least SDN nodes, and to optimize the deployment efficiency of intercept access points and improve the intercept link coverage, so as to optimize the performance of the whole intercepting system. According to RMVCA, we analyze the effect of different SDN node ratios on the intercept link coverage and the minimum vertex coverage set. Considering the intercept link coverage and the minimum vertex coverage set, we found a suitable SDN node ratio for deploying intercept access points reasonably, namely, to intercept the whole hybrid SDN with the least SDN nodes.

Based RMVCA, we put forward three approaches PA, RA, and HA for experiments, and compared the three experimental approaches. The experimental results show that HA is the best approach, which can significantly optimize the efficiency of deploying intercept access points (i.e., optimize TTC) and improve the intercept link coverage of the whole hybrid SDN.

By the way, we analyzed the influence of three SDN interception models on various performance metrics of lawful interception system using three real-world topologies. The simulation results reveal that the three SDN interception models have little effect on the intercept link coverage, and T model and ECMP-T model have no effect on user's traffic transmission quality. Compared with T model and ECMP-T model, Fermat-point model is the best interception model for the reason that Fermat-point model can make MILC, TOC, NFD the lowest by sacrificing a small part of user's traffic transmission quality (UQoS) and deployment time (TTC), intercepting the whole hybrid SDN at dramatically lower costs.

This paper has not considered the traffic bottleneck (link capacity) problem but has proposed the deployment and optimization strategy of intercept access points that pave the way for the future work that joint deployment of IAPs and LEAs in H-SDNs based on the consideration of the traffic bottleneck problem.

Author Contributions: X.X. and W.-K.J. conceived and designed the study. X.X. performed the simulations. X.X. wrote the paper. Y.W. and X.W. reviewed and edited the manuscript. All authors have read and agreed to the published version of the manuscript.

Funding: This research was financially supported by the National Natural Science Foundation of China No. U1805262, No. 61871131, and No. 61701118 in part by the Natural Science Foundation of Fujian Province, China No.2018J05101, 2018H6007, and Special Fund for Marine Economic Development of Fujian Province (ZHHY-2020-3).

Institutional Review Board Statement: Not applicable.

Informed Consent Statement: Not applicable.

Data Availability Statement: The data are not publicly available due to their containing information that could compromise the privacy of research participants.

Acknowledgments: The authors would like to thank the anonymous reviewers for their useful comments and careful reading of the manuscript.

Conflicts of Interest: The authors declare no conflict of interest.

Appendix A

Based on Proactive Approach (PA) or Hybrid Approach (HA), the details of pseudo code of T, ECMP-T or Fermat-point model can be presented respectively in Algorithm A1 or Algorithm A2.

Algorithm A1 Proactive Approach—T or ECMP-T or Fermat-point Model

Input: $P; S; D; L$
Output: I
1: **if** $P \neq \emptyset$ **then**
2:    $I = \emptyset$
3:    **for** node i in P **do**
:   ......
4: **return** I

Algorithm A2 Hybrid Approach—T or ECMP-T or Fermat-point Model

Input: $N_{SDN}; P; S; D; L$
Output: I
1: $I = \emptyset$
: ......
2:    $SP \leftarrow Select(N_{S,D}, P)$ or $Select(N_{S,D,L}, P)$
3: **if** the SDN node-set $SP \neq \emptyset$ **then**
4:    **for** node i in SP **do**
:   ......
5: **else**
6:    $Algo3(N_{SDN}, S, D, L)$ or $Algo4(N_{SDN}, S, D, L)$
7: **end if**
8: **return** I

References

1. Fundation, O.N. Software-Defined Networking: The New Norm for Networks. *ONF White Paper* **2012**, *2*, 11.
2. Kadhim, A.J.; Seno, S.A.H. Maximizing the Utilization of Fog Computing in Internet of Vehicle Using SDN. *IEEE Commun. Lett.* **2019**, *23*, 140–143. [CrossRef]
3. Görkemli, B.; Tatlıcıoğlu, S.; Tekalp, A.M.; Civanlar, S.; Lokman, E. Dynamic Control Plane for SDN at Scale. *IEEE J. Sel. Areas Commun.* **2018**, *36*, 2688–2701. [CrossRef]
4. McKeown, N.; Anderson, T.; Balakrishnan, H.; Parulkar, G.; Peterson, L.; Rexford, J.; Turner, J. OpenFlow: Enabling innovation in campus networks. *ACM Sigcomm. Comput. Commun. Rev.* **2008**, *38*, 69–74. [CrossRef]
5. Cheng, T.Y.; Jia, X. Compressive Traffic Monitoring in Hybrid SDN. *IEEE J. Sel. Areas Commun.* **2018**, *36*, 2731–2743. [CrossRef]
6. Dijkstra, E.W. A Note on Two Problems in Connexion with Graphs. *Numer. Math.* **1959**, *1*, 269–271. [CrossRef]
7. Li, G. *An Improvement of Dijkstra Algorithm*; Computer Development & Applications: New York, NY, USA, 2009.
8. Wang, S.; Zhao, X. The improved Dijkstra's shortest path algorithm. In Proceedings of the 2011 Seventh International Conference on Natural Computation, Shanghai, China, 26–28 July 2011; pp. 2313–2316. [CrossRef]
9. Kadry, S.; Abdallah, A.; Joumaa, C. On The Optimization of Dijkstras Algorithm. In *Informatics in Control, Automation and Robotics*; Springer: Berlin/Heidelberg, Germany, 2012.
10. Zhang, X.; Chen, Y.; Li, T. Optimization of logistics route based on Dijkstra. In Proceedings of the 2015 6th IEEE International Conference on Software Engineering and Service Science (ICSESS), Beijing, China, 23–25 September 2015; pp. 313–316. [CrossRef]
11. Bozyiğit, A.; Alankuş, G.; Nasiboğlu, E. Public transport route planning: Modified dijkstra's algorithm. In Proceedings of the 2017 International Conference on Computer Science and Engineering (UBMK), Antalya, Turkey, 5–8 October 2017; pp. 502–505. [CrossRef]
12. Makariye, N. Towards shortest path computation using Dijkstra algorithm. In Proceedings of the 2017 International Conference on IoT and Application (ICIOT), Nagapattinam, India, 19–20 May 2017; pp. 1–3. [CrossRef]
13. Wei, K.; Gao, Y.; Zhang, W.; Lin, S. A Modified Dijkstra's Algorithm for Solving the Problem of Finding the Maximum Load Path. In Proceedings of the 2019 IEEE 2nd International Conference on Information and Computer Technologies (ICICT), Kahului, HI, USA, 14–17 March 2019; pp. 10–13. [CrossRef]
14. Wenzheng, L.; Junjun, L.; Shunli, Y. An Improved Dijkstra's Algorithm for Shortest Path Planning on 2D Grid Maps. In Proceedings of the 2019 IEEE 9th International Conference on Electronics Information and Emergency Communication (ICEIEC), Beijing, China, 12–14 July 2019; pp. 438–441. [CrossRef]
15. Levin, D.; Canini, M.; Schmid, S.; Schaffert, F.; Feldmann, A. Panopticon: Reaping the benefits of incremental SDN deployment in enterprise networks. In Proceedings of the 2014 {USENIX} Annual Technical Conference, Philadelphia, PA, USA, 19–20 June 2014; pp. 333–345.
16. Hong, D.K.; Ma, Y.; Banerjee, S.; Mao, Z.M. Incremental deployment of SDN in hybrid enterprise and ISP networks. In Proceedings of the Symposium on SDN Research, Santa Clara, CA, USA, 14–15 March 2016.

17. Poularakis, K.; Iosifidis, G.; Smaragdakis, G.; Tassiulas, L. One step at a time: Optimizing SDN upgrades in ISP networks. In Proceedings of the IEEE INFOCOM 2017-IEEE Conference on Computer Communications, Atlanta, GA, USA, 1–4 May 2017.
18. Zhou, H.; Zhang, D.; Xie, K.; Wang, X. Data reconstruction in internet traffic matrix. *China Commun.* **2014**, *11*, 1–12. [CrossRef]
19. Chen, J.; Kanj, I.A. Constrained minimum vertex cover in bipartite graphs: Complexity and parameterized algorithms. *J. Comput. Syst. Sci.* **2003**, *67*, 833–847. [CrossRef]
20. Behsaz, B.; Hatami, P.; Mahmoodian, E.S. On minimum vertex cover of generalized Petersen graphs. *Australas. J. Comb.* **2010**, *40*, 253–264.
21. Madhavi, L.; Maheswari, B. Edge Cover Domination in Mangoldt Graph. *Momona Ethiop. J. Sci.* **2011**, *3*, 37–51. [CrossRef]
22. Wu, Y.; Yu, Q. A Characterization of Graphs with Equal Domination Number and Vertex Cover Number. *Bull. Malays. Math. Soc. Ser.* **2012**, *3*, 803–806.
23. Volkmann, L. On graphs with equal domination and covering numbers. *Discret. Appl. Math.* **1994**, *51*, 211–217. [CrossRef]
24. Grandoni, F.; Könemann, J.; Panconesi, A.; Sozio, M. A Primal-dual Bicriteria Distributed Algorithm for Capacitated Vertex Cover. *Siam J. Comput.* **2008**, *38*, 825–840. [CrossRef]
25. Zhang, N.; Sheng, Z. *Improved Greedy Algorithm for the Minimum Vertex Cover Problem*; Mongolia Normal University: Hohhot, China, 2012.
26. Tomar, D. An Improved Greedy Heuristic for Unweighted Minimum Vertex Cover. In Proceedings of the 2014 International Conference on Computational Intelligence and Communication Networks, Bhopal, India, 14–16 November 2014; pp. 618–622. [CrossRef]
27. Angel, D. A breadth first search approach for minimum vertex cover of grid graphs. In Proceedings of the 2015 IEEE 9th International Conference on Intelligent Systems and Control (ISCO), Coimbatore, India, 9–10 January 2015; pp. 1–4. [CrossRef]
28. Fayaz, M.; Arshad, S.; Zaman, U.; Ahmad, A. A Simple, Fast and Near Optimal Approximation Algorithm for Optimization of Un-Weighted Minimum Vertex Cover. In Proceedings of the International Conference on Frontiers of Information Technology IEEE, Islamabad, Pakistan, 19–21 December 2017.
29. Çınaroğlu, S.; Bodur, S. A new hybrid approach based on genetic algorithm for minimum vertex cover. In Proceedings of the 2018 Innovations in Intelligent Systems and Applications (INISTA), Thessaloniki, Greece, 3–5 July 2018; pp. 1–5. [CrossRef]
30. Zhou, Y.; Qiu, C.; Wang, Y.; Fan, M.; Yin, M. An Improved Memetic Algorithm for the Partial Vertex Cover Problem. *IEEE Access* **2019**, *7*, 17389–17402. [CrossRef]
31. Branch, P.A. Lawful Interception of the Internet. *Int. J. Emerg. Technol. Soc.* **2003**, *1*, 38–51.
32. Csikor, L.; Szalay, M.; Rétvári, G.; Pongrácz, G.; Pezaros, D.P.; Toka, L. Transition to SDN is HARMLESS: Hybrid Architecture for Migrating Legacy Ethernet Switches to SDN. *IEEE/ACM Trans. Netw.* **2020**, *28*, 275–288. [CrossRef]
33. Bshouty, N.H.; Burroughs, L. Massaging a linear programming solution to give a 2-approximation for a generalization of the vertex cover problem. In *Symposium on Theoretical Aspects of Computer Science*; Springer: Berlin/Heidelberg, Germany, 1998. [CrossRef]
34. Cao, W.; Feng, X.; Jia, J.; Zhang, H. Characterizing the Structure of the Railway Network in China: A Complex Weighted Network Approach. *J. Adv. Transp.* **2019**, *2019*, 3928260. [CrossRef]
35. De Maesschalck, S.; Colle, D.; Lievens, I.; Pickavet, M.; Demeester, P.; Mauz, C.; Jaeger, M.; Inkret, R.; Mikac, B.; Derkacz, J. Pan-European optical transport networks. *Photonic Netw. Commun.* **2003**, *5*, 203–225. [CrossRef]
36. Claffy, K.C.; Polyzos, G.C.; Braun, H. Traffic characteristics of the T1 NSFNET backbone. In Proceedings of the IEEE INFOCOM '93 The Conference on Computer Communications, San Francisco, CA, USA, 28 March–1 April 1993; Volume 2, pp. 885–892. [CrossRef]

Review

A Comprehensive Study of Anomaly Detection Schemes in IoT Networks Using Machine Learning Algorithms

Abebe Diro [1], Naveen Chilamkurti [2], Van-Doan Nguyen [2,*] and Will Heyne [3]

1 College of Business and Law, RMIT University, Melbourne 3001, Australia; abebe.diro3@rmit.edu.au
2 Department of Computer Science and I.T., La Trobe University, Melbourne 3086, Australia; n.chilamkurti@latrobe.edu.au
3 BAE Systems Australia, Adelaide 5000, Australia; will.heyne@baesystems.com
* Correspondence: o.nguyen@latrobe.edu.au

Citation: Diro, A.; Chilamkurti, N.; Nguyen, V.-D.; Heyne, W. A Comprehensive Study of Anomaly Detection Schemes in IoT Networks Using Machine Learning Algorithms. *Sensors* **2021**, *21*, 8320. https://doi.org/10.3390/s21248320

Academic Editors: Zihuai Lin and Wei Xiang

Received: 8 November 2021
Accepted: 8 December 2021
Published: 13 December 2021

Publisher's Note: MDPI stays neutral with regard to jurisdictional claims in published maps and institutional affiliations.

Copyright: © 2021 by the authors. Licensee MDPI, Basel, Switzerland. This article is an open access article distributed under the terms and conditions of the Creative Commons Attribution (CC BY) license (https://creativecommons.org/licenses/by/4.0/).

Abstract: The Internet of Things (IoT) consists of a massive number of smart devices capable of data collection, storage, processing, and communication. The adoption of the IoT has brought about tremendous innovation opportunities in industries, homes, the environment, and businesses. However, the inherent vulnerabilities of the IoT have sparked concerns for wide adoption and applications. Unlike traditional information technology (I.T.) systems, the IoT environment is challenging to secure due to resource constraints, heterogeneity, and distributed nature of the smart devices. This makes it impossible to apply host-based prevention mechanisms such as anti-malware and anti-virus. These challenges and the nature of IoT applications call for a monitoring system such as anomaly detection both at device and network levels beyond the organisational boundary. This suggests an anomaly detection system is strongly positioned to secure IoT devices better than any other security mechanism. In this paper, we aim to provide an in-depth review of existing works in developing anomaly detection solutions using machine learning for protecting an IoT system. We also indicate that blockchain-based anomaly detection systems can collaboratively learn effective machine learning models to detect anomalies.

Keywords: cybersecurity; anomaly detection; the Internet of Things; machine learning; deep learning; blockchain

1. Introduction

The IoT consists of myriad smart devices capable of data collection, storage, processing, and communication. The adoption of the IoT has brought about tremendous innovation opportunities in industries, homes, the environment, and businesses, and it has enhanced the quality of life, productivity, and profitability. However, infrastructures, applications, and services associated with the IoT introduced several threats and vulnerabilities as emerging protocols and workflows exponentially increased attack surfaces [1]. For instance, the outbreak of the Mirai botnet exploited IoT vulnerabilities and crippled several websites and domain name systems [2].

It is challenging to secure IoT devices as they are heterogeneous, traditional security controls are not practical for these resource-constrained devices, and the distributed IoT networks fall out of the scope of perimeter security, and existing solutions such as the cloud suffer from centralisation and high delay. Another reason for this challenge is that IoT device vendors commonly overlook security requirements due to a rush-to-market mentality. Furthermore, the lack of security standards has added another dimension to the complexity of securing IoT devices. These challenges and the nature of IoT applications call for a monitoring system such as anomaly detection at device and network levels beyond the organisational boundary.

An anomaly is a pattern or sequence of patterns in IoT networks or data that significantly deviate from the normal behaviour. Anomalies can be contextual and collective points based on the sources of anomalies [3]. Point anomaly represents a specific data point that falls outside the norm, and it indicates random irregularity, extremum, or deviation with no meaning, often known as outliers. The contextual anomaly denotes a data point that deviates from the norm in a specific context such as in a time window. It means that the same normal observation in a given context can be abnormal in a different context. The contextual anomaly is driven by contextual features such as time and space and behavioural features such as the application domain. A collection of related data points, specifically in sequential, spatial, and graph data, that fall outside of normal behaviour forms collective anomalies. It is denoted as a group of interconnected, correlated, or sequential instances, where individuals of the group are not anomalous themselves; the collective sequence is anomalous. Anomalous events rarely occur; however, these events bring about dramatic negative impacts in businesses and governments using IoT applications [4].

As for protecting IoT and I.T. applications, intrusion detection systems (I.D.S.s) that alert abnormal events or suspicious activities that might lead to an attack have been developed. I.D.S.s can be divided into two main categories: anomaly-based and signature-based. With anomaly-based I.D.S.s, unidentified attacks or zero-day attacks can be detected as deviations from normal activities [5]. However, signature-based I.D.S cannot identify unknown attacks until the vendors release updated versions consisting of the new attack signatures [5]. This indicates that anomaly-based I.D.S.s are strongly positioned to secure IoT devices better than signature-based I.D.S.s. Moreover, there is a large amount of raw data generated by IoT devices, which leads to the process of identifying suspicious behaviour from data suffering from high computation cost due to included noise. Hence, lightweight distributed anomaly-based I.D.S.s play a significant role in thwarting cyber-attacks in the IoT network.

In recent years, using machine learning techniques to develop anomaly-based I.D.S.s to protect the IoT system has produced encouraging results as machine learning models are trained on normal and abnormal data and then used to detect anomalies [1,2]. However, building effective and efficient anomaly detection modules is a challenging task as machine learning has the following drawbacks:

- First, machine learning models, specifically with classical algorithms, are shallow to extract features that can truly represent underlying data to discriminate anomaly events from normal ones.
- Second, running machine learning models can consume extensive resources, making it challenging to deploy such models on resource-constrained devices.
- Third, it requires massive data for training machine learning models to archive high accuracy in anomaly detection. Therefore, machine learning models may not capture all of the cyber-attacks or suspicious events due to training data. This means that machine learning suffers from both false positives and false negatives in some circumstances.

However, with the advancement in hardware such as GPU and neural networks such as deep learning, machine learning has constantly improved. This makes it promising for anomaly detection emerging platforms such as blockchain.

This paper aims to provide an in-depth review of current works in developing anomaly detection solutions using machine learning to protect an IoT system, which can help researchers and developers design and implement new anomaly-based I.D.S.s. Our contributions are summarised as follows: first, we present the significance of anomaly detection in the IoT system (Section 2); then, we identify the challenges of applying anomaly detection to an IoT system (Section 3); after that, we describe the state-of-the-art machine learning techniques for detecting anomalies in the system (Section 4); finally, we analyse the use of machine learning techniques for IoT anomaly detection (Section 5). In particular, this paper also covers the federated learning technique that helps to collaboratively train effective machine learning models to detect anomalies (Section 4) and indicates that the use of blockchain for anomaly detection is a novel contribution as the inherent characteristics of a distributed ledger is an ideal solution to defeat adversarial learning systems (Section 5).

2. Significance of Anomaly Detection in the IoT

Over the years, anomaly-based I.D.S.s have been applied in a wide range of IoT applications, as illustrated in Table 1. This section will focus on the important roles of anomaly detection systems in industries, smart grids, and smart cities.

Table 1. Anomaly-Based I.D.S.s according to Anomaly Types and Applications.

		ANOMALY TYPES		
		Points	Contextual	Collective
APPLICATIONS	Generic	[6] [9]	[7]	[8] [10] [11] [12] [13] [14] [15]
	Flights		[16]	
	Industries	[17] [18] [19]		
	Health		[20]	
	Smart Cities	[21]		
	Smart Grids		[22]	
	Smart Home	[23]		[24] [25] [26]
	Unmanned Aerial Vehicles		[27]	

Industrial IoT is one of the beneficiaries of anomaly detection tools. Anomaly detection has been leveraged for industrial IoT applications such as power systems, health monitoring [28], heating ventilation and air conditioning system fault detection [29], production plant maintenance scheduling [30], and manufacturing quality control systems [31]. In [32], machine learning approaches such as linear regression have been applied to sensor readings of engine-based machines to learn deviations from normal system behaviours. The study demonstrated that anomaly detection plays a significant role in preventive maintenance by detecting machine failures and inefficiencies. In another study, autoencoder (A.E.)-based outlier detection was investigated in audio data using reconstruction error [33]. The study showed that early detection of anomalies could be used as responsive maintenance for machine failures, thereby reducing downtime. Furthermore, water facilities used IoT anomaly detection [34] to monitor and identify certain chemical concentration levels as a reactive alerting mechanism. These studies show that IoT anomaly detection provides mechanisms of improving efficiency and system up-time for industry machines by monitoring machine health.

The power sector including existing smart grids has also attracted anomaly detection systems to identify power faults and outages. The study in [35] utilised statistical methods to develop an anomaly detection framework using smart meter data. The authors argue that hierarchical network data can be used to model anomaly detection for power systems. The other study [36] employed high-frequency signals to detect anomalies in power network faults. The article concludes that local anomaly detection depends more on network size than topology. In [37], big data analysis schemes were explored to detect and localise failures and faults in power systems. The study showed that the compensation theorem in circuit theory could be applied to event detection in power networks. Physical attacks on smart grids such as energy theft can also be detected by using anomaly detection systems, as shown in [38]. It is compelling that anomaly detection plays a paramount role in detecting failures and faults in power systems, enhancing system reliability and efficiency.

Abnormality detection can be used for smart city facilities such as roads and buildings. Road surface anomalies were studied in [39]. It has been indicated that damage to private vehicles can be reduced if the road surface is monitored for anomalies so that timely measures such as maintenance are taken before road incidents. In the study undertaken in [40], pollution monitoring and controlling were modelled as an anomaly to enable policymaker decisions in health, traffic, and environment. Similarly, assisted living can also benefit from IoT-based anomaly detection as deviations from normal alert caregivers as studied in [41]. Thus, it can be summed up that abnormal situations in smart cities and buildings can be detected using anomaly detection systems, and these can be provided to policymakers for decision-making purposes.

3. Challenges in IoT Anomaly Detection Using Machine Learning

The development of anomaly detection schemes in the IoT environment is challenging due to several factors such as (1) scarcity of IoT resources; (2) profiling normal behaviours; (3) the dimensionality of data; (4) context information; and (5) the lack of resilient machine learning models [15]. These factors will be explained in this section.

3.1. Scarcity of IoT Resources

The leverage of device-level IoT anomaly detection can be hindered by the constraints in storage, processing, communication, and power resources. To compensate for this, the cloud can be adopted as a data collection, storage, and processing platform. However, the remoteness of the cloud can introduce high latency due to resource scheduling and round trip time. This delay may not be acceptable for real-time requirements of IoT suspicious events [15]. It is also evident that the scale of traffic in the IoT may degrade the detection performance of the anomaly detection system if it exceeds the capacity of the devices. A better solution is to offload certain storage and computations from devices to edge nodes or to send aggregated data to the cloud. Sliding window techniques can also offer reduced storage benefits by withholding only certain data points, though the anomaly detection system may require patterns/trends [26].

3.2. Profiling Normal Behaviours

The success of an anomaly detection system depends on gathering sufficient data about normal behaviours; however, defining normal activities is challenging. Due to their rare occurrence, anomalous behaviours might be collected within normal behaviours. There is a lack of datasets representing both IoT normal and abnormal data, making supervised learning impractical, specifically for massively deployed IoT devices. This drives the need to model IoT anomaly detection systems in unsupervised or semi-supervised schemes, where data deviating from those collected in normal operations are taken as anomalous [3].

3.3. Dimensionality of Data

IoT data can be univariate as key-value x_t or multivariate as temporally correlated univariate $x_t = [x_t^1, \ldots, x_t^n]$. The IoT anomaly detection using univariate series compares current data against historical time series. In contrast, multivariate-based detection provides historical stream relationships and relationships among attributes at a given time. Thus, choosing a specific anomaly detection mechanism in IoT applications depends on data dimensionality due to associated overheads in processing [3,29]. Furthermore, multivariate data introduces the complexity of processing for models, which needs dimension reduction techniques using principal components analysis (P.C.A.) and A.E.s. On the other hand, univariate data may not represent finding patterns and correlations that enhance machine learning performance.

3.4. Context Information

The distributed nature of IoT devices caters to context information for anomaly detection. However, the challenge is to capture the temporal input at a time t_1 is related to input at a time t_n and spatial contexts in large IoT deployments where some IoT devices are mobile in their operations. This means that introducing context enriches anomaly detection systems, but increases complexity if the right context is not captured [3].

3.5. Lack of Machine Learning Models Resiliency against Adversarial Attacks

The lack of a low false-positive rate of existing machine learning models and the vulnerability to adversarial attacks during training and detection call for both accurate algorithms and resilient models. On the other hand, the massive deployment of IoT devices could be leveraged for collective anomaly detection as most of the devices in the network exhibit similar characteristics. This large number of devices helps to utilise the power of cooperation against cyber-attacks such as malware [42]. Model poisoning and evasion can decrease the utility of machine learning models as adversaries can introduce fake data to train or tamper the model.

4. Machine Learning Techniques for Detecting Anomalies in the IoT

Several aspects of IoT anomaly detection using machine learning must be considered. Learning algorithm methods can be categorised into three groups: supervised, unsupervised, and semi-supervised. The technique to train the learning algorithms across many decentralised IoT devices is known as federated learning. In addition, anomaly detection can be seen in terms of extant data dimension, leading to univariate-and multivariate-based approaches. In the rest of this section, we will present the anomaly detection schemes based on (1) machine learning algorithms; (2) federated learning; and (3) data sources and dimensions.

4.1. Detection Schemes Based on Machine Learning Algorithms

Supervised algorithms, known as discriminative algorithms, are classification-based learning through labelled instances. These algorithms consist of classification algorithms such as the K-nearest neighbour (K.N.N.), support vector machine (SVM), Bayesian network, and neural network (N.N.) [43,44]. K.N.N. is one of the distance-based algorithms of anomaly detection where the distances of anomalous points from the majority of the dataset are greater than a specific threshold. Calculating the distances is computationally complex; it seems impossible to provide on-device anomaly detection using this algorithm. On the other hand, SVM provides a hyperplane that divides data points for classification. As in the case of K.N.N., it is so resource-intensive that the applicability to IoT anomaly detection is impractical. As the Bayesian network may not require the prior knowledge of neighbour nodes for anomaly detection, it can be adopted for resource-constrained devices through low accuracy. Finally, N.N. algorithms have been extensively used to train on normal data so that anomalous data can be detected as the deviation from normal. The resource requirements of N.N. algorithms make it challenging to adapt to the IoT environment. Hence, supervised algorithms are the least applicable for IoT anomaly detection systems for their labelled dataset requirements and extensive resource requirements.

Commonly known as generative algorithms, unsupervised algorithms use unlabelled data to learn hierarchical features. Clustering-based algorithms such as K-means and density-based spatial clustering of applications with noise (D.B.S.C.A.N.) are unsupervised techniques that apply similarity and density attributes to classify data points into clusters [43,44]. Abnormal points are small data points significantly far from the dense area, while normal points are either close to or within the clusters. Usually, clustering algorithms are used with classification algorithms to enhance anomaly detection accuracy. Because of resource usage, most of the clustering algorithms cannot be directly applied to IoT devices for anomaly detection. Another unsupervised learning technique involves dimension-reduction approaches such as P.C.A. and A.E. to remove noise and redundancy from data to reduce the dimension of original data [44,45]. P.C.A. has been extensively applied to anomaly detection, but it fails in the dynamic IoT environment. A.E. has produced promising results in IoT anomaly detection in reducing data sizes and in reconstructing errors to identify anomalous points. However, these techniques have been used extensively as a part of feature extraction for classification algorithms. The dimensionality reduction algorithms in unsupervised learning can be adapted to IoT anomaly detection. Semi-supervised algorithms combine discriminative and generative algorithms by providing normal data instances so that deviation from normal behaviour is seen as abnormal behaviour. Hence, anomaly detection in IoT is geared toward unsupervised or semi-supervised algorithms where normal system profiling is utilised as a baseline environment [46].

Table 2 shows the state-of-the-art machine learning algorithms according to three anomaly types.

Table 2. Learning Algorithms According to Anomaly Types and Machine Learning Schemes.

		ANOMALY TYPES		
		Points	Contextual	Collective
MACHINE LEARNING SCHEMES	Supervised	RF [21] DL [17]	RL [16] LSTM [22]	CNN [24] GNN [8] Multiple [10] AE-ANN [11] LSTM [12] AE-CNN [13] Ensemble [14]
	Unsupervised	AE-CNN [6] AE [18]	Subspace [27]	AE [25] Self-learning [26]
	Semi-Supervised	TCN [23]	AE-LSTM [20] DBN [7]	DNN [15]

4.2. Training Detection Schemes Based on Federated Learning Algorithms

Federated learning, also known as collaborative learning, allows IoT devices to train machine learning models locally and send the trained models, not the local data, to the server for aggregation [47,48]. This training method is different from the standard machine learning training approaches that require centralising the training data in one place such as a server or data centre.

The federating learning method consists of four main steps. First, the server initialises a global machine learning model for anomaly detection and selects a subset of IoT devices to send the initialised model. Second, each selected IoT device will train the model by using its local data, then send the trained model back to the server. Next, the server will aggregate received models to form the global model. Finally, the server will send the final model to all IoT devices to detect anomalies. Note that the server can repeat the tasks of selecting a sub-set of IoT devices, sending the global model, receiving the trained models, and aggregating the received models multiple times, as some devices may not be available at the time of federated computation or some may have dropped out during each round.

By using federated learning, data in the IoT system is decentralised, and data privacy is protected. The other advantages of federated learning include lower latency, less network load, less power consumption, and can be applied across multiple organisations. However, federated learning also suffers from some drawbacks such as inference attacks [49] and model poisoning [50].

4.3. Detection Mechanisms Based on Data Sources and Dimensions

Univariate IoT data consists of data representation from a single IoT device over time. In reality, anomaly detection systems utilise data from multiple IoT devices deployed in complex environments. These multivariate multi-sources feed richer contexts by providing noise-tolerant temporal and spatial information than a single source.

4.3.1. Univariate Using Non-Regressive Scheme

In the non-regressive scheme, threshold-based mechanisms can be leveraged by setting low and high thresholds of observations on univariate stationary data to flag anomalies if a data point falls outside the boundary. More advanced mechanisms such as mean and variance thresholds produced over historical data can replace this min–max approach. Another similar approach is using a box plot to split data distribution into a range of small categories where new data points are compared against the boxes. These non-regressive approaches are ideal in saving resources such as processors and memories for IoT devices. However, being distributed techniques over univariate observations, the range-based schemes fail to detect contextual and collective anomalies due to the lack of the ability to capture temporal relationships [3].

N.N.s such as A.E.s, recurrent neural networks (R.N.N.), and long short-term memory (L.S.T.M.) can be used as non-regressive models to solve the problem of anomaly detection in the IoT ecosystem using univariate time series data. A.E. is used to reconstruct data symmetrically from the input to the output layer, and a high reconstruction error probably indicates abnormality [13]. A.E. can also be applied to resource-constrained IoT devices for conserving resources and battery power. On the other hand, R.N.N. provides memory in the network by affecting neurons from previous outputs through feedback loops. This enables the capture of temporal contexts over time. The vanishing gradient problem in R.N.N. makes it unsuitable for large IoT networks. L.S.T.M. can provide semi-supervised learning on normal time series data to identify anomaly sequences from reconstruction to solve this error problem. Hence, it seems that combining A.E. and L.S.T.M. can bring about resource-saving and accuracy requirements of the IoT anomaly detection tasks.

4.3.2. Univariate Using Regressive Scheme

Predictive approaches, known as regressive schemes, enable identifying anomalies by comparing predicted value to actual value in time series data. Parametric models such as autoregressive moving average (A.R.M.A.) are popular techniques despite seasonality or mean shift problems in non-stationary datasets. However, these problems can be solved by using enhanced variants of A.R.M.A. such as autoregressive integrated moving average (A.R.I.M.A.) and seasonal A.R.M.A. As another approach to predictive IoT anomaly detection, NN-based predictive models such as M.L.P., R.N.N., L.S.T.M., and others can be applied to capture the dynamics of a time series on complex univariate data [46]. For instance, R.N.N., L.S.T.M., and G.R.U. models can represent the variability in time series data to predict the expected values for time sequences. Recently, attention-based models have been applied to IoT anomaly detection in complex long sequential data. Similar to the non-regressive scheme, sequential models can boost the accuracy of IoT anomaly detection if dimensional reduction algorithms can be used in feature extraction.

4.3.3. Multivariate Using Regressive Scheme

As the additional variables increase data sizes, dimensionality reduction techniques such as P.C.A., A.E., and others can be employed to decrease overall data size. P.C.A. can capture the interdependence of variables for multivariate sources. It reduces the data size by decomposing multivariate data into a reduced set. The linearity and computational complexity of P.C.A. can limit its usage for IoT anomaly detection. A.E. works like P.C.A. and can discover anomalies in multivariate time series data using reconstruction error, the same way as in univariate cases. The promising aspect of A.E. is its low resource usage and its non-linear feature extraction. Similar to predictive and non-predictive models on univariate data, schemes using L.S.T.M., CNN, DBN, and others can also be applied to identifying anomalies in multi-source IoT systems. Specifically, CNN and L.S.T.M. algorithms can be preceded by A.E. for important feature extraction and resource savings. These deep learning schemes can learn spatio-temporal aspects of multivariate IoT data [12].

Clustering mechanisms are another approach to detect anomalies in multivariate data. In addition, graph networks can be used to learn models about variable or sequence relationships where the weakest weight between graph nodes is considered anomalous.

5. Analysis of Machine Learning for IoT Anomaly Detection

Anomaly detection systems have proven their capabilities of defending traditional networks by detecting suspicious behaviours. However, the standalone anomaly detection systems in classical systems do not fit the architecture of distributed IoT networks. In such systems, a single node compromise could damage the entire network. By collecting traffic from various spots, a collaborative anomaly detection framework plays a paramount role in thwarting cyber threats. However, the trust relationship and data sharing form two major challenges [42,51]. In this massive network, insider attacks can be a serious issue.

Furthermore, as most anomaly detection systems apply machine learning, nodes may not be willing to share normal profiles for training or performance optimisation due to privacy issues. The trust problem can be solved by implementing a central server that handles trust computation and data sharing. However, this approach could lead to a single point of failure and security, specifically for the large-scale deployment of IoT devices. Recently, blockchain has attracted much interest in financial sectors for its capability of forming trust among mistrusting entities using contracts and consensus. Blockchain could provide an opportunity to solve the problem of collaborative anomaly detection by providing trust management and a data-sharing platform. In the remainder of this section, we will focus on analysing (1) the collaborative architecture for IoT anomaly detection using blockchain; (2) datasets and algorithms for IoT anomaly detection; and (3) resource requirements of IoT anomaly detection.

5.1. Collaborative Architecture for IoT Anomaly Detection

Blockchain is a decentralised ledger that provides immutability, trustworthiness, authenticity, and accountability mechanisms for the maintained records based on majority consensus. Though it was originally applied to digital currency systems, blockchain can be applied in various fields. With the power of public-key cryptography, strong hash functions, and consensus algorithms, participating nodes in a blockchain can verify the formation of new blocks. A block typically consists of a group of records, timestamp, previous block hash, nonce, and a block's hash. Thus, the change in a record or group of records will be reflected in the next block's previous hash field, which makes it immune to adversarial change [42].

The powerful attributes of blockchain could provide a solid foundation for anomaly detection in distributed networks such as the IoT. IoT devices can collaboratively develop a global anomaly detection model from local models without adversarial attacks using blockchain architecture. As IoT needs mutual trust to share local models in a secure and tamper-proof way, consensus algorithms and decentralised blockchain storage make it challenging for malicious actors to manipulate the network. However, the successful Bitcoin consensus algorithms in financial areas such as proof-of-work require extensive storage and processing capabilities. Etherium has applied proof-of-stake where the participants' stakes determine consensus. It uses smart contracts, and is less computationally intensive. Hyperledger Fabric is another customisable blockchain platform that applies smart contracts in distributed systems rather than cryptocurrencies. As it relies on central service to enable participants to endorse transactions, endorsing participants must agree on the value of a transaction to reflect changes in the local participant ledger. These three popular blockchain systems do not seem to solve resource-constrained IoT devices [51].

Blockchain-based security solutions have been discussed in a mix of traditional and IoT systems [52,53]. In these studies, a resource-rich device was connected to IoT devices, where the device acts as a proxy to connect IoT devices to the blockchain. A similar study was conducted in [54]. The main advantages of these approaches lie in resource savings, but they may also create a central point of failure. In [55], the author's utilised smart contracts to integrate IoT devices into blockchain for communication integrity and authenticity through the resource requirement issues that may not make it practical. The most promising result has been achieved on distributed and collaborative IoT anomaly detection [51]. The study uses a self-attestation mechanism to establish a dynamic trusted model against which nodes compare to detect anomalous behaviour. The model is cooperatively updated by majority consensus before being distributed to peers.

5.2. Datasets and Algorithms for IoT Anomaly Detection

The lack of labelled realistic datasets has hampered anomaly detection research in the IoT. The existing data suffer from lacking realistic representation for IoT traffic patterns and lack capture of the full range of anomalies that may occur in the IoT. Class imbalance between normal traffic and anomalous patterns also manifests, which makes classification systems inefficient. Most IoT traffic can be represented as normal behaviour while it dynamically changes over time. As contextual information such as time, environment, and neighbour nodes profile rich information to improve anomaly detection in the IoT, it seems that multivariate data plays a significant role. The challenges associated with the absence of truly representative, realistic, and balanced datasets favour an anomaly detection scheme that profiles normal behaviours to detect anomalous points that deviate from the normal data [56]. Table 3 shows the common datasets that have been commonly used in some recent studies in this research area. As can be seen, most datasets are not specific to the IoT system; however, they are still suitable for training and evaluating anomaly-based I.D.S.s because they contain both normal and abnormal data.

Table 3. Common Datasets for Anomaly Detection in the IoT System (Adapted from [1]).

Dataset	Published Year	IoT Specific	Dimensions	Normal Instances	Abnormal Instances
N-BaIoT [57]	2018	Yes	115	555,932	6,545,967
CICIDS 2017 [58]	2017	No	80	2,273,097	557,646
AWID [59]	2015	No	155	530,785	44,858
UNSW-NB15 [60]	2015	No	49	2,218,761	321,283
NLS-KDD [61]	2009	No	43	77,054	71,463
Kyoto [62]	2006	No	24	50,033,015	43,043,255
KDD CUP 1999 [63]	1999	No	43	1,033,372	4,176,086

The initial deployment of the IoT anomaly detection system lacks historical data that specify normal and anomalous points. This absence and the rare nature of anomalies challenge the usage of traditional machine learning schemes. Though several techniques of solving imbalanced data have been proposed, such methods cannot maintain the temporal context of anomalies. In addition, supervised algorithms capture only known anomalies while failing to detect novel attacks. Thus, unsupervised or semi-supervised approaches can be used to solve the limitations of supervised algorithms [54].

While several techniques have been used in IoT anomaly detection, most of the approaches have failed to satisfy the resource and power requirements of IoT devices [54]. Though there is no single best anomaly detection approach, deep learning techniques, specifically A.E. and CNN, have shown promising results in both delivering better resource-saving and accuracy, respectively [64]. While algorithms such as CNN and L.S.T.M. can boost detection accuracy, A.E. can be used to reduce the dimension of data and extract representative features by eliminating noise. Specifically, L.S.T.M. can be applied to dynamic and complex observations within time-series IoT data over a long sequence. Thus, it suggests that these techniques or combinations could be further explored to detect anomalies in the IoT ecosystem [65].

5.3. Resource Requirements of IoT Anomaly Detection

The resource-constrained nature of IoT devices prohibits the deployment of traditional host-based intrusion detection such as anti-malware and anti-virus. As traffic analysis consumes huge computational resources during anomaly detection, incremental approaches such as sliding windows can reduce the processing and storage requirements for IoT devices. It is also critical that the anomaly detection engine of the IoT system should operate in near real-time for reliable detection. This indicates that adaptive techniques help to improve the detection model over time without major retraining. However, offline training may be applied for initial deployment.

6. Conclusions

The IoT environment's massive number, heterogeneity, and resource constraints have hindered cyber-attack prevention and detection capabilities. These characteristics attract monitoring IoT devices at the network level as on-device solutions are not feasible. To this end, anomaly detection is better positioned to protect the IoT network. To protect the system, anomaly detection is considered to be an important tool as it helps identify and alert abnormal activities in the system. Machine learning has been applied for anomaly detection systems in I.T. and IoT systems. However, the applications of anomaly detection systems using machine learning in I.T. systems have been better than the IoT ecosystem due to their resource capabilities and in-perimeter location. Nevertheless, the existing machine learning-based anomaly detection is vulnerable to adversarial attacks. This article has presented a comprehensive survey of anomaly detection using machine learning in the IoT system. The significance of anomaly detection, the challenges when developing anomaly detection systems, and the analysis of the used machine learning algorithms are provided. Finally, it has been recommended that blockchain technology can be applied to mitigate model corruption by adversaries where IoT devices can collaboratively produce a single model using blockchain consensus mechanisms. In the future, we plan to implement a blockchain-based anomaly detection system for protecting high-end IoT devices such as Raspberry Pi. The system can be built on a python-based machine learning platform such as TensorFlow and a blockchain platform such as Hyperledger Fabric, where Raspberry Pi devices act as distributed nodes.

Author Contributions: Conceptualization: A.D. and N.C.; methodology: A.D. and V.-D.N.; formal analysis: V.-D.N.; investigation: V.-D.N.; resources: N.C.; data curation: V.-D.N.; writing—original draft preparation: A.D. and V.-D.N.; writing—review and editing: A.D., V.-D.N., W.H. and N.C.; supervision: N.C.; project administration: N.C. and W.H.; funding acquisition: N.C. and W.H. All authors have read and agreed to the published version of the manuscript.

Funding: This work was supported by the SmartSat C.R.C., whose activities are funded by the Australian Government's C.R.C. Program.

Conflicts of Interest: The authors declare no conflict of interest in this research.

References

1. Alsoufi, M.A.; Razak, S.; Siraj, M.M.; Nafea, I.; Ghaleb, F.A.; Saeed, F.; Nasser, M. Anomaly-Based Intrusion Detection Systems in IoT Using Deep Learning: A Systematic Literature Review. *Appl. Sci.* **2021**, *11*, 8383. [CrossRef]
2. Njilla, L.; Pearlstein, L.; Wu, X.; Lutz, A.; Ezekiel, S. Internet of Things Anomaly Detection using Machine Learning. In Proceedings of the 2019 IEEE Applied Imagery Pattern Recognition Workshop (A.I.P.R.), Washington, DC, USA, 15–17 October 2019; pp. 1–6.
3. Cook, A.A.; Mısırlı, G.; Fan, Z. Anomaly Detection for IoT Time-Series Data: A Survey. *IEEE Internet Things J.* **2020**, *7*, 6481–6494. [CrossRef]
4. Cauteruccio, F.; Cinelli, L.; Corradini, E.; Terracina, G.; Ursino, D.; Virgili, L.; Savaglio, C.; Liotta, A.; Fortino, G. A Framework for Anomaly Detection and Classification in Multiple IoT Scenarios. *Future Gener. Comput. Syst.* **2021**, *114*, 322–335. [CrossRef]
5. Doshi, R.; Apthorpe, N.; Feamster, N. Machine Learning DDoS Detection for Consumer Internet of Things Devices. In Proceedings of the 2018 IEEE Security and Privacy Workshops (S.P.W.), San Francisco, CA, USA, 24 May 2018; pp. 29–35.
6. Hwang, R.H.; Peng, M.C.; Huang, C.W.; Lin, P.C.; Nguyen, V.L. An Unsupervised Deep Learning Model for Early Network Traffic Anomaly Detection. *IEEE Access* **2020**, *8*, 30387–30399. [CrossRef]

7. Manimurugan, S.; Al-Mutairi, S.; Aborokbah, M.M.; Chilamkurti, N.; Ganesan, S.; Patan, R. Effective Attack Detection in Internet of Medical Things Smart Environment Using a Deep Belief Neural Network. *IEEE Access* **2020**, *8*, 77396–77404. [CrossRef]
8. Protogerou, A.; Papadopoulos, S.; Drosou, A.; Tzovaras, D.; Refanidis, I. A Graph Neural Network Method for Distributed Anomaly Detection in IoT. *Evol. Syst.* **2021**, *12*, 19–36. [CrossRef]
9. Cauteruccio, F.; Fortino, G.; Guerrieri, A.; Liotta, A.; Mocanu, D.C.; Perra, C.; Terracina, G.; Torres Vega, M. Short-long term anomaly detection in wireless sensor networks based on machine learning and multi-parameterized edit distance. *Inf. Fusion* **2019**, *52*, 13–30. [CrossRef]
10. Hasan, M.; Islam, M.M.; Zarif, M.I.I.; Hashem, M. Attack and Anomaly Detection in IoT Sensors in IoT Sites Using Machine Learning Approaches. *Internet Things* **2019**, *7*, 100059. [CrossRef]
11. AL-Hawawreh, M.; Moustafa, N.; Sitnikova, E. Identification of Malicious Activities in Industrial Internet of Things Based on Deep Learning Models. *J. Inf. Secur. Appl.* **2018**, *41*, 1–11. [CrossRef]
12. Shukla, R.; Sengupta, S. Scalable and Robust Outlier Detector using Hierarchical Clustering and Long Short-Term Memory (L.S.T.M.) Neural Network for the Internet of Things. *Internet Things* **2020**, *9*, 100167. [CrossRef]
13. Yin, C.; Zhang, S.; Wang, J.; Xiong, N.N. Anomaly Detection Based on Convolutional Recurrent Autoencoder for IoT Time Series. *IEEE Trans. Syst. Man Cybern. Syst.* **2020**, 1–11. [CrossRef]
14. Tsogbaatar, E.; Bhuyan, M.H.; Taenaka, Y.; Fall, D.; Gonchigsumlaa, K.; Elmroth, E.; Kadobayashi, Y. SDN-Enabled IoT Anomaly Detection Using Ensemble Learning. I.F.I.P. In *International Conference on Artificial Intelligence Applications and Innovations*; Springer International Publishing: Cham, Switzerland, 2020; pp. 268–280.
15. Diro, A.A.; Chilamkurti, N. Distributed Attack Detection Scheme Using Deep Learning Approach for Internet of Things. *Future Gener. Comput. Syst.* **2018**, *82*, 761–768. [CrossRef]
16. Farshchi, M.; Weber, I.; Della Corte, R.; Pecchia, A.; Cinque, M.; Schneider, J.G.; Grundy, J. Contextual Anomaly Detection for a Critical Industrial System Based on Logs and Metrics. In Proceedings of the 2018 14th European Dependable Computing Conference (E.D.C.C.), Iasi, Romania, 10–14 September 2018; pp. 140–143.
17. Ferrari, P.; Rinaldi, S.; Sisinni, E.; Colombo, F.; Ghelfi, F.; Maffei, D.; Malara, M. Performance Evaluation of Full-Cloud and Edge-Cloud Architectures for Industrial IoT Anomaly Detection Based on Deep Learning. In Proceedings of the 2019 II Workshop on Metrology for Industry 4.0 and IoT (MetroInd4.0 IoT), Naples, Italy, 4–6 June 2019; pp. 420–425.
18. Bhatia, R.; Benno, S.; Esteban, J.; Lakshman, T.V.; Grogan, J. Unsupervised Machine Learning for Network-Centric Anomaly Detection in IoT. In Proceedings of the 3rd A.C.M. CoNEXT Workshop on Big DAta, Machine Learning and Artificial Intelligence for Data Communication Networks, Orlando, FL, USA, 9 December 2019; pp. 42–48.
19. Savic, M.; Lukic, M.; Danilovic, D.; Bodroski, Z.; Bajovic, D.; Mezei, I.; Vukobratovic, D.; Skrbic, S.; Jakovetic, D. Deep Learning Anomaly Detection for Cellular IoT With Applications in Smart Logistics. *IEEE Access* **2021**, *9*, 59406–59419. [CrossRef]
20. Ngo, M.V.; Luo, T.; Chaouchi, H.; Quek, T.S. Contextual-Bandit Anomaly Detection for IoT Data in Distributed Hierarchical Edge Computing. In Proceedings of the 2020 IEEE 40th International Conference on Distributed Computing Systems (I.C.D.C.S.), Singapore, 29 November–1 December 2020; pp. 1227–1230.
21. Alrashdi, I.; Alqazzaz, A.; Aloufi, E.; Alharthi, R.; Zohdy, M.; Ming, H. AD-IoT: Anomaly Detection of IoT Cyberattacks in Smart City Using Machine Learning. In Proceedings of the 2019 IEEE 9th Annual Computing and Communication Workshop and Conference (C.C.W.C.), Las Vegas, NV, USA, 7–9 January 2019; pp. 305–310.
22. Utomo, D.; Hsiung, P.A. Anomaly Detection at the IoT Edge using Deep Learning. In Proceedings of the 2019 IEEE International Conference on Consumer Electronics—Taiwan (ICCE-TW), Yilan, Taiwan, 20–22 May 2019; pp. 1–2.
23. Cheng, Y.; Xu, Y.; Zhong, H.; Liu, Y. Leveraging Semisupervised Hierarchical Stacking Temporal Convolutional Network for Anomaly Detection in IoT Communication. *IEEE Internet Things J.* **2021**, *8*, 144–155. [CrossRef]
24. Han, N.; Gao, S.; Li, J.; Zhang, X.; Guo, J. Anomaly Detection in Health Data Based on Deep Learning. In Proceedings of the 2018 International Conference on Network Infrastructure and Digital Content (IC-NIDC), Guiyang, China, 22–24 August 2018; pp. 188–192.
25. Chalapathy, R.; Toth, E.; Chawla, S. Group Anomaly Detection Using Deep Generative Models. In *Machine Learning and Knowledge Discovery in Databases*; Springer International Publishing: Cham, Switzerland, 2019; pp. 173–189.
26. Nguyen, T.D.; Marchal, S.; Miettinen, M.; Fereidooni, H.; Asokan, N.; Sadeghi, A.R. DÏoT: A Federated Self-learning Anomaly Detection System for IoT. In Proceedings of the 2019 IEEE 39th International Conference on Distributed Computing Systems (I.C.D.C.S.), Dallas, TX, USA, 7–10 July 2019; pp. 756–767.
27. He, Y.; Peng, Y.; Wang, S.; Liu, D.; Leong, P.H.W. A Structured Sparse Subspace Learning Algorithm for Anomaly Detection in UAV Flight Data. *IEEE Trans. Instrum. Meas.* **2018**, *67*, 90–100. [CrossRef]
28. Himeur, Y.; Ghanem, K.; Alsalemi, A.; Bensaali, F.; Amira, A. Artificial Intelligence Based Anomaly Detection of Energy Consumption in Buildings: A Review, Current Trends and New Perspectives. *Appl. Energy* **2021**, *287*, 116601. [CrossRef]
29. Piscitelli, M.S.; Brandi, S.; Capozzoli, A.; Xiao, F. A Data Analytics-Based Tool for The Detection and Diagnosis of Anomalous Daily Energy Patterns in Buildings. *Build. Simul.* **2021**, *14*, 131–147. [CrossRef]
30. Kim, D.; Yang, H.; Chung, M.; Cho, S.; Kim, H.; Kim, M.; Kim, K.; Kim, E. Squeezed Convolutional Variational AutoEncoder for Unsupervised Anomaly Detection in Edge Device Industrial Internet of Things. In Proceedings of the 2018 International Conference on Information and Computer Technologies (I.C.I.C.T.), DeKalb, IL, USA, 23–25 March 2018; pp. 67–71.

31. Kanaway, A.; Sane, A. Machine Learning for Predictive Maintenance of Industrial Machines Using IoT Sensor Data. In Proceedings of the 2017 8th IEEE International Conference on Software Engineering and Service Science (I.C.S.E.S.S.), Beijing, China, 24–26 November 2017; pp. 87–90.
32. Shah, G.; Tiwari, A. Anomaly Detection in IIoT: A Case Study Using Machine Learning. In Proceedings of the The A.C.M. India Joint International Conference on Data Science and Management of Data. Association for Computing Machinery, Goa, India, 11–13 January 2018; pp. 295–300.
33. Oh, D.Y.; Yun, I.D. Residual Error Based Anomaly Detection Using Auto-Encoder in S.M.D. Machine Sound. *Sensors* **2018**, *18*, 1308. [CrossRef]
34. Giannoni, F.; Mancini, M.; Marinelli, F. Anomaly Detection Models for IoT Time Series Data. *arXiv* **2018**, arXiv:abs/1812.00890.
35. Moghaddass, R.; Wang, J. A Hierarchical Framework for Smart Grid Anomaly Detection Using Large-Scale Smart Meter Data. *IEEE Trans. Smart Grid* **2018**, *9*, 5820–5830. [CrossRef]
36. Passerini, F.; Tonello, A.M. Smart Grid Monitoring Using Power Line Modems: Anomaly Detection and Localization. *IEEE Trans. Smart Grid* **2019**, *10*, 6178–6186. [CrossRef]
37. Farajollahi, M.; Shahsavari, A.; Mohsenian-Rad, H. Location Identification of Distribution Network Events Using Synchrophasor Data. In Proceedings of the 2017 North American Power Symposium (NAPS), Morgantown, WV, USA, 17–19 September 2017; pp. 1–6.
38. Yip, S.C.; Tan, W.N.; Tan, C.; Gan, M.T.; Wong, K. An Anomaly Detection Framework for Identifying Energy Theft and Defective Meters in Smart Grids. *Int. J. Electr. Power Energy Syst.* **2018**, *101*, 189–203. [CrossRef]
39. El-Wakeel, A.S.; Li, J.; Rahman, M.T.; Noureldin, A.; Hassanein, H.S. Monitoring Road Surface Anomalies Towards Dynamic Road Mapping for Future Smart Cities. In Proceedings of the 2017 IEEE Global Conference on Signal and Information Processing (GlobalSIP), Montreal, QC, Canada, 14–16 November 2017; pp. 828–832.
40. Kong, X.; Song, X.; Xia, F.; Guo, H.; Wang, J.; Tolba, A. LoTAD: Long-Term Traffic Anomaly Detection Based on Crowdsourced Bus Trajectory Data. *World Wide Web* **2018**, *21*, 825–847. [CrossRef]
41. Bakar, U.A.B.U.A.; Ghayvat, H.; Hasanm, S.F.; Mukhopadhyay, S.C. Activity and Anomaly Detection in Smart Home: A Survey. In *Next Generation Sensors and Systems*; Springer International Publishing: Cham, Switzerland, 2016; pp. 191–220.
42. Alexopoulos, N.; Vasilomanolakis, E.; Ivánkó, N.R.; Mühlhäuser, M. Towards Blockchain-Based Collaborative Intrusion Detection Systems. In *Critical Information Infrastructures Security*; Springer International Publishing: Cham, Switzerland, 2018; pp. 107–118.
43. Hastie, T.; Tibshirani, R.; Friedman, J. *The Elements of Statistical Learning: Data Mining, Inference and Prediction*, 2nd ed.; Springer: New York, NY, USA, 2009.
44. Murphy, K.P. Machine Learning: A Probabilistic Perspective. MIT Press: Cambridge, MA, USA, 2013.
45. Chadha, G.S.; Islam, I.; Schwung, A.; Ding, S.X. Deep Convolutional Clustering-Based Time Series Anomaly Detection. *Sensors* **2021**, *21*, 5488. [CrossRef]
46. Jiang, J.; Han, G.; Liu, L.; Shu, L.; Guizani, M. Outlier Detection Approaches Based on Machine Learning in the Internet-of-Things. *IEEE Wirel. Commun.* **2020**, *27*, 53–59. [CrossRef]
47. Mothukuri, V.; Khare, P.; Parizi, R.M.; Pouriyeh, S.; Dehghantanha, A.; Srivastava, G. Federated Learning-based Anomaly Detection for IoT Security Attacks. *IEEE Internet Things J. (Early Access)* **2021**. [CrossRef]
48. Liu, Y.; Garg, S.; Nie, J.; Zhang, Y.; Xiong, Z.; Kang, J.; Hossain, M.S. Deep Anomaly Detection for Time-Series Data in Industrial IoT: A Communication-Efficient On-Device Federated Learning Approach. *IEEE Internet Things J.* **2021**, *8*, 6348–6358. [CrossRef]
49. Lee, H.; Kim, J.; Ahn, S.; Hussain, R.; Cho, S.; Son, J. Digestive neural networks: A novel defense strategy against inference attacks in federated learning. *Comput. Secur.* **2021**, *109*, 102378. [CrossRef]
50. Wang, C.; Chen, J.; Yang, Y.; Ma, X.; Liu, J. Poisoning attacks and countermeasures in intelligent networks: Status quo and prospects. *Digit. Commun. Netw. (Early Access)* **2021**. [CrossRef]
51. Meng, W.; Tischhauser, E.W.; Wang, Q.; Wang, Y.; Han, J. When Intrusion Detection Meets Blockchain Technology: A Review. *IEEE Access* **2018**, *6*, 10179–10188. [CrossRef]
52. Novo, O. Blockchain Meets IoT: An Architecture for Scalable Access Management in IoT. *IEEE Internet Things J.* **2018**, *5*, 1184–1195. [CrossRef]
53. Dorri, A.; Kanhere, S.S.; Jurdak, R. Towards an Optimized BlockChain for IoT. In Proceedings of the 2017 IEEE/ACM Second International Conference on Internet-of-Things Design and Implementation (IoTDI), Pittsburgh, PA, USA, 18–21 April 2017; pp. 173–178.
54. Özyılmaz, K.R.; Yurdakul, A. Work-in-Progress: Integrating low-Power IoT Devices to a Blockchain-Based Infrastructure. In Proceedings of the 2017 International Conference on Embedded Software (E.M.S.O.F.T.), Seoul, Korea, 15–20 October 2017; pp. 1–2.
55. Huh, S.; Cho, S.; Kim, S. Managing IoT Devices Using Blockchain Platform. In Proceedings of the 2017 19th International Conference on Advanced Communication Technology (I.C.A.C.T.), PyeongChang, Korea, 19–22 February 2017; pp. 464–467.
56. Alsoufi, M.A.; Razak, S.; Siraj, M.M.; Ali, A.; Nasser, M.; Abdo, S. Anomaly Intrusion Detection Systems in IoT Using Deep Learning Techniques: A Survey. In *Innovative Systems for Intelligent Health Informatics*; Springer International Publishing: Cham, Switzerland, 2021; pp. 659–675.
57. Meidan, Y.; Bohadana, M.; Mathov, Y.; Mirsky, Y.; Shabtai, A.; Breitenbacher, D.; Elovici, Y. N-BaIoT—Network-Based Detection of IoT Botnet Attacks Using Deep Autoencoders. *IEEE Pervasive Comput.* **2018**, *17*, 12–22. [CrossRef]

58. Sharafaldin, I.; Lashkari, A.H.; Ghorbani, A.A. Toward Generating a New Intrusion Detection Dataset and Intrusion Traffic Characterization. In Proceedings of the 4th International Conference on Information Systems Security and Privacy (I.C.I.S.S.P. 2018), Funchal, Portugal, 22–24 January 2018; pp. 108–116.
59. Kolias, C.; Kambourakis, G.; Stavrou, A.; Gritzalis, S. Intrusion Detection in 802.11 Networks: Empirical Evaluation of Threats and a Public Dataset. *IEEE Commun. Surv. Tutor.* **2016**, *18*, 184–208. [CrossRef]
60. Moustafa, N.; Slay, J. UNSW-NB15: A comprehensive data set for network intrusion detection systems (UNSW-NB15 network data set). In Proceedings of the 2015 Military Communications and Information Systems Conference (MilCIS), Canberra, Australia, 10–12 November 2015; pp. 1–6.
61. Tavallaee, M.; Bagheri, E.; Lu, W.; Ghorbani, A.A. A detailed analysis of the KDD CUP 99 data set. In Proceedings of the 2009 IEEE Symposium on Computational Intelligence for Security and Defense Applications, Ottawa, ON, Canada, 8–10 July 2009; pp. 1–6.
62. Malaiya, R.K.; Kwon, D.; Suh, S.C.; Kim, H.; Kim, I.; Kim, J. An Empirical Evaluation of Deep Learning for Network Anomaly Detection. *IEEE Access* **2019**, *7*, 140806–140817. [CrossRef]
63. Stolfo, S.; Fan, W.; Lee, W.; Prodromidis, A.; Chan, P. Cost-based modeling for fraud and intrusion detection: Results from the J.A.M. project. In Proceedings of the DARPA Information Survivability Conference and Exposition, Hilton Head, SC, USA, 25–27 January 2000; Volume 2, pp. 130–144.
64. Kamat, P.; Sugandhi, R. Anomaly Detection for Predictive Maintenance in Industry 4.0-A Survey. In Proceedings of the E3S Web of Conferences, Pune City, India, 18–20 December 2019; p. 02007.
65. Bovenzi, G.; Aceto, G.; Ciuonzo, D.; Persico, V.; Pescapé, A. A Hierarchical Hybrid Intrusion Detection Approach in IoT Scenarios. In Proceedings of the GLOBECOM 2020—2020 IEEE Global Communications Conference, Virtual Event, Taiwan, 7–11 December 2020; pp. 1–7.

Article

Intelligent Electromagnetic Sensors for Non-Invasive Trojan Detection

Ethan Chen *, John Kan, Bo-Yuan Yang, Jimmy Zhu and Vanessa Chen

Department of Electrical and Computer Engineering, Carnegie Mellon University, Pittsburgh, PA 15213, USA; johnkan@andrew.cmu.edu (J.K.); boyuany@andrew.cmu.edu (B.-Y.Y.); jzhu@cmu.edu (J.Z.); vanessachen@cmu.edu (V.C.)
* Correspondence: ethanchen@cmu.edu

Abstract: Rapid growth of sensors and the Internet of Things is transforming society, the economy and the quality of life. Many devices at the extreme edge collect and transmit sensitive information wirelessly for remote computing. The device behavior can be monitored through side-channel emissions, including power consumption and electromagnetic (EM) emissions. This study presents a holistic self-testing approach incorporating nanoscale EM sensing devices and an energy-efficient learning module to detect security threats and malicious attacks directly at the front-end sensors. The built-in threat detection approach using the intelligent EM sensors distributed on the power lines is developed to detect abnormal data activities without degrading the performance while achieving good energy efficiency. The minimal usage of energy and space can allow the energy-constrained wireless devices to have an on-chip detection system to predict malicious attacks rapidly in the front line.

Keywords: hardware security; electromagnetic sensing; machine learning; real time

Citation: Chen, E.; Kan, J.; Yang, B.-Y.; Zhu, J.; Chen, V. Intelligent Electromagnetic Sensors for Non-Invasive Trojan Detection. *Sensors* **2021**, *21*, 8288. https://doi.org/10.3390/s21248288

Academic Editors: Zihuai Lin and Wei Xiang

Received: 31 October 2021
Accepted: 8 December 2021
Published: 11 December 2021

Publisher's Note: MDPI stays neutral with regard to jurisdictional claims in published maps and institutional affiliations.

Copyright: © 2021 by the authors. Licensee MDPI, Basel, Switzerland. This article is an open access article distributed under the terms and conditions of the Creative Commons Attribution (CC BY) license (https://creativecommons.org/licenses/by/4.0/).

1. Introduction

The rapid growth of sensors and the Internet of Things (IoT) has the potential to transform society, the economy, and the quality of life. Many devices at the extreme edge collect and transmit sensitive information wirelessly for remote computing. The sensitive information can be leaked from side channels, including power consumption and electromagnetic (EM) emissions. Some devices are simply controlled by a simple wake-up signal to activate data transmission without two-way authentication. Moreover, the wireless charging techniques that allow energy constrained devices and electric cars to stay connected and operate continuously provide another entry point to exploit the sensitive information and the vulnerability in the power domain, as shown in Figure 1a. The vulnerability of those wireless devices to hacking or exploitation has emerged as a major concern on both security and public safety. For instance, because the electronic devices may continue receiving and transmitting signals while they are being wirelessly powered, the data activities are exposed to the energy source. Nevertheless, state-of-the-art cybersecurity approaches are mainly focused on software and digital modules. Security measures are not integrated in the analog/radio frequency (RF) domain to verify signal and power sources or to suppress the side-channel emissions in real time. To bridge the gap, this study presents a self-testing approach incorporating nanoscale EM sensing devices and learning algorithms to detect threats directly at the RF and analog front-end. As shown in Figure 1b, the EM sensors are integrated into the RF/analog front-end through post processing to monitor the EM emissions from power wires and critical signal nodes. Machine-learning modules were developed to analyze the sensed data for threat and vulnerability detection. Combing emerging material, device, circuit, and system concepts, this study developed a built-in threat detection approach in the RF/analog domain without degrading the performance while achieving good energy efficiency.

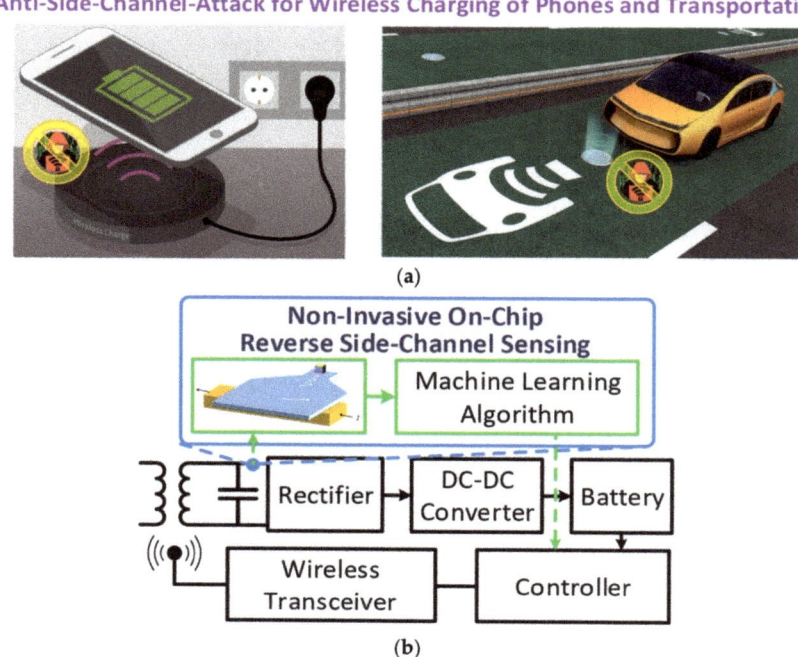

Figure 1. (**a**) Security vulnerability of electromagnetic emissions and wireless charging (figure credit for wireless charging: Infineon and PowerElectronics.com Available online: https://www.powerelectronics.com/markets/automotive/article/21864097/wireless-charging-of-electric-vehicles accessed on 9 December 2021), and (**b**) the proposed non-invasive on-chip sensing system.

2. Relevant Work

Wireless power transfer technology relies on the principle of electromagnetic induction, which falls into two categories, near field and far field. Near-field techniques utilize inductive coupling between coils of wire or capacitive coupling between metal electrodes [1–3]. Inductive coupling for power transfer over a short distance through magnetic fields is one of the most widely used wireless powering technologies. Considering the conversion efficiency and the safety criteria, radiative wireless power transfer is the most popular far-field technique to remotely power mobile devices over a long distance for low-power devices [4–6] and wireless sensor networks (WSNs) [7]. Nondirective antennas can be used to energize sensors, but the efficiency is low. On the other hand, directive transmitting antennas are more efficient to increase the maximum distances that can be remotely powered [8]. Many radiative wireless power transfer techniques have been discussed with different operation frequencies, average chargeable distances, beamforming techniques, and overall system complexity. Depending on different transfer schemes, specific wireless power transfer patterns can be adopted to charge the devices without interrupting the operations. For instance, the electric vehicles can be dynamically charged as they ride on the wireless power lane [9]. Hence, the wireless power source is becoming a new shared input to the devices and vehicles as they are all connected on the wireless charging platform, so the data activities are exposed to the energy source. Especially when the devices and vehicles with hardware trojans need to be recharged, its battery can be very low and the existing security functions may not work intermittently, which raises critical concerns of safety and security. However, the security vulnerabilities for pervasive devices accessing the shared wireless charging platform have not been addressed.

2.1. Survey of Hardware Trojans

Much attention has been focused on hardware trojan taxonomy, development, and detection in the past two decades, especially since the Defense Science Board of US Department of Defense released a report in 2005 on the security of the supply of high-performance integrated circuits (ICs) which highlighted the need for "secure and authentic hardware" [10]. The resulting research produced numerous publications [11–27] which not only provide insight into existing hardware trojans, but also develop a general framework of hardware trojan understanding. This section will first briefly review hardware trojan taxonomy and detection methods, point out the lack of literature related to analog trojan development, taxonomy, and detection, and then present a number of trojans scenarios that can be possible in the analog/RF domain, specifically attacking a Class-E amplifier in the later sections.

2.1.1. Hardware Trojan Taxonomy and Insertion

Hardware trojans defined by [12] are an intentional and malicious modification of a circuit that is designed to alter the circuit's behavior in order to accomplish a specific objective. It also makes a distinction between such trojans and design bugs and defects by defining such a bug as "an unintentional problem (i.e., error) that is unknowingly introduced into the circuit during its design and development phases" and a defect as "unintentional physical phenomenon (e.g., imperfection) that occurs during the circuit's fabrication, assembly, and testing phases". Trojans, as they are malicious, seek to tamper with the function of the integrated circuit and avoid detection, whereas flaws and bugs are generally discoverable via conventional models of testing and verification.

Reference [11] provides an excellent description of a general view of hardware trojan taxonomy. Furthermore, it details the supply chain and hardware development layout for ASICs and FPGAs. Other resources, such as reference [12], also provide excellent insight into the taxonomy of hardware trojans and the section of the design process into which hardware trojans can be inserted. Hence, it is clear from the various literature that the number of trojans, their triggers, payloads, and development can be inserted in numerous areas of the design phase. These publications indicate not only the type of trojan trigger and payload, but also the potential points at which the trojan can be inserted [1,4,17] which all developed trojans that can be inserted by an untrusted foundry, or which insert the trojan post-fabrication [11,12].

However, while a number of papers indicate hardware trojans, most focus on the digital domain, even while utilizing "analog" trojans. A search of literature generally yields results in which papers such as [15,18] developed "analog" trojans. However, the in these papers, the circuits they are trying to attack generally are in the realm of digital integrated circuits. In [12] the authors indicate there are at least four types of trojans, side channel, semiconductor, analog, and digital, and includes a catch-all category of other for those that do not directly fall into those other categories. Side-channel attacks generally leak information out of an analog channel [12,14]. Semiconductor trojans generally tamper with the dopant polarity [17], analog trojans seek to insert some sort of analog device or additional circuit that will affect some sort of change in the operation of the circuit, either immediate or over time [14,20–23], such as the capacitor trojan threat identified in [18]. Digital trojans generally try to cause issues with the finite state machines (FSM) [12] to cause the overall FSM to end up in a "don't-care" state, if available, that would contain a trojan payload.

Whilst there is a great deal of literature seeking to define hardware trojans, all of them, however, seek to attack circuits that remain in the digital domain. All the literature mentioned above, even those considered "analog" or even "semiconductor" trojans, seek to attack either microprocessors, digital circuits, etc. Additionally, some developed trojans for "RF applications", but their trojans generally remain in the realm of different transmitter input termination [20], and do not cover the full spectrum of possible hardware trojan attack vectors in the RF domain. Hence, we seek to contribute not only to this prior research, but also to provide some initial steps at the lack of research into trojans that can occur in the analog domain.

2.1.2. Analog Circuit Trojans

Hence, after noting the expansive literature focusing much needed attention on hardware trojans in the digital circuit design domain, the following papers attempt to address trojans in the analog circuit design domain. Quite a few indicate the noticeable lack of research in this field [23,25]. These trojans are more difficult to obfuscate or deceptively insert than larger, more topographically complex digital circuits that can feasibly hide a trojan, digital, analog or semiconductor, these trojans can still exist [25]. The class of power/area/architecture and signature transparent (PAAST) trojans impact the fundamental operation of analog circuits, such as amplifiers, but do not require extra components, area, or semiconductor changes [24]. The study states that by changing the high side supply bus to an amplifier or oscillator circuit, it is possible to cause the circuit to go into a trojan state without adding extra components. It also proves that even without having to physically change the circuit, hardware trojans within analog IC development exist. The research into PAAST trojans has generally focused on the determination of the possible trojan states [26].

Other trojans in the analog/RF front-end domain have also been detected such as changing the termination to the entrance of the power amplifier (PA) [20], or inserting a trojan on a mm-wave RF PA matching network to leak information [27]. Hence, while hardware trojans are possible, there has been little focus on detecting and defending the RF front end from inserted hardware trojans.

2.2. Hardware Trojan Detection

Hardware trojan detection methods are broadly categorized as destructive and nondestructive [11,12]. Destructive methods can be applied on a smaller scale, but also provide a way to find and form golden models for verification [12]. Nondestructive methods include techniques such as side-channel analysis, formal verification, simulation, and logic testing. Many nondestructive methods require a golden model, and thus, together, these nondestructive and destructive methods form a complementary approach to hardware trojan detection. Other recently studied methods include an optical analysis of various ICs [28].

2.2.1. Side-Channel Detection

Side-channel analysis is a well-studied detection method, with physical side channels such as temporal (propagation delay), thermal, and electrical (current, EMI, voltage, charge). Side-channel attack (SCA) analysis utilizes the hardware runtime characteristic, such as power, of a cryptographic device to evaluate if it leaks secret information or reveals encryption behaviors. Unlike exploiting software bugs, such attacks on hardware components are not due to buggy hardware. Side-channel attacks can be categorized in a simple power analysis [29], differential power analysis [29], and correlation power analysis [30]. Since a correlation power analysis requires far fewer traces for recovering the key than a simple power analysis or differential power analysis, a correlation power analysis that retrieves the key through analyzing the correlation between the computing data and the measured power consumption, has become the most popular way for side-channel attacks to crack many cryptographic implementations [31,32]. Among many kinds of targets, an awareness of the potential of the EM side-channel attacks is developing [33–39]. The attacker is typically interested in emanations resulting from data processing operations, such as state changes and current flows in the CMOS circuits. These currents result in EM emanations, sometimes, in unintended ways. Such emanations carry information about the data or clock rates. The emanations provide multiple views of events unfolding within the device at each clock cycle because each active component produces and induces various types of emanations, increasing their vulnerability to hacking or exploitation. However, much of the literature on the utilization of current and EM side channels is generally not isolated from the side-channel under test [20], and it requires components to be placed in the circuit itself to detect changes in the waveform. In the case of an EM side-channel analysis, the unit under test must be within a particular test setup in order for those verifying the chip to discover the trojan, and once it leaves that setting, if the trojan goes undetected, it cannot be discovered until malicious events occur. Hence, this study proposes IC power interconnect the EM side-channel analysis via magnetic tunnel junction sensors.

2.2.2. IC Current Sensing for Hardware Trojan Detection

Various IC current sensing methods have been previously proposed, including built-in current sensors (BICSs) [40] and magnetoresistance sensors [41,42]. Previously, BICS were employed and shown to be able to detect trojans [40]. Other research indicated that MTJ sensors can be utilized for anomaly detection [43]. In many current sensing schemes, the conventional methods utilize invasive series components, such as a series resistor, a power MOSFET (to observe on-resistance), and even an integrator [44–47]. These schemes cause high-power dissipation and have many limitations, including process dependency, control difficulty and high complexity. The major issue is that the inserted components change the characteristics of the overall circuits unless a small resistance component is inserted in the loop. Although using a small resistance component can reduce the risk of degrading the performance, it increases the difficulty to sense the signal accurately. In this study, novel on-chip non-invasive EM sensors will be exploited to collect EM emanations for (1) observing if the device reveals detectable patterns; (2) monitoring if the device is under attack, which may result in unusual activities. To enable the on-chip security detection function for mobile devices, we propose an EM-sensing system to monitor critical signals with non-invasive sensors that can avoid inserting new components in the signal path, so that the system characteristics will not be modified by the sensing circuits.

This study proposes to utilize MTJ sensors for current sensing, and machine learning models to develop an on-chip, isolated current sensor that will enable hardware trojan detection for protection of RF transceivers. Thus, this study not only focuses on the physics of hardware transceivers, computationally light-weight machine learning models, and MTJ sensors, but also develops a number of potential hardware trojans that cover the vulnerabilities pointed out in previous research, such as added components and injected noise.

3. On-Chip Magnetic Tunnel Junction (MTJ) Based Sensors for Instant Device Power/Current and EM Emission Monitoring

The basic MTJ structure consists of two ferromagnetic layers separated by the insulator layer. The pinned layer has the fixed magnetization direction, while the magnetization direction can be changed in the free layer. Conventionally, the MTJ devices have been used as oscillators or memory [48–52]. The MTJ devices can be fabricated monolithically over CMOS circuits. An e-beam-based nanofabrication process was developed to fabricate the MTJ-based spin torque oscillator over the Metal-4 layer of CMOS circuits. In this study we directly changed the resistance of MTJ devices with an external magnetic field. Hence, the MTJ devices are utilized as a non-invasive current sensor, which resistance is a function of the external magnetic field. The MTJ devices can be exploited as EM sensors placed near the critical signal paths.

Figure 2a shows the on-chip non-invasive current sensor that consists of the magnetic flux guide and concentrator along with the magnetic tunnel junction to convert magnetization rotation into a voltage change. A current along the power line of the chip generates magnetization rotation in the above magnetic layer with its rotation magnitude linearly proportional to the current amplitude. The patterned planar funnel-shaped magnetic film will amplify the rotation angle as the magnetization flux travels along the strip. An MTJ is placed at the end of the strip with its free layer exchange coupled to the flux guide. An MgO-based tunnel barrier is used to obtain high magnetoresistance ratio (MR) of larger than 300%. The reference magnetic layer on the other side of the tunnel barrier has its magnetization pinned in the direction orthogonal to the flux propagation direction by using an antiferromagnetic layer deposited above. The resistance of the MTJ depends on the relative magnetization orientation of the two magnetic layers sandwiching the tunnel barrier, i.e., the angle q in the figure on the left. The resistance can be computed by $R(\theta) = \frac{R_\perp}{1+p^2 \cos\theta}$ where $R\_\perp$ is the resistance when $q = 90°$ and p is polarization factor. The maximum and minimum resistance can be calculated as $R_{min} = R(0°) = \frac{R_\perp}{1+p^2}$ and $R_{max} = R(180°) = \frac{R_\perp}{1-p^2}$. Therefore, $MR = (R_{max} - R_{min})/R_{min} = \left(\frac{R_\perp}{1-p^2} - \frac{R_\perp}{1+p^2}\right) / \frac{R_\perp}{1+p^2} = \frac{2p^2}{1-p^2}$. For today's typical MTJ, p is equal to 0.70~0.75 and MR is equal to ~200%. The resulting resistance-area product $(R_\perp A)$ is around $1 \text{ k}\Omega \cdot \mu m^2 \sim 1 \text{ M}\Omega \cdot \mu m^2$. The analysis shows that millivolts level signal voltage is expected for a milliampere-level current change. Here, a bridge sensing structure [53] in Figure 2b is used to eliminate any response to the external stray field disturbance, such as the earth field effect.

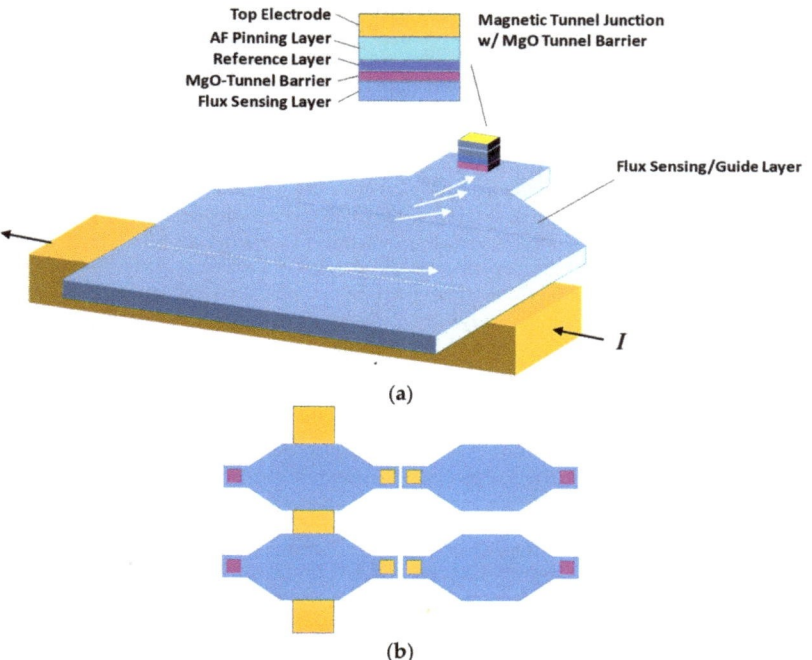

Figure 2. (a) The on-chip non-invasive current sensor that consists of the magnetic flux guide and concentrator along with magnetic tunnel junction, and (b) the bridge sensing structure to eliminate any response to field disturbance.

The entire MTJ-based sensor structure can be directly fabricated on top of the top metal layer of the semiconductor chip/circuit with two potential methods. The first one is the chemical mechanical polishing (CMP) process that will be performed over the top metal layer with deposition of the magnetic flux guide and MTJ film stack using the sputtering technique. An e-beam/optical lithography with an ion-mill process will be employed to fabricate the sensor structure along with contacting pads and connection to the circuit underneath. The other method is to adopt the dry etch to remove the top passivation layers for chip protection from the electrode areas. The silicon dioxide can be further thinned down by an optional dielectric reactive etch in order to enhance the coupling efficiency and the minimum detectable resolution.

4. Machine Learning Algorithms for Real-Time Threat-and-Vulnerability Detection

A typical side-channel signal analysis involves pre-processing to diminish dimensionality, where the measured traces are compared with predicted leakage using distinguishing algorithms. The most common technique is correlation computation [11]. For example, the Pearson correlation coefficient, ρ, for the information component, t, of all measured traces between predicted leakage, L_p, and measured leakage, $L_m(t)$, is defined as follows: $\rho(t) = Cov(L_p, L_m(t)) / \sqrt{Var(L_p) \cdot Var(L_m(t))}$, where Cov is covariance and Var defines variance. Pre-processing is adopted to diminish the set of points in the trace to remove high-order signals. However, it is still computationally expensive to realize pre-processing and correlation computation on energy-constrained RF/analog devices. To eliminate the need of pre-processing the data, Bayesian neural networks (BNNs) are exploited to directly process data and extract the features in the proposed research.

4.1. Bayesian Neural Networks

Bayesian neural networks (BNNs) have been investigated as a computationally lightweight yet robust approach to the classification of electrical signals. In particular, a previous work [26] investigated the use of BNNs as a way to classify power amplifiers (PAs) based upon variational differences due to process corners. This study also investigated classifying side-channel signals sensed from the MTJ sensors, such as integrated circuit (IC) supply current, through Bayesian neural networks. BNNs are based upon Bayes' probability theorem which states that the probability for a hypothesis from a given set of data D to be true is equal to the probability that D is true given a hypothesis h multiplied by the probability the hypothesis is true divided by the probability of D.

$$P(h|D) = \frac{P(D|h)P(h)}{P(D)} \qquad (1)$$

In this case, $P(h|D)$ is the posterior probability of h because it reflects the confidence that h holds after seeing D. Bayes concept learning is based upon some main assumptions, that is, that the BNN is trained utilizing a sequence of training examples (D), consisting of a set of instances x, which are mapped to a label, y such that:

$$D = [(X_n, y_n) | n = 1, 2, \ldots, N] \qquad (2)$$

For some n, X_n is a vector of a set of points corresponding to an IC current signal sensed through a MTJ resistive sensor and y_n is a vector of assigned class labels, corresponding to a set of K classes. Given a model with parameters θ, and prior distribution $Pr(\theta)$ the posterior distribution for the parameters is as follows:

$$Pr(\theta | X_{tr}, y_{tr}) = \frac{Pr(\theta) Pr(y_{tr} | X_{tr}, \theta)}{\int Pr(\theta) Pr(y_{tr} | X_{tr}, \theta) d\theta} \qquad (3)$$

In classifying a test set X_{new}, the predictive distribution of the classification set Y_{new} becomes:

$$Pr(Y_{new} | X_{new}, X_{tr}, y_{tr}) = \int Pr(Y_{new} | X_{new}, \theta) Pr(\theta | X_{tr}, y_{tr}) d\theta \qquad (4)$$

Due to the intractable nature of the integral in Equation (4), various numerical methods, such as the computationally heavy Markov Chain Monte Carlo method, must be applied to estimate the predictive distribution. In this study, the comparatively lighter computational method variational inference [54] is used to estimate the integral. The variational posterior is assumed to be a Gaussian distribution, where the samples of the weights are obtained by shifting and scaling unit Gaussian variables with mean μ and standard deviation σ, where $\sigma = \ln(1 + \exp(\rho))$. Thus, each sample of the weights can be expressed as:

$$w = \mu + \sigma \circ \epsilon \qquad (5)$$

where " ∘ " denotes an element-wise multiplication and ϵ is a vector of Gaussian normal distribution $N(0,1)$ to introduce variance to the weights for the Bayesian neural network as in Figure 3.

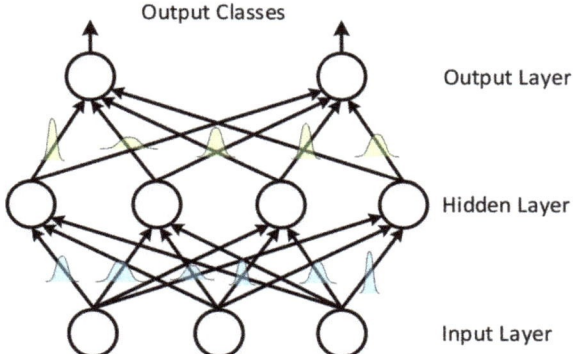

Figure 3. The weights of the Bayesian neural network weights are sampled from probability distributions.

4.2. BNN Architecture and Optimization

The BNN for this study was a network of two hidden layers with thirty-two nodes per layer, as shown in Figure 4. The BNN was trained utilizing the Python library Pytorch. The Cadence simulation data were quantized and classified to train the BNN. The BNN was trained and tested with the data from a number of different hardware trojans. We first tested the ability of the system to classify individual trojans, the details of which will be discussed in a further section. For each of these cases, certain trojans were easier to detect than others, with some of the particular trojans being able to be detected with nearly 100% accuracy. Furthermore, we also trained and tested the BNN with all the different hardware trojans combined into one dataset. We equally trained the BNN with normal and abnormal data, and noticed that due to similarities between the normal and abnormal data, the accuracy for the overall combined dataset was around 90%. We determined the exact structure of our BNN in order to maximize the accuracy for the total trojan dataset and we found that utilizing 32 hidden neurons per layer produced nearly 6% higher accuracy after 1000 training epochs than 16 neurons, but more statistically insignificant accuracy depreciation than a network with 64 neurons in the same amount of time. Thus, we decided to utilize 32 neurons to minimize resource usage and accuracy.

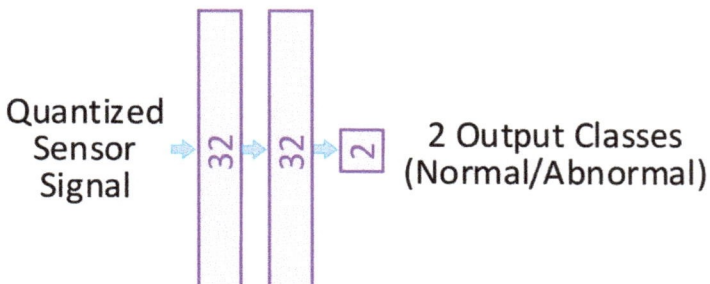

Figure 4. The optimized architecture of the lightweight Bayesian neural network for classification of the sensed EM signals.

5. Experimental Results

5.1. Fabracation of the MTJ Sensors

This work aims to develop the reliable methods for design and fabricating the novel MTJ sensor on the CMOS circuits. The planar funnel-shaped magnetic film was developed to efficiently amplify and convert the sensed magnetic field to a voltage change. The behavior of the MTJ sensor was characterized and modeled for the Cadence simulations.

5.1.1. Fabrication

To fabricate MTJ on top of top metal layer, first, we used chemical-mechanical planarization (CMP) to polish and planarize the passivation layer. Then we deposited the bottom electrode and MTJ stack at room temperature by magnetron sputtering with base pressure $<2 \times 10^{-8}$ Torr. The film structure, as shown in Figure 5, is Ta/Ru multilayer (30)/CoFeB(2)/MgO(1.5)/CoFeB(2)/Ta(0.5)/CoFe(1)/Ru(0.85)/CoFe(2.5)/IrMn(8)/Ta(1.5)/RU(7) (in nm). After deposition, the film is post-annealed at 330 °C for 10 min with a 5000 Oe magnetic field applied along in the plane direction. The deposited film is processed into elliptical pillars by e-beam lithography and carefully controlled ion milling. The long and short axis of the pillars are 300 nm and 70 nm, respectively. We deposited a SiN layer on top of the nanopillars for passivation followed by low angle ion milling for planarization. The trench and via were defined by photolithography and etched by reactive ion etching (RIE). Finally, Ti(5)/Au(100) (in nm) was deposited for the top electrode. Figure 5 also shows the cross-section image of the MTJ device.

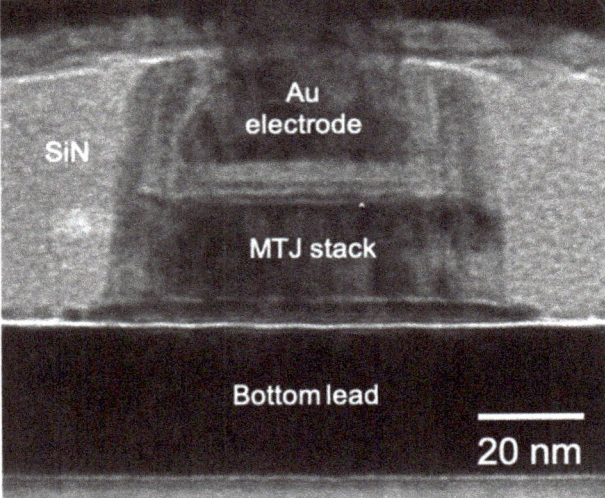

Figure 5. Cross-section TEM images of the MTJ pillar and the film structure.

5.1.2. MTJ Measurement Results

The MTJ sensor is characterized by applying magnetic field along the axis of the pillar and measured the resistance change corresponding to the magnitude of magnetic field. Figure 6 shows the typical tunneling magnetoresistance (TMR) curves. The sensor is in the low-resistance state when the two CoFeB layers' magnetizations are aligned in the parallel state by a large magnetic field. While in the range of a small magnetic field, the magnetization of the sensing CoFeB layer with respect to the reference CoFeB changes gradually with the field intensity and eventually reaches a high-resistance state when they are in the antiparallel state.

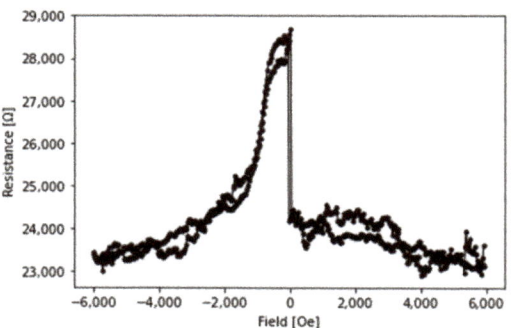

Figure 6. Typical tunneling magnetoresistance (TMR) curves. This graph illustrates the H-field-resistance curve for one MTJ sensor developed for this study. Two sweeps of the H-field are separate tests of the sensor. For the purposes of our modeling, we used the greater resistance response.

5.1.3. Modeling for Cadence Simulation

The measured results discussed in Section 5.1.2 were utilized to find the relationship between the magnetic field and resistance for this particular MTJ resistor. Once the magnetic field-resistance relationship was characterized, then electromagnetism, in particular Ampere's law, could be used to find the current-resistance relationship of the sensor.

The numerical analysis indicated the relationship between the magnetic field and resistance was piecewise linear in two different regions of interest:

$$R = \begin{cases} 0.849 * H(t) + 15979.24 \ [\Omega], & H \geq -425 \frac{A}{m} \\ 2.629 * H(t) + 16755.52 \ [\Omega], & H < -425 \frac{A}{m} \end{cases} \quad (6)$$

In order to determine the magnitude of the magnetic field sensed at the sensor, a simplified electromagnetic analysis of the system was conducted. First, the current density was assumed to be equally distributed over the entire surface area of the interconnect. Next, the equation was determined for the magnetic field from an infinitely thin, finite width plate a distance h beneath the sensor. Next, based on the principle of superposition, the magnetic fields due to a number of plates that were an equal distance apart from each other within the depth of the interconnect, $H_{\text{interconnect}}$ in Figure 7, were computed. For each plate, it was assumed the current density was equal and a proportional current equal to the current magnitude divided by the number of plates. Thus, the H field could be determined mathematically, as shown in the following equations:

$$\int_C B \cdot dl = \oint J \cdot dS \quad (7)$$

$$H = \frac{B}{\mu_0} \quad (8)$$

$$\int_C H \cdot dl = I \quad (9)$$

Using polar coordinates we analyze the H-field from a single plate.

$$dH = \frac{I}{2 * \pi * r} \hat{\theta}, \quad r = \text{sqrt}\left(y^2 + h^2\right) \quad (10)$$

$$\hat{\theta} = -\sin\theta \hat{y} + \cos\theta \hat{z} = -\frac{h}{r}\hat{y} + \frac{y}{r}\hat{z} \quad (11)$$

$$dH = \frac{I}{2 * \pi * r} * \frac{1}{r}(-h\hat{y} + y\hat{z}) \quad (12)$$

$$dH = \frac{I}{2*\pi*r^2} * (-h\hat{y} + y\hat{z}) \quad (13)$$

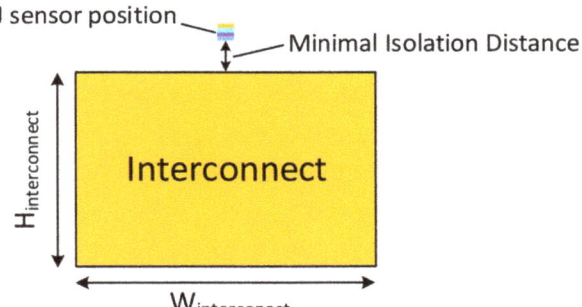

Figure 7. Current carrying interconnect and sensor location.

Integrating in the y dimension of I yields:

$$\int dH = \int \frac{I}{2*\pi*y^2 + h^2}(-h\hat{y}) \quad (14)$$

$$H = \int \frac{1}{2\pi}\frac{I}{w}\frac{h}{h^2+y^2}dy \quad (15)$$

$$H = \frac{I}{\pi*w}\tan^{-1}\left(\frac{w}{2h}\right) \quad (16)$$

H depends not only on the distance away from the interconnect, but also on the width of the wire, a finding found to be true in [43].

To find the total estimated current in the interconnect, a large, but finite, number of plates were integrated with varying distances from the first plate, which was at a distance h from the sensor to the depth of the entire interconnect, $H_{interconnect}$. Using the minimal h distance illustrated in Figure 8, 50 nm from the surface of the interconnect, the estimated magnetic field over varying currents could then be determined.

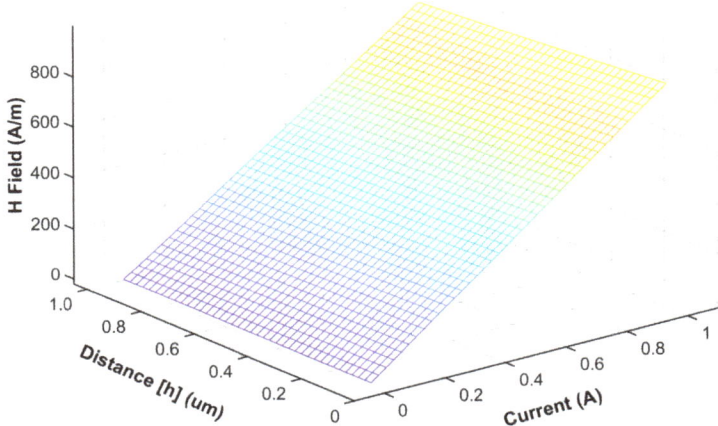

Figure 8. Magnetic field and current at varying sensor distances.

Fitting the H-I relationship of the sensor 50 nm above the interconnect, the H-I relationship in Figure 9 was found to be approximated by the linear function:

$$R = \begin{cases} 0.849 * H(t) + 15979.24 \ [\Omega], & H \geq -425 \frac{A}{m} \\ 2.629 * H(t) + 16755.52 \ [\Omega], & H < -425 \frac{A}{m} \end{cases} \quad (17)$$

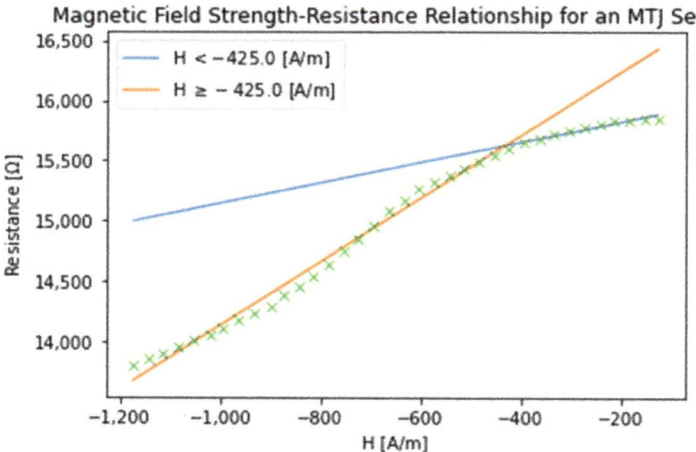

Figure 9. Magnetic field-resistance relationship for an MTJ Sensor.

Although empirical evidence concerning the bandwidth of our MTJ sensor was not collected, the authors of [53] indicated that while theoretical MTJ sensors have a wide bandwidth of GHz, in practice the bandwidth is closer to 100 MHz. Hence, for the simulation, we filtered the H-field at about a 100 MHz cutoff frequency.

$$H = 906.81 * I(t), \ -400\text{mA} \leq I < 400\text{mA} \quad (18)$$

For this particular application, DC currents on the IC trace are approximately in the range of ±400 mA. Hence, the maximum of change in resistance of this particular sensor will be ±307 Ω according to the following equations, leading to a sensitivity of around 1.9%.

$$\Delta R \approx (\pm 400 \text{ mA} * 906.81 * 0.849) \quad (19)$$

$$\Delta R = \pm 307 \ \Omega \quad (20)$$

$$\therefore \frac{\Delta R}{R} \approx \frac{307}{15979} \approx 1.9\% \quad (21)$$

With the small 1.9 % change in resistance, it should be noted that accurate measurements will be difficult, with potential sensing voltages through the utilization of a Wheatstone bridge of approximately 3 mV peak to peak. Because the application requires low-power ADC and a tolerable resolution, some way to boost the signal, either through an amplifier circuit or through a current-to-frequency converter or a voltage-to-time converter might be utilized to accurately measure the changing resistance, and hence, the changes in the measured current for accurate classification of the BNN. Such small-signal measurement techniques have been developed for MTJ sensor networks in the past [55]. Based on these calculations and considerations, a Verilog-A model was created to simulate the current sensing capabilities of the MTJ sensor.

5.2. Attacker Models and Evalution Results

The attacker scenarios developed based in the analog domain, and thus, can be detected using a reverse side-channel analysis. Hardware trojans are well-researched in literature [11] and detection with methods such as a side-channel analysis is also widely researched. However, there is little available information concerning analog hardware trojans. Hence, this study created three classes of power amplifier stage hardware trojans. The goal for these trojans was either decreasing efficiency, shutting off the device, or to inject noise in the amplification stage. Furthermore, these trojans were developed to appear to be like components found on an actual power amplifier IC and thus increase the probability of being detected. Most importantly, no matter the trojan, all were able to be classified using the lightweight BNN.

5.2.1. Power Amplifier Designs

A single-ended cascoded Class-E PA was designed and simulated for demonstration of the proposed system. By reducing the overlapping time of the transistor's output voltage and current, power dissipation at the transistor of the switching mode PAs is minimized. Hence, supply power can be delivered to the output load more efficiently. Figure 10 shows the PA schematic that exploits the switching Class-E operation to achieve high efficiency to reflect the stringent power consumption requirements of IoT applications, and their prominent nonlinearity. A cascode transistor was added to prevent the device from breaking down. The harmonic content at the transistor drain is a result of the soft switching effect generated by C_{sw} and L_{sw}. In schematic-level simulation, an output power of 18.7 to 20 dBm and a drain efficiency of 40 to 44% across process and temperature corners is achieved.

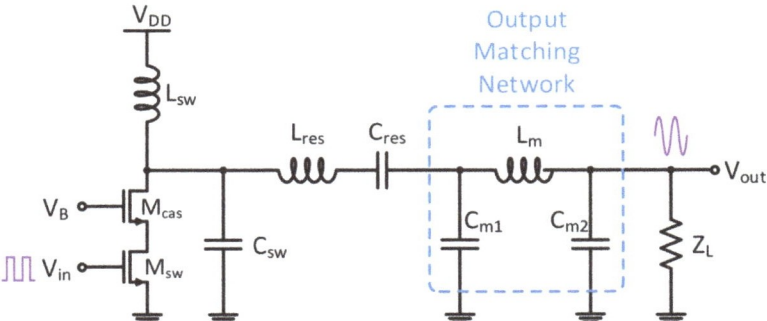

Figure 10. The designed Class-E PA that was used for demonstration of the proposed system.

5.2.2. Attacker Models on PA Designs

The attacker models are segmented into three main categories: shut-off, parasitic capacitance, and noise injection. Few trojan models are available in literature due to the pernicious nature of trojans. Figure 11 illustrates the main areas identified in this study that can be targeted by attackers. The first area is the active device, the switching FET controlled by the input signal. Two different attacks can be carried out here, a source switch turn-off attack and a noise injection attack. A third attack can be at the output of the drain of the cascaded MOSFET, increasing the parasitic capacitance through an injected trojan capacitance circuit.

The development of trojans consisted of focusing on inserting trojans into various regions of the device and the impact of these insertions on the PA efficiency. The most obvious trojan is one that completely disables the PA. To disable the amplifier, a switch can be placed at the source of lower MOSFET that, when triggered, will cause the active devices (the transistors) to be shifted from the operation region to the off state, limiting the ability of the device to operate, as shown in Figure 12.

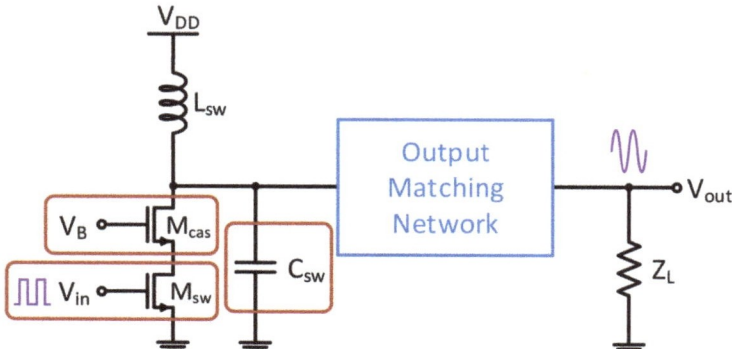

Figure 11. The main areas of attacker models are segmented into three categories: shut-off, parasitic capacitance, and noise injection.

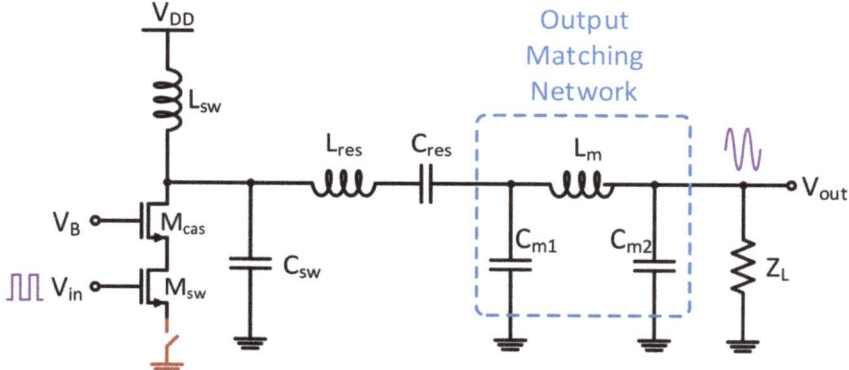

Figure 12. The killer switch that is used to shut off the operation of the power amplifier.

The second trojan studied was the parasitic capacitance trojan, impacting the matching network Q factor of the circuit. By increasing the capacitance on the output of the drain of the cascaded MOSFET (Figure 13), an attacker can easily cause the system to become less efficient in transmitting the input signals. The efficiency of the power amplifier determined the voltage-current relationship of the switching circuit. A Class-E amplifier is tuned to be most efficient, and thus, the drain output capacitance magnitude is carefully selected. Hence, a capacitance with a switch that can be triggered by an attacker can plausibly be fabricated on the device and be thus activated to limit the efficiency and increase the power consumption of the PA.

The third trojan studied was an AC-coupled noise source at the input of the cascaded MOSFET in the Class-E topology as in Figure 14. The radio frequency (RF) circuit designers usually set tolerances at which the amplifier can work, and hence, in moving outside of that range, the power amplifier will be less effective by coupling a noise source at the input, especially outside of that tolerance range. In this study, we analyzed noise voltages of 10% of the DC voltage and higher at frequencies equal to the input frequency.

Finally, the last trojan studied was noise injection at the input signal of the amplifier. A two-toned input, added through an RF power combiner in Figure 15, which vector-adds two analog signals together, not only causing an issue with the gain of the circuit, but also with noise in the side-band channels. By inserting a noise signal in the sideband of the of the desired signal, with the signal large enough to interfere with standard specifications, in this case the Bluetooth specification, the attacker can not only change the power consumption of the PA, but also interfere with signals in other channels as well.

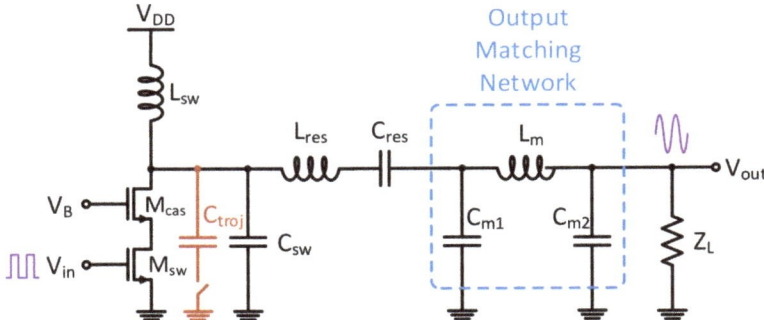

Figure 13. The added parasitic capacitance degrades the output efficiency and increases the power consumption of the power amplifier.

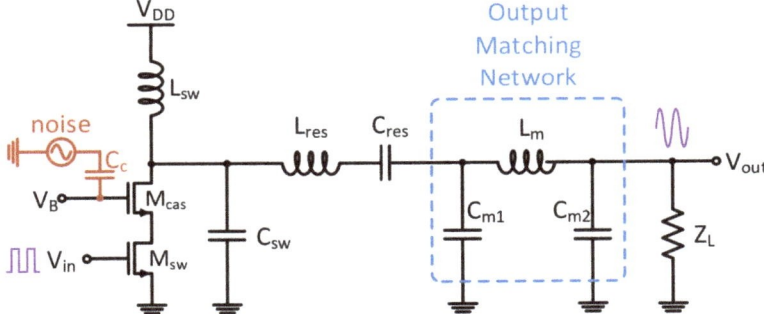

Figure 14. The AC-coupled noise source at the input of the cascaded MOSFET.

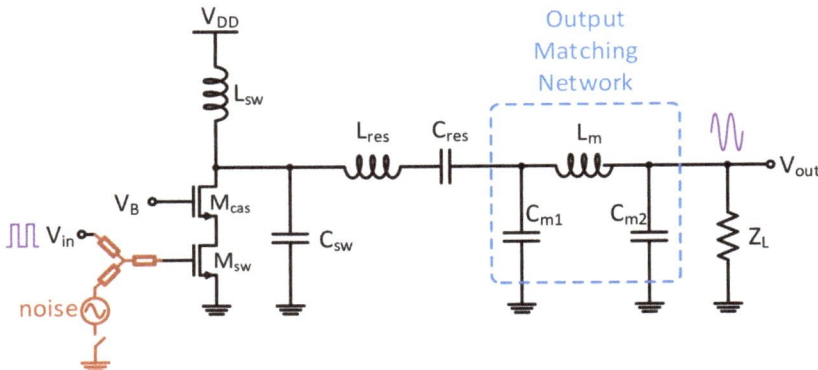

Figure 15. An RF power combiner to mix the noise source into the input signal.

Hence, all of these PA trojans were developed to change the operating ability of the Class-E power amplifier.

5.2.3. Evaluation Results

All the of presented trojans and the sensor model were simulated in Cadence. The MTJ model was written in Verilog-A utilizing the current to H-field and H-field for resistance equations earlier mentioned in this study. Based on previous literature search, we also included a low-pass filter on the current-to-H-field equations at 100 MHz to approximately model the actual frequency response of the sensor. Simulated with the Class-E PA with trojans, tests were performed with a Wheatstone bridge configuration as suggested in [53], and the output went to an amplifier to allow for the determination of optimal gain for this sensing configuration.

The simulation results then were used as the input signals for the proposed BNN to classify the results. The dataset for evaluating the BNN classifier was generated by simulating the PA with various trojans in Cadence. The trojans themselves were tested by utilizing non-ideal switches that would be cycled on and off in the simulation. Each simulation was run at 1.5 V with process variations in fast-fast (FF), slow-slow (SS), and typical-typical (TT) process variations. Furthermore, the data were generated for temperatures of -40 °C, 27 °C, 60 °C, and 125 °C. Thus, for each trojan that was run, there were 12 different tests at different temperatures and process variations. For each trojan besides the switched trojan, we tested various configurations of the trojans to determine the precision of the classifier. The switched trojan only had one configuration (on and off), while the voltage tolerance trojan was swept from 0.1 V to 0.5 V in 0.1 V increments, the parasitic capacitance trojan was tested with 10 fF, 100 fF, 1 pF, and 10 pF capacitors, and the power combiner trojan was tested with combined signals of 0.024 GHz, 0.24 GHz and 2.4 GHz. All of these data for the process technologies and temperatures were combined together for each trojan in the following way: the trojan region for the source switch trojan was determined, the same length of data for that trojan was taken for each of the trojans and then were quantized at different quantization levels (4, 6, 8, 10, 12, 14, 16, and 24) between ± 0.8 to produce eight different quantized test sets, and those points were then added to an overall trojan vector test vector for each quantization level that included all the different process and temperature for that particular trojan (e.g., switch, pcap 10 f, pcomb 2.4 GHz, etc.). These vectors were then added to an overall test vector that included normal operation data and all of the variations in trojans. The training sets for all the training used 20,000 points and a test set of 1000 points. Furthermore, to determine how well the classifier can resolve individual trojans, vectors that included only normal operation with a particular trojan were included. Note that in training, the dataset included an equal number of "normal" operation and "trojan" operation sets to avoid over-training the model on trojan data.

In determining the difference between a source switch circuit, a power combiner circuit, and a parasitic capacitance, the BNN performed well over all different quantization levels. The BNN was able to determine a source switch trojan with 96% accuracy over all quantization levels, a power combiner trojan with different frequencies from 200 MHz through 2.4 GHz with nearly 100% accuracy, and an approximately 85% accuracy for the parasitic capacitance trojan over all the quantization levels. When all the different types of trojans were put into the same class and compared against the "typical" signal, there was greater than 95% testing accuracy for the BNN over the various quantization levels. Table 1 summarizes the accuracies of the different type trojans.

Table 1. Accuracy summary of the different type trojans.

Trojan Type	Accuracy
Source switch trojan	96%
Parasitic capacitance trojan	85%
Noise trojan	100%
Combined trojans of all types	95%

6. Conclusions

Novel non-invasive sensors were developed to collect data for analysis of analog, mixed-signal, power, and EM signal behavior. To sense small changes in magnetic fields and inform the machine learning circuits, the nanoscale heterostructure was developed to be able to monolithically integrate CMOS circuits with novel spin-torque devices that can be utilized as robust high-fidelity sensors and embedded into interconnects. Lightweight learning algorithms were developed for fast threat detection at the front-end of resource-constraint devices in real time. The MTJ sessors were fabricated, measured and modeled for Cadence simulations together with the presented attacker models. The results show that the proposed system achieves 95% of the accuracy to recognize the attacker with all trojan types applied.

Author Contributions: Conceptualization, E.C., J.Z. and V.C.; methodology, E.C., J.K. and B.-Y.Y.; software, J.K.; validation, J.K. and B.-Y.Y.; formal analysis, E.C., J.K. and B.-Y.Y.; investigation, E.C., J.K. and B.-Y.Y.; resources, J.Z. and V.C.; data curation, E.C. and J.K.; writing—original draft preparation, E.C., J.K. and B.-Y.Y.; writing—review and editing, J.Z. and V.C.; visualization, E.C., J.K. and B.-Y.Y.; supervision, J.Z. and V.C.; project administration, J.Z. and V.C.; funding acquisition, J.Z. and V.C. All authors have read and agreed to the published version of the manuscript.

Funding: This work was supported by the National Science Foundation under Grant No. 1952907, 1953801, and 2028893, and the Data Storage Systems Center (DSSC) at Carnegie Mellon University.

Institutional Review Board Statement: Not applicable.

Informed Consent Statement: Not applicable.

Data Availability Statement: This study did not report any data.

Conflicts of Interest: The authors declare no conflict of interest.

References

1. Erfani, R.; Marefat, F.; Sodagar, A.M.; Mohseni, P. Modeling and experimental validation of a capacitive link for wireless power transfer to biomedical implants. *IEEE Trans. Circuits Syst. II Express Briefs* **2018**, *65*, 923–927. [CrossRef]
2. Erfani, R.; Marefat, F.; Sodagar, A.M.; Mohseni, P. Modeling and characterization of capacitive elements with tissue as dielectric material for wireless powering of neural implants. *IEEE Trans. Neural Syst. Rehabil. Eng.* **2018**, *26*, 1093–1099. [CrossRef] [PubMed]
3. Erfani, R.; Marefat, F.; Sodagar, A.M.; Mohseni, P. Transcutaneous capacitive wireless power transfer (C-WPT) for biomedical implants. In Proceedings of the 2017 IEEE International Symposium on Circuits and Systems (ISCAS), Baltimore, MD, USA, 28–31 May 2017.
4. Popovic, Z.; Falkenstein, E.A.; Costinett, D.; Zane, R. Low-power far-field wireless powering for wireless sensors. *Proc. IEEE* **2013**, *101*, 1397–1409. [CrossRef]
5. Huang, K.; Lau, V.K.N. Enabling wireless power transfer in cellular networks: Architecture, modeling and deployment. *IEEE Trans. Wirel. Commun.* **2014**, *13*, 902–912. [CrossRef]
6. Huang, K.; Zhou, X. Cutting the last wires for mobile communications by microwave power transfer. *IEEE Commun. Mag.* **2015**, *53*, 86–93. [CrossRef]
7. Xie, L.; Shi, Y.; Hou, Y.T.; Lou, A. Wireless power transfer and applications to sensor networks. *IEEE Wirel. Commun.* **2013**, *20*, 140–145.
8. Massa, A.; Oliveri, G.; Viani, F.; Rocca, P. Array designs for long-distance wireless power transmission: State-of-the-art and innovative solutions. *Proc. IEEE* **2013**, *101*, 1464–1481. [CrossRef]
9. Futuristic Roads May Make Recharging Electric Cars A Thing of the Past. Available online: https://www.nbcnews.com/mach/mach/futuristic-roads-may-make-recharging-electric-cars-thing-past-ncna766456 (accessed on 31 October 2021).
10. Defense Science Board, Department of Defense. *Report of the Defense Science Board Task Force on High Performance Microchip Supply*; Defense Science Board, Department of Defense: Washington, DC, USA, 2005.
11. Krieg, C.; Dabrowski, A.; Hobel, H.; Krombholz, K.; Weippl, E. *Hardware Malware*; Morgan & Claypool Publishers: San Rafael, CA, USA, 2013.
12. Vosatka, J. Introduction to hardware trojans. In *The Hardware Trojan War: Attacks, Myths, and Defenses*; Bhunia, S., Tehranipoor, M.M., Eds.; Springer International Publishing: Cham, Switzerland, 2018; pp. 15–51.
13. Chakraborty, R.S.; Narasimhan, S.; Bhunia, S. Hardware Trojan: Threats and emerging solutions. In Proceedings of the 2009 IEEE International High Level Design Validation and Test Workshop, San Francisco, CA, USA, 4–6 November 2009.
14. Lin, L.; Burleson, W.; Paar, C. MOLES: Malicious off-chip leakage enabled by side-channels. In Proceedings of the 2009 IEEE/ACM International Conference on Computer-Aided Design, San Jose, CA, USA, 2–5 November 2009.

15. Narasimhan, S.; Wang, X.; Du, D.; Chakraborty, R.S.; Bhunia, S. TeSR: A robust temporal self-referencing approach for hardware trojan detection. In Proceedings of the 2011 IEEE International Symposium on Hardware-Oriented Security and Trust, San Diego, CA, USA, 5–6 June 2011.
16. Tehranipoor, M.; Koushanfar, F. A survey of hardware trojan taxonomy and detection. *IEEE Des. Test Comput.* **2010**, *27*, 10–25. [CrossRef]
17. Becker, G.T.; Regazzoni, F.; Paar, C.; Burleson, W.P. Stealthy dopant-level hardware trojans: Extended version. *J. Cryptogr. Eng.* **2014**, *4*, 19–31. [CrossRef]
18. Yang, K.; Hicks, M.; Dong, Q.; Austin, T.; Sylvester, D. A2: Analog malicious hardware. In Proceedings of the 2016 IEEE Symposium on Security and Privacy (SP), San Jose, CA, USA, 23–25 May 2016.
19. Subramani, K.S.; Helal, N.; Antonopoulos, A.; Nosratinia, A.; Makris, Y. Amplitude-modulating analog/rf hardware trojans in wireless networks: Risks and remedies. *IEEE Trans. Inf. Forensics Secur.* **2020**, *15*, 3497–3510. [CrossRef]
20. Subramani, K.S.; Antonopoulos, S.; Abotabl, A.A.; Nosratinia, A.; Makris, Y. ACE: Adaptive channel estimation for detecting analog/RF trojans in WLAN transceivers. In Proceedings of the 2017 IEEE/ACM International Conference on Computer-Aided Design (ICCAD), Irvine, CA, USA, 12–16 November 2017.
21. Chang, D.; Bakkaloglu, B.; Ozev, S. Enabling unauthorized RF transmission below noise floor with no detectable impact on primary communication performance. In Proceedings of the 2015 IEEE 33rd VLSI Test Symposium (VTS), Napa, CA, USA, 27–29 April 2015.
22. Liu, Y.; Jin, Y.; Nosratinia, A.; Makris, Y. Silicon demonstration of hardware trojan design and detection in wireless cryptographic ICs. *IEEE Trans. Very Large Scale Integr. (VLSI) Syst.* **2017**, *25*, 1506–1519. [CrossRef]
23. Jin, Y.; Maliuk, D.; Makris, Y. Hardware trojan detection in analog/rf integrated circuits. In *Secure System Design and Trustable Computing*; Chang, C.-H., Potkonjak, M., Eds.; Springer International Publishing: Cham, Switzerland, 2016; pp. 241–268.
24. Wang, Q.; Chen, D.; Geiger, R.L. Transparent side channel trigger mechanism on analog circuits with PAAST hardware trojans. In Proceedings of the 2018 IEEE International Symposium on Circuits and Systems (ISCAS), Florence, Italy, 27–30 May 2018.
25. Wang, W.; Geiger, R.L.; Chen, D. Hardware trojans embedded in the dynamic operation of analog and mixed-signal circuits. In Proceedings of the 2015 IEEE National Aerospace and Electronics Conference (NAECON), Dayton, OH, USA, 16–19 June 2015.
26. Wang, Y.-T.; Wang, W.; Chen, D.; Geiger, R.L. Hardware trojan state detection for analog circuits and systems. In Proceedings of the 2014 IEEE National Aerospace and Electronics Conference (NAECON), Dayton, OH, USA, 24–27 June 2014.
27. Gungor, B.; Yazici, M.; Salman, E.; Gurbuz, Y. Establishing a covert communication channel in rf and mm-wave circuits. In Proceedings of the 2020 IEEE 63rd International Midwest Symposium on Circuits and Systems (MWSCAS), Springfield, MA, USA, 9–12 August 2020.
28. Zhou, B.; Aksoylar, A.; Vigil, K.; Adato, R.; Tan, J.; Goldberg, B.; Ünlü, M.S.; Joshi, A. Hardware trojan detection using backside optical imaging. *IEEE Trans. Comput.-Aided Des. Integr. Circuits Syst.* **2021**, *40*, 24–37. [CrossRef]
29. Kocher, P.; Jaffe, J.; Jun, B. Differential power analysis. Advances in Cryptology—Crypto'99. In Proceedings of the 19th Annual International Cryptology Conference, Santa Barbara, CA, USA, 15–19 August 1999.
30. Brier, E.; Clavier, C.; Olivier, F. Correlation power analysis with a leakage model. In Proceedings of the 6th International Workshop on Cryptographic Hardware and Embedded Systems (CHES), Boston, MA, USA, 11–13 August 2004.
31. Kocher, P.C. Timing attacks on implementations of Diffie-Hellman, RSA, DSS, and other systems. In Proceedings of the 16th Annual International Cryptology Conference, Santa Barbara, CA, USA, 18–22 August 1996.
32. Kocher, P.; Jaffe, J.; Jun, B.; Rohatgi, P. Introduction to differential power analysis. *J. Cryptogr. Eng.* **2011**, *1*, 5–27. [CrossRef]
33. NSA Tempest Series. Available online: http://cryptome.org/#NSA--TS (accessed on 31 October 2021).
34. Gandolfi, K.; Mourtel, C.; Olivier, F. Electromagnetic attacks: Concrete results. In *Cryptographic Hardware and Embedded Systems—CHES 2001*; Springer: Berlin/Heidelberg, Germany, 2001; Volume 2162, pp. 251–261.
35. Quisquater, J.-J.; Samyde, D. ElectroMagnetic Analysis (EMA): Measures and counter-measures for smart cards. *Smart Card Program. Secur.* **2001**, *2140*, 200–210.
36. Balasch, J.; Gierlichs, B.; Verbauwhede, I. Electromagnetic circuit fingerprints for Hardware Trojan detection. In Proceedings of the IEEE International Symposium on Electromagnetic Compatibility (EMC), Dresden, Germany, 16–22 August 2015.
37. Du, D.; Narasimhan, S.; Chakraborty, R.S.; Bhunia, S. Self-referencing: A scalable side-channel approach for hardware trojan detection. In Proceedings of the 12th International Workshop on Cryptographic Hardware and Embedded Systems, Santa Barbara, CA, USA, 17–20 August 2010.
38. Wang, L.; Xie, H.; Luo, H. Malicious circuitry detection using transient power analysis for IC security. In Proceedings of the 2013 International Conference on Quality, Reliability, Risk, Maintenance and Safety Engineering (QR2MSE), Chengdu, China, 15–18 July 2013.
39. Agrawal, D.; Baktir, S.; Karakoyunlu, D.; Rohatgi, P.; Sunar, B. Trojan detection using IC fingerprinting. In Proceedings of the 2007 IEEE Symposium on Security and Privacy (SP'07), Berkeley, CA, USA, 20–23 May 2007.
40. Cimino, M.; Lapuyade, H.; De Matos, M.; Taris, T.; Deval, Y.; Begueret, J.B. A Robust 130nm-CMOS built-in current sensor dedicated to rf applications. In Proceedings of the Eleventh IEEE European Test Symposium (ETS'06), Southampton, UK, 21 May 2006.
41. Le Phan, K.; Boeve, H.; Vanhelmont, F.; Ikkink, T.; Talen, W. Geometry optimization of TMR current sensors for on-chip IC testing. *IEEE Trans. Magn.* **2005**, *41*, 3685–3687. [CrossRef]

42. Dąbek, M.; Wiśniowski, P.; Kalabińskia, P.; Wrona, J.; Moskaltsova, A.; Cardoso, S.; Freitas, P.P. Tunneling magnetoresistance sensors for high fidelity current waveforms monitoring. *Sens. Actuators A Phys.* **2016**, *251*, 142–147. [CrossRef]
43. Le Phan, K.; Boeve, H.; Vanhelmont, F.; Ikkink, T.; de Jong, F.; de Wilde, H. Tunnel magnetoresistive current sensors for IC testing. *Sens. Actuators A Phys.* **2006**, *129*, 69–74. [CrossRef]
44. Lee, C.F.; Mok, P.K.T. On-chip current sensing technique for cmos monolithic switch-mode power converter. In Proceedings of the IEEE International Symposium on Circuits and Systems (ISCAS), Scottsdale, AZ, USA, 26–29 May 2002.
45. Lee, C.F.; Mok, P.K.T. A Monolithic current-mode CMOS DC-DC converter with on-chip current-sensing technique. *IEEE J. Solid-State Circuits* **2004**, *39*, 3–14. [CrossRef]
46. Somerville, T.A. High Side Current Sense Amplifier. U.S. Patent 5,627,494, 6 May 1997.
47. Marschalkowski, E.; Malcolm, J. Current Sensing Circuit for DC/DC Buck Converters. U.S. Patent 6,992,473, 31 January 2006.
48. Bromberg, D.M.; Sumbul, H.E.; Zhu, J.-G.; Pileggi, L. All-magnetic magnetoresistive random access memory based on four terminal mCell device. *J. Appl. Phys.* **2015**, *117*, 17B510. [CrossRef]
49. Bromberg, D.M.; Moneck, M.T.; Sokalski, V.M.; Zhu, J.; Pileggi, L.; Zhu, J.-G. Experimental demonstration of four-terminal magnetic logic device with separate read- and write-paths. In Proceedings of the 2014 IEEE International Electron Devices Meeting, San Francisco, CA, USA, 15–17 December 2014.
50. Bromberg, D.M.; Morris, D.H.; Pileggi, L.; Zhu, J.-G. Novel STT-MTJ device enabling all-metallic logic circuits. *IEEE Trans. Magn.* **2012**, *48*, 3215–3218. [CrossRef]
51. Sokalski, V.; Bromberg, D.M.; Moneck, M.T.; Yang, E.; Zhu, J.-G. Increased perpendicular TMR in FeCoB/MgO/FeCoB magnetic tunnel junctions by seedlayer modifications. *IEEE Trans. Magn.* **2013**, *49*, 4383–4385. [CrossRef]
52. Morris, D.H.; Bromberg, D.M.; Zhu, J.-G.; Pileggi, L. mLogic: Ultra-low voltage non-volatile logic circuits using STT-MTJ devices. In Proceedings of the 49th Annual ACM Design Automation Conference, San Francisco, CA, USA, 3–7 June 2012.
53. Reig, C.; Cardoso, S.; Mukhopadhyay, S.C. *Giant Magnetoresistance (GMR) Sensors from Basis to State-of-the-Art Applications*; Springer: Berlin/Heidelberg, Germany, 2013.
54. Xu, J.; Shen, Y.; Chen, E.; Chen, V. Bayesian neural networks for identification and classification of radio frequency transmitters using power amplifiers' nonlinearity signatures. *IEEE Open J. Circuits Syst.* **2021**, *2*, 457–471. [CrossRef]
55. Takenaga, T.; Tsuzaki, Y.; Furukawa, T.; Yoshida, C.; Yamazaki, Y.; Hatada, A.; Nakabayashi, M.; Iba, Y.; Takahashi, A.; Noshiro, H.; et al. Scalable sensing of interconnect current with magnetic tunnel junctions embedded in Cu interconnects. In Proceedings of the 2013 IEEE International Electron Devices Meeting, Washington, DC, USA, 11–15 December 2013.

Article

Device-to-Device (D2D) Multi-Criteria Learning Algorithm Using Secured Sensors

Khalid Haseeb [1], Amjad Rehman [2], Tanzila Saba [2], Saeed Ali Bahaj [3] and Jaime Lloret [4,5,*]

1. Department of Computer Science, Islamia College Peshawar, Peshawar 25000, Pakistan; khalid.haseeb@icp.edu.pk
2. Artificial Intelligence and Data Analytics (AIDA) Lab, CCIS Prince Sultan University, Riyadh 11586, Saudi Arabia; arkhan@psu.edu.sa or rkamjad@gmail.com (A.R.); tsaba@psu.edu.sa or drstanzila@gmail.com (T.S.)
3. MIS Department College of Business Administration, Prince Sattam Bin Abdulaziz University, Alkharj 16278, Saudi Arabia; s.bahaj@psau.edu.sa
4. Instituto de Investigación para la Gestión Integrada de Zonas Costeras, Universitat Politecnica de Valencia, 46379 Gandia, València, Spain
5. School of Computing and Digital Technologies, Staffordshire University, Stoke ST4 2DE, UK
* Correspondence: jlloret@dcom.upv.es

Abstract: Wireless networks and the Internet of things (IoT) have proven rapid growth in the development and management of smart environments. These technologies are applied in numerous research fields, such as security surveillance, Internet of vehicles, medical systems, etc. The sensor technologies and IoT devices are cooperative and allow the collection of unpredictable factors from the observing field. However, the constraint resources of distributed battery-powered sensors decrease the energy efficiency of the IoT network and increase the delay in receiving the network data on users' devices. It is observed that many solutions are proposed to overcome the energy deficiency in smart applications; though, due to the mobility of the nodes, lots of communication incurs frequent data discontinuity, compromising the data trust. Therefore, this work introduces a D2D multi-criteria learning algorithm for IoT networks using secured sensors, which aims to improve the data exchange without imposing additional costs and data diverting for mobile sensors. Moreover, it reduces the compromising threats in the presence of anonymous devices and increases the trustworthiness of the IoT-enabled communication system with the support of machine learning. The proposed work was tested and analyzed using broad simulation-based experiments and demonstrated the significantly improved performance of the packet delivery ratio by 17%, packet disturbances by 31%, data delay by 22%, energy consumption by 24%, and computational complexity by 37% for realistic network configurations.

Keywords: wireless systems; mobile sensors; D2D; technological development; Internet of things

1. Introduction

IoT-based technologies have gained numerous growth in the development of smart cities and support to real-time communication systems [1–3]. Wireless sensor networks (WSN) enable low-power deployments, and they have become the dominating choice in the composition of IoT devices. The technology of WSN is broadly utilized in various applications, such as precision agriculture, healthcare, vehicle transportation, smart cities, etc. [4–6]. It is comprised of tiny and battery-powered nodes with limited memory and transmission power, and such constraints restrict the amount of computation in facilitating the network users [7–9]. In large network domains, most of the solutions prefer the multi-hop paradigm rather than the single hop, which improves the connectivity among IoT networks and supports the connected devices for transmitting their collected data. However, with increases in data traffic, frequent changes occur in communication channels and most of the network systems degrade their performance in terms of resource management and security. Many solutions have used the techniques of machine learning to make an intelligent decision and forward the monitoring data with nominal overhead [10–12]. However, most of the existing solutions are not able to cope with malicious threats in the existence of mobile IoT devices. Accordingly, it is easy to degrade the performance of smart cities while transferring the gathered data among the D2D communication system over the unreliable source of network channels [13–15]. Furthermore, due to the constraints in devices for resources, authentication, integrity, and security are other significant parameters in transmitting the smart data from the critical field [16–18]. Therefore, network devices must be protected from unauthorized access and maintain their accuracy in terms of privacy and trustworthiness.

This study aims to propose a D2D multi-criteria reinforcement learning algorithm using secured and mobile IoT devices. It offers an efficient way of collecting sensors' data and utilizes the technique of machine learning with the least overheads on sensors and provides intelligent methods for ensuring low-latency data routing. Additionally, unlike most of the existing solutions, the proposed algorithm increases the robustness of the mobile network for security, given as follows. (i) Mobile devices are authenticated in their transmission radius before connecting to the route discovery process, and after mutual verification, the proposed algorithm allows them to become involved in the data routing. (ii) It offers a trusted analysis of security measurements and improves enhancement in terms of privacy with data integrity in mobile networks. (iii) Moreover, it provides the security flow between three stages i.e., mobile devices, gateways, and the sink node. The significant contributions of the proposed work are as follows:

i. D2D authentication algorithm is developed for a mobile network to ensure the authenticity and trustworthiness between partial and fully-connected nodes.
ii. Using a multi-criteria process, the reinforcement learning technique is applied and the network system is trained using realistic conditions. The proposed algorithm offers the selection of optimal neighbors using the computation of rank value that is comprised of energy, speed, link cost, and radio coverage. Accordingly, it reduces the sizes of routing tables and avoids excessive routing intervals.
iii. Moreover, the proposed algorithm protects devices and attains uncompromised data against security attacks.
iv. The simulations are performed to verify the improvement of the proposed algorithm in the comparison of existing work.

The rest of the article is divided into the following subsections: The related work and problem statement are described in Section 2. Section 3 explains the proposed algorithm with flow diagrams. Simulation configuration and experimental results are discussed in Section 4, and Section 5 provides the conclusion with suggestions for future work.

2. Related Work

IoT is one of the most promising technologies of the current era and interacts with sensors for observing the physical world [19–21]. These technologies have expanded into the real-time environment and support the applications to govern their operations. Recently, many solutions have presented to optimize the transmissions and increase the accuracy of online data retrieval systems. The authors of [22] determined the required resources of energy at the BS for IoT-enabled systems. Numerous agricultural sensors have utilized in precision agriculture for continually monitoring the field and communicating with the smart nodes. They presented a unique product density model for estimating the energy requirements for BS. Additionally, a method for Improved Duty Cycling was provided that makes use of the residual energy parameter. The proposed routing protocol [23] employs a region-based static clustering technique to efficiently cover the agricultural area while utilizing threshold-sensitive hybrid routing to send sensed data to the base station. In addition, the proposed protocol uses fuzzy logic to select the optimal cluster head (CH) among all sensor nodes in a given round, minimizing node energy usage during each data transmission period. The suggested energy-efficient protocol is compared to establish benchmark protocols, such as energy-efficient heterogeneous clustering (EEHC), developed distributed energy-efficient clustering (DDEEC), and region-based hybrid routing (RBHR). The research and testing findings indicate that user-defined transmission thresholds substantially decrease the data transmission rate. Furthermore, the balanced employment of fuzzy logic, static clustering, and hybrid routing effectively reduces the energy consumption of sensor nodes throughout each data transmission round, therefore extending the network's total lifetime. In [24], the authors proposed PAwCOR to develop a distributed method for the selection of CH by using node energy, latency, and congestion characteristics. Energy saving is accomplished via the use of nodes that are selected depending on sensing inaccuracy. PAwCOR enables the application of periodic data with the least amount of delay possible via the use of various routing routes. Additionally, it fulfills the need for non-delay-tolerant applications by utilizing service differentiation to prioritize time-critical data transmission. By allocating at least one route for both essential and routine data transfers, the suggested method improves performance compared to current protocols. It improved the performance in terms of latency, average energy consumption, packet delivery ratio, and average residual energy to attain reliable transmission. The authors of [25] proposed CTEER, an energy-efficient routing protocol based on cluster trees, to address the fast energy loss experienced by ordinary nodes while using the conventional static routing tree method. This protocol is a rendezvous-based method with a low-delay characteristic. As a result, the protocol is well-suited for time-critical applications, such as network live broadcast systems, automated railway operation systems, ticketing software, and intelligent home systems. It creates a cross-routing tree in which the mobile sink serves as the central node. Clustering algorithms are used to group the ordinary nodes and aggregate the data packets based on the routing tree. The suggested approach outperforms RRP in terms of the network lifecycle, energy consumption, and data latency. The authors of [26] proposed a deep-reinforcement learning-based quality-of-service (QoS)-aware secure routing protocol (DQSP). It aims to ensure the QoS and extract knowledge from traffic history by cooperating with the observing environment. Moreover, the proposed protocol optimizes the policies of routing. It performs significant improvement under different network metrics and has proven high convergence and effectiveness. The authors of [27] presented QL-MAC based on Q-learning, which iteratively tweaks the MAC parameters through a trial-and-error process and attains energy-efficient communication. It offers minimization problems without predetermining the system model, and also provides a self-adaptive protocol in case of topological or any external events. It readjusts the duty cycle of nodes and explicitly minimizes the energy consumption. The large-scale simulation experiments demonstrate its efficacy over other schemes.

It was noticed that technologies of IoT and sensors are performing an extraordinary role in the development of smart communication. The sensors are widely used in different applications, including remote operations to observe the data and respond with a timely reaction [28–30]. However, they are bound in terms of resources and limit the online services for IoT networks. Moreover, transporting sensitive data from network devices towards the data centers is another important characteristic for any IoT-enabled system. It has also been seen that different solutions are discussed to improve the energy consumption and QoS parameters by using artificial intelligence and machine learning techniques for D2D communication; however, most of the reinforcement learning solutions lack the optimal consumption of resources, especially in the routing phase for mobile devices. In addition, they are not able to cope with the dynamic evaluation of routing links, and in such cases, sensors' data were frequently dropped. Moreover, it was also observed that a few solutions are still vulnerable to external attacks and not able to cope with data security under mobile nodes. Such solutions could not provide a robust mutual authentication system, and as a result, communication performance is non-collaborative and uncertain. Table 1 summarizes the discussion of the existing solutions.

Table 1. Summary of related discussion.

Comparative Approaches	Pros and Cons
Existing solutions	Most of the existing solutions have proposed for efficient utilization of energy consumption with constraint devices and improved the performance of data delivery. However, it is noticed that some solutions can tackle mobile devices at the cost of frequent data lost and overloaded wireless channels. Although machine learning techniques are explored by different researchers for IoT networks; however, it was seen that they overlooked security threats such as privacy, integrity, and authentication for mobile devices. Such a solution affects the reliability of smart cities and compromised communication system in the presence of unknown machines.
Proposed D2D multi-criteria learning algorithm using secured sensors technologies	An algorithm is developed for smart cities using reinforcement learning techniques based on devices and packets' reception information. It supports gathering real-time data by imposing security restrictions for mobile devices against malicious actions. Moreover, mobile devices are verified first, and afterward, they are allowed to be involved in the data-gathering phase. It also supports data encryption with a session-oriented function and leads to lightweight complexity for the mobile network.

3. Proposed Multi-Criteria Learning Algorithm Using Secured Sensors

Sensors integrated with IoT objects are utilized in different domains to gather data and support the community using a smart communication system. IoT network provides the processes of data collection and assists the end-users in observing and optimizing the transmission based on environmental conditions. In this section, we present the details of the proposed algorithm and its working flow.

The proposed algorithm is comprised of two stages. In the first stage, D2D authentication is performed, and afterward, using the machine learning approach the optimal forwarding tables are established. The forwarding tables are updated based on the network conditions, which decreases the overheads in determining the optimal routes. The second stage provides the trustworthiness forwarding in terms of privacy and integrity from the observing field to network applications. In this stage, the proposed algorithm ensures the accuracy of the collected data and eliminates the number of attacks from unknown devices. Additionally, the proposed algorithm imposes the lowest computing cost and data diverting for ensuring security between mobile devices, with nominal communication delays. Figure 1 illustrates the development flow of the proposed algorithm.

The contributions of the proposed algorithm are as follows:

i. The first component is D2D authentication and key distribution. It consists of mobile devices and is associated with the inline gateway for obtaining the secret keys. Additionally, gateways are directly associated with the sink node for forwarding the network information to data centers.

ii. Reinforcement learning is executed in the second component by fetching the nodes' statistics from the constructed forwarding tables along with information of packets' reception. The forwarding tables are updated iteratively, thus the proposed algorithm converges the desired outcomes optimally. Using the machine learning technique, the proposed algorithm imposes lower routing overhead on the constraint devices and informs about the latest information to mobile nodes by exploring network parameters.

iii. The third component is secured IoT communication and accomplishing sustainable routing with the support of a D2D session-oriented system. It provides authentic and verifiable sessions between devices, gateways, and sink nodes with low-security costs.

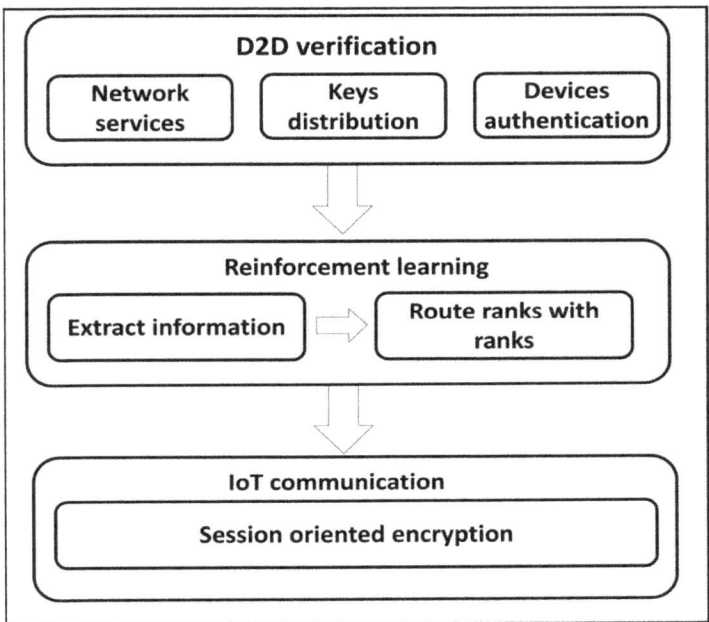

Figure 1. Development flow of the proposed algorithm.

3.1. D2D Authentication with Multi-Criteria Reinforcement Learning

In the beginning, the devices build a table containing their neighbor information, which is saved in their memories. We consider that the devices are mobile, and they advertise their current address when they are away from their home network. In the table, each device maintains the neighbors' information, such as identity id, distance, d_i, residual energy, e_i, and radio coverage limit, CR_i, to next-level nodes. Moreover, as the devices are mobile, the proposed algorithm initiates the process of authentication using gateways, w_i, by utilizing the session keys, K_s. All the nodes are required to distribute the tokens, T_k, at the beginning of data forwarding, which consists of identity, timestamp, and positioning coordinates. Additionally, the token is encrypted using the obtained session key, from device x to device y. The session keys are temporary for a specific authentication process, and when the positioning coordinates of the devices are changed, the generated keys are revoked. Afterward, device x has to obtain a new session key from the proximity gateway for communication with its other peer devices. Each device generates a request with its id to the nearest gateway for mutual communication with a peer device. Upon receiving this information, the gateway constructs a record inside its table and generates a symmetric key sK for the peer devices over the secured channel. Later, both devices perform an encryption function, e, to securely transmit the data packets m_i as defined in Equations (1) and (2):

$$w_i \rightarrow x : e_{sK}(m_i) + d' \qquad (1)$$

$$w_i \rightarrow y : e_{sK}(m_i) + d' \qquad (2)$$

where d' shows the digital signatures. On the other hand, the devices first verify the validity of the encryption blocks using digital signature, and afterward, the peer nodes perform a decryption function to recover the data packets. In the proposed algorithm, each device updates the information in the constructed table and makes an entry of the authorized device as well. In case any device is found faulty, then its entry is removed from the table by the source device.

Most of the solutions [31,32] utilize multiple parameters for data aggregation and route the data in the network system. The proposed algorithm uses the concept of multi-criteria evaluation for data aggregation and optimizing the learning procedure in terms of constraint resources. The learning procedure also makes use of radio coverage, nodes' mobility, and link cost to attain an energy-efficient and stable end-to-end communication system. In the proposed algorithm, each node obtains the information of the neighbor and utilizes the reinforcement learning technique for optimizing the intelligence process with nominal resources' consumption. The source node initiates the process for the selection of the next-hop based on the highest rank. This route rank, $R(i)$, denotes the most optimal neighbor, i, for decreasing the communication delay, energy consumption, and data disturbance, as defined in Equation (3):

$$R(i) = re_i + \left(\frac{1}{s_i}\right) + CR_i + 1/lcost_{i,j} \qquad (3)$$

where re_i is residual energy, s_i is speed, CR_i is radio coverage, and $lcost_i$ denotes the link cost from node i to node j. $lcost_{i,j}$ is the integration of packet reception ratio, PRR, and average delay time, ave_{dtime}. To compute this, the source node distributes n number of probes' packets in a fixed time interval, t, and as a result, the neighboring node j determines the value of $lcost_{i,j}$ for node i, as defined in Equations (4) and (5):

$$lcost_i = \left(PRR_{(i,j)} + \frac{1}{ave_{dtime}}\right) + 1/d_{err} \qquad (4)$$

$$ave_{dtime} = \frac{(p_n - p_i)}{t} \qquad (5)$$

where p_n and p_i denote the reception time for the first and last probe packets, t is the given time interval, and d_{err} is the data error, used to measure the number of retransmissions.

The proposed algorithm utilizes reinforcement learning [33] for computing and selecting the routing states using network conditions and experiences. The reinforcement algorithm is comprised of agents, states, S, and a set of actions, A, per state. Using reinforcement learning, node i exploits the $R(i)$ values and selects the next hop using energy, speed, radio coverage, and link cost metrics. On receiving the data, the next-hop performs the re-computation of the $R(i)$ value and forwards the data through its selected routing states. This process is continued for each neighbor selection until network data are received at the sink node. Additionally, when device i needs to route the data at the time t_0, it performs a set of actions and selects the neighbor node based on the computed route rank. The value of route rank is dynamically changed by evaluating the network and nodes' statistics. Later, the device i gains a reward, Rwd, and enters the next state, i.e., (S, a, Rwd). A node has only a single reward value at any time. If any node has no reward value at any moment, then it will not be allowed to participate in the routing phase. On entering into the next state, the device i updates its forwarding table by adding the value of the reward. Moreover, the preceding device retrieves the updated information of device i. This practice of reinforcement learning is exploited by the proposed algorithm for finding the most optimal routes for forwarding the IoT data towards the sink node. At the end of the learning period, the entries of forwarding tables are converged to a numeric value that indicates the optimal route from the source device to the sink node. Converged forwarding tables with computation of route rank not only decreases the unnecessary data diverting but also increases the packets reception ratio over the communication channels in the existence of malicious nodes. Figure 2 illustrates the flow of reinforcement learning by exploiting the computed route rank. It uses the multi-criteria of the nodes to determine its rank value and accordingly assign the reward. Based on the updated forwarding tables and reward values the proposed algorithm offers convergence results and increases the route lifetime in terms of energy, speed, and link cost. The convergence levels depend on the number of iterations until end-to-end routes are established with the efficient distribution of constraint resources. Figure 3 shows the message flow for the selection of the next-hop between the source node and its neighbors. The source node floods the route request packet in its radio coverage and identifies the nearest neighbors. In a case when no reply has been received, then it resends the request packet. Once it has found the list of neighbors, then the process of data discovery is initiated, utilizing the node-level table to fetch the statistics. Based on the fetched data, the proposed algorithm computes the route rank using a multi-criteria process and the assigned reward value by exploiting reinforcement learning. Thus, selected nodes advertise their status for the connection in the routing phase, and sensors' data is forwarded to the sink node.

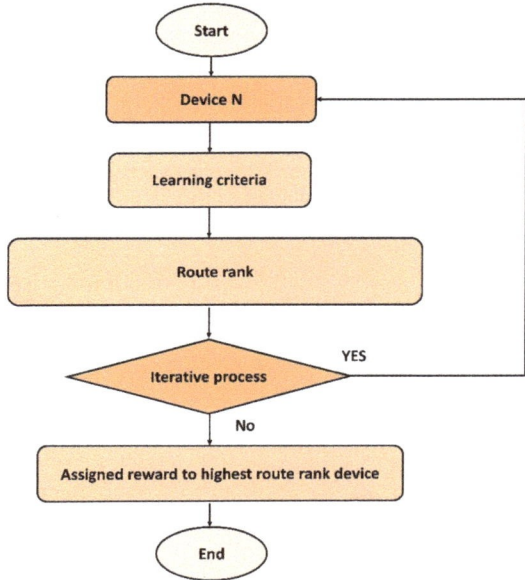

Figure 2. Route rank using reinforcement learning.

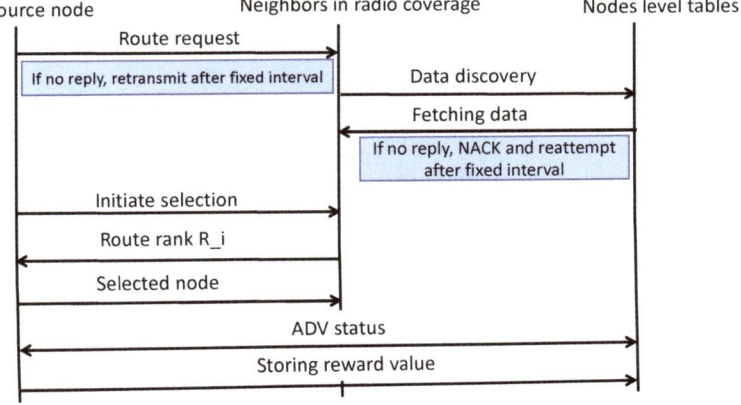

Figure 3. Next-hop selection procedure.

The format of the node-level table is presented in Table 2.

Table 2. Node level information.

1 Byte	1 Byte	1 Byte	2 Bytes	1 Byte	2 Bytes	1 Byte
Identity, id	Energy, e_i	Distance, d_i	Link cost, $lcost_i$	Radio Coverage, CR_i	Route rank, $R(i)$	Reward, Rwd

3.2. Secured Data Transmission Using a Secured Session-Oriented Scheme

The proposed algorithm offers secure IoT-enabled smart data routing by utilizing the interaction of session keys between devices, gateways, and the sink node. This process is comprised of two levels. In the first level, the devices and gateways exchange their session keys and obtain the cipher information over the insecure channel. In the second level, the session keys are shared among the gateway and the sink node. Furthermore, session keys have an expiration time and are revoked after the completion of this time. However, we consider that the devices are mobile, so it might be a case that the device moves to another communication range, thus the session key is also revoked, and it sends a new request to the nearest gateway for providing the new session key and executes the authentication process. The session keys are encrypted using the public key. Let us consider that $(ks_i)^n$ denotes the set of session keys. Then, data encryption, E, from the mobile network device i to the gateway j can be obtained as shown in Equation (6). Before this, device i to the gateway j performs an authentication function to validate the session key, as defined in Equation (6):

$$i \to j : E(ks_i, [n_i, t_i]) \quad (6)$$

where t_i is a timestamp and n_i is a nonce, also known as a random number. It is encrypted using the symmetric key of mobile device i. On receiving the encrypted session key, the gateway j includes its nonce, n_j, along with the timestamp, t_j, and sends back the confirmation message, as defined in Equation (7):

$$j \to i : E(ks_i, [n_j, t_j]) \quad (7)$$

Accordingly, both devices on the network authenticate themselves, and now the network messages, m_i, can be ciphered using the encryption function, as provided in Equation (8):

$$i \to j : xor\,(m_i, ks_i) \quad (8)$$

Finally, when data are received by gateways, they establish separate sessions with sink nodes using Equations (6) and (7). Afterward, the device data, M, are forwarded to the sink node, $sink$, including the digital signature, MAC, of the gateway with its private key, R_j, and ciphered data, $E[m_i, ks_i]$, as shown in Equation (9):

$$M(j, sink) = MAC(R_j, E[m_i, ks_i]) \quad (9)$$

Figure 4a,b describes the flowcharts of the proposed algorithm. Initially, the network services and mobile devices gather the network data from the smart environment. Network keys are generated for D2D authentications, and after their verification, they can be a part of the routing. The proposed algorithm determines the value of route rank based on the multi-criteria and updates the nodes' tables. Afterward, it utilizes reinforcement learning to assign rewards for the nodes. These rewards significantly improve the training process for the devices to extract the optimal neighbors from the set of choices, and accordingly, offer energy-efficient, least error rate delivery paths. Moreover, the proposed algorithm also secures the sessions among the gateway and the sink node for data transfer. Both the gateway and the sink node established secure sessions for their direct communication and are valid for a fixed time interval. After the mutual authentication, the gateways interact with the sink node for forwarding the network data with nominal communication costs.

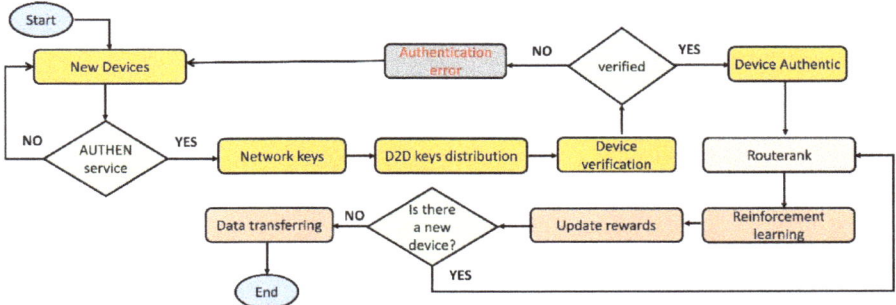

(**a**) D2D authentication and routing computation

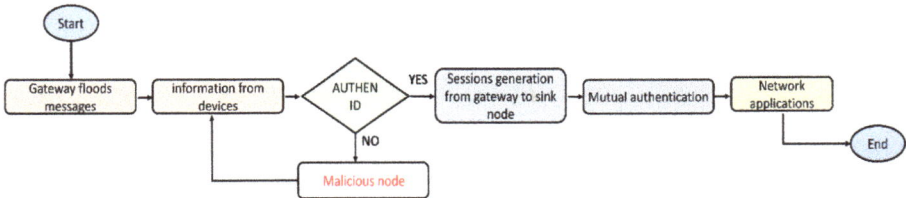

(**b**) Session generation with mutual forwarding among the gateway and the sink node

Figure 4. Flowchart of the proposed algorithm.

Figure 5 shows the flow of messages between the gateway and sink node for the establishment of a secure session with encrypted data transfer. In the beginning, the gateway device transmits the route request packet along with its *id* towards the sink node. Upon successful verification, the sink node acknowledges it, and later the gateway device requests the session key. If the time expires, the gateway device resends the request for the session key. Once the sink node receives the request for the session key, it generates the key and sends it towards the gateway device in encrypted form. The gateway device decrypts it and sends an acknowledgment message to the sink node that it has received the session key. The sink node confirms the acknowledgment message and afterward, both devices use the same session key for data encryption and decryption.

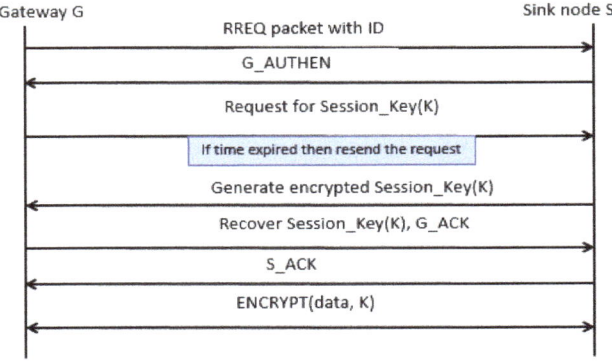

Figure 5. Message flow between the gateway and the sink node.

Algorithm 1 explains the pseudocode for the proposed work. It has two main components: one for the authentication of mobile devices with the reinforcement learning technique to assign the rewards, and the other for session-oriented data encryption from mobile sensors towards the sink node. After the successful verification of mobile sensors, the proposed algorithm evaluates the route rank for the neighbors using multiple parameters, along with the link cost. Accordingly, the neighbor with the highest route rank is assigned a reward value and selected as a forwarder. Moreover, the proposed algorithm also established a secure session from mobile sensors towards the sink node using gateway services. In this case, only that node is allowed to send the route request to the sink node that has a valid session key. The secure session key is utilized by both the mobile sensor and the sink node for data encryption and decryption, respectively.

Algorithm 1: Multi-criteria learning algorithm with secured devices.

Input: SN: Sensor nodes
RREQ: Route request
ID: Identity
K: Session key
S: Sink node
G: Gateway nodes
Output: Authentic devices, Dynamic routes, R, Secure transmission, Sec

1. **for** i = 1:N
2.   initiate Authen_service
3.   **if** Authen_service = True
4.   call keys_gen_process
5.   **else**
6.   node is declared as faulty
7. **end for**
8.   **for** j = 1:G
9.   **if** dist(j) closest to S
10.   mutual_authen service
11.   encrypt(data, K)
12.   **else**
13.   execute keys_gen_process
14.   mutual_authen service
15.   e = encrypt(data, K)
16.   **end if**
17. **end for**
18. **If** destination == S
19.   Recover K
20.   decrypt(e, K)
21. **end if**

4. Simulation Setup

This section presents the simulation configuration to evaluate the performance of the proposed algorithm. We experimented with the proposed algorithm, CTEER [25], and QL-MAC [27] solutions in terms of energy consumption, packet delivery ratio, packet disturbance, data latency, and computational complexity. The experiments were performed under varying rounds and the varying number of nodes using NS-3. Initially, nodes have homogeneous energy levels of 5 joules. The transmission range was set to 10 m. We deployed varying sensor nodes in the field of 300 × 300 m with a static sink. Sensor nodes are mobile with an installed GPS. Additionally, we assumed the number of malicious nodes to be 10. The data traffic between connected devices is a type of Constant Bit Rate (CBR).

We assumed the energy model as discussed in [34,35]. Equations (10) and (11) define the energy consumption by exploiting the transmitted and received data bits:

$$E_{tx}(k,d) = \begin{cases} E_{elect} * k + k * E_{fs} * d^2 \text{ if } d < d_0 \\ E_{elect} * k + k * E_{amp} * d^4 \text{ if } d \geq d_0 \end{cases} \quad (10)$$

$$E_{rx}(k) = E_{elect} * k \quad (11)$$

where E_{tx} and E_{rx} are the transmitting and receiving energy, k is data bits, d is the distance among sensor nodes, E_{elect} is the amount of consumed energy per data bit, and the energy of the transmitting amplifier is denoted by E_{fs}. Table 3 illustrates the parameters for simulation configuration.

Table 3. Simulation configuration.

Parameter	Value
Simulation area	300 × 300 m
Deployment	Random
Propagation Model	Two Ray Ground
Node speed	5 m/s
Pause time	20 s
Malicious nodes	10
Simulations	10
Regular nodes	100–500
Initial energy	5 j
Transmission range	10 m
MAC layer	IEEE 802.11 b
Mobility model	Random waypoint
Simulation rounds	500–2500 s
Data traffic	CBR

Results and Discussion

In this section, we first evaluate the security test of our proposed algorithm against different possible attacks. In the proposed algorithm, the D2D communication is based on the authenticity and verification of devices. Once on the communication channels, the devices are verified, and then they generate session keys. Using the sessional keys, the device initiates the sharing of data over secure links. It might be possible that the session keys are compromised, so the proposed algorithm utilizes the encrypted procedure to forward the security keys. Additionally, session keys are automatically revoked, and each node has to send a new request to the gateway for issuing the session key. In our proposed algorithm, the devices are assumed as mobile, so when any node shifts to other coverage limits, then the obtained session key will not work and it generates a request packet to the new gateway for mutual association and further authentication. Table 4 shows the general network attacks and the procedures of the proposed algorithm used to avoid them.

In Figure 6a,b, the performance evaluation of the proposed algorithm is evaluated with other solutions for the packet delivery ratio. It can be computed as the ratio of the number of delivered packets to the total number of transmitted packets from the source node to the destination node. It is seen that with a varying number of nodes and rounds, the proposed algorithm increases the packet delivery ratio by an average of 18% and 17%. This is because the proposed algorithm utilizes the route rank function to estimate the aggregate condition of the devices. Unlike QL-MAC and CTEER solutions, the proposed algorithm periodically judges the situation of the mobile devices in terms of speed, coverage ratio, and packets information, and supports robust IoT-based routing development. Accordingly, it offers the most reliable neighbors for the selection of routing states. Moreover, the proposed algorithm utilizes the position of mobile sensors to balance the load among devices, thus ultimately increasing the delivery performance towards the sink node. Additionally, the sensing range is exploited by the route rank function to incorporate the high node density option in routing decisions. The proposed algorithm makes use of multi-hop mode for forwarding the network data rather than single-hop mode, and gateways perform the intermediate roles among mobile sensors and the sink node. Such an approach efficiently exploits the coverage option and robust connectivity among deployed devices and network applications.

Table 4. Security attacks and their related procedures.

Security Attacks	Proposed Procedures
Device authentication	Unique ID Session keys
Session key security	Encryption
Verification	Decryption using symmetric key
Confidentiality	Ciphered data using the session-oriented encryption
Malicious nodes regenerate request packet for session key	ID and session key expire automatically
Storage overload	Distributed data chunks and diffusion
Connectivity loss	Reinforcement learning
Additional resources' consumption	Computing route rank
Network load	Distributed forwarding
Data originality	MAC, Digital hashes

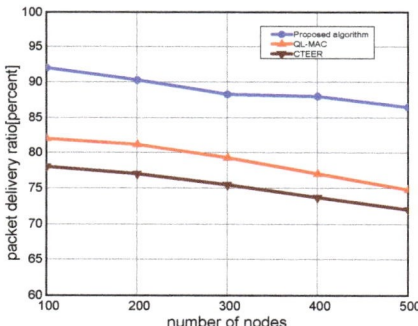

(a) Number of nodes with packet delivery ratio

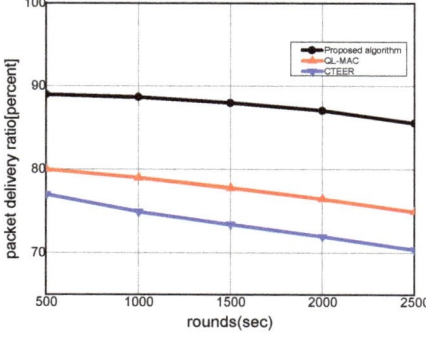

(b) Simulation rounds with packet delivery ratio

Figure 6. The performance evaluation of the proposed algorithm compared to QL-MAC and CTEER for packet delivery ratio.

Figure 7a,b illustrates the performance evaluation of the proposed algorithm for packet disturbance with existing solutions. It can be determined as a ratio of the number of packets lost to the data packets transmitted in a communication system. Compared to other work, the proposed algorithm improves the packet disturbance ratio by an average of 34% and 28% under a varying number of nodes and rounds. This is because the proposed algorithm does not overlook the constraint resource of the nodes and uniformly distributes the communication load among devices using reinforcement learning. Unlike QL-MAC and CTEER, the proposed algorithm utilizes network conditions in terms of multiple criteria and balances the data distribution on transmission links with the evaluation of the link cost. The proposed algorithm utilizes the PRR and response time factors in determining the optimal neighbors from the set of nodes. Additionally, with better utilization of the energy consumption of data forwarders, the proposed algorithm increases the strength of routes and prolongs the stability of the transmission system. Moreover, the D2D multi-criteria reinforcement learning-based routing decision avoids the chance of selecting faulty and overloaded links for the forwarding of IoT data. Based on the reward value, the proposed algorithm increases the efficiency for learning and offers a stable routing performance. Unlike QL-MAC and CTEER, the authentication and verification process of the proposed algorithm offers trustworthy communication among devices and supports improved packets' distribution over the links.

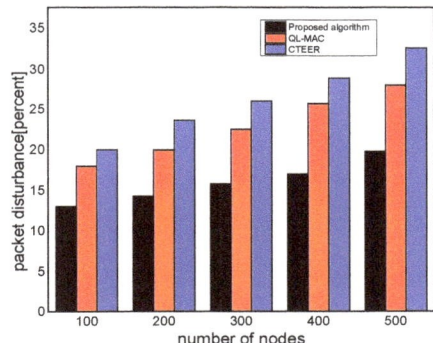

(a) Number of nodes with packet disturbance

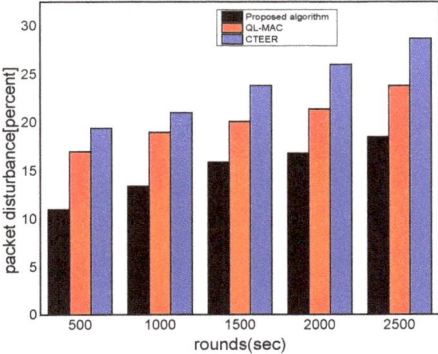

(b) Simulation rounds with packet disturbance

Figure 7. The performance evaluation of the proposed algorithm compared to QL-MAC and CTEER for packet disturbance.

Figure 8a,b illustrates the experimental results of the proposed algorithm compared to the existing solution. It measures the round-trip time of forwarding the network data towards its destination in the communication system. It was observed that the proposed algorithm significantly decreased the data delay by an average of 20% and 23% for varying nodes and rounds. The proposed algorithm utilizes the concept of mobile sensors that rapidly shift their positions for the observation and forwarding of network data. In addition, the multi-criteria parameters in the forwarding scheme explicitly achieve optimal performance for constraint devices. Unlike most of the other proposed reinforcement learning schemes, the proposed algorithm assigns the appropriate rewards to nodes, decreasing the response time and data delay for smart mobile devices. Moreover, it uses the error rate metrics in determining the loss ratio, and thus only optimal neighbors whose link cost is not congested are chosen for data routing. The D2D direct authentication and verification in the routing of data packets also decrease the involvement of unauthorized nodes. Such an approach improves the transmission path, avoiding unnecessary delays and retransmissions. Unlike other solutions that impose overheads for securing the data forwarding and lead to a high delay rate, our proposed algorithm offers lightweight session keys based on secure routing, which explicitly minimizes the latency ratio on communication paths.

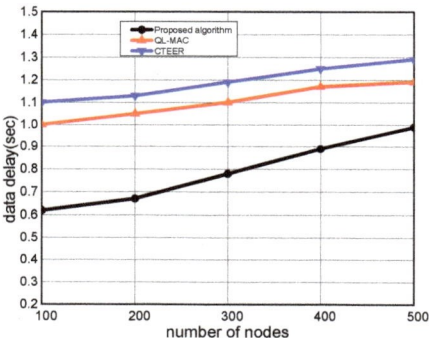

(**a**) Number of nodes with data delay metrics

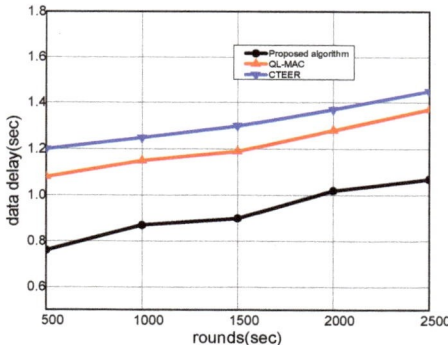

(**b**) Simulation rounds with data delay

Figure 8. The performance evaluation of the proposed algorithm compared to QL-MAC and CTEER for data delay.

In Figure 9a,b, the performance analysis of the proposed algorithm is evaluated compared to other solutions in terms of energy consumption. It is computed as a ratio of depleted energy to the total network energy in data sensing, receiving, and transmitting. It was found that the proposed algorithm minimized the energy consumption by an average of 22% and 27% under a varying number of nodes and rounds. This is because of the uniform load distribution among forwarders based on the machine learning technique. The reward value significantly trains the source node to fetch the previous information of the selected neighbors from its node-level table and optimize the performance for the constraint network. Moreover, it also decreases the extra energy consumption in sending the data from agricultural land using mobile sensors, which near uniformly balances the load on nodes. Additionally, only those nodes that fall into the coverage range exchange their information to proceed with the data routing. In case no node is found, then the next inline gateway device is assigned the responsibility of achieving the routing process. In all these processes, the proposed algorithm decreases the load on the mobile nodes and explicitly optimizes the consumption of energy resources.

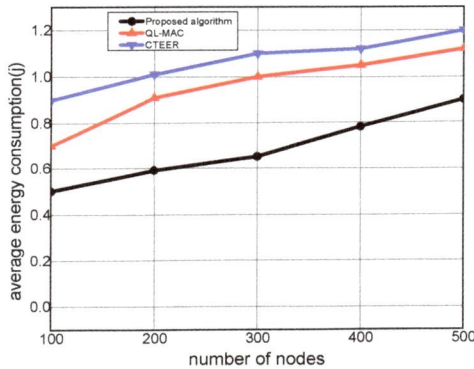

(a) Number of nodes with average energy consumption

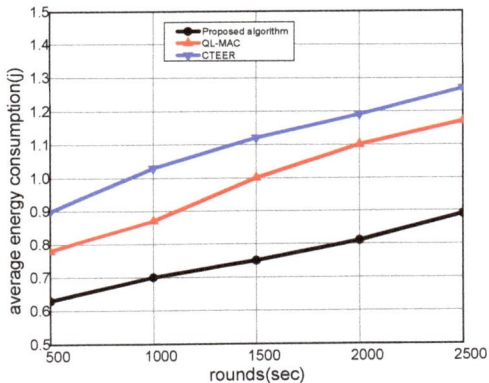

(b) Simulation rounds with average energy consumption

Figure 9. The performance evaluation of the proposed algorithm compared to QL-MAC and CTEER for average energy consumption.

Figure 10a,b demonstrate the experimental results of the proposed algorithm compared to the existing solution for computational complexity. The results determined the number of processing overheads it takes to execute the proposed algorithm. It was seen that the proposed algorithm minimized the computational complexity by 36% and 39% for varying numbers of nodes and rounds. In computing the computational time, the proposed algorithm measures the number of route requests and route response packets, especially in the presence of malicious nodes. Moreover, it also considers the number of retransmissions in computing the computational time of the proposed algorithm. Based on the security function, the proposed algorithm efficiently identified the false requester, which significantly decreases the ratio for a computational time as compared to other solutions. Furthermore, using the reinforcement learning technique, balanced the resources' consumption among the nodes and decreased the communication complexity by minimizing the least distance towards the sink node. The gateways perform the role of local supervision and reduced the cost of D2D communication by utilizing the method of coverage limit. Unlike QL-MAC and CTEER, the proposed algorithm supports the authentication and verification phase for mobile sensors and avoids the chance of malicious nodes generating excessive false traffic. Additionally, such security methods of the proposed algorithm impose the least communication complexity on constraint devices in the presence of malicious nodes, with affordable data retransmissions. Moreover, the link cost function identifies the more appropriate trusted links by utilizing the information of packets' reception and data error.

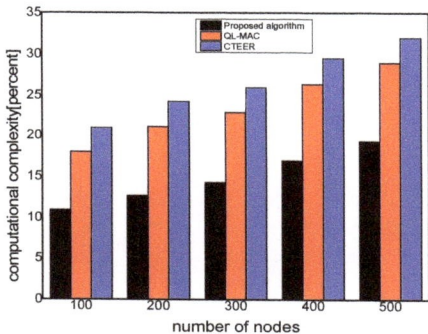

(a) Number of nodes with network overhead

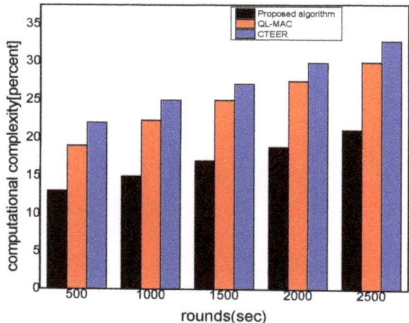

(b) Simulation rounds with network overhead

Figure 10. The performance evaluation of the proposed algorithm compared to QL-MAC and CTEER for computational complexity.

5. Conclusions

IoT technology and sensor networks are widely utilized for monitoring, data collection, and analysis of smart environments using the wireless communication system. However, due to the constraints of resources of the nodes, most of the solutions are unable to balance the routing load on the selected routes and incur rapid data losses in the presence of security attacks. In this paper, we presented a D2D multi-criteria reinforcement learning algorithm with secured IoT infrastructure for smart cities. It offers a more authentic and verified solution for directly connected devices and increases the trustworthiness of transmission. Using multi-criteria reinforcement learning, the proposed algorithm offers intelligent methods for sensing the coverage area and efficiently distributing the energy load between mobile devices. The proposed algorithm can be used for smart buildings to interconnect various operations and for security surveillance using mobile IoT devices and sensors technologies. Our proposed algorithm makes it possible to gather the real-time data from the smart building and timely transmit the data towards network applications for further analysis and appropriate actions.

However, the proposed algorithm still suffers from link disruption with the high exchange of control packets, and thus in the future, we aim to utilize the deep learning model and real-time dataset to train the network nodes and cope with network anomalies. Additionally, we would like to introduce the concept of multi-clouds in the proposed algorithm for high scalability and parallel processing.

Author Contributions: Conceptualization, K.H. and A.R.; methodology, K.H. and A.R.; software, T.S. and S.A.B.; validation, J.L. and T.S.; formal analysis, A.R., J.L., and K.H.; investigation, J.L. and S.A.B.; resources, J.L. and A.R.; data curation, A.R. and T.S.; writing—original draft preparation, K.H. and A.R.; writing—review and editing, T.S. and J.L.; visualization, S.A.B. and T.S.; supervision, J.L. and T.S.; project administration, J.L. and A.R.; funding acquisition, J.L. and A.R. All authors have read and agreed to the published version of the manuscript.

Funding: This research received no external funding.

Institutional Review Board Statement: Not applicable.

Informed Consent Statement: Not applicable.

Data Availability Statement: All data are available in the manuscript.

Acknowledgments: This research was technically supported by the Artificial Intelligence and Data Analytics Lab (AIDA), CCIS Prince Sultan University, Riyadh, Saudi Arabia. All authors are thankful for the technical support.

Conflicts of Interest: The authors declare no conflict of interest.

References

1. Khelifi, F. *Monitoring System Based in Wireless Sensor Network for Precision Agriculture, in Internet of Things (IoT)*; Springer: Berlin/Heidelberg, Germany, 2020; pp. 461–472.
2. Kumar, K.A.; Jayaraman, K. Irrigation control system-data gathering in WSN using IOT. *Int. J. Commun. Syst.* **2020**, *33*, e4563. [CrossRef]
3. Gharaei, N.Y.D.; Al-Otaibi, S.A.; Butt, G.; Sahar; Rahim, S. Energy-efficient and coverage-guaranteed unequal-sized clustering for wireless sensor networks. *IEEE Access* **2019**, *7*, 157883–157891. [CrossRef]
4. Sharma, A.; Jain, A.; Gupta, P.; Chowdary, V. Machine learning applications for precision agriculture: A comprehensive review. *IEEE Access* **2020**, *2020*, 4843–4873. [CrossRef]
5. Haseeb, K.; Saba, T.; Rehman, A.; Ahmed, I.; Lloret, J. Efficient data uncertainty management for health industrial internet of things using machine learning. *Int. J. Commun. Syst.* **2021**, *34*, e4948. [CrossRef]
6. Abbasi, Z.A.; Islam, N.; Shaikh, Z.A. A review of wireless sensors and networks' applications in agriculture. *Comput. Stand. Interfaces* **2014**, *36*, 263–270.
7. Malik, N.N.; Alosaimi, W.; Uddin, M.I.; Alouffi, B.; Alyami, H. Wireless Sensor Network Applications in Healthcare and Precision Agriculture. *J. Health Eng.* **2020**, *2020*, 8836613. [CrossRef]
8. Saba, T.; Haseeb, K.; Din, I.U.; Almogren, A.; Altameem, A.; Fati, S.M. EGCIR: Energy-Aware Graph Clustering and Intelligent Routing Using Supervised System in Wireless Sensor Networks. *Energies* **2020**, *13*, 4072. [CrossRef]
9. Rahman, G.M.; Wahid, K.A. LDAP: Lightweight Dynamic Auto-Reconfigurable Protocol in an IoT-Enabled WSN for Wide-Area Remote Monitoring. *Remote Sens.* **2020**, *12*, 3131. [CrossRef]
10. Saba, T.; Haseeb, K.; Shah, A.A.; Rehman, A.; Tariq, U.; Mehmood, Z. A Machine-Learning-Based Approach for Autonomous IoT Security. *IT Prof.* **2021**, *23*, 69–75. [CrossRef]

11. Mazzia, V.; Comba, L.; Khaliq, A.; Chiaberge, M.; Gay, P. UAV and machine learning based refinement of a satellite-driven vegetation index for precision agriculture. *Sensors* **2020**, *20*, 2530. [CrossRef]
12. Liakos, G.K.; Busato, P.; Moshou, D.; Pearson, S.; Bochtis, D. Machine learning in agriculture: A review. *Sensors* **2018**, *18*, 2674. [CrossRef] [PubMed]
13. Haseeb, K.; Din, I.U.; Almogren, A.; Islam, N. An Energy Efficient and Secure IoT-Based WSN Framework: An Application to Smart Agriculture. *Sensors* **2020**, *20*, 2081. [CrossRef] [PubMed]
14. Song, J.; Zhong, Q.; Wang, W.; Su, C.; Tan, Z.; Liu, Y. FPDP: Flexible privacy-preserving data publishing scheme for smart agriculture. *IEEE Sens. J.* **2020**, *21*, 17430–17438. [CrossRef]
15. De Araujo Zanella, A.R.; da Silva, E.; Albini, L.C.P. Security challenges to smart agriculture: Current state, key issues, and future directions. *Array* **2020**, *8*, 100048. [CrossRef]
16. Ali, R.; Pal, A.K.; Kumari, S.; Karuppiah, M.; Conti, M. A secure user authentication and key-agreement scheme using wireless sensor networks for agriculture monitoring. *Future Gener. Comput. Syst.* **2018**, *84*, 200–215. [CrossRef]
17. Banerjee, A.; Mitra, A.; Biswas, A. Wiley Online Library. Available online: https://onlinelibrary.wiley.com/doi/abs/10.1002/9781119769231.ch9 (accessed on 10 December 2021).
18. Haseeb, K.; Islam, N.; Saba, T.; Rehman, A.; Mehmood, Z. LSDAR: A Light-weight Structure based Data Aggregation Routing Protocol with Secure Internet of Things Integrated Next-generation Sensor Networks. *Sustain. Cities Soc.* **2019**, 101995. [CrossRef]
19. Rehman, A.; Haseeb, K.; Saba, T.; Kolivand, H. M-SMDM: A model of security measures using Green Internet of Things with Cloud Integrated Data Management for Smart Cities. *Environ. Technol. Innov.* **2021**, *24*, 101802. [CrossRef]
20. Shafique, K.; Khawaja, B.A.; Sabir, F.; Qazi, S.; Mustaqim, M. Internet of things (IoT) for next-generation smart systems: A review of current challenges, future trends and prospects for emerging 5G-IoT scenarios. *IEEE Access* **2020**, *8*, 23022–23040. [CrossRef]
21. Garcia, M.; Bri, D.; Sendra, R.; Lloret, J. Available online: http://citeseerx.ist.psu.edu/viewdoc/summary?doi=10.1.1.681.7101 (accessed on 10 December 2021).
22. Agrawal, H.; Dhall, R.; Iyer, K.; Chetlapalli, V. An improved energy efficient system for IoT enabled precision agriculture. *J. Ambient. Intell. Humaniz. Comput.* **2020**, *11*, 2337–2348. [CrossRef]
23. Maurya, S.; Jain, V.K. Fuzzy based energy efficient sensor network protocol for precision agriculture. *Comput. Electron. Agric.* **2016**, *130*, 20–37. [CrossRef]
24. Agarkhed, J.; Dattatraya, P.Y.; Patil, S. Precision agriculture with cluster-based optimal routing in wireless sensor network. *Int. J. Commun. Syst.* **2021**, *34*, e4800. [CrossRef]
25. Lu, J.; Hu, K.; Yang, X.; Hu, C.; Wang, T. A cluster-tree-based energy-efficient routing protocol for wireless sensor networks with a mobile sink. *J. Supercomput.* **2021**, *77*, 6078–6104. [CrossRef]
26. Guo, X.; Lin, H.; Li, Z.; Peng, M. Deep-reinforcement-learning-based QoS-aware secure routing for SDN-IoT. *IEEE Internet Things J.* **2019**, *7*, 6242–6251. [CrossRef]
27. Savaglio, C.; Pace, P.; Aloi, G.; Liotta, A.; Fortino, G. Lightweight reinforcement learning for energy efficient communications in wireless sensor networks. *IEEE Access* **2019**, *7*, 29355–29364. [CrossRef]
28. Gharaei, N.; Malebary, S.J.; Bakar, K.A.; Hashim, S.Z.M.; Butt, S.A.; Sahar, G. Energy-efficient mobile-sink sojourn location optimization scheme for consumer home networks. *IEEE Access* **2019**, *7*, 112079–112086. [CrossRef]
29. Ullo, L.S.; Sinha, G. Advances in smart environment monitoring systems using IoT and sensors. *Sensors* **2020**, *20*, 3113. [CrossRef]
30. Rehman, A.; Haseeb, K.; Fati, S.M.; Lloret, J.; Peñalver, L. Reliable Bidirectional Data Transfer Approach for the Internet of Secured Medical Things Using ZigBee Wireless Network. *Appl. Sci.* **2021**, *11*, 9947. [CrossRef]
31. Mahdi, A.O.; Wahab, A.W.A.; Idris, M.Y.I.; Znaid, A.A.; Al-Mayouf, Y.R.B.; Khan, S. WDARS: A weighted data aggregation routing strategy with minimum link cost in event-driven WSNs. *J. Sens. 2016*, *2016*, 1–12. [CrossRef]
32. Sennan, S.; Balasubramaniyam, S.; Luhach, A.K.; Ramasubbareddy, S.; Chilamkurti, N.; Nam, Y. Energy and delay aware data aggregation in routing protocol for Internet of Things. *Sensors* **2019**, *19*, 5486. [CrossRef]
33. Kaelbling, P.L.; Littman, M.L.; Moore, A.W. Reinforcement learning: A survey. *J. Artif. Intell. Res.* **1996**, *4*, 237–285. [CrossRef]
34. Wang, J.; Gao, Y.; Liu, W.; Wu, W.; Lim, S.-J. An asynchronous clustering and mobile data gathering schema based on timer mechanism in wireless sensor networks. *Comput. Mater. Contin.* **2019**, *58*, 711–725. [CrossRef]
35. Wang, J.; Gao, Y.; Liu, W.; Sangaiah, A.K.; Kim, H.-J. Energy efficient routing algorithm with mobile sink support for wireless sensor networks. *Sensors* **2019**, *19*, 1494. [CrossRef] [PubMed]

Review

Machine Learning Enhances the Performance of Bioreceptor-Free Biosensors

Kenneth E. Schackart III [1] and Jeong-Yeol Yoon [1,2,*]

[1] Department of Biosystems Engineering, The University of Arizona, Tucson, AZ 85721, USA; schackartk@email.arizona.edu
[2] Department of Biomedical Engineering, The University of Arizona, Tucson, AZ 85721, USA
* Correspondence: jyyoon@arizona.edu

Abstract: Since their inception, biosensors have frequently employed simple regression models to calculate analyte composition based on the biosensor's signal magnitude. Traditionally, bioreceptors provide excellent sensitivity and specificity to the biosensor. Increasingly, however, bioreceptor-free biosensors have been developed for a wide range of applications. Without a bioreceptor, maintaining strong specificity and a low limit of detection have become the major challenge. Machine learning (ML) has been introduced to improve the performance of these biosensors, effectively replacing the bioreceptor with modeling to gain specificity. Here, we present how ML has been used to enhance the performance of these bioreceptor-free biosensors. Particularly, we discuss how ML has been used for imaging, Enose and Etongue, and surface-enhanced Raman spectroscopy (SERS) biosensors. Notably, principal component analysis (PCA) combined with support vector machine (SVM) and various artificial neural network (ANN) algorithms have shown outstanding performance in a variety of tasks. We anticipate that ML will continue to improve the performance of bioreceptor-free biosensors, especially with the prospects of sharing trained models and cloud computing for mobile computation. To facilitate this, the biosensing community would benefit from increased contributions to open-access data repositories for biosensor data.

Keywords: label-free biosensor; machine learning; support vector machine; artificial neural network; principal component analysis

1. Introduction

The field of biosensing has exploded into nearly all areas of research, from medical applications [1] to environmental monitoring [2]. Some of the greatest appeals of biosensors are their specificity and sensitivity. These properties are primarily due to bioreceptors, which are selected for their inherent specificities such as enzymes [3], antibodies [4], and aptamers [5]. However, the very aspect that makes biosensors so specific and sensitive can also limit the sensor stability due to the degradation of the bioreceptor [6]. Additionally, as the bioreceptor is specific to an individual analyte, the particular sensor's scope is limited to the specific analyte to which the bioreceptor can bind.

To obviate these issues, many nature-inspired sensors have emerged that are bioreceptor-free. Some of the most notable examples that have made great progress include the electronic nose (Enose) [7–11] and electronic tongue (Etongue) [12–16]. Additionally, surface enhanced Raman spectroscopy (SERS)-based sensors have demonstrated incredible chemosensing ability [17–21]. Without a bioreceptor, however, there is the risk of significantly compromised biosensor performance including the limit of detection (LOD) and specificity. Researchers have introduced machine learning (ML) to bioreceptor-free biosensors to bridge this trade-off gap, improving the LOD and specificity [22]. In a sense, ML can be used to take the place of a bioreceptor by reintroducing specificity during data analysis. This is made possible by powerful ML techniques capable of detecting subtle patterns in sensor responses.

While this approach has demonstrated success, there are still several challenges that these systems must overcome. A major challenge being faced is model generalizability. Since many models rely on subtle patterns in the data, they can be quite sensitive to underlying data changes. This can make the models susceptible to error when faced with sensor drift or replacing parts of the system [14].

Since the scope of this review is quite large and covers all bioreceptor-free biosensors that utilize ML, there are a few points to clarify. Many subsets of our scope have received thorough attention and review. For instance, the use of ML for Enose and Etongue [23–27] and SERS-based biosensors [28] have previously been described. Since the literature is rich in these areas, we realize that all recent original research cannot be adequately covered here. Rather, our intent is to provide a unified discussion of the relevant methods and challenges to give a bigger picture. We also would like to acknowledge that there is a complementary review in the literature discussing the use of ML in biosensing in general [29], but not for biosensors that are bioreceptor-free.

In this review, we will give the current state of using ML to enhance the performance of bioreceptor-free biosensors. Section 2 briefly introduces the types of biosensors that have most benefited from ML. Section 3 provides some background on machine learning algorithms and how their performance can be assessed. Section 4 covers electrochemical biosensors, with particular emphasis on Enose and Etongue. Successful methods are discussed as well as some of the challenges and how they are being addressed with ML. Section 5 discusses optical biosensors, notable for imaging- and SERS-based biosensors. Additional considerations and future perspectives are discussed in Section 6 including what currently prevents many of these systems from being commercialized and what directions may be taken. We also present some considerations on best practices for ML in biosensing, especially regarding communication of methods and reproducibility.

2. How Biosensors Can Benefit from Machine Learning

Biosensors in the classic definition are sensors that utilize a bioreceptor such as antibody, enzyme, peptide, nucleic acid, etc. A bioreceptor binds to a target biological molecule and generates a signal when coupled with a transducer. Biosensors have evolved to a wide range of transducer types including electrochemical, optical, and spectroscopic biosensors. Traditionally, it is the bioreceptor that provides specificity and sensitivity to the biosensor. Increasingly, however, researchers are developing biosensors that lack a specific bioreceptor. A typical example is a semi-specific chemical sensor array, termed Enose (from gas), or Etongue (in solution). Since such a sensor's specificity is not provided by the bioreceptor, a fingerprinting technique is used to recognize signal patterns indicative of a particular analyte. Frequently, machine learning techniques are employed to detect these patterns and provide specificity.

The use of machine learning to enhance the performance (e.g., specificity, sensitivity, and LOD) of bioreceptor-free biosensors is not limited to chemical sensor arrays. It has been employed in various biosensor mechanisms. Some of the most famous examples aside from Enose and Etongue are imaging-based biosensors and SERS-based biosensors. Additionally, the use of machine learning for biosensors is not limited to those that lack bioreceptors. Cui et al. [29] cover several examples of traditional biosensors employing machine learning to enhance performance.

Table 1 provides an overview of the tasks for which machine learning has been applied, the specific algorithms used, and the relevant papers. More information on the algorithms themselves can be found in Section 3. Additionally, Table 2 gives a comparison of each of the major biosensing mechanisms including data type and appropriate feature engineering and ML methods. All information in Table 2 comes from Table 1 and serves as a higher-level summary.

Table 1. Machine learning tasks and algorithms used in biosensing.

Biosensing Mechanism	Task	Target	Algorithm	Ref.
ELECTROCHEMICAL				
CV	Regression	Maleic hydrazide	ANN	[30]
CV	Classification	Industrial chemicals	LSTM, CNN	[31]
Enose	Feature extraction	Harmful gases	PCA	[32]
	Classification		DT, RF, SVM	
	Regression		SVR	
Enose	Regression	Formaldehyde	BPNN	[33]
Enose	Classification	Chinese wines	BPNN	[34]
	Target task change	Chinese liquors	Transfer learning	
Enose	Sensor drift compensation for classification	Gases	JDA	[35]
			DTBLS	[36]
			TrLightGBM	[37]
			ELM	[38]
Enose	Sensor drift compensation & noise reduction	Bacteria	ELM	[39]
EIS	Classification	Breast tissue	ELM + SVM	[40]
EIS	Classification	Milk adulteration	k-NN	[41]
EIS	Classification	Breast tissue	RBFN	[42]
EIS	Feature extraction	Avocado ripeness	PCA	[43]
	Classification		SVM	
EIS & EIT	Classification	Prostatic tissue	SVM	[44]
Etongue	Taste classification	Tea storage time	CNN	[45]
	Increase generalizability		Transfer learning	
Etongue	Feature Extraction	Beverages	t-SNE	[46]
	Classification		k-NN	
Etongue	Classification	Cava wine age	LDA	[47]
Etongue	Regression	Black tea theaflavin	Si-CARS-PLS	[48]
OPTICAL				
Colorimetric	Classification	Plant disease VOCs (blight)	PCA	[49]
Diff. contrast microscopy	Digital staining & domain adaptation	Leukocytes	GAN	[50]
Fluorescence imaging	Classification	Microglia	ANN	[51]
FTIR imaging	Digital staining	H&E stain	Deep CNN	[52]
Lens-free imaging	Image reconstruction	Blood & tissue	CNN	[53,54]
		Herpes		[55]
Lens-free imaging	Image reconstruction & classification	Bioaerosol	CNN	[56]
Multi-modal multi-photon microscopy	Digital staining & modal mapping	Liver tissue	DNN	[57]
Multispectral imaging	Classification	Pollen species	CNN	[58]
Quantitative phase imaging	Digital staining	Skin, kidney & liver tissue	GAN	[59]
Raman spectroscopy	Feature extraction	Thyroid dysfunction biomarker	PCA	[60]
	Classification		SVM	

Table 1. Cont.

Biosensing Mechanism	Task	Target	Algorithm	Ref.
TLC-SERS	Feature extraction	Histamine	PCA	[61]
	Quantification		SVR	
SERS	Exploratory analysis	Malachite green & crystal violet	PCA	[37,62]
	Quantification		PLSR	
SERS	Quantification	Methotrexate	PLSR	[63]
SERS	Classification	Oil vs lysate spectra Leukemia cell lysate	k-means clustering	[64]
	Dimension reduction		PCA	
	Classification		SVM	
SERS	Dimension reduction	Levofloxacin	PCA	[38,65]
	Regression		PLSR	
SERS	Quantification	Potassium sorbate & sodium benzoate	PLSR	[66]
SERS	Dimension reduction & regression	Congo red	PCR	[39,67]
SERS	Dimension reduction	Mycobacteria	PCA	[40,68]
	Classification		LDA	
SERS	Quantification	Biofilm formation	PLSR	[41,69]
SERS	Feature extraction	Non-structural protein 1	PCA	[70,71]
	Classification		BPNN, ELM	
SERS	Exploratory analysis	Pollen species	PCA, HCA	[72]
	Classification		ANN	
SERS	Feature extraction	Human serum	KPCA	[73]
	Classification		SVM	

Note. CV = cyclic voltammetry; ANN = artificial neural network; LSTM = Long short-term memory; PCA = principal component analysis; DT = decision tree; RF = random forest; SVM = support vector machine; SVR = support vector regression; BPNN = back-propagation neural network; JDA = joint distribution adaptation; DTBLS = domain transfer broad learning system; GBM = gradient boost machine; ELM = extreme learning machine; EIS = electrical impedance spectroscopy; EIT = electrical impedance tomography; k-NN = k-nearest neighbor; RBFN = radial basis function network; CNN = convolutional neural network; t-SNE = t-distributed stochastic neighbor embedding; Si-CARS-PLS = synergy interval partial least square with competitive adaptive reweighted sampling; FTIR = Fourier transform infrared; VOC = volatile organic compound; GAN = generative adversarial network; DNN = deep neural network; TLC = thin layer chromatography; SERS = surface enhance Raman spectroscopy; PLSR = partial least squares regression; PCR = principal component regression; LDA = linear discriminant analysis; HCA = hierarchical cluster analysis; KPCA = kernel principal component analysis.

Table 2. Summary of data types and useful ML methods for biosensing mechanisms.

Biosensing Mechanism	Description of Data	Feature Extraction	ML Model
CV	Cyclic voltammogram		ANN, LSTM, CNN
EIS	Nyquist plot	PCA	k-NN, ELM, SVM, RBFN
Enose	Multivariate	PCA	DT, RF, ELM, SVM, BPNN
Etongue	Multivariate	PCA, t-SNE	LDA, k-NN, CNN, PLS
Lens-free imaging	Image		CNN
Digital staining	Image		Deep learning, GAN
SERS	Spectrum	PCA, KPCA	PLSR, LDA, SVM, SVR, BPNN, ELM

3. A Brief Tour of Machine Learning

In simple terms, machine learning aims to learn patterns in data to make predictions on new data. Generally, this prediction is either categorical classification (into one of a set of classes) or regression (continuous numerical output). In machine learning terms, the data used for prediction (i.e., biosensor data) are termed features or predictors. The set of features associated with one "observation" (e.g., biosensor data from one sample) is termed the feature vector.

3.1. Feature Engineering

Frequently, the predictor variables (feature vector) are not the raw biosensor data. One of the most challenging parts of using machine learning is the construction of the feature vector from the raw data. This process is termed feature engineering and mostly entails finding the relevant information from the data to aid the machine learning algorithm's performance. Common feature engineering steps include denoising, normalization, and rescaling.

One of the most powerful feature engineering processes is dimension reduction. This reduces a large number of features to a smaller number of features while minimizing information loss. Perhaps the most common method of dimension reduction is principal component analysis (PCA) [74], which reduces the original set of variables to a smaller set of independent variables termed principal components (PCs). The effectiveness of PCA to represent the data can be assessed by the amount of variance in the data explained by the PCs. Since PCA determines the PCs based on the eigenvectors' directions in the feature space, data must first be centered and rescaled to avoid bias toward those variables with a larger magnitude. Another common dimension reduction algorithm is linear discriminant analysis (LDA), which also produces a smaller number of variables but is supervised and optimally maximizes class separation [75]. Other more complex dimension reduction methods exist including artificial neural networks (ANN), as discussed in Section 3.3. ANN is typically used as a supervised machine learning method, while it has occasionally also been used for dimension reduction.

3.2. Unsupervised vs. Supervised

The two broad categories of machine learning algorithms are unsupervised and supervised [76]. In unsupervised methods, data labels are not provided during model training, while in supervised methods, they are. An example of an unsupervised algorithm is cluster analysis, used to group similar data. Unsupervised methods are less common in biosensing since we generally know what kind of prediction(s) we would like the model to make. A notable exception is PCA, as mentioned in Section 3.1. While PCA may be considered an unsupervised machine learning method, its use has recently been limited to dimension reduction (one of feature engineering processes) prior to supervised machine learning analyses.

3.3. Classification Algorithms

Among the supervised methods, classification algorithms are some of the most well-known. Classification gives prediction in the form of a class label (e.g., which bacteria species is present), thus, the output is inherently categorical. Briefly, some of the most common classification algorithms are presented in the following.

k-nearest neighbors (k-NN): One of the simplest classification algorithms, *k*-NN is a distance-based classifier. Class is predicted as the most common class of the *k*-nearest neighbors in the feature space [77]. In the example shown in Figure 1, the feature space is two dimensional (with variables x_1 and x_2) and the value of *k* is 4. In *k*-NN, the number of neighbors used for assignment, *k*, is a hyperparameter (i.e., a model parameter that is not optimized during the training process itself). As with most ML models, hyperparameter selection may strongly influence performance [78].

Support vector machine (SVM) is a non-probabilistic, binary, linear classifier [79]. SVM relies on the construction of hyper-plane boundaries in the feature space to separate data of different classes. Although SVM itself only accounts for linear separation of classes (i.e., hyper-plane boundaries must be "flat"), the data may be mapped to a higher-dimensional feature-space using the "kernel trick" [80]. Some of the most common kernels are radial basis function and Gaussian. When the hyperplane boundaries are projected back into the original feature space, they allow for non-linear boundaries, as shown in Figure 1. Additionally, there are methods allowing SVM to be used for multi-class prediction [81]. The placement of hyperplanes is determined by minimizing the distance between the hyperplane and several of the points closest to the boundary between classes. SVM's robustness against outliers is improved by a soft margin. This allows for a certain quantity of misclassifications, which are presumably outliers, to improve the separation of the other observations [82]. While SVM shows resilience against outliers and performs well in high-dimension feature spaces, it is prone to over-fitting, especially when using non-linear kernels [83]. Overfitting is when the model performs well on training data but performs poorly when generalized to unseen data.

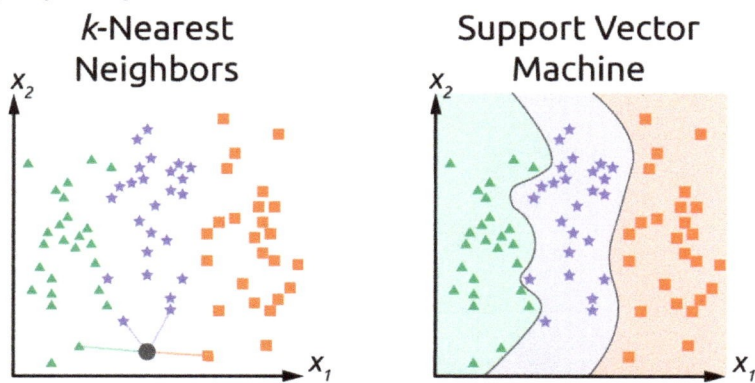

Figure 1. Comparison of classification technique using *k*-NN and SVM. In *k*-NN, four nearest neighbors are shown contributing to the gray point's assignment. Classification of the gray point is the blue star class. In hypothetical SVM with nonlinear kernel, new data are classified in which region the point lies. In both examples, the feature space consists of two dimensions. Classification could be, for example, bacterial species like *E. coli*, *Salmonella* spp., *Pseudomonas* spp., *Staphylococcus* spp., *Enterococcus* spp., etc. In practical applications, the feature space has many more dimensions, where decision boundaries for SVM are hyperplanes in the $(n-1)$ dimension for an n-dimensional feature vector.

Linear discriminant analysis (LDA): In addition to dimension reduction, LDA can be used for classification. Other related algorithms allow for non-linear classification such as quadratic discriminant analysis (QDA) [84]. One of the limitations of LDA and its relatives is that they assume the data are normally distributed.

Decision tree (DT) and random forest (RF): In tree-based models such as decision tree (DT), the feature vector starts at the tree's "trunk," and at each branching point a decision is made based on the learned decision rules. The end classification would then be at the terminal or "leaf" node that the instance results. DTs can be used for classification and regression [85]. When the target variable is categorical, it is referred to as a classification tree; when the target variable is numerical and continuous, it is referred to as a regression tree [86]. Random forest (RF) is so called because it can be considered a forest of decision trees (Figure 2) [87]. There are many RF architectures, but in all instances, the classification from each decision tree contributes to the overall classification for an observation.

Artificial neural network (ANN) draws inspiration from biological neural networks (i.e., neurons in the brain) and is composed of a collection of connected nodes called artificial neurons (see Figure 3). ANNs can be used for classification and regression. As mentioned earlier, ANN can be used for dimension reduction prior to supervised machine learning. There are a large variety of ANN structures such as (1) recurrent neural network (RNN) [88], (2) extreme learning machine (ELM) [89], and (3) deep learning algorithms such as the convolutional neural network (CNN) [90], deep belief network [91], and back-propagation neural network (BPNN) [92]. "Deep" indicates several hidden layers. ANN architectures have many hyperparameters such as the number of hidden layers, connectedness, and activation functions [93].

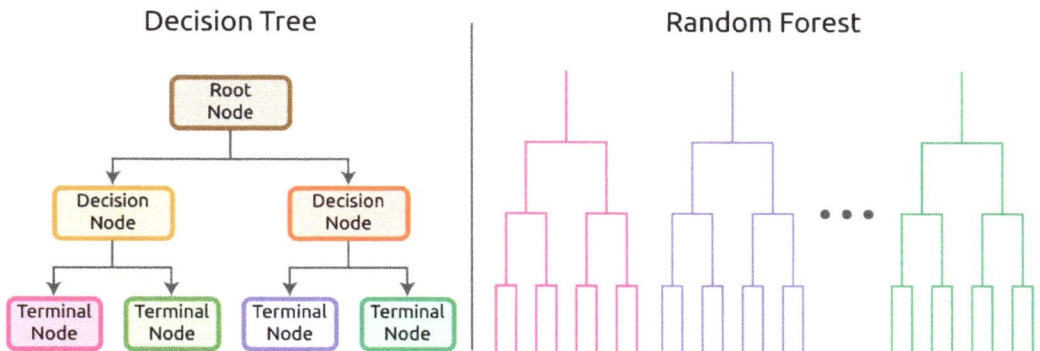

Figure 2. Decision tree (DT) showing nodes at which binary decisions are made on features. Terminal node dictates model prediction. Actual DTs have many more nodes than shown here. Random forest (RF) shown as a series of distinct decision trees.

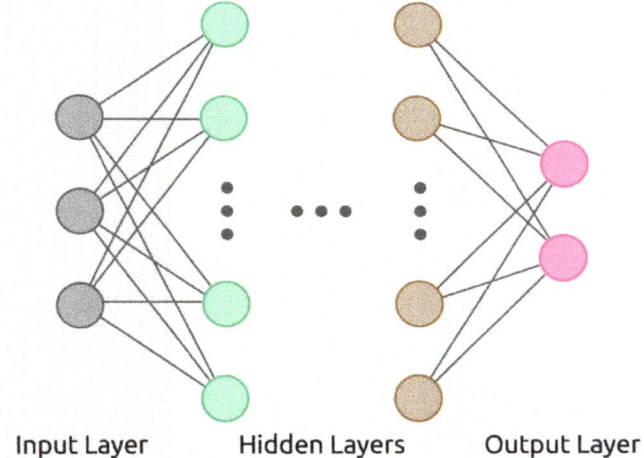

Figure 3. Artificial neural network (ANN) showing nodes of the input, hidden, and output layers.

One of the aspects that makes ANN so powerful is that features do not need to be well-defined real numbers. This allows them to excel at working with data such as images for which extracting numerical features would be difficult and inefficient. One limitation of ANNs is that they require a large amount of data for effective training. In some settings, training data sparsity can be mitigated through a generative adversarial network (GAN) using back propagation [94].

Common classification model performance metrics are accuracy, precision, sensitivity (also known as recall), specificity, and F1. For binary classification with labels "positive" and "negative", they are defined as follows:

$$accuracy = \frac{TP + TN}{TP + TN + FP + FN} \quad (1)$$

$$precision = \frac{TP}{TP + FP} \quad (2)$$

$$sensitivity = \frac{TP}{TP + FN} \quad (3)$$

$$specificity = \frac{TN}{TN + FP} \quad (4)$$

$$F1 = \frac{2 \times precision \times sensitivity}{precision + sensitivity} \quad (5)$$

where TP is true positive, TN is true negative, FP is false positive, and FN is false negative.

3.4. Regression Algorithms

In contrast to classification, the prediction made by a regression algorithm is a numeric value from a continuous scale (e.g., glucose concentration in blood). A simple regression example fits a linear model of the form $y = mx + b$, where a model is built for the prediction of the output variable y based on the input variable x, and the coefficients m and b are "learned" from the data. The learning is typically done by the least-squares regression approach, minimizing the sum of the squared residuals. The following are some of the most common regression algorithms.

Multilinear regression (MLR) is a simple regression model, which expands the above linear model example, accounting for multiple input variables. This model shows how it can be difficult to determine when an algorithm becomes sophisticated enough to be considered "machine learning".

Support vector regression (SVR) is an adaptation of SVM used for regression problems. Like SVM, SVR can utilize kernels to allow for non-linear regression. An advantage of SVR over traditional regression is that one need not assume a model that might not be accurate. For instance, with linear regression, there is an assumption that the data distribution is linear. SVR does not require such pre-determined assumptions [95].

Regression tree is an adaptation of DT for regression. Regression tree has the advantage that it is non-parametric, implying that no assumptions are made about the underlying distribution of values of the predictors [86].

Artificial neural network (ANN) is also widely used for regression problems, and many varieties exist, some of which were mentioned previously.

A large variety of metrics exist for regression model performance. Since there are too many to define here, for further reading, we suggest the study by Hoffman et al. [96] to learn more. Some of the most common metrics are briefly presented here. Root mean squared error (RMSE) and mean absolute error (MAE) have the benefit that their units are the same as the output (predicted) variables, but this makes the metrics less universally understandable. Normalized root mean squared error (NRMSE) partially resolves that. Coefficient of determination, R^2, on the other hand, is unitless and $R^2 \leq 1$, where a value near 1 is generally considered good performance (although this is a bit oversimplified).

3.5. Model Performance Assessment

Frequently, researchers will try various models and compare their performance. The value of the performance metrics listed above can be treated as random variables and statistical analyses can be used to test hypotheses regarding which model is better [96]. While this sounds simple, it can be nuanced: for instance, when working with a classification model, which metric is most important for your application? In some cases, specificity may be more important than accuracy, for instance. Additionally, when using statistical tests to compare model performances, certain assumptions are made, and their validity should be assessed such as when using NRMSE, as it is assumed that noise affecting the output is random and normally distributed.

The best practice for model selection, tuning, and performance assessment is to split the data into 3 sets: training, testing, and validation. For example, if the database consists of 1000 observations, 100 (10%) are assigned to the validation set and the remaining 900 (90%) are split between the training and test sets as 810 (90%) for training, 90 (10%) for testing. The model is then trained on the labeled training set. Model selection and hyperparameter tuning is conducted based on model performance when challenged using the test set. In addition to train–test splitting, cross-fold validation can be used on the training set when tuning hyperparameters or comparing models [97]. Train–test splitting and cross-validation are most important when you intend to generalize the model to predict new, unseen data [96]. Final model performance validation is conducted on the validation set, which should not be used until all model selection and hyperparameter tuning have been completed.

4. Electrochemical Bioreceptor-Free Biosensors

Since their inception, electrochemical biosensors have become extremely popular. In traditional electrochemical biosensors, the bioreceptor interacts with the target to generate a signal at the electrical interface. A widespread scheme is an enzyme (e.g., glucose dehydrogenase or glucose oxidase) interacting directly with the target analyte (e.g., glucose), catalyzing a redox reaction that generates a signal at the electrical interface [98]. Electrical interfaces include metal electrodes, nanoparticles, nanowires, and field-effect transistors (FET) [99].

It is also possible to eliminate the biorecognition element (=bioreceptor, e.g., an enzyme) in electrochemical biosensors. Voltametric sensors described in Section 4.1 can detect biomolecules based on direct interaction with the electrical interface [30]. Electrical impedance spectroscopic biosensors can also detect subtle differences in a solution or material's electrical impedance, as discussed in Section 4.2. Alternatively, we can use an array of chemical or physical sensors varying the electrical interface to create multi-dimensional data. Machine learning-based pattern recognition is used to identify the target analyte. Two of the most common sensor arrays are termed Enose and Etongue, which are covered in Section 4.3.

4.1. Cyclic Voltammetry (CV)

Voltammetry sensors apply electric potential to a "working" electrode and measure the current response, which is affected by analyte oxidation or reduction [100]. Cyclic voltammetry (CV) is a specific voltammetry technique in which the potential is swept across a range of values, and current response is recorded. These CV curves (cyclic voltammograms) can serve as a fingerprint of the sensor response. A typical CV curve is shown in Figure 4A.

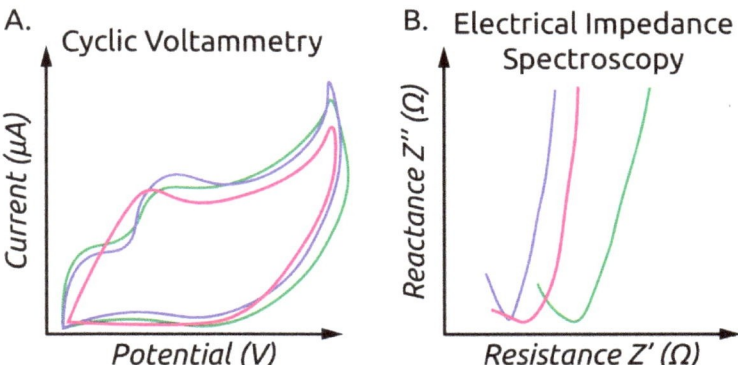

Figure 4. (**A**) Hypothetical cyclic voltammograms for three samples. (**B**) Hypothetical Nyquist plot obtained through EIS showing curves for three samples.

CV biosensors often employ bioreceptors to provide specificity in the interaction between target analyte and electrode surface. However, there has also been research on utilizing more complex electrode surface structures and modifications to allow for semi-specific interaction with the target analyte without the need of a bioreceptor. Sheng et al. [30] describe a compound electrode utilizing Cu/PEDOT-4-COOH particles for CV detection of the phytoinhibitor maleic hydrazide. They found that several regression models had poor performance for modeling the sensor current response with respect to target concentration. However, they employed an ANN with great success for the same regression task. The result is that their detection range is broader than comparable methods by an order of magnitude at each extreme (detection range = 0.06–1000 µM and LOD = 0.01 µM).

4.2. Electrical Impedance Spectroscopy (EIS)

Electrical or electrochemical impedance spectroscopy (EIS) is an analytical technique that provides a fingerprint of the electrical properties of a material. EIS is performed by applying a sinusoidal electric potential to a test sample and recording the impedance (both resistance and reactance expressed in a complex number) over a range of frequencies [101]. Frequently, an equivalent circuit model is fitted to EIS data to provide a fingerprint of the material properties [101]. Figure 5 shows an equivalent circuit diagram for EIS being performed on a single cell suspension. An example EIS spectrum is shown in Figure 4B. It is the classification and regression on such fingerprints that machine learning tends to be well suited.

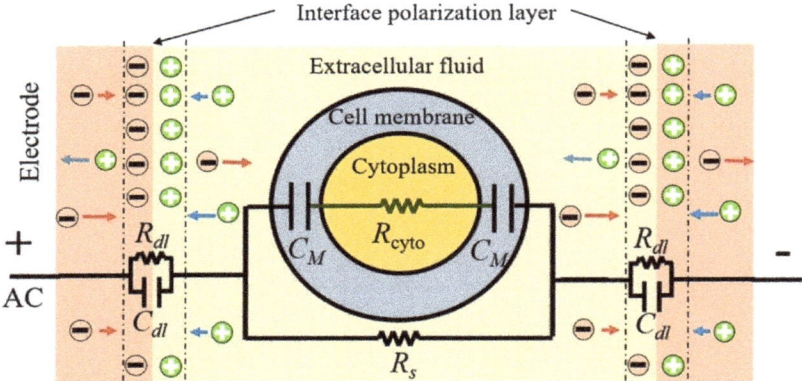

C_M : Cell membrane capacitance
R_{cyto}: Cytoplasmic resistance
C_{dl} : Electric double layer capacitor
R_{dl} : Electric double layer resistor
R_s : Extracellular fluid resistance

Figure 5. Equivalent circuit diagram of single cell suspension. Reproduced with permission from [102] without modification. Copyright 2020 John Wiley and Sons.

A simple example of this is the use of *k*-NN on EIS data for the detection of adulteration in milk [41]. In this work, the feature space was composed of resistance at a certain temperature and pH. They demonstrated good accuracy of 94.9%. However, the data were highly imbalanced, and in the example classification plot [41], one of the three unadulterated samples were misclassified, a 66% specificity.

More robust classification has been performed using SVM. One example is for the assessment of avocado ripeness [43]. This work describes using PCA for feature extraction, resulting in two PCs that explain >99.3% of the variance. SVM is then used for classification based on the first two PCs. SVM for EIS was also described by Murphy et al. [44] for classification of malignant and benign prostatic tissue. However, instead of using PCA for feature extraction, equivalent electrical circuit model parameters were used as predictors. The feature vector size was 2160, consisting of four electrical features for each of eleven frequencies across multiple electrode configurations. Classification was also performed on electrical impedance tomography (EIT) data from the same samples using SVM. Both showed good classification performance, though the authors mention that EIT may be preferable since the measurements are not dependent on probe electrical properties, and thus can be compared more easily to other studies.

While SVM is renowned for its tolerance of outliers, this is a trade-off in that data points not near the boundary between classes do not contribute to defining class attributes. However, ANNs preserve more of this information for prediction. When the number of observations or predictors are small, this can lead to overfitting. However, with sufficient data size, ANNs can preserve predictive information and be robust against outliers and overfitting. These attributes have been utilized for EIS based classification of breast tissue [40,42]. Both works use the same publicly available dataset of EIS measurements from freshly excised breast tissue [103], made available on the University of California, Irvine (UCI) Machine Learning Repository [104]. The dataset contains nine spectral features from EIS. Daliri [40] describes using three ELMs, each with different numbers of nodes, and feeding the output of the three ELMs (extreme learning machines) into SVM for classification. This method showed improved performance over previous methods for the same dataset such as LDA [105]. Helwan et al. [42] compared both BPNN and radial basis function network (RBFN) for the same task. Both methods showed an improvement over ELM-SVM as described by Daliri [40], with RBFN performing better than the BPNN including improved generalizability (i.e., classification performance on new data).

It is seen that in the case of EIS classification, node-based models have shown improved performance over other models. This can be seen most clearly when comparing classification accuracy for those methods that utilized the same dataset. The RBFN and BPNN had the highest classification accuracy, with 93.39% and 94.33%, respectively [42]. The next best performance was achieved by the ELM-SVM, achieving 88.95% accuracy [40]. These results show marked performance increase over LDA [105]. Model performance is greatest in those models that do not utilize distance for classification (i.e., SVM and LDA). While distance-based classifiers are robust to outliers, in these EIS datasets, performance benefitted by node-based classification.

4.3. Enose and Etongue

Enose and Etongue are named in analogy to their respective animal organs. Both sensor types rely on an array of semi-specific sensors, each of which interacts to a different degree with a wide range of analytes. Figure 6 shows a comparison between Enose and Etongue alongside the analogy to their respective biological systems [27,106]. The sensor arrays can be composed of any variety of sensors. The following chemical gas sensors have been used in Enose systems: metal oxide (MOX) gas sensor, surface or bulk acoustic wave (SAW and BAW) sensors, piezoelectric sensor, metal oxide semiconductor field-effect transistor (MOSFET) sensor, and conducting polymer (CP) based sensor [107]. Similarly, a variety of sensors can be employed in Etongue systems such as ion-selective field-effect transistor (ISFET) and light-addressable potentiometric sensor (LAPS) [108].

Analyte presence, or a more general attribute such as odor or taste, is detected through pattern recognition of the sensor array response. For pattern recognition on this naturally high-dimensional data, machine learning techniques are an obvious choice. Scott et al. provided a relevant and succinct paper on data analysis for Enose systems [23]. As discussed in Section 3 of this review, feature engineering is critical in any machine learning pipeline. Yan et al. [24] provide a review article on the feature extraction methods for Enose data. For non-linear feature extraction of Etongue data, Leon-Medina et al. [46] give a great comparison of seven manifold learning methods.

A vast number of papers exist detailing such systems and their use of machine learning. As such, it would be infeasible to cover all of them adequately. For this review, a higher-level analysis is presented by looking at the conclusions reached in the review papers covering this topic as well as a few notable examples of specific papers. Of particular interest is which algorithms had the most success with Enose and Etongue sensors or applications.

A common task of Enose is the prediction of "scent", which is a classification problem. Before the application of the classification algorithm, it is common to perform dimension reduction. PCA is the most common choice for this task, although independent component analysis (ICA, a generalization of PCA) has shown success [25]. PCA has been shown to improve the performance over classification algorithms alone for the piezoelectric Enose [25]. The two classifiers most commonly in use are SVM [109,110] and various ANN methods [25,111]. In addition to classification problems, Enose may be used for analyte concentration prediction. One example is the use of MOS (metal oxide semiconductor) gas sensors for formaldehyde concentration assessment. In this case, the back-propagated neural network (BPNN) outperformed radial basis function network (RBFN) and SVR [33]. In another instance, with the single nickel oxide (NiO) gas sensor, PCA with SVR was utilized for harmful gas classification and quantification [32]. In cases where the amount of data are not large, SVM may be advantageous over node-based models (ANNs) for its resilience against outliers and overfitting.

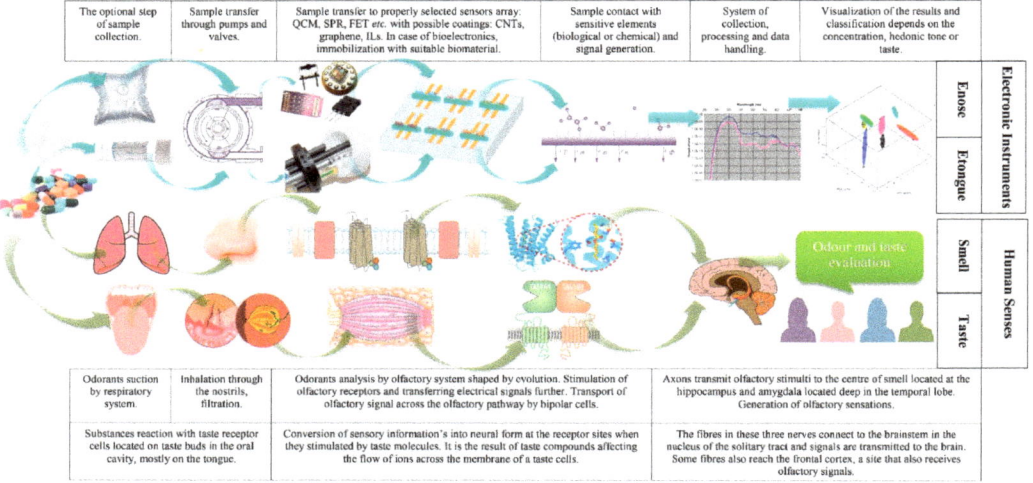

Figure 6. Comparison of operation principle of Enose and Etongue, and the analogy to the biological systems. Reproduced with permission from [27] without modification. Copyright 2019 Elsevier.

While Enose and Etongue systems have shown great promise for non-destructive analytical devices, there are challenges that have limited their use in commercial settings. Several challenges involve changes in the sensor data, which affect the performance of the trained model. A common phenomenon is when the sensor array response changes over time or upon prolonged expose under identical conditions. Such change in sensor response is referred to as sensor drift and can greatly affect the trained models' performance [14]. Another way in which the sensor response may change is if a sensor in the array becomes defective and must be replaced, as it is difficult to replace it with one that responds identically, largely due to variability in manufacturing [112,113]. For both challenges, time consuming and computationally expensive recalibration may be necessary.

The issue of needing retraining due to underlying data distribution changes is commonly addressed through transfer learning in many machine learning settings. Transfer learning is a computational method for minimizing the need for retraining when either the data distributions change (e.g., sensor array response to an analyte) or the task changes (e.g., new classes of analytes are being detected).

Transfer learning has been extensively employed to counter Enose sensor drift and reduce the need for complete retraining [35–38]. It has also been used to reduce the deleterious effect of background interference [39,114]. Although several of the above papers [35,36,38,39] demonstrate the efficacy of their approach on a shared sensor drift dataset shown in Figure 7 [115], ranking of the methods is difficult due to inconsistent benchmarking metrics. As mentioned previously, the data distribution may also change due to replacing a sensor with a new sensor, or when attempting to apply a trained model to a theoretically identical array with differences due to manufacturing variability. Transfer learning, specifically using ANN, has demonstrated decent recalibration [116].

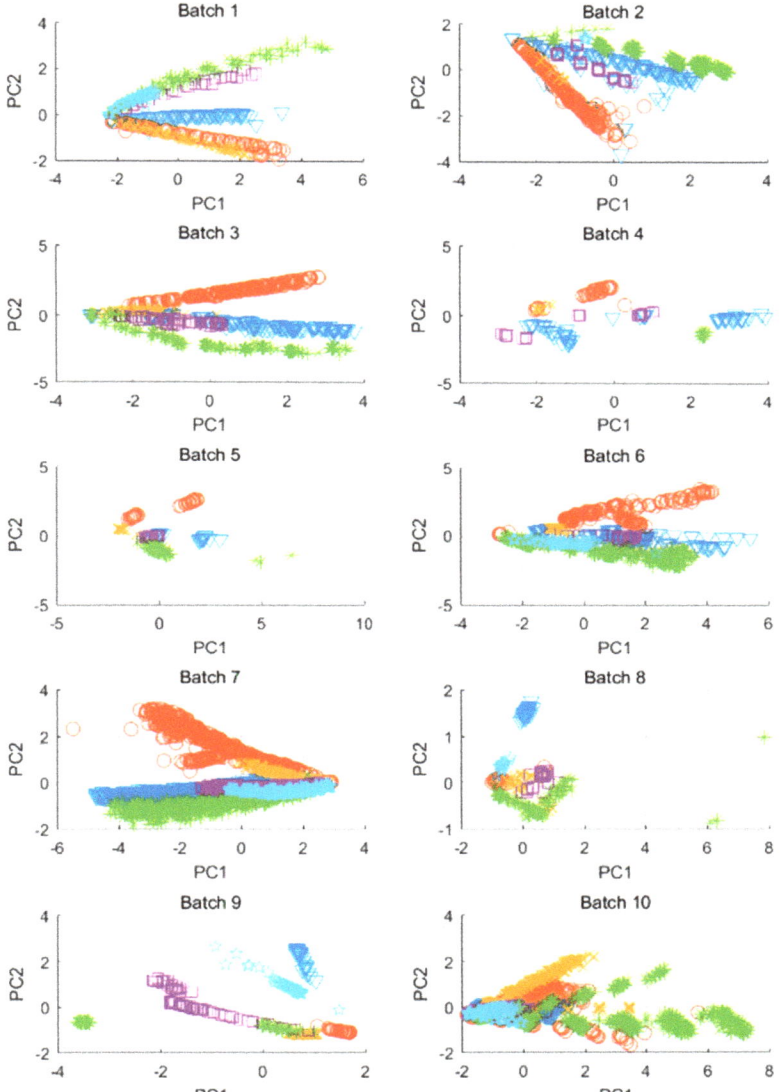

Figure 7. Gas sensor drift dataset from [36]. Each color represents a different gas. Each panel represents a measurement "batch" at various times spanning 36 months. Reproduced from [36] without modification, under Creative Commons Attribution 4.0 License.

One instance of utilizing transfer learning for target task change was demonstrated by Yang et al. by training an Enose classifier on wines (source task) and applying it to classify Chinese liquors (target task) while only retraining the output layer [34]. Interestingly, transfer learning has been used much less commonly for Etongue systems, although they also face sensor drift. However, Yang et al. utilized transfer learning to improve the generalizability of their Etongue [45]. In this work, they demonstrate the superiority of their transfer learning trained CNN over other methods such as BPNN, ELM, and SVM for tea age classification.

A trend that has been gaining traction is data fusion to combine Enose and Etongue systems. The value of this can again be appreciated in how closely the senses of smell and taste are linked in animals [117], complementing each other to provide the most accurate assessment. Similarly, by using information from both Enose and Etongue, better analysis can be conducted. As illustrated in Figure 8, data fusion can be performed at three levels: low, mid, and high [118]. Recently, mid-level fusion schemes have shown promising results for fusion of Enose and Etongue data [119,120], especially when performing PCA on the two systems and using those features for fusion before model training [121–123]. Such systems have also benefitted from the inclusion of a computer vision system in data fusion [121,124].

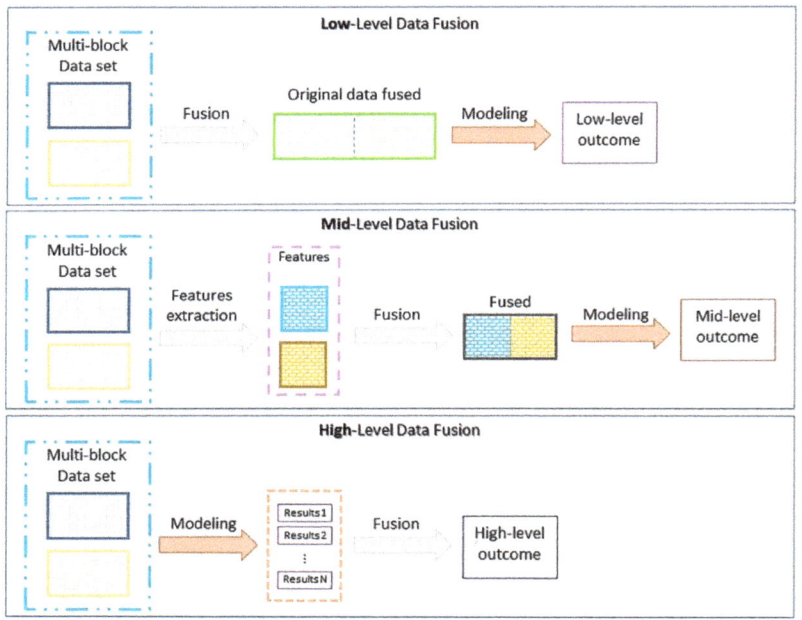

Figure 8. General scheme depicting the main differences among low-, mid-, and high-level data fusion. Reproduced with permission from [118] without modification. Copyright 2019 Elsevier.

Currently, another class of systems exist with the same goals as Enose and Etongue that utilize biochemical recognition elements, termed bioelectronic nose (bEnose) and tongue (bEtongue). These devices utilize biological elements such as taste receptors, cells, or even tissues for sensing [106,125]. These systems show impressive selectivity and sensitivity, especially when coupled with nanomaterials to aid in signal transduction from the biochemical recognition element [106,126]. Their major challenges, as with most biosensors, is stability and reproducibility of the biological element [106]. For these reasons, Enose and Etongue remain popular for their sensor stability. Continued efforts are necessary to improve sensitivity closer to their bioelectronic counterparts, especially regarding sensor design and feature extraction methods.

With such a large variety of sensors in use for Enose and Etongue systems, data processing can vary significantly. Of particular interest is finding appropriate feature extraction methods [23,24]. A huge variety of machine learning classification and regression methods have been employed, both on unsupervised dimensionally reduced feature vectors and classically extracted features. Transfer learning methods have been successful in allowing target task change with minimal retraining, especially when using node-based models. However, the challenges posed by sensor drift and manufacturing variability are still significant and will likely remain a focus for researchers over the next several years.

4.4. Summary of Electrochemical Bioreceptor-Free Biosensing

Many electrochemical bioreceptor-free biosensors employ chemical or physical sensor arrays coupled with machine learning. These are most obvious in Enose and Etongue systems, inspired by nature (humans and animals). Other systems generate multivariate spectral data also coupled with machine learning. In both cases, machine learning models can aid in analyte classification or quantification. Especially when using distance-based models, choice of feature extraction method is important to optimally capture the features relevant to the task (i.e., classification or regression). Node-based models, primarily ANNs often require less feature extraction pre-processing as this step is built into the model learning. Additionally, node-based models offer a great solution to target task change and noise elimination through transfer learning, often aided by integration through the back-propagation step so that only the final layer needs to be refined [34].

5. Optical Bioreceptor-Free Biosensors

The mechanisms of optical detection in biosensing are diverse. A classic example is the colorimetric lateral flow assay [127–129]. Mechanisms beyond colorimetry include fluorescence [130–132], luminescence [133], surface plasmon resonance [134], and light scattering [135,136].

Machine learning has been widely employed in optical biosensors. An example with similarities to Enose and Etongue is the bacterial bioreporter panel. Each bacterial bioreporter responds to target analytes in a semi-specific manner. Machine learning is used to discover patterns in the bioreporter panel response and relate them to analyte presence or concentration [137,138]. However, this review's focus is to discuss cases in which the bioreceptor is absent, so such sensors are not covered in detail.

Another prevalent use of machine learning for analyzing images as biosensor data is for image processing, especially segmentation [139–142]. The literature is rich in reviews on machine learning for image segmentation, and this technology is in no way specific to biosensors, so this review will not discuss those examples. However, the topic is essential to many biosensors, so it must be mentioned.

5.1. Imaging

Imaging sensors utilize an array of optical sensors such as a CMOS array (complementary metal-oxide-semiconductor array; the most used image sensor for digital cameras). Images of the specimen can be used to identify the target presence and concentration as the molecules exhibit different coloration, fluorescence, or light scattering, with varying morphology and spatial distribution. In this manner, several imaging biosensors have been developed to eliminate the need for labels and bioreceptors.

A growing field of imaging-based biosensors utilizes lens-free imaging techniques [143,144]. Since the images from lens-free imaging are not in focus, computational techniques are needed for image reconstruction, the most common of which is deep learning (mostly based on ANN with "deeper" layers) [53,54,145]. Lens-free imaging may be used to detect the aggregation of particles caused by bioreceptor–analyte interaction [55] (Figure 9). However, an exciting application is the direct, label-free classification of particles by lensless holography. Wu et al. [56] presented a lensless holography biosensor for classifying pollen and spores. As with many of these systems, a CNN was used for image reconstruction. In this work, another CNN was used to classify the particles, yielding > 94% accuracy.

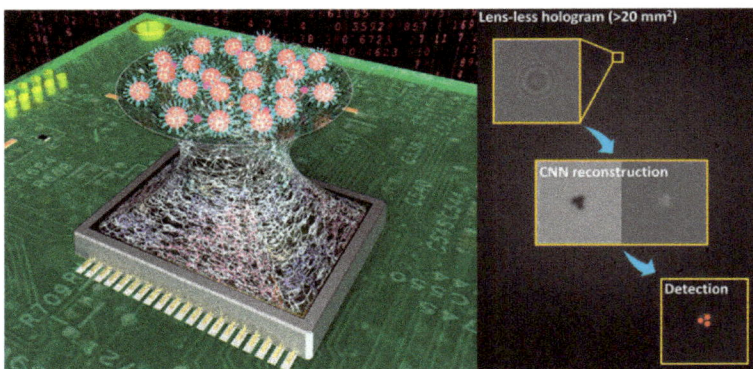

Figure 9. Lensless hologram reconstruction via CNN for particle aggregation detection. Reprinted with permission from [55] without modification. Copyright 2019 American Chemical Society.

Another work on the imaging classification of pollen utilizes multispectral imaging [58]. Again, a CNN was trained for classification, and a species-averaged accuracy of 96% was achieved for 35 plant species.

Artificial neural networks (ANNs) have also found great success in the developing field of digital staining. Hematoxylin and eosin (H&E) stain is the most common stain for histology [146]. However, the quality of tissue staining is subject to many factors that can affect the diagnosis. Digital staining is an alternative in which tissue sections are imaged unstained, and a trained model generates an image simulating stained tissue (Figure 10). Deep learning has been applied for digital staining on images acquired from a variety of methods including quantitative phase imaging [59], Fourier transformed infrared spectroscopy (FTIR) [52], and multi-modal multi-photon microscopy [57]. To overcome the issue of data scarcity and overfitting, researchers have frequently employed generative adversarial neural network (GAN) for medical imaging [147], which has shown promising results for digital staining model training [148]. Additionally, transfer learning has improved the model's generalizability to multiple domains [50].

Fluorescence-based imaging biosensors are also worthy of mention. Sagar et al. [51] presented a microglia classification based on fluorescence lifetime utilizing ANN.

The applications of imaging biosensors are extensive. Indeed, the scope is too large to analyze all papers in this review. However, of particular importance to imaging biosensors is the ANN, especially the CNN. This preference is expected since CNN has shown exceedingly good performance in a variety of image classification contexts [149,150].

5.2. Colorimetry

One class of optical biosensors is the colorimetric biosensor. Currently, the applications of machine learning to enhance the performance of bioreceptor-free colorimetric biosensors are limited. This limitation is because the colorimetric biosensors (most notably lateral flow assays) mostly utilize bioreceptors (e.g., antibodies, enzymes, and aptamers) [98]. One example of such a bioreceptor-free biosensor is non-invasive plant disease diagnosis by Li et al. [49]. They utilized an array of plasmonic nanocolorants and chemo-responsive organic dyes that interact with volatile compounds from the plant. Their technique is similar to Enose and Etongue since it is a fingerprinting approach to the array response for classification. They used PCA, but do not cite an actual classifier, although they give performance metrics such as accuracy. At this time, it is unclear how the classification was performed on the PCA-transformed data.

Most colorimetric biosensors do not require machine learning due to their simplicity for readout. However, the arrays of bioreceptor-free (semi-specific) colorimetric sensors require machine learning-based classification in a way similar to Enose and Etongue.

In these instances, they will likely benefit from the same treatment, namely dimension reduction by PCA and SVM classification.

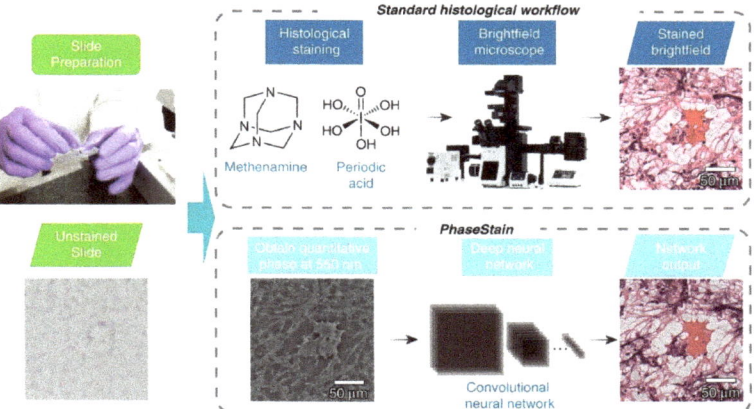

Figure 10. A quantitative phase image of a label-free specimen is virtually stained by a deep neural network, bypassing the standard histological staining procedure that is used as part of clinical pathology. Reproduced from [59] without modification under Creative Commons Attribution 4.0 License.

5.3. Spectroscopy

Of the spectroscopic biosensing techniques, surface-enhanced Raman spectroscopy (SERS) has shown great success [151,152]. SERS is a vibrational surface sensing technique that enhances Raman scattering based on surface characteristics. Briefly, SERS utilizes incident laser light to induce inelastic scattering (Raman scattering) from the target analyte. The intensity of the Raman scattering is enhanced by interaction with the conduction electrons of metal nanostructures (SERS substrate). The enhancement of the Raman scattering is what makes SERS so sensitive. Researchers have reported enhancement factors of up to ten or eleven orders of magnitude [153]. Figure 11 illustrates a SERS sensor for the analysis of breath volatile organic compound (VOC) biomarkers [154]. Due to the complex nature of the obtained spectral signal, various machine learning algorithms have been used to process SERS data in multiple contexts [28].

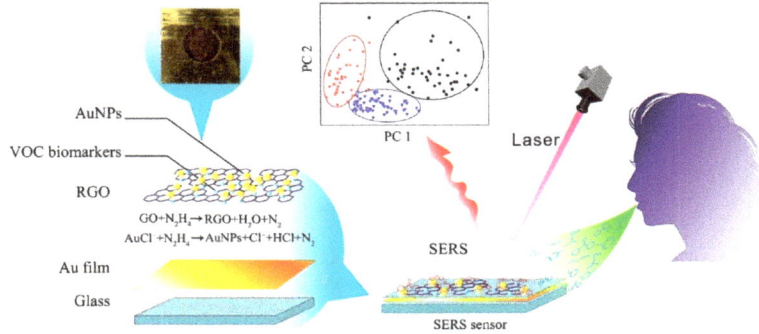

Figure 11. SERS sensor for analysis of breath VOC biomarkers utilizing AuNPs. Reprinted with permission from [154] without modification. Copyright 2016 American Chemical Society.

Although bioreceptors may be used to allow for specific binding of the target analyte to the SERS sensing surface [155,156], direct detection is also possible. Robust classification and regression algorithms can bring specificity and sensitivity to these biosensors. A simple yet effective method for SERS based quantification is partial least squares regression (PLSR). PLSR has been used for a variety of quantification applications such as biofilm formation monitoring [69], blood serum methotrexate concentration [63], aquaculture toxins [62], and food antiseptics [66]. PLSR has the advantage of model simplicity with well-defined parameters, but it may be insufficient in modeling data with significant sources of noise.

Since the spectra have high dimensionality, dimension reduction is a frequent preprocessing step (Figure 12). PCA is again popularly used as a dimension reduction or feature extraction step [60,61,64,65,68,70,71,73], or for exploratory analysis [62,72,157]. Once the spectra are remapped using PCA, a classifier or regression model is employed such as an extreme learning machine (ELM) [71], LDA [68], SVM [60,64,73], PLSR [65], or ANN [70]. An alternative to dimension reduction is utilizing the high dimensionality spectral data directly with a node-based algorithm such as ANN [72,158,159] and CNN [160,161].

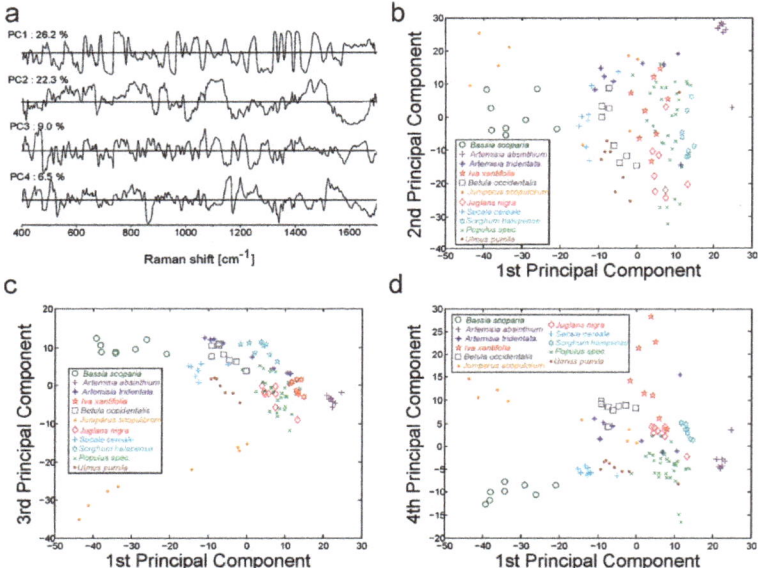

Figure 12. PCA results using the spectral range of 400–1700 cm^{-1} of 112 average SERS spectra from 14 different commercially available pollen species. Loadings of the first four PCs (**a**) as well as the scores of the first and second (**b**), first and third (**c**), and first and fourth PC (**d**) are shown. PCA was done with standardized first derivatives of the mean spectra of 500 vector-normalized spectra. Reprinted with permission from [72] without modification. Copyright 2016 John Wiley and Sons.

The reusability and generalizability of the trained models are often limited. Spectral response is affected not just by analyte presence but surface structure. Therefore, for the model to be reused on a new SERS biosensing dataset, the surface characteristics must be very similar. In terms of transfer learning, this is an issue of changes in the underlying data distributions. However, if the surface structure methods are well documented and reproducible, transfer learning could be employed on a spectral library [28]. Ideally, researchers could contribute to this library in an open-access manner and use these spectra for model training. In this case, the quality of the attached metadata would be a crucial factor.

Clearly, machine learning has been used extensively in the context of SERS sensors. The most common pipeline is to perform unsupervised dimensionality reduction/feature extraction for which PCA is generally the preferred method. Less consistency is seen in the algorithms used for classification and regression. Alternatively, ANNs can be used directly on the data, and the advantage of one approach over the other is not clearly illustrated in the literature. We anticipate, however, that like in the case of electrochemical sensors, node-based models would allow for more efficient transfer learning to accommodate target task change.

5.4. Summary of Optical Bioreceptor-Free Biosensing

A variety of optical sensing methods have benefited from machine learning techniques, with the preferred method being dependent on the data type. For image type data, CNN is the most obvious choice for its ability to detect features as well as reconstruct images obtained by lensless systems. For spectral data, the approach is similar to spectral data obtained with electrochemical sensors. In those instances, dimensionality reduction coupled with a classification/regression algorithm may perform nearly as well as node-based methods. Indeed, they may be preferable in instances where the quantity of training data is small.

6. Considerations and Future Perspectives

Biosensor research has shown great success and promise. For both systems with and without bioreceptor, ML has demonstrated huge success in going from large, complex sensor datasets to getting meaningful measurements and classification of analytes. However, in many of these systems, a key challenge is consistency in device manufacturing. This manifests itself regarding sensor reproducibility for Enose and Etongue, or as substrate reproducibility for SERS. Since the models used to process these data often rely on subtle signals in the data, even small changes in sensor response characteristics can lead to poor performance. These issues have effectively limited widespread commercial adoption of these technologies. There has been some success in accommodating these inconsistencies through computational methods, notably with transfer learning for Enose. More work, both from a manufacturing and computational standpoint, needs to be done before many of these systems are robust enough for widespread adoption.

One area in which these systems have pushed to increase commercial potential is through miniaturization and modularity. There have been efforts with several of the methods presented here to develop compact standalone devices that rival their bulkier counterparts in terms of performance [16,47,162–166]. We believe that cloud computing may be a key element to the success of these endeavors. Some of the models in use, especially for image-based sensors, are computationally expensive. By offloading the computational work to cloud computing, the device footprint imposed by processing and memory needs is greatly reduced.

A central question is what the relative advantages and disadvantages are between systems that utilize a bioreceptor and those that do not. A key advantage of those that eliminate the bioreceptor addresses one of the barriers to commercialization—manufacture variability. By eliminating the bioreceptor, device manufacture is simplified, and may decrease manufacture variability. Additionally, sensor longevity is generally improved because the long-term stability of the bioreceptor is often limited [6]. However, to match LOD and specificity of bioreceptors, improvements must be made. Nanomaterials show promise for improving device performance [167].

There have been studies that attempt to gain the advantages of both systems by creating artificial bioreceptors, notably nanomaterials with enzymatic properties referred to as nanozymes [168,169]. While exciting progress has been made in this field, current nanozyme-based biosensors have inferior catalytic activity and specificity to their biological alternatives [170,171]. Nanozyme catalytic activity is also currently limited to oxidase-like activity [171]. If researchers can broaden nanozyme activity and improve selectivity, these biosensors may become a competitive alternative for biological bioreceptors.

In addition to device considerations, there are computational challenges to consider. Although some ML algorithms have been in use for decades such as PCA and SVM, the field of ML is advancing rapidly with new algorithms being described frequently. While many areas are quick to adopt the new methods, improper usage is common and certainly not limited to biosensing. Some common mistakes are inappropriate data splitting, hidden variables serving as bad predictors, and mistaking the objective of the model [172]. Great emphasis must be placed on the importance of reporting appropriate performance metrics. A great example of a misleading metric is reporting accuracy on highly imbalanced data such as in Durante et al. [41]. It can often be difficult to determine if the proper pre-processing and model assumption checks are being performed. This may be centering and re-scaling prior to PCA, or normality checks for LDA.

Some of these issues can be solved with better methods reporting, especially regarding computational methods. Certain key details are frequently left out, making critical evaluation difficult and reproducibility impossible [173]. One of the most striking examples from the literature described herein is reporting classification metrics, without reporting what classifier was used on PCA processed data [49]. Perhaps the best way to make methods clear and reproducible is to release all associated code, preferably publicly.

Increased availability in general can greatly improve this field. More open access repositories of training sets may allow researchers to improve model robustness by exposing them to more diverse datasets [16]. Some examples currently exist such as the gas sensor drift dataset [115] and the EIS breast tissue dataset [103], both available in the UCI Machine Learning repository [104]. One vision would be to have large repositories of gas sensor responses to many analytes under various experimental conditions. Models could be trained on such repositories to improve generalizability. Ideally, with such repositories and improved manufacturing consistency, trained models could be shared directly and need only minimal recalibration.

7. Conclusions

In this review we have explored the ways in which bioreceptor-free biosensors can benefit from ML methods. Robust ML models bring specificity and accuracy to array-based biosensors such as Enose and Etongue by learning the patterns in the sensor responses. Notably, PCA has shown great performance as a feature extraction technique for these systems. Similar power of PCA has been demonstrated for optical biosensors that generate spectra such as Raman spectra or SERS. ANNs using deep learning generate impressive results for imaging-based sensors including lensless holography and digital staining. ML has also been used in creative ways such as for data fusion of multiple biosensors, and transfer learning for noise correction, sensor drift compensation, and domain adaptation.

However, many practical challenges still exist. Many of the methods presented here are not widely used in commercial settings. This is due to many reasons including variability in manufacturing and the ability to make compact versions of the biosensors while maintaining performance. ML models that can adapt to differences in sensor response are at an advantage, and transfer learning shows promise to be part of the solution.

In recent years, ML has garnered strong research interest in many fields including biosensing, as evidenced in this review. If this review has inspired interest to learn more about how machine learning is being used for one of the methods presented here, we encourage you to seek more specific reviews for the subject. There are great reviews in the literature, many of which were referenced, that take a closer look at the methods presented in this review.

Author Contributions: Conceptualization, K.E.S.III and J.-Y.Y.; Methodology, K.E.S.III; Formal analysis, K.E.S.III; Investigation, K.E.S.III; Data curation, K.E.S.III; Writing—original draft preparation, K.E.S.III; Writing—review and editing, J.-Y.Y.; Supervision, J.-Y.Y.; Project administration, J.-Y.Y.; Funding acquisition, K.E.S.III and J.-Y.Y. All authors have read and agreed to the published version of the manuscript.

Funding: This work was supported by the U.S. National Institutes of Health under the grant T32GM132008.

Institutional Review Board Statement: Not applicable.

Informed Consent Statement: Not applicable.

Data Availability Statement: This study did not report any data.

Acknowledgments: The authors would like to thank Lane E. Breshears for her contribution to the collection of papers and for discussions on how to organize this manuscript. The authors would also like to thank Kattika Kaarj (now at Mahidol University) for helpful discussions and proofreading this manuscript.

Conflicts of Interest: The authors declare no conflict of interest. The funders had no role in the design of the study; in the collection, analyses, or interpretation of data; in the writing of the manuscript, or in the decision to publish the results.

References

1. Metkar, S.K.; Girigoswami, K. Diagnostic biosensors in medicine—A review. *Biocatal. Agric. Biotechnol.* **2019**, *17*, 271–283. [CrossRef]
2. Justino, C.I.L.; Duarte, A.C.; Rocha-Santos, T.A.P. Recent progress in biosensors for environmental monitoring: A review. *Sensors* **2017**, *17*, 2918. [CrossRef] [PubMed]
3. Nguyen, H.H.; Lee, S.H.; Lee, U.J.; Fermin, C.D.; Kim, M. Immobilized enzymes in biosensor applications. *Materials* **2019**, *12*, 121. [CrossRef] [PubMed]
4. Hock, B. Antibodies for immunosensors a review. *Anal. Chim. Acta* **1997**, *347*, 177–186. [CrossRef]
5. Lim, Y.C.; Kouzani, A.Z.; Duan, W. Aptasensors: A review. *J. Biomed. Nanotechnol.* **2010**, *6*, 93–105. [CrossRef] [PubMed]
6. Massah, J.; Asefpour Vakilian, K. An intelligent portable biosensor for fast and accurate nitrate determination using cyclic voltammetry. *Biosyst. Eng.* **2019**, *177*, 49–58. [CrossRef]
7. Esfahani, S.; Shanta, M.; Specht, J.P.; Xing, Y.; Cole, M.; Gardner, J.W. Wearable IoT electronic nose for urinary incontinence detection. In Proceedings of the 2020 IEEE Sensors, Virtual Conference, Rotterdam, The Netherlands, 25–28 October 2020; IEEE: Piscataway, NJ, USA, 2020; pp. 1–4. [CrossRef]
8. Pelosi, P.; Zhu, J.; Knoll, W. From gas sensors to biomimetic artificial noses. *Chemosensors* **2018**, *6*, 32. [CrossRef]
9. Wilson, A.D. Noninvasive early disease diagnosis by electronic-nose and related VOC-detection devices. *Biosensors* **2020**, *10*, 73. [CrossRef]
10. Szulczyński, B.; Armiński, K.; Namieśnik, J.; Gębicki, J. Determination of odour interactions in gaseous mixtures using electronic nose methods with artificial neural networks. *Sensors* **2018**, *18*, 519. [CrossRef]
11. Zambotti, G.; Soprani, M.; Gobbi, E.; Capuano, R.; Pasqualetti, V.; Di Natale, C.; Ponzoni, A. Early detection of fish degradation by electronic nose. In Proceedings of the 2019 IEEE International Symposium on Olfaction and Electronic Nose (ISOEN), Fukuoka, Japan, 26–29 May 2019; IEEE: Piscataway, NJ, USA, 2019; pp. 1–3. [CrossRef]
12. Podrażka, M.; Bączyńska, E.; Kundys, M.; Jeleń, P.S.; Witkowska Nery, E. Electronic tongue—A tool for all tastes? *Biosensors* **2018**, *8*, 3. [CrossRef]
13. Chen, X.; Xu, Y.; Meng, L.; Chen, X.; Yuan, L.; Cai, Q.; Shi, W.; Huang, G. Non-parametric partial least squares—Discriminant analysis model based on sum of ranking difference algorithm for tea grade identification using electronic tongue data. *Sens. Actuat. B Chem.* **2020**, *311*, 127924. [CrossRef]
14. Kovacs, Z.; Szöllősi, D.; Zaukuu, J.-L.Z.; Bodor, Z.; Vitális, F.; Aouadi, B.; Zsom-Muha, V.; Gillay, Z. Factors influencing the long-term stability of electronic tongue and application of improved drift correction methods. *Biosensors* **2020**, *10*, 74. [CrossRef] [PubMed]
15. Guedes, M.D.V.; Marques, M.S.; Guedes, P.C.; Contri, R.V.; Kulkamp Guerreiro, I.C. The use of electronic tongue and sensory panel on taste evaluation of pediatric medicines: A systematic review. *Pharm. Dev. Technol.* **2020**, *26*, 119–137. [CrossRef] [PubMed]
16. Ross, C.F. Considerations of the use of the electronic tongue in sensory science. *Curr. Opin. Food Sci.* **2021**, *40*, 87–93. [CrossRef]
17. Guerrini, L.; Alvarez-Puebla, R.A. Chapter 19—Surface-enhanced Raman scattering chemosensing of proteins. In *Vibrational Spectroscopy in Protein Research*; Ozaki, Y., Baranska, M., Lednev, I.K., Wood, B.R., Eds.; Academic Press: London, UK, 2020; pp. 553–567. [CrossRef]
18. Feng, J.; Hu, Y.; Grant, E.; Lu, X. Determination of thiabendazole in orange juice using an MISPE-SERS chemosensor. *Food Chem.* **2018**, *239*, 816–822. [CrossRef]
19. Langer, J.; Jimenez de Aberasturi, D.; Aizpurua, J.; Alvarez-Puebla, R.A.; Auguié, B.; Baumberg, J.J.; Bazan, G.C.; Bell, S.E.J.; Boisen, A.; Brolo, A.G.; et al. Present and future of surface-enhanced Raman scattering. *ACS Nano* **2020**, *14*, 28–117. [CrossRef]
20. Zheng, X.S.; Jahn, I.J.; Weber, K.; Cialla-May, D.; Popp, J. Label-free SERS in biological and biomedical applications: Recent progress, current challenges and opportunities. *Spectrochim. Acta A Mol. Biomol. Spectrosc.* **2018**, *197*, 56–77. [CrossRef] [PubMed]
21. Krafft, C.; Osei, E.B.; Popp, J.; Nazarenko, I. Raman and SERS spectroscopy for characterization of extracellular vesicles from control and prostate carcinoma patients. *Proc. SPIE* **2020**, *11236*, 1123602. [CrossRef]
22. Sang, S.; Wang, Y.; Feng, Q.; Wei, Y.; Ji, J.; Zhang, W. Progress of new label-free techniques for biosensors: A review. *Crit. Rev. Biotechnol.* **2016**, *36*, 465–481. [CrossRef]
23. Scott, S.M.; James, D.; Ali, Z. Data analysis for electronic nose systems. *Microchim. Acta* **2006**, *156*, 183–207. [CrossRef]

24. Yan, J.; Guo, X.; Duan, S.; Jia, P.; Wang, L.; Peng, C.; Zhang, S. Electronic nose feature extraction methods: A review. *Sensors* **2015**, *15*, 27804–27831. [CrossRef] [PubMed]
25. Hotel, O.; Poli, J.-P.; Mer-Calfati, C.; Scorsone, E.; Saada, S. A review of algorithms for SAW sensors Enose based volatile compound identification. *Sens. Actuat. B Chem.* **2018**, *255*, 2472–2482. [CrossRef]
26. Da Costa, N.L.; da Costa, M.S.; Barbosa, R. A review on the application of chemometrics and machine learning algorithms to evaluate beer authentication. *Food Anal. Meth.* **2021**, *14*, 136–155. [CrossRef]
27. Wasilewski, T.; Migoń, D.; Gębicki, J.; Kamysz, W. Critical review of electronic nose and tongue instruments prospects in pharmaceutical analysis. *Anal. Chim. Acta* **2019**, *1077*, 14–29. [CrossRef]
28. Lussier, F.; Thibault, V.; Charron, B.; Wallace, G.Q.; Masson, J.-F. Deep learning and artificial intelligence methods for Raman and surface-enhanced Raman scattering. *TrAC-Trends Anal. Chem.* **2020**, *124*, 115796. [CrossRef]
29. Cui, F.; Yue, Y.; Zhang, Y.; Zhang, Z.; Zhou, H.S. Advancing biosensors with machine learning. *ACS Sens.* **2020**, *5*, 3346–3364. [CrossRef] [PubMed]
30. Sheng, Y.; Qian, W.; Huang, J.; Wu, B.; Yang, J.; Xue, T.; Ge, Y.; Wen, Y. Electrochemical detection combined with machine learning for intelligent sensing of maleic hydrazide by using carboxylated PEDOT modified with copper nanoparticles. *Microchim. Acta* **2019**, *186*, 543. [CrossRef] [PubMed]
31. Dean, S.N.; Shriver-Lake, L.C.; Stenger, D.A.; Erickson, J.S.; Golden, J.P.; Trammell, S.A. Machine learning techniques for chemical identification using cyclic square wave voltammetry. *Sensors* **2019**, *19*, 2392. [CrossRef] [PubMed]
32. Tonezzer, M.; Le, D.T.T.; Iannotta, S.; Van Hieu, N. Selective discrimination of hazardous gases using one single metal oxide resistive sensor. *Sens. Actuat. B Chem.* **2018**, *277*, 121–128. [CrossRef]
33. Xu, L.; He, J.; Duan, S.; Wu, X.; Wang, Q. Comparison of machine learning algorithms for concentration detection and prediction of formaldehyde based on electronic nose. *Sens. Rev.* **2016**, *36*, 207–216. [CrossRef]
34. Yang, Y.; Liu, H.; Gu, Y. A model transfer learning framework with back-propagation neural network for wine and Chinese liquor detection by electronic nose. *IEEE Access* **2020**, *8*, 105278–105285. [CrossRef]
35. Leon-Medina, J.X.; Pineda-Muñoz, W.A.; Burgos, D.A.T. Joint distribution adaptation for drift correction in electronic nose type sensor arrays. *IEEE Access* **2020**, *8*, 134413–134421. [CrossRef]
36. Liu, B.; Zeng, X.; Tian, F.; Zhang, S.; Zhao, L. Domain transfer broad learning system for long-term drift compensation in electronic nose systems. *IEEE Access* **2019**, *7*, 143947–143959. [CrossRef]
37. Wang, X.; Gu, Y.; Liu, H. A transfer learning method for the protection of geographical indication in China using an electronic nose for the identification of Xihu Longjing tea. *IEEE Sens. J.* **2021**, *21*, 8065–8077. [CrossRef]
38. Yi, R.; Yan, J.; Shi, D.; Tian, Y.; Chen, F.; Wang, Z.; Duan, S. Improving the performance of drifted/shifted electronic nose systems by cross-domain transfer using common transfer samples. *Sens. Actuat. B Chem.* **2021**, *329*, 129162. [CrossRef]
39. Liang, Z.; Tian, F.; Zhang, C.; Sun, H.; Song, A.; Liu, T. Improving the robustness of prediction model by transfer learning for interference suppression of electronic nose. *IEEE Sens. J.* **2018**, *18*, 1111–1121. [CrossRef]
40. Daliri, M.R. Combining extreme learning machines using support vector machines for breast tissue classification. *Comput. Meth. Biomech. Biomed. Eng.* **2015**, *18*, 185–191. [CrossRef]
41. Durante, G.; Becari, W.; Lima, F.A.S.; Peres, H.E.M. Electrical impedance sensor for real-time detection of bovine milk adulteration. *IEEE Sens. J.* **2016**, *16*, 861–865. [CrossRef]
42. Helwan, A.; Idoko, J.B.; Abiyev, R.H. Machine learning techniques for classification of breast tissue. *Proc. Comput. Sci.* **2017**, *120*, 402–410. [CrossRef]
43. Islam, M.; Wahid, K.; Dinh, A. Assessment of ripening degree of avocado by electrical impedance spectroscopy and support vector machine. *J. Food Qual.* **2018**, *2018*, 4706147. [CrossRef]
44. Murphy, E.K.; Mahara, A.; Khan, S.; Hyams, E.S.; Schned, A.R.; Pettus, J.; Halter, R.J. Comparative study of separation between ex vivo prostatic malignant and benign tissue using electrical impedance spectroscopy and electrical impedance tomography. *Physiol. Meas.* **2017**, *38*, 1242–1261. [CrossRef] [PubMed]
45. Yang, Z.; Miao, N.; Zhang, X.; Li, Q.; Wang, Z.; Li, C.; Sun, X.; Lan, Y. Employment of an electronic tongue combined with deep learning and transfer learning for discriminating the storage time of Pu-erh tea. *Food Control* **2021**, *121*, 107608. [CrossRef]
46. Leon-Medina, J.X.; Anaya, M.; Pozo, F.; Tibaduiza, D. Nonlinear feature extraction through manifold learning in an electronic tongue classification task. *Sensors* **2020**, *20*, 4834. [CrossRef]
47. Giménez-Gómez, P.; Escudé-Pujol, R.; Capdevila, F.; Puig-Pujol, A.; Jiménez-Jorquera, C.; Gutiérrez-Capitán, M. Portable electronic tongue based on microsensors for the analysis of Cava wines. *Sensors* **2016**, *16*, 1796. [CrossRef]
48. Ouyang, Q.; Yang, Y.; Wu, J.; Liu, Z.; Chen, X.; Dong, C.; Chen, Q.; Zhang, Z.; Guo, Z. Rapid sensing of total theaflavins content in black tea using a portable electronic tongue system coupled to efficient variables selection algorithms. *J. Food Compos. Anal.* **2019**, *75*, 43–48. [CrossRef]
49. Li, Z.; Paul, R.; Ba Tis, T.; Saville, A.C.; Hansel, J.C.; Yu, T.; Ristaino, J.B.; Wei, Q. Non-invasive plant disease diagnostics enabled by smartphone-based fingerprinting of leaf volatiles. *Nat. Plants* **2019**, *5*, 856–866. [CrossRef]
50. Tomczak, A.; Ilic, S.; Marquardt, G.; Engel, T.; Forster, F.; Navab, N.; Albarqouni, S. Multi-task multi-domain learning for digital staining and classification of leukocytes. *IEEE Trans. Med. Imaging* **2020**. [CrossRef]
51. Sagar, M.A.K.; Cheng, K.P.; Ouellette, J.N.; Williams, J.C.; Watters, J.J.; Eliceiri, K.W. Machine learning methods for fluorescence lifetime imaging (FLIM) based label-free detection of microglia. *Front. Neurosci.* **2020**, *14*, 931. [CrossRef]

52. Lotfollahi, M.; Berisha, S.; Daeinejad, D.; Mayerich, D. Digital staining of high-definition Fourier transform infrared (FT-IR) images using deep learning. *Appl. Spectrosc.* **2019**, *73*, 556–564. [CrossRef]
53. Rivenson, Y.; Zhang, Y.; Günaydın, H.; Teng, D.; Ozcan, A. Phase recovery and holographic image reconstruction using deep learning in neural networks. *Light Sci. Appl.* **2018**, *7*, 17141. [CrossRef]
54. Rivenson, Y.; Wu, Y.; Ozcan, A. Deep learning in holography and coherent imaging. *Light Sci. Appl.* **2019**, *8*, 85. [CrossRef]
55. Wu, Y.; Ray, A.; Wei, Q.; Feizi, A.; Tong, X.; Chen, E.; Luo, Y.; Ozcan, A. Deep learning enables high-throughput analysis of particle-aggregation-based biosensors imaged using holography. *ACS Photon.* **2019**, *6*, 294–301. [CrossRef]
56. Wu, Y.; Calis, A.; Luo, Y.; Chen, C.; Lutton, M.; Rivenson, Y.; Lin, X.; Koydemir, H.C.; Zhang, Y.; Wang, H.; et al. Label-free bioaerosol sensing using mobile microscopy and deep learning. *ACS Photon.* **2018**, *5*, 4617–4627. [CrossRef]
57. Borhani, N.; Bower, A.J.; Boppart, S.A.; Psaltis, D. Digital staining through the application of deep neural networks to multi-modal multi-photon microscopy. *Biomed. Opt. Express* **2019**, *10*, 1339–1350. [CrossRef]
58. Dunker, S.; Motivans, E.; Rakosy, D.; Boho, D.; Mäder, P.; Hornick, T.; Knight, T.M. Pollen analysis using multispectral imaging flow cytometry and deep learning. *New Phytol.* **2021**, *229*, 593–606. [CrossRef]
59. Rivenson, Y.; Liu, T.; Wei, Z.; Zhang, Y.; de Haan, K.; Ozcan, A. PhaseStain: The digital staining of label-free quantitative phase microscopy images using deep learning. *Light Sci. Appl.* **2019**, *8*, 23. [CrossRef] [PubMed]
60. Zheng, X.; Lv, G.; Du, G.; Zhai, Z.; Mo, J.; Lv, X. Rapid and low-cost detection of thyroid dysfunction using Raman spectroscopy and an improved support vector machine. *IEEE Photon. J.* **2018**, *10*, 1–12. [CrossRef]
61. Tan, A.; Zhao, Y.; Sivashanmugan, K.; Squire, K.; Wang, A.X. Quantitative TLC-SERS detection of histamine in seafood with support vector machine analysis. *Food Control.* **2019**, *103*, 111–118. [CrossRef]
62. Chen, X.; Nguyen, T.H.D.; Gu, L.; Lin, M. Use of standing gold nanorods for detection of malachite green and crystal violet in fish by SERS. *J. Food Sci.* **2017**, *82*, 1640–1646. [CrossRef] [PubMed]
63. Fornasaro, S.; Marta, S.D.; Rabusin, M.; Bonifacio, A.; Sergo, V. Toward SERS-based point-of-care approaches for therapeutic drug monitoring: The case of methotrexate. *Faraday Discuss.* **2016**, *187*, 485–499. [CrossRef] [PubMed]
64. Hassoun, M.; Rüger, J.; Kirchberger-Tolstik, T.; Schie, I.W.; Henkel, T.; Weber, K.; Cialla-May, D.; Krafft, C.; Popp, J. A droplet-based microfluidic chip as a platform for leukemia cell lysate identification using surface-enhanced Raman scattering. *Anal. Bioanal. Chem.* **2018**, *410*, 999–1006. [CrossRef]
65. Hidi, I.J.; Jahn, M.; Pletz, M.W.; Weber, K.; Cialla-May, D.; Popp, J. Toward levofloxacin monitoring in human urine samples by employing the LoC-SERS technique. *J. Phys. Chem. C* **2016**, *120*, 20613–20623. [CrossRef]
66. Hou, M.; Huang, Y.; Ma, L.; Zhang, Z. Quantitative analysis of single and mix food antiseptics basing on SERS spectra with PLSR method. *Nanoscale Res. Lett.* **2016**, *11*, 296. [CrossRef]
67. Kämmer, E.; Olschewski, K.; Stöckel, S.; Rösch, P.; Weber, K.; Cialla-May, D.; Bocklitz, T.; Popp, J. Quantitative SERS studies by combining LOC-SERS with the standard addition method. *Anal. Bioanal. Chem.* **2015**, *407*, 8925–8929. [CrossRef]
68. Mühlig, A.; Bocklitz, T.; Labugger, I.; Dees, S.; Henk, S.; Richter, E.; Andres, S.; Merker, M.; Stöckel, S.; Weber, K.; et al. LOC-SERS: A promising closed system for the identification of mycobacteria. *Anal. Chem.* **2016**, *88*, 7998–8004. [CrossRef]
69. Nguyen, C.Q.; Thrift, W.J.; Bhattacharjee, A.; Ranjbar, S.; Gallagher, T.; Darvishzadeh-Varcheie, M.; Sanderson, R.N.; Capolino, F.; Whiteson, K.; Baldi, P.; et al. Longitudinal monitoring of biofilm formation via robust surface-enhanced Raman scattering quantification of *Pseudomonas aeruginosa*-produced metabolites. *ACS Appl. Mater. Interfaces* **2018**, *10*, 12364–12373. [CrossRef] [PubMed]
70. Othman, N.H.; Lee, K.Y.; Radzol, A.R.M.; Mansor, W. PCA-SCG-ANN for detection of non-structural protein 1 from SERS salivary spectra. In *Intelligent Information and Database Systems*; Nguyen, N.T., Tojo, S., Nguyen, L.M., Trawiński, B., Eds.; Springer: Cham, Switzerland, 2017; pp. 424–433. [CrossRef]
71. Othman, N.H.; Yoot Lee, K.; Mohd Radzol, A.R.; Mansor, W.; Amanina Yusoff, N. PCA-polynomial-ELM model optimal for detection of NS1 adulterated salivary SERS spectra. *J. Phys. Conf. Ser.* **2019**, *1372*, 012064. [CrossRef]
72. Seifert, S.; Merk, V.; Kneipp, J. Identification of aqueous pollen extracts using surface enhanced Raman scattering (SERS) and pattern recognition methods. *J. Biophotonics* **2016**, *9*, 181–189. [CrossRef]
73. Sun, H.; Lv, G.; Mo, J.; Lv, X.; Du, G.; Liu, Y. Application of KPCA combined with SVM in Raman spectral discrimination. *Optik* **2019**, *184*, 214–219. [CrossRef]
74. Wold, S.; Esbensen, K.; Geladi, P. Principal component analysis. *Chemom. Intell. Lab. Syst.* **1987**, *2*, 37–52. [CrossRef]
75. Cunningham, P. Dimension reduction. In *Machine Learning Techniques for Multimedia: Case Studies on Organization and Retrieval, Cognitive Technologies*; Cord, M., Cunningham, P., Eds.; Springer: Berlin/Heidelberg, Germany, 2008; pp. 91–112. [CrossRef]
76. Alpaydin, E. *Introduction to Machine Learning*; MIT Press: Cambridge, MA, USA, 2020.
77. Peterson, L.E. K-nearest neighbor. *Scholarpedia* **2009**, *4*, 1883. [CrossRef]
78. Balaprakash, P.; Salim, M.; Uram, T.D.; Vishwanath, V.; Wild, S.M. DeepHyper: Asynchronous hyperparameter search for deep neural networks. In Proceedings of the 2018 IEEE 25th International Conference on High Performance Computing (HiPC), Bengaluru, India, 17–20 December 2018; IEEE: Piscataway, NJ, USA, 2019; pp. 42–51. [CrossRef]
79. Boser, B.E.; Guyon, I.M.; Vapnik, V.N. A training algorithm for optimal margin classifiers. In Proceedings of the Fifth Annual Workshop on Computational Learning Theory, COLT'92, Pittsburgh, PA, USA, 27–29 July 1992; Association for Computing Machinery: New York, NY, USA, 1992; pp. 144–152. [CrossRef]
80. Zhang, J. A complete list of kernels used in support vector machines. *Biochem. Pharmacol.* **2015**, *4*, 2. [CrossRef]

81. Mathur, A.; Foody, G.M. Multiclass and binary SVM classification: Implications for training and classification users. *IEEE Geosci. Remote Sens. Lett.* **2008**, *5*, 241–245. [CrossRef]
82. Shawe-Taylor, J.; Cristianini, N. On the generalization of soft margin algorithms. *IEEE Trans. Inf. Theory* **2002**, *48*, 2721–2735. [CrossRef]
83. Han, H.; Jiang, X. Overcome support vector machine diagnosis overfitting. *Cancer Inform.* **2014**, *13*, 145–158. [CrossRef]
84. Ghojogh, B.; Crowley, M. Linear and quadratic discriminant analysis: Tutorial. *arXiv* **2019**, arXiv:1906.02590. Available online: https://arxiv.org/abs/1906.02590v1 (accessed on 2 August 2021).
85. Myles, A.J.; Feudale, R.N.; Liu, Y.; Woody, N.A.; Brown, S.D. An introduction to decision tree modeling. *J. Chemom.* **2004**, *18*, 275–285. [CrossRef]
86. Lewis, R.J. An introduction to classification and regression tree (CART) analysis. In *The 2000 Annual Meeting of the Society for Academic Emergency Medicine in San Francisco, California*; Society for Academic Emergency Medicine: Des Plaines, IL, USA, 2000; Available online: http://citeseerx.ist.psu.edu/viewdoc/download?doi=10.1.1.95.4103&rep=rep1&type=pdf (accessed on 4 August 2021).
87. Criminisi, A.; Shotton, J.; Konukoglu, E. Decision forests: A unified framework for classification, regression, density estimation, manifold learning and semi-supervised learning. *Found. Trends Comput. Graph. Vis.* **2012**, *7*, 81–227. [CrossRef]
88. Sherstinsky, A. Fundamentals of recurrent neural network (RNN) and long short-term memory (LSTM) network. *Phys. D Nonlinear Phenom.* **2020**, *404*, 132306. [CrossRef]
89. Huang, G.B.; Zhu, Q.Y.; Siew, C.-K. Extreme learning machine: Theory and applications. *Neurocomput. Neural Netw.* **2006**, *70*, 489–501. [CrossRef]
90. Fukushima, K. Neocognitron: A hierarchical neural network capable of visual pattern recognition. *Neural Netw.* **1998**, *1*, 119–130. [CrossRef]
91. Hinton, G.E.; Salakhutdinov, R.R. Reducing the dimensionality of data with neural networks. *Science* **2006**, *313*, 504–507. [CrossRef]
92. Hecht-Nielsen, R. Theory of the backpropagation neural network. In *Neural Networks for Perception, Volume 2: Computation, Learning, and Architectures*; Academic Press: Boston, MA, USA, 1992; pp. 65–93. [CrossRef]
93. Alto, V. Neural Networks: Parameters, Hyperparameters and Optimization Strategies. Available online: https://towardsdatascience.com/neural-networks-parameters-hyperparameters-and-optimization-strategies-3f0842fac0a5 (accessed on 2 August 2021).
94. Goodfellow, I.J.; Pouget-Abadie, J.; Mirza, M.; Xu, B.; Warde-Farley, D.; Ozair, S.; Courville, A.; Bengio, Y. Generative adversarial networks. *arXiv* **2014**, arXiv:1406.2661. Available online: https://arxiv.org/abs/1406.2661 (accessed on 13 April 2021). [CrossRef]
95. Zhang, F.; O'Donnell, L.J. Chapter 7—Support vector regression. In *Machine Learning*; Mechelli, A., Vieira, S., Eds.; Academic Press: London, UK, 2020; pp. 123–140. [CrossRef]
96. Hoffmann, F.; Bertram, T.; Mikut, R.; Reischl, M.; Nelles, O. Benchmarking in classification and regression. *WIREs* **2019**, *9*, e1318. [CrossRef]
97. Shao, J. Linear model selection by cross-validation. *J. Am. Stat. Assoc.* **1993**, *88*, 486–494. [CrossRef]
98. Yoon, J.-Y. *Introduction to Biosensors: From Electric Circuits to Immunosensors*, 2nd ed.; Springer: New York, NY, USA, 2016. [CrossRef]
99. Grieshaber, D.; MacKenzie, R.; Vörös, J.; Reimhult, E. Electrochemical biosensors—Sensor principles and architectures. *Sensors* **2008**, *8*, 1400–1458. [CrossRef] [PubMed]
100. Dhanjai; Sinha, A.; Lu, X.; Wu, L.; Tan, D.; Li, Y.; Chen, J.; Jain, R. Voltammetric sensing of biomolecules at carbon based electrode interfaces: A review. *TrAC—Trends Anal. Chem.* **2018**, *98*, 174–189. [CrossRef]
101. Grossi, M.; Riccò, B. Electrical impedance spectroscopy (EIS) for biological analysis and food characterization: A review. *J. Sens. Sens. Syst.* **2017**, *6*, 303–325. [CrossRef]
102. Yao, J.; Wang, L.; Liu, K.; Wu, H.; Wang, H.; Huang, J.; Li, J. Evaluation of electrical characteristics of biological tissue with electrical impedance spectroscopy. *Electrophoresis* **2020**, *41*, 1425–1432. [CrossRef]
103. Jossinet, J. Variability of impedivity in normal and pathological breast tissue. *Med. Biol. Eng. Comput.* **1996**, *34*, 346–350. [CrossRef]
104. Dua, D.; Graff, C. UCI Machine Learning Repository (https://archive.ics.uci.edu/); University of California, Irvine, School of Information and Computer Sciences: Irvine, CA, USA, 2017.
105. Estrela da Silva, J.; Marques de Sá, J.P.; Jossinet, J. Classification of breast tissue by electrical impedance spectroscopy. *Med. Biol. Eng. Comput.* **2000**, *38*, 26–30. [CrossRef]
106. Wasilewski, T.; Kamysz, W.; Gębicki, J. Bioelectronic tongue: Current status and perspectives. *Biosens. Bioelectron.* **2020**, *150*, 111923. [CrossRef] [PubMed]
107. Röck, F.; Barsan, N.; Weimar, U. Electronic nose: Current status and future trends. *Chem. Rev.* **2008**, *108*, 705–725. [CrossRef] [PubMed]
108. Ha, D.; Sun, Q.; Su, K.; Wan, H.; Li, H.; Xu, N.; Sun, F.; Zhuang, L.; Hu, N.; Wang, P. Recent achievements in electronic tongue and bioelectronic tongue as taste sensors. *Sens. Actuat. B: Chem.* **2015**, *207*, 1136–1146. [CrossRef]
109. Chen, C.Y.; Lin, W.C.; Yang, H.Y. Diagnosis of ventilator-associated pneumonia using electronic nose sensor array signals: Solutions to improve the application of machine learning in respiratory research. *Respir. Res.* **2020**, *21*, 45. [CrossRef]

110. Tan, J.; Xu, J. Applications of electronic nose (Enose) and electronic tongue (Etongue) in food quality-related properties determination: A review. *Artif. Intell. Agric.* **2020**, *4*, 104–115. [CrossRef]
111. Sanaeifar, A.; ZakiDizaji, H.; Jafari, A.; de la Guardia, M. Early detection of contamination and defect in foodstuffs by electronic nose: A review. *TrAC—Trends Anal. Chem.* **2017**, *97*, 257–271. [CrossRef]
112. Zhang, L.; Tian, F.; Kadri, C.; Xiao, B.; Li, H.; Pan, L.; Zhou, H. On-line sensor calibration transfer among electronic nose instruments for monitoring volatile organic chemicals in indoor air quality. *Sens. Actuat. B Chem.* **2011**, *160*, 899–909. [CrossRef]
113. Ciosek, P.; Wróblewski, W. Sensor arrays for liquid sensing—Electronic tongue systems. *Analyst* **2007**, *132*, 963–978. [CrossRef]
114. Liu, H.; Li, Q.; Li, Z.; Gu, Y. A suppression method of concentration background noise by transductive transfer learning for a metal oxide semiconductor-based electronic nose. *Sensors* **2020**, *20*, 1913. [CrossRef]
115. Vergara, A.; Vembu, S.; Ayhan, T.; Ryan, M.A.; Homer, M.L.; Huerta, R. Chemical gas sensor drift compensation using classifier ensembles. *Sens. Actuat. B Chem.* **2012**, *166–167*, 320–329. [CrossRef]
116. Zhang, L.; Tian, F.; Peng, X.; Dang, L.; Li, G.; Liu, S.; Kadri, C. Standardization of metal oxide sensor array using artificial neural networks through experimental design. *Sens. Actuat. B Chem.* **2013**, *177*, 947–955. [CrossRef]
117. DeVere, R. Disorders of taste and smell. *Continuum* **2017**, *23*, 421. [CrossRef]
118. Biancolillo, A.; Boqué, R.; Cocchi, M.; Marini, F. Chapter 10—Data fusion strategies in food analysis. In *Data Handling in Science and Technology*; Cocchi, M., Ed.; Elsevier: Amsterdam, The Netherlands, 2019; Volume 31, pp. 271–310. [CrossRef]
119. Banerjee, M.B.; Roy, R.B.; Tudu, B.; Bandyopadhyay, R.; Bhattacharyya, N. Black tea classification employing feature fusion of E-nose and E-tongue responses. *J. Food Eng.* **2019**, *244*, 55–63. [CrossRef]
120. Men, H.; Shi, Y.; Fu, S.; Jiao, Y.; Qiao, Y.; Liu, J. Mining feature of data fusion in the classification of beer flavor information using E-tongue and E-nose. *Sensors* **2017**, *17*, 1656. [CrossRef] [PubMed]
121. Buratti, S.; Malegori, C.; Benedetti, S.; Oliveri, P.; Giovanelli, G. E-nose, E-tongue and E-eye for edible olive oil characterization and shelf life assessment: A powerful data fusion approach. *Talanta* **2018**, *182*, 131–141. [CrossRef]
122. Dai, C.; Huang, X.; Huang, D.; Lv, R.; Sun, J.; Zhang, Z.; Ma, M.; Aheto, J.H. Detection of submerged fermentation of *Tremella aurantialba* using data fusion of electronic nose and tongue. *J. Food Proc. Eng.* **2019**, *42*, e13002. [CrossRef]
123. Tian, X.; Wang, J.; Ma, Z.; Li, M.; Wei, Z. Combination of an E-nose and an E-tongue for adulteration detection of minced mutton mixed with pork. *J. Food Qual.* **2019**, *2019*, e4342509. [CrossRef]
124. Di Rosa, A.R.; Leone, F.; Cheli, F.; Chiofalo, V. Fusion of electronic nose, electronic tongue and computer vision for animal source food authentication and quality assessment—A review. *J. Food Eng.* **2017**, *210*, 62–75. [CrossRef]
125. Cave, J.W.; Wickiser, J.K.; Mitropoulos, A.N. Progress in the development of olfactory-based bioelectronic chemosensors. *Biosens. Bioelectron.* **2019**, *123*, 211–222. [CrossRef]
126. Ahn, S.R.; An, J.H.; Song, H.S.; Park, J.W.; Lee, S.H.; Kim, J.H.; Jang, J.; Park, T.H. Duplex bioelectronic tongue for sensing umami and sweet tastes based on human taste receptor nanovesicles. *ACS Nano* **2016**, *10*, 7287–7296. [CrossRef] [PubMed]
127. Edachana, R.P.; Kumaresan, A.; Balasubramanian, V.; Thiagarajan, R.; Nair, B.G.; Thekkedath Gopalakrishnan, S.B. Paper-based device for the colorimetric assay of bilirubin based on in-vivo formation of gold nanoparticles. *Microchim. Acta* **2019**, *187*, 60. [CrossRef]
128. Mutlu, A.Y.; Kılıç, V.; Kocakuşak Özdemir, G.; Bayram, A.; Horzum, N.; Solmaz, M.E. Smartphone-based colorimetric detection via machine learning. *Analyst* **2017**, *142*, 2434–2441. [CrossRef]
129. Solmaz, M.E.; Mutlu, A.Y.; Alankus, G.; Kılıç, V.; Bayram, A.; Horzum, N. Quantifying colorimetric tests using a smartphone app based on machine learning classifiers. *Sens. Actuat. B Chem.* **2018**, *255*, 1967–1973. [CrossRef]
130. Lin, B.; Yu, Y.; Cao, Y.; Guo, M.; Zhu, D.; Dai, J.; Zheng, M. Point-of-care testing for streptomycin based on aptamer recognizing and digital image colorimetry by smartphone. *Biosens. Bioelectron.* **2018**, *100*, 482–489. [CrossRef] [PubMed]
131. Song, E.; Yu, M.; Wang, Y.; Hu, W.; Cheng, D.; Swihart, M.T.; Song, Y. Multi-color quantum dot-based fluorescence immunoassay array for simultaneous visual detection of multiple antibiotic residues in milk. *Biosens. Bioelectron.* **2015**, *72*, 320–325. [CrossRef] [PubMed]
132. Wang, L.; Chen, W.; Ma, W.; Liu, L.; Ma, W.; Zhao, Y.; Zhu, Y.; Xu, L.; Kuang, H.; Xu, C. Fluorescent strip sensor for rapid determination of toxins. *Chem. Commun.* **2011**, *47*, 1574–1576. [CrossRef] [PubMed]
133. Chen, X.; Lan, J.; Liu, Y.; Li, L.; Yan, L.; Xia, Y.; Wu, F.; Li, C.; Li, S.; Chen, J. A paper-supported aptasensor based on upconversion luminescence resonance energy transfer for the accessible determination of exosomes. *Biosens. Bioelectron.* **2018**, *102*, 582–588. [CrossRef] [PubMed]
134. Guo, X. Surface plasmon resonance based biosensor technique: A review. *J. Biophoton.* **2012**, *5*, 483–501. [CrossRef] [PubMed]
135. Heinze, B.C.; Yoon, J.-Y. Nanoparticle immunoagglutination Rayleigh scatter assay to complement microparticle immunoagglutination Mie scatter assay in a microfluidic device. *Colloids Surf. B Biointerfaces* **2011**, *85*, 168–173. [CrossRef]
136. Park, T.S.; Li, W.; McCracken, K.E.; Yoon, J.-Y. Smartphone quantifies Salmonella from paper microfluidics. *Lab Chip* **2013**, *13*, 4832–4840. [CrossRef] [PubMed]
137. Elad, T.; Benovich, E.; Magrisso, S.; Belkin, S. Toxicant identification by a luminescent bacterial bioreporter panel: Application of pattern classification algorithms. *Environ. Sci. Technol.* **2008**, *42*, 8486–8491. [CrossRef]
138. Jouanneau, S.; Durand, M.-J.; Courcoux, P.; Blusseau, T.; Thouand, G. Improvement of the identification of four heavy metals in environmental samples by using predictive decision tree models coupled with a set of five bioluminescent bacteria. *Environ. Sci. Technol.* **2011**, *45*, 2925–2931. [CrossRef] [PubMed]

139. Gou, T.; Hu, J.; Zhou, S.; Wu, W.; Fang, W.; Sun, J.; Hu, Z.; Shen, H.; Mu, Y. A new method using machine learning for automated image analysis applied to chip-based digital assays. *Analyst* **2019**, *144*, 3274–3281. [CrossRef] [PubMed]
140. Ker, J.; Wang, L.; Rao, J.; Lim, T. Deep learning applications in medical image analysis. *IEEE Access* **2018**, *6*, 9375–9389. [CrossRef]
141. Uslu, F.; Icoz, K.; Tasdemir, K.; Yilmaz, B. Automated quantification of immunomagnetic beads and leukemia cells from optical microscope images. *Biomed. Signal Proc. Control.* **2019**, *49*, 473–482. [CrossRef]
142. Zeng, N.; Wang, Z.; Zhang, H.; Liu, W.; Alsaadi, F.E. Deep belief networks for quantitative analysis of a gold immunochromatographic strip. *Cogn. Comput.* **2016**, *8*, 684–692. [CrossRef]
143. Roy, M.; Seo, D.; Oh, S.; Yang, J.-W.; Seo, S. A review of recent progress in lens-free imaging and sensing. *Biosens. Bioelectron.* **2017**, *88*, 130–143. [CrossRef]
144. Wu, Y.; Ozcan, A. Lensless digital holographic microscopy and its applications in biomedicine and environmental monitoring. *Methods* **2018**, *136*, 4–16. [CrossRef]
145. Greenbaum, A.; Luo, W.; Su, T.W.; Göröcs, Z.; Xue, L.; Isikman, S.O.; Coskun, A.F.; Mudanyali, O.; Ozcan, A. Imaging without lenses: Achievements and remaining challenges of wide-field on-chip microscopy. *Nat. Methods* **2012**, *9*, 889–895. [CrossRef]
146. Slaoui, M.; Fiette, L. Histopathology procedures: From tissue sampling to histopathological evaluation. In *Drug Safety Evaluation: Methods and Protocols*; Gautier, J.-C., Ed.; Humana Press: Totowa, NJ, USA, 2011; pp. 69–82. [CrossRef]
147. Yi, X.; Walia, E.; Babyn, P. Generative adversarial network in medical imaging: A review. *Med. Image Anal.* **2019**, *58*, 101552. [CrossRef]
148. Rana, A.; Yauney, G.; Lowe, A.; Shah, P. Computational histological staining and destaining of prostate core biopsy RGB images with generative adversarial neural networks. In *2018 17th IEEE International Conference on Machine Learning and Applications (ICMLA)*; IEEE: Piscataway, NJ, USA, 2018; pp. 828–834. [CrossRef]
149. Affonso, C.; Rossi, A.L.D.; Vieira, F.H.A.; de Leon Ferreira de Carvalho, A.C.P. Deep learning for biological image classification. *Expert Syst. Appl.* **2017**, *85*, 114–122. [CrossRef]
150. Russakovsky, O.; Deng, J.; Su, H.; Krause, J.; Satheesh, S.; Ma, S.; Huang, Z.; Karpathy, A.; Khosla, A.; Bernstein, M.; et al. ImageNet large scale visual recognition challenge. *Int. J. Comput. Vis.* **2015**, *115*, 211–252. [CrossRef]
151. Yang, T.; Guo, X.; Wang, H.; Fu, S.; Wen, Y.; Yang, H. Magnetically optimized SERS assay for rapid detection of trace drug-related biomarkers in saliva and fingerprints. *Biosens. Bioelectron.* **2015**, *68*, 350–357. [CrossRef]
152. Zhang, D.; Huang, L.; Liu, B.; Ni, H.; Sun, L.; Su, E.; Chen, H.; Gu, Z.; Zhao, X. Quantitative and ultrasensitive detection of multiplex cardiac biomarkers in lateral flow assay with core-shell SERS nanotags. *Biosens. Bioelectron.* **2018**, *106*, 204–211. [CrossRef] [PubMed]
153. Schlücker, S. Surface-enhanced Raman spectroscopy: Concepts and chemical applications. *Angew. Chem. Int. Ed.* **2014**, *53*, 4756–4795. [CrossRef]
154. Chen, Y.; Zhang, Y.; Pan, F.; Liu, J.; Wang, K.; Zhang, C.; Cheng, S.; Lu, L.; Zhang, W.; Zhang, Z.; et al. Breath analysis based on surface-enhanced Raman scattering sensors distinguishes early and advanced gastric cancer patients from healthy persons. *ACS Nano* **2016**, *10*, 8169–8179. [CrossRef]
155. Banaei, N.; Moshfegh, J.; Mohseni-Kabir, A.; Houghton, J.M.; Sun, Y.; Kim, B. Machine learning algorithms enhance the specificity of cancer biomarker detection using SERS-based immunoassays in microfluidic chips. *RSC Adv.* **2019**, *9*, 1859–1868. [CrossRef]
156. Guselnikova, O.; Postnikov, P.; Pershina, A.; Svorcik, V.; Lyutakov, O. Express and portable label-free DNA detection and recognition with SERS platform based on functional Au grating. *Appl. Surf. Sci.* **2019**, *470*, 219–227. [CrossRef]
157. Goodacre, R.; Graham, D.; Faulds, K. Recent developments in quantitative SERS: Moving towards absolute quantification. *TrAC—Trends Anal. Chem.* **2018**, *102*, 359–368. [CrossRef]
158. Guselnikova, O.; Trelin, A.; Skvortsova, A.; Ulbrich, P.; Postnikov, P.; Pershina, A.; Sykora, D.; Svorcik, V.; Lyutakov, O. Label-free surface-enhanced Raman spectroscopy with artificial neural network technique for recognition photoinduced DNA damage. *Biosens. Bioelectron.* **2019**, *145*, 111718. [CrossRef] [PubMed]
159. Thrift, W.J.; Cabuslay, A.; Laird, A.B.; Ranjbar, S.; Hochbaum, A.I.; Ragan, R. Surface-enhanced Raman scattering-based odor compass: Locating multiple chemical sources and pathogens. *ACS Sens.* **2019**, *4*, 2311–2319. [CrossRef] [PubMed]
160. Thrift, W.J.; Nguyen, C.Q.; Wang, J.; Kahn, J.E.; Dong, R.; Laird, A.B.; Ragan, R. Improved regressions with convolutional neural networks for surface enhanced Raman scattering sensing of metabolite biomarkers. *Proc. SPIE* **2019**, *11089*, 1108907. [CrossRef]
161. Thrift, W.J.; Ragan, R. Quantification of analyte concentration in the single molecule regime using convolutional neural networks. *Anal. Chem.* **2019**, *91*, 13337–13342. [CrossRef]
162. Cheng, L.; Meng, Q.H.; Lilienthal, A.J.; Qi, P.F. Development of compact electronic noses: A review. *Meas. Sci. Technol.* **2021**, *32*, 062002. [CrossRef]
163. Jiang, L.; Hassan, M.M.; Ali, S.; Li, H.; Sheng, R.; Chen, Q. Evolving trends in SERS-based techniques for food quality and safety: A review. *Trends Food Sci. Technol.* **2021**, *112*, 225–240. [CrossRef]
164. Pang, S.; Yang, T.; He, L. Review of surface enhanced Raman spectroscopic (SERS) detection of synthetic chemical pesticides. *TrAC—Trends Anal. Chem.* **2016**, *85*, 73–82. [CrossRef]
165. Barreiros dos Santos, M.; Queirós, R.B.; Geraldes, Á.; Marques, C.; Vilas-Boas, V.; Dieguez, L.; Paz, E.; Ferreira, R.; Morais, J.; Vasconcelos, V.; et al. Portable sensing system based on electrochemical impedance spectroscopy for the simultaneous quantification of free and total microcystin-LR in freshwaters. *Biosens. Bioelectron.* **2019**, *142*, 111550. [CrossRef]

166. Huang, X.; Li, Y.; Xu, X.; Wang, R.; Yao, J.; Han, W.; Wei, M.; Chen, J.; Xuan, W.; Sun, L. High-precision lensless microscope on a chip based on in-line holographic imaging. *Sensors* **2021**, *21*, 720. [CrossRef]
167. Zhang, R.; Belwal, T.; Li, L.; Lin, X.; Xu, Y.; Luo, Z. Nanomaterial-based biosensors for sensing key foodborne pathogens: Advances from recent decades. *Compr. Rev. Food Sci. Food Saf.* **2020**, *19*, 1465–1487. [CrossRef]
168. Wang, Q.; Wei, H.; Zhang, Z.; Wang, E.; Dong, S. Nanozyme: An emerging alternative to natural enzyme for biosensing and immunoassay. *TrAC—Trends Anal. Chem.* **2018**, *105*, 218–224. [CrossRef]
169. Zhu, X.; Liu, P.; Ge, Y.; Wu, R.; Xue, T.; Sheng, Y.; Ai, S.; Tang, K.; Wen, Y. MoS2/MWCNTs porous nanohybrid network with oxidase-like characteristic as electrochemical nanozyme sensor coupled with machine learning for intelligent analysis of carbendazim. *J. Electroanal. Chem.* **2020**, *862*, 113940. [CrossRef]
170. Mahmudunnabi, G.R.; Farhana, F.Z.; Kashaninejad, N.; Firoz, S.H.; Shim, Y.-B.; Shiddiky, M.J.A. Nanozyme-based electrochemical biosensors for disease biomarker detection. *Analyst* **2020**, *145*, 4398–4420. [CrossRef]
171. Zhang, X.; Wu, D.; Zhou, X.; Yu, Y.; Liu, J.; Hu, N.; Wang, H.; Li, G.; Wu, Y. Recent progress in the construction of nanozyme-based biosensors and their applications to food safety assay. *TrAC—Trends Anal. Chem.* **2019**, *121*, 115668. [CrossRef]
172. Riley, P. Three pitfalls to avoid in machine learning. *Nature* **2019**, *572*, 27–29. [CrossRef] [PubMed]
173. Stodden, V.; McNutt, M.; Bailey, D.H.; Deelman, E.; Gil, Y.; Hanson, B.; Heroux, M.A.; Ioannidis, J.P.A.; Taufer, M. Enhancing reproducibility for computational methods. *Science* **2016**, *354*, 1240–1241. [CrossRef] [PubMed]

Article

IoTactileSim: A Virtual Testbed for Tactile Industrial Internet of Things Services

Muhammad Zubair Islam, Shahzad, Rashid Ali, Amir Haider * and Hyungseok Kim *

School of Intelligent Mechatronics Engineering, Sejong University, Seoul 05006, Korea;
zubair@sju.ac.kr (M.Z.I.); shahzad@sju.ac.kr (S.); rashidali@sejong.ac.kr (R.A.)
* Correspondence: amirhaider@sejong.ac.kr (A.H.); hyungkim@sejong.ac.kr (H.K.)

Abstract: With the inclusion of tactile Internet (TI) in the industrial sector, we are at the doorstep of the tactile Industrial Internet of Things (IIoT). This provides the ability for the human operator to control and manipulate remote industrial environments in real-time. The TI use cases in IIoT demand a communication network, including ultra-low latency, ultra-high reliability, availability, and security. Additionally, the lack of the tactile IIoT testbed has made it more severe to investigate and improve the quality of services (QoS) for tactile IIoT applications. In this work, we propose a virtual testbed called IoTactileSim, that offers implementation, investigation, and management for QoS provisioning in tactile IIoT services. IoTactileSim utilizes a network emulator Mininet and robotic simulator CoppeliaSim to perform real-time haptic teleoperations in virtual and physical environments. It provides the real-time monitoring of the implemented technology parametric values, network impairments (delay, packet loss), and data flow between operator (master domain) and teleoperator (slave domain). Finally, we investigate the results of two tactile IIoT environments to prove the potential of the proposed IoTactileSim testbed.

Keywords: 5G/6G; URLLC; tactile Internet; industrial IoT; network emulator; robotic simulator; virtual testbed

1. Introduction

The rapid development of communication technologies from First-Generation (1G) to Sixth-Generation (6G) has gained enormous attention due to its emerging services like human-to-human (H2H), machine-to-machine (M2M), and human-to-machine (H2M) communication. These emerging services are induced by drivers like mobile Internet, Internet of Things (IoT), and tactile Internet (TI). The IoT envisions to fill the gap between the cyber and physical world [1]. It is defined as to interrelate every existing computing object around us such as, mobile devices, sensors, and actuators, over the Internet. Moreover, IoT technology provides data sharing and communication in the M2M environment. Recently, the TI, with the aim to enable haptic communications, has shifted the IoT paradigm to real-time interaction between H2M and revolutionized the next-generation communication technologies [2,3]. The TI is envisioned to empower H2M communication where a human can interact with machines in a virtual and physical environment, while experiencing the haptic sensations (touch and forces) in addition to traditional audio-video data [4]. Figure 1 depicts the technological evolution of the communication trends from 1G to 6G wireless communication.

Several international standard organizations, such as the international telecommunication union, the Third-Generation Partnership Project (3GPP), and the Institute of Electrical and Electronics Engineering (IEEE), are working to enable the existing and develop new network architectures to carry haptic data over the communication in real-time. The TI standard working group IEEE P1819.1 has already initiated and defined reference architecture, technical functions, and the definition of the TI [5]. Moreover, it also described

standard use cases of the TI and corresponding strict requirements, including teleoperation, automotive, immersive virtual/augmented reality, internet of drones, interpersonal communication, live haptic broadcast, and cooperative automated driving. However, these use cases demand near real-time connectivity (ultra-reliable and ultra-responsive) for M2M and H2M communication. This type of real-time connectivity is termed as ultra-reliable and low latency communication (URLLC). The URLLC is one of the key services of the Fifth-Generation (5G) networks, along with enhanced mobile broadband and massive machine-type communication. Moreover, 3GPP has introduced the 5G new radio to increase reliability and minimize end-to-end (E2E) communication latency for the URLLC services. In Release 15, 3GPP describes the URLLC requirement with the reliability of 99.9% for a single 32-byte packet under 1ms latency [6]. Conclusively, 5G URLLC services are one of the potential enablers for the extreme requirements of the TI.

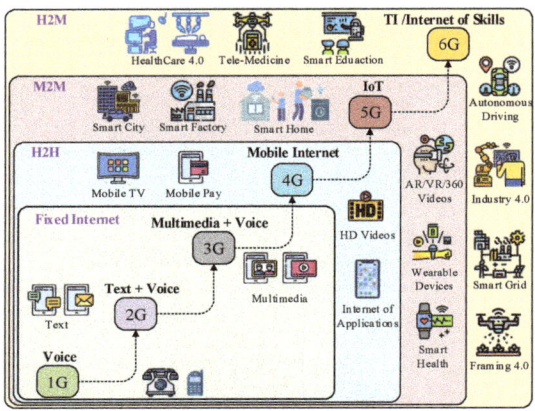

Figure 1. A taxonomy of the different emerging communication trends.

Moreover, these requirements become more critical for loss-intolerant and delay-sensitive TI industrial and medical applications. For example, remote industrial management and the automation of industrial robots (sensitivity of control circuits) demand latency between 0.25–10 ms with a packet loss of $\leq 10^{-9}$ [7]. Therefore, supporting next-generation industrial applications, including immersive reality, holographic, and haptic/tactile communication, demands a 5G network with new physical and upper layer techniques to guarantee quality of service (QoS) and quality of experience (QoE) provisioning. Furthermore, the 6G technology paradigm promises to break the 5G network limitations and enable them to virtualize human skills and transfer them from one place to another within 1ms through 6G native artificial intelligence (AI) network architecture. In-depth work on 5G URLLC services, beyond the 5G and 6G communication network, is presented in these articles [8–12]. Table 1 compares the connectivity requirements of the traditional and emerging tactile IIoT applications (adapted from [7]). The relationships between emerging technologies such as IoT, IIoT, Industrial Internet, Internet of Everything (IoE), TI, tactile IoT, tactile IIoT, Industry 4.0 and 5.0 are presented in Figure 2.

An in-depth discussion on conventional and emerging industrial is presented in [7], where the authors investigated the role of TI in the industrial environment, along with technical connectivity requirements of the TI industrial services. One of the vital use cases of the TI in the industrial domain is the bilateral/multilateral haptic-driven teleoperation systems. A teleoperation system consists of a human operator (master), teleoperator (slave), and a communication network that link the master to a slave domain, and enable the operator to interact with the teleoperator in the distant and inaccessible remote environment to perform complex tasks. The TI-based network provides bilateral communication to manage touch and actuation in real-time between the master and slave domain with a focus to ensure QoS and QoE requirements. Haptic-enabled teleoperation systems have numerous applications in Industry 4.0, such as robotic automation, smart manufacturing, smart logistic, the mining industry, food industry, healthcare industry, and industrial management (controlling and monitoring). Contrary to the traditional application, Haptic-enabled industrial applications demand high QoS and QoE, and depend on the nature of the application.

Table 1. Summary of the connectivity requirements for traditional IIoT and emerging tactile IIoT services.

	Applications/Requirements	Latency (ms)	Reliability (%)	Scalability	Data Rate (Mbps)
Conventional	Monitoring	50–100	99.9–99.99	100–1000	0.1–0.5
	Safety control	10	99.99–99.999	10–20	0.1–1
	Motion control	0.5–2	99.9999–99.99999	10–50	1–5
	Closed-loop control	100–150	99.99–99.999	100–150	1–5
Emerging	Remote monitoring and maintenance	20–50	99.99–99.999	500–1000	1–2
	Remote operation (teleoperations)	2–10	99.999–99.99999	1–5	100–200
	Mobile workforce	5–10	99.999–99.9999	50–100	10–50
	Augmented reality	10	99.99–99.999	10–20	500–1000

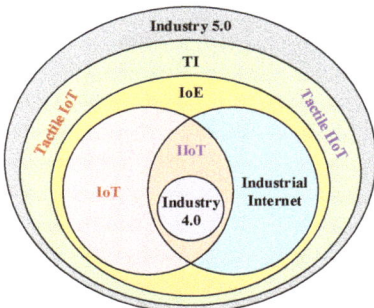

Figure 2. An overview of the relation between IoT, IIoT, tactical IoT, tactile IIoT, Industry 4.0, and Industry 5.0.

One of the effective ways to investigate the tactile IIoT application requirements, performance, and testing the new solutions to ensure QoS and QoE, is to set up a virtual testbed similar to the real network. The testbed must allow us to utilize and maintain hardware and software virtually on a standard computer without purchasing them. In the literature, several recent articles have proposed testbeds to overcome the above-mentioned challenges. The work in [13] proposed a haptic system testbed to characterize and validate E2E haptic communication of different use cases of TI. The authors introduce a framework comprised of multiple sub-blocks that can be re-configured based on the nature of use cases, with a focus on minimizing cost and evaluation time. It also provides an option to integrate the testbed with the simulation platform through a connector interface to perform testing. Commonly, it is intended to offer an extensive range of haptic hardware, including sensors, actuators, and tactile interface boards. A testbed for tactile and kinesthetic data coding was proposed in [14] aligned with IEEE P1918.1 TI standard working group to improve and standardize haptic codec. The proposed haptic coding testbed is considered as a reference testbed with the aim to develop optimal data compression schemes to exchange tactile and kinesthetic information and enable human-in-the-loop TI services. The authors also provide some reference tactile data traces, software, and hardware to evaluate newly developed kinesthetic and tactile codecs.

In [15], a framework for tactile cyber physical systems was proposed, which is specifically for physical remote environments and based on network simulator NS3. It provides an interface for robotic experiments, along with haptic communication modules. However, the authors ignored the extensibility of the proposed testbed for other haptic-driven applications. Similarly, the authors in [16,17] proposed a generic testbed framework for different TI use cases. A data-driven experiment setup was proposed in [16] to provide a common playground for testing haptic applications. The proposed haptic communication testbed at the Otto-von-Guericke University of Magdeburg (OVGU-HC) focused on providing experiment testbed for long-distance haptic-enabled teleoperation systems, in addition to small scale wireless haptic-driven applications. The OVGU-HC presents experiment automation and data collection utilizing experiment description language (DES-Cript). The proposed OVGU-HC did not work standalone, is a part of the MIoT-Lab, and is just used to gather hepatic experiment information. Moreover, it utilized domain-specific language DES-Cript [18], and did not provide an open-source facility to the research and development community.

The study in [17] presents a two-level classification of the TI applications based on controlled environment and master-slave integrations to develop a generic testbed, with a focus to ensure compatibility for all these classified applications, which is named as TI- eXtensible Testbed (XT). To demonstrate the potential of the TIXT, they discuss H2M haptic communication in the virtual and physical environment. However, they ignored the explanation on how to characterize the network impairments (delay, jitter, and packet losses) and investigate the performance of the haptic-driven IIoT application. Therefore, there is a strong need for a testbed that offers flexibility, scalability, open-source availability, tailored to examine network impairments, communication flow, and extensible for TI IIoT use cases. In this regard, we proposed a virtual testbed called IoTactileSim to investigate tactile IIoT services from QoS and QoE perspectives. The IoTactileSim employs Software Define Network (SDN) and edge computing at the core network to tactile industrial application. The following section presents the main contribution of the proposed IoTactileSim testbed.

1.1. Research Contributions

The primary contributions of this work are summarized below as:

- We presented the details of TI in the context of various industrial environments and discussed some emerging applications of the tactile IIoT.
- A hybrid virtual testbed, IoTactileSim, is proposed by combining a network emulator and an industrial robotic simulator to simulate tactile IIoT applications and investigate their performance.
- We designed the IoTactileSim by adopting a hierarchical approach, where the network is divided into two parts; a core and an edge layer. The core layer consists of SDN routers to perform intelligent routing, while the edge layer performs as an intelligent support engine for tactile IIoT services.
- The proposed IoTactileSim identifies the challenges imposed by the tactile IIoT and their strict QoS/QoE requirements. Moreover, it focuses on investigating the communication network parameters (latency, reliability) and other configurations corresponding to the identified requirements.
- We conduct two different experiments in the tactile IIoT environment to evaluate the performance and present the potential of the proposed IoTactileSim testbed.

1.2. Paper Organization

As illustrated in Figure 3, the rest of this paper is organized as follows. Section 2 discusses the proposed IoTactileSim structure and setup, along with the topological view.

Section 3 presents the scenarios and case studies to demonstrate the potential of the proposed testbed. Finally, Section 3 concludes the paper and offers some future avenues.

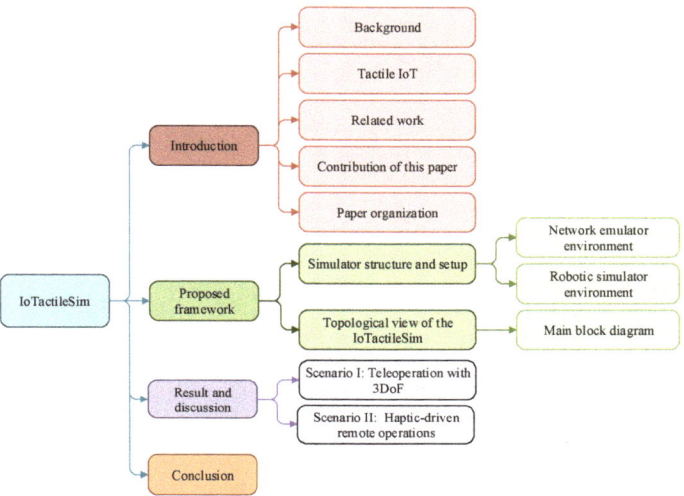

Figure 3. Diagrammatic view of the structure of the paper.

2. Proposed Framework

In this section, we describe the proposed IoTactileSim testbed to support a broad range of tactile IIoT services. At first, we will present the network emulator to mimic the real-world communication network, followed by a detailed discussion on the industrial robotic simulator. Finally, we present the topologic view of the proposed simulator, along with the basic parametric settings.

2.1. Simulator Structure and Setup

The structure of the proposed testbed IoTactileSim as depicted in Figure 4, following the IEEE P1918.1 TI standard architecture. In general, the TI use cases are comprised of three key domains: master domain, network domain, and slave or controlled domain [5]. The master domain consists of operators (human or control algorithms) that exploit haptic devices. The slave domain deals with the slave robots or teleoperators that the master side operator directly controls via control signals. The network domain connects the master and slave sides to enable bi-directional communication. To control the slave side teleoperator, the master side sends the control signal, and in return receives the feedback information including haptic and audio-visual signals. The master and slave domain creates a global control loop over communication network infrastructure. To maintain stability for the tactile IIoT services and provide a real-time haptic sensation to the users, this global control loop demands a haptic packet sampling rate of ≥ 1 kHz, a packet loss rate between 10^{-3}–10^{-5}, and latency ranging from 1–10 ms. The proposed IoTactileSim helps the users to investigate these strict requirements and evaluate their newly developed strategies for emerging industrial applications.

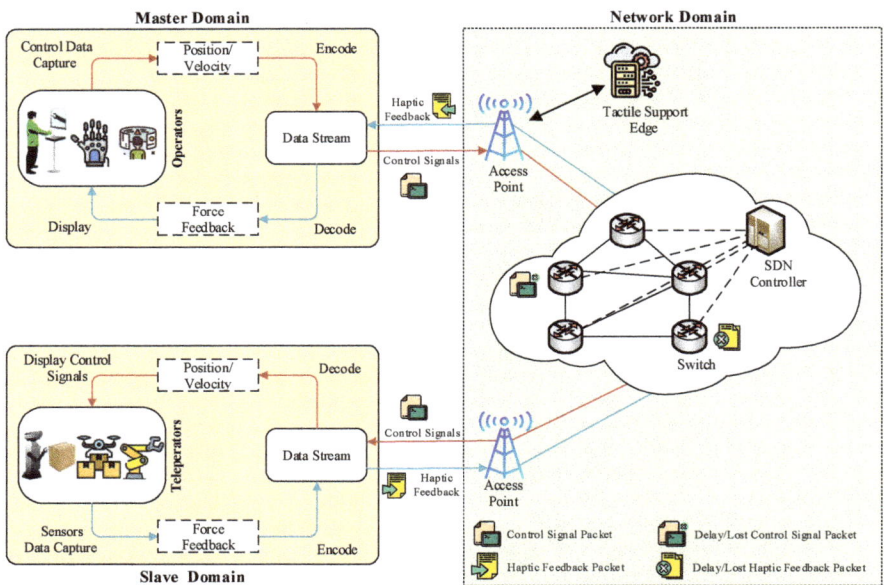

Figure 4. Indepth overview of the proposed virtual testbed IoTactileSim.

2.1.1. Network Emulator Environment

As can be seen in Figure 4, the control signal is captured through the controller on the master side and forwarded to the network domain in a specifically encoded format. SDN, network function virtualization (NFV), and mobile edge computing (MEC) are employed with 5G technology in the core network of the proposed testbed to overcome the 1 ms latency challenge and to provide support for the next-generation industrial applications [19]. The network domain receives these packets and forwards them to the slave domain to perform the required task. The feedback in form of the haptic data is sent from the slave teleoperator to the master side human operator. The proposed IoTactileSim utilizes the Mininet emulator for the network design, resembling a real-world network operations and hardware in a virtual environment [20]. Mininet employs process-level virtualization to develop a virtual communication network with virtual hosts and connects them via virtual Ethernet pairs.

The proposed IoTactileSim enables the evaluation of large custom topologies with actual application traffic traces by deploying them into the physical network. It also enables the utilization of emerging technologies such as SDN, NFV, and MEC. In the SDN framework, control planes are separated from the forward plane in the network. The emulator is written in python language and freely available at the Mininet official website (http://mininet.org/, accessed on 10 November 2021). An overview of the basic architecture of the Mininet with open-source virtual switches Open vSwitch (OVS) and SDN standard protocol OpenFlow is depicted in Figure 5a. Mininet by itself is a network emulator that allows users to mimic real network topologies. It also enables users to build such network topologies in SDN architecture. This is what Mininet is capable of in a nutshell. It does not provide any support for integrating tactile Input/Output (I/O) modules or any other modules for that matter. It only provides us with a virtual environment where all network nodes are present (virtually) on a single physical device. The contribution of the IoTactileSim over Mininet is defined as:

- IoTactileSim allows users to integrate several tactile I/O modules with the Mininet environment.

- IoTactileSim enables users to implement each network module on a separate physical device.
- IoTactileSim also has an embedded tactile support engine which is not present in Mininet. This support engine can be modified by the user based on their use cases.

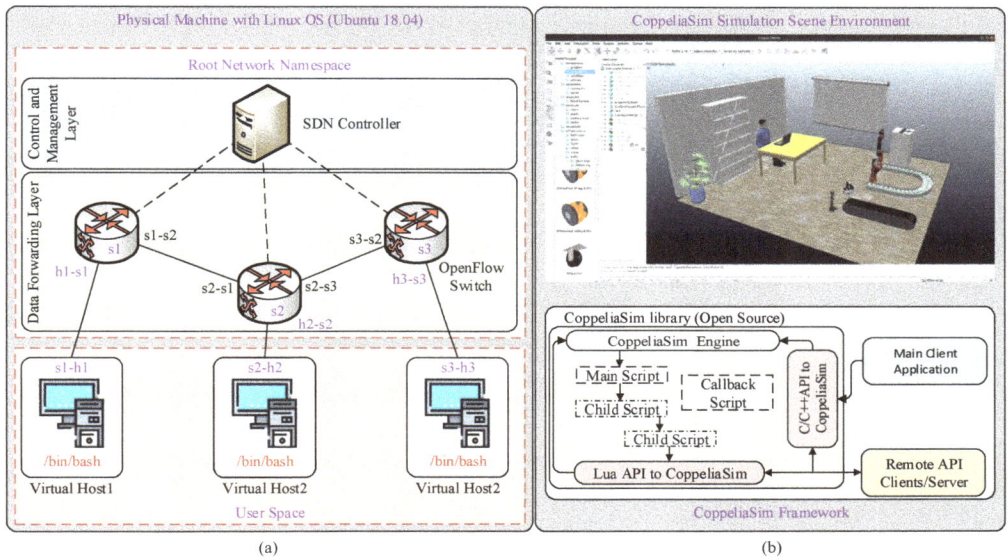

Figure 5. Overview of the basic framework for Mininet and CoppeliaSim. (**a**) represents the Mininet network emulator with Open vSwitch and SDN based OpenFlow protocol, (**b**) illustrates the CoppeliaSim simulator scene environment along with control architecture.

2.1.2. Robotic Simulator Environment

To develop the smart industry with human-in-the-loop and human-robot interaction haptic-driven teleoperation use cases like remote maintenance, inspection, industrial management, we utilized one of the famous industrial robotic simulators CoppeliaSim. This industrial robotic simulators CoppeliaSim is formerly known as (V-REP: Virtual Robot Experimentation Platform) [21]. The reason to use the CoppeliaSim is that it provides a range of emerging industrial applications, including factory automation, remote monitoring, safety monitoring, telerobotic operations, etc. Figure 5b illustrates the basic control architecture and scene environment of the CoppeliaSim simulator. As can be seen from Figure 5b, the simulation loop consists of the main and child script. The main script controls all child scripts attached to the specific object in the simulation environment. The remote Application Programming Interface (API) allows the user to interact with the simulator from outside the system through socket communication. The remote API client and server are responsible for providing these services through different programming languages like C/C++, Python, Matlab, Java, etc.

In the proposed IoTactileSim, we utilized the python remote API client to interact with the smart industrial application in the CoppeliaSim environment through socket communication. The control code of the developed IIoT applications is executed in the same computing machine where the network emulator was employed. The CoppeliaSim simulator is connected with a network to represent network-slave interaction that was designed utilizing Mininet. Additionally, the network is linked with the master domain and makes the master-network relationship. The overview of the connection between master, network, and slave domain using Mininet and CoppeliaSim in a single computing machine (personal computer) is illustrated in Figure 6, and an in-depth discussion on each module is presented in the next section.

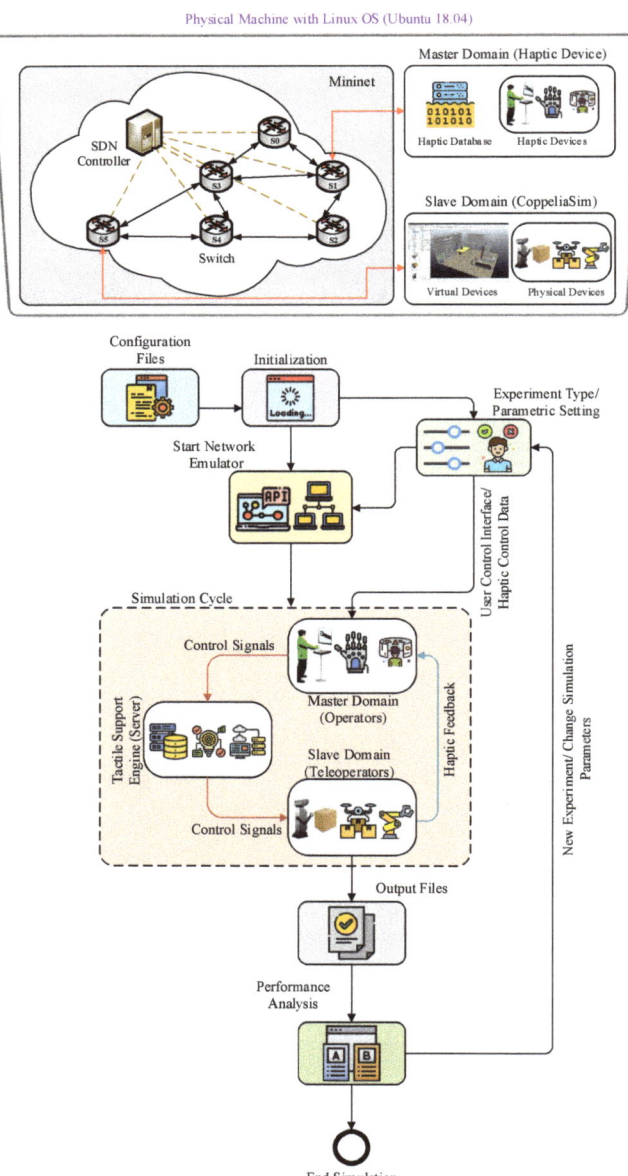

Figure 6. The flowchart illustrating the overall flow of the IoTactileSim testbed dynamics and relationship between different parts.

2.2. Topological View of the IoTactileSim

This section focuses on the architecture of the proposed IoTactileSim testbed as depicted in Figure 6. At first, we discuss the parameter initialization, settings, and user interface to interact with the proposed testbed. Second, the topological view of the core network architecture is presented. Third, an in-depth discussion on application-agnostic design with the application and network connectivity is reported.

Initialization: In the initialization module, the simulator reads the parametric configuration files. It sets packet size, packet rate, Internet Protocol (IP) suite, IP address, link bandwidth, and link latencies. This module facilitates the users to provide the parameter settings as per their experiment need. If the user does not provide the parametric settings, then it automatically uses the default values of the parameters, as defined in Table 2.

Table 2. Summary of parameters and settings used for Simulation.

Parameters	Settings Used
Simulation environment	
Operation system	Linux (Ubuntu 18.04)
Programming language	Python 3.8
Network emulator	Mininet 3.6.9
Industrial robotic simulator	CoppeliaSim 4.2
Network emulator	
Network topology	Mesh network of switches
IP suite	User datagram protocol
Software switch type	Open vSwitch 2.9.8
SDN controller	OVS-controller
Interface protocol for controller	OpenFlow
Link latency	Shortest route 1.2 ms Longest route 1.8 ms
Link bandwidth	100 Mbps
No. of packets	10, 100, 1000, 10,000
Packet sampling Rate	1 kHz
Industrial robotic simulator	
Remote API	Python legacy remote API client
Simulation mode	Real-time simulation
Execution techniques	Same machine with the same thread
Interaction network	Socket communication
Simulation scene model	Custom design (Teleoperation)

Start Network Emulator: In order to design a real-world network, a Mininet emulator creates a custom topology with five OVSs and three hosts. The hosts act as a standard computing machine and are responsible for the master domain, slave domain, and tactile support engine. The OVSs are connected with a single SDN controller. The SDN controller decides to handle the data plane and allows the network operator to control and manage the whole network via API provided by the Mininet. Therefore, we utilized the Mininet Python API in IoTactileSim, so that the users can change the network settings as per their experiment demands to evaluate their newly developed approaches. The users can change the parametric values from the configuration files as discussed earlier in the initialization step.

Simulation Cycle: After creating the network topology with three hosts that work as a master, slave, and tactile support engine, the simulator enters into the simulation cycle. In the simulation cycle, the actual experiments are performed as per defined conditions by the users through parametric settings or default values. The network host that acts as the master side, connects to the haptic device that the human operator uses to send the control signals to the salve side manipulator. The other host acting as a slave, connects to the CoppeliaSim simulator for executing the control commands in the designed virtual/physical tactile IIoT applications. After conducting a desired control experiment, the haptic feedback is sent to the human teleoperator at the master domain. This loop runs until the simulator reaches the defined threshold values like 1000 packets, etc. Finally, at the end of the simulation cycle, the simulator stores the experimental results into a file to investigate various network impairments.

Performance Analysis: This module receives the stored experimental data file and compiles result graphs to understand the strengths and weaknesses of the conducted experiments from the QoE and QoS perspectives. After representing the experimental result in the visual form, it stores these results, and enables the users to perform new experiments or to exit the simulator. The list of the default parameter values and settings used in the proposed IoTactileSim are summarized in Table 2. The proposed IoTactileSim is publicly available at Github (https://github.com/zubair1811/IoTactileSimV1.git, accessed on 30 November 2021) to the interested researchers to conduct extensive experiments to evaluate their suggested approaches.

3. Result and Discussion

In this section, we demonstrate the effectiveness of the proposed IoTactileSim with two different tactile industrial scenarios using the simulation environment and parameters setting defined in Table 2. These two realistic applications belonging to tactile industrial use cases define as follows

- Scenario I: Teleoperation with 3 Degree-of-Freedom (3DoF)
- Scenario II: Haptic-driven remote operations

Moreover, these scenarios can be classified into two categories, like offline and real-time applications. On the one hand, offline (teleoperation with 3DoF) experimentation means that we already have a static dataset of some real-world teleoperation applications and utilize previously collected data for analytical analysis. On the other hand, real-time or online (haptic-driven remote operations) experiments indicate that interaction data between operator and teleoperator are collected in real-time to make a suitable decision to ensure stability and transparency. The real-time scenario is complex compared to offline because it deals with more data under time constraints. Most of the existing studies on industrial testbeds just utilized the offline methodology, while the proposed IoTactileSim considered both scenarios. The discussion considering offline and real-time scenarios is presented in detail in the following subsections.

3.1. Scenario I: Teleoperation with 3DoF

Scenario I considers the offline experimentation, where publicly available 3DoF haptic traces in [22] on teleoperations were utilized. To record the haptic traces, a human operator employs a haptic device (Phantom Omni) at the master to interact with the virtual environment, which acts as a slave domain. The virtual environment is comprised of a rigid movable cube lying on a wooden, smooth surface. The human operator makes an interaction (static and dynamic) with the rigid cube via the haptic device and receives force feedback. Figure 7 illustrates the 3DoF position and velocity control by the human operator via the haptic device and the received force feedback of the used haptic dataset [22] for experimentation. In the proposed IoTactileSim testbed, we transmit the control signals (position/velocity) from the master side to the slave side and get the haptic feedback (force). To minimize the E2E communication delay, a common practice is to transmit the haptic traffic packets instantly after receiving sensors' data, resembling a real-time tactile industrial IoT application.

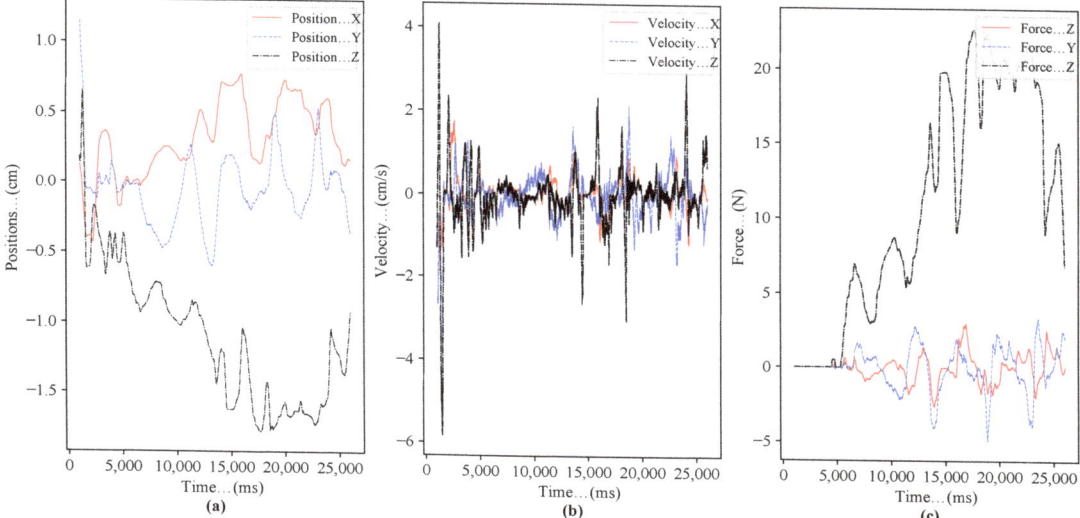

Figure 7. Dynamic interaction of the human operator with virtual application via haptic device. (**a**) positions of the human operator's hand at master side device, (**b**) velocity traces of the operator (**c**) force data traces of the teleopertor x, y and z-axis.

We adopt the same method in our experiments; after reading the sensors data, the system makes packets of the haptic traces as following:

$$PacketSize = Ethernet/UDP/IP/8 \times N_{DoF} \qquad (1)$$

where $Ethernet/UDP/IP$ indicates the header of the ethernet, user datagram protocol, and internet protocol layer, respectively. N_{DoF} is the number of DoF in the experimental data. We are using 3DoF, so the formulation can be evaluated as:

$$Packet\ Size = 14/8/20/8 \times 3$$
$$Packet\ Size = 14 + 8 + 20 + 24 = 78$$

The interface of the IoTactileSim during communication between master and slave domain is depicted in Figure 8. At the master domain, control signals from utilized 3DoF haptic traces are selected and transmitted to the slave domain through the network domain. Similarly, after receiving specific control signals, the slave domain returns the corresponding force feedback to the master domain. The overview of the data flow interfaces between master and slave for the scenario I is depicted in Figure 8.

The performance analysis for a scenario I in terms of round trip delay is presented in Figure 9. To investigate the effect of the number of haptic data packets on round trip delay for IoT applications, the scenario I was simulated for the number of haptic data traces = 10 to 10,000. The latency investigation using IoTactileSim for a scenario I with the number of packets = 10, 100, 1000, 10,000 is depicted in Figure 9a–d, respectively. It can be seen clearly from the results, the tendency with an increase in the number of haptic data packets the round trip delay decreases from 5 to 2 ms. In Figure 9a, with the number of packet = 1, the packet latency approaches 6 ms as compared to Figure 9b–d, where packet latency is below 5 ms. To elaborate this latency decrement in detail, Figure 10 illustrates the packet delay histogram for a scenario I.

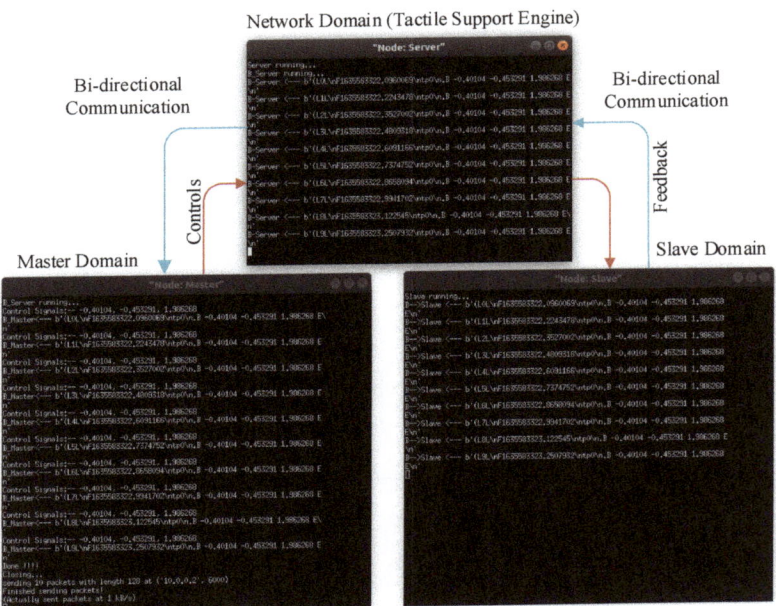

Figure 8. IoTactileSim interface for scenario I experiment.

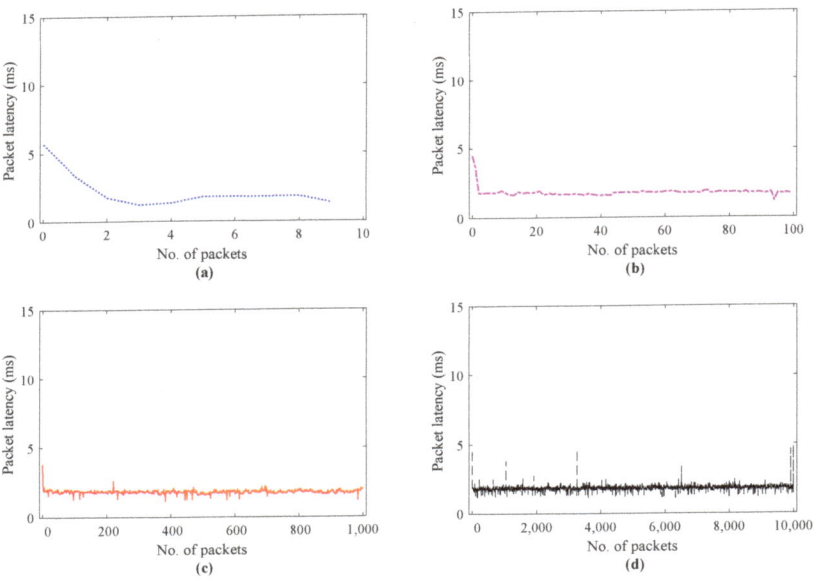

Figure 9. Packet latency investigation for scenario I haptic data transmission; (**a**) data packets = 10, (**b**) data packets = 100, (**c**) data packets = 1000, (**d**) data packets = 10,000.

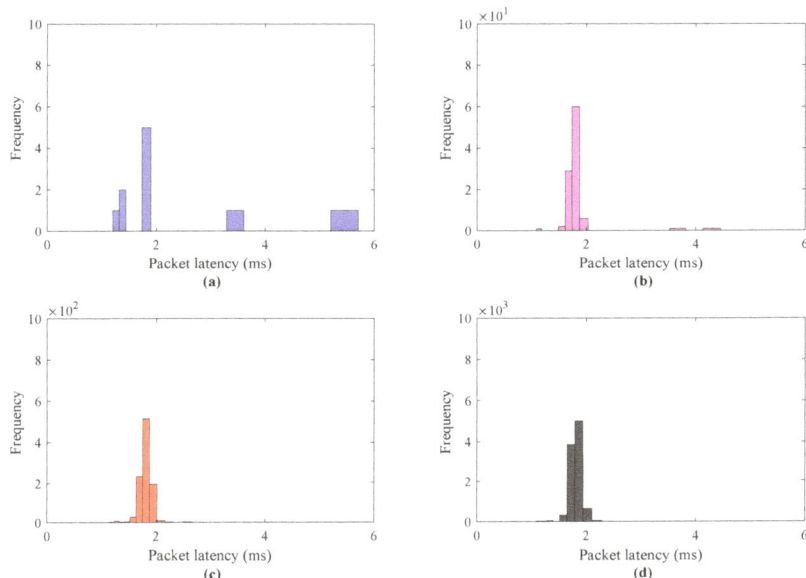

Figure 10. Packet latency histogram for scenario I haptic data transmission; (**a**) data packets = 10, (**b**) data packets = 100, (**c**) data packets = 1000, (**d**) data packets = 10,000.

Contrary to Figure 9, the results in Figure 10 reveal the latencies of the most frequent haptic data packets. Similar to the results presented in Figure 10a–d, the simulation results in Figure 10a–d also indicate the decrease in packet latencies from 5.8 to 2.1 ms. From Figure 10a–d, it can be seen clearly that most of the haptic traces latencies centered between 1 to 2 ms, which is one of the stringent requirements for the tactile IoT services. Figure 10b–d, depicts the improvement in the communications network impairments (delay, jitter) with the number of packets = 100, 1000, 10,000. The reason behind this is that at the beginning, the proposed testbed IoTactileSim understands and fine-tunes the simulation parameters to support delay-sensitive and loss-intolerant applications. The efficacy of the scenario I regarding reliability characterization is summarized in Table 3. The reliability of the transmitted haptic data packets is evaluated in terms of delayed/lost and out-of-ordered packets.

Table 3. Summary of the reliability characterization for haptic datset and real-time haptic drive teleoperation experiment.

Experiments		Packet Statistics	
		Dropped/Delayed (%)	Out-of-Order (%)
		Haptic data transmission	
Data Packets	10	20.0 (2 Packets)	11.1
	100	2.00 (2 Packets)	1.00
	1000	0.10 (1 Packets)	0.00
	10,000	0.10 (7 Packets)	0.00

Table 3. Cont.

Experiments		Packet Statistics	
		Dropped/Delayed (%)	Out-of-Order (%)
		Haptic-driven remote operations	
Data Packets	10	100 (10 Packets)	100
	100	44.0 (44 Packets)	44.4
	1000	3.10 (31 Packets)	3.00
	10,000	0.40 (39 Packets)	0.30

3.2. Scenario II: Haptic-Driven Remote Operations

In this section, we will present the real-time control of the teleoperator in the virtual environment to mimic the real-world tactile industrial remote operations. Similar to the scenario I, II also consists of the master, slave, and network domain where virtual teleoperator developed in CoppeliaSim acts as salve domain. In the master domain, physical haptic devices (haptic computer mouse, glove, and hapkit) are used to interact with the virtual environment, as illustrated in Figure 11. These haptic devices are easy to develop because their supplementary material is available publicly for the research community. The tactile computer mouse was presented in [14], while the study in [23,24] provide the design and development detail on a haptic glove and Hapkit, respectively. However, in this experiment, we only employed the (computer mouse and glove) to interact with the teleoperator as slave side. We mapped the physical computer mouse and glove X and Y direction to the XY coordinates of the developed virtual teleoperator in CoppeliaSim. The key focus of this experiment is to investigate the communication network parameters (latency, reliability) that affect the TI services. Additionally, it also demonstrates the potential of the proposed IoTactileSim to provide TI services under TI QoS/QoE requirements (1–5 ms). Human operators in the master domain use the haptic device to interact with the teleoperator at the slave side and receive haptic feedback. The interface of the IoTactileSim during direct controlling of the teleoperator in the virtual environment is depicted in Figure 11.

To observe the effect of the number of data packets on latency for scenario II, the simulation results are summarized in Figure 12. These results also demonstrate that with the increase in the number of packets from 10 to 10,000, the network communication latency tends to decrease. In this experiment, we directly control the teleoperator in the virtual environment via a computer mouse in real-time. The control signals from the computer mouse are packetized as defined in (1), sampled as per haptic system requirement, and transmitted to the teleoperator using default parameters values as listed in Table 2. The teleoperator at slave side receives the control commands to perform the required task and backward the force feedback to the human operator. Figure 12a–d, indicates the results of the packet latency for the number of data packets 10, 100, 1000, 10,000, respectively. In Figure 12a with the number of packets = 10, the value of latency lies between 9–7 ms.

In Figure 12b, up to 40 packets, latency value remains higher than 5 ms, and after that, the system gets convergence round trip latency around 2.5 ms. Similarly, the results in Figure 12c, gain a minimum latency value of 2.5 ms from the 20th data packet to the 1000th packet. To continue on a similar line as mentioned above, Figure 12d, exhibits a quick decreasing trend in round trip latency from 6 to 2.2∼2.0 ms , as the number of data packets increases from 10, to 10,000.

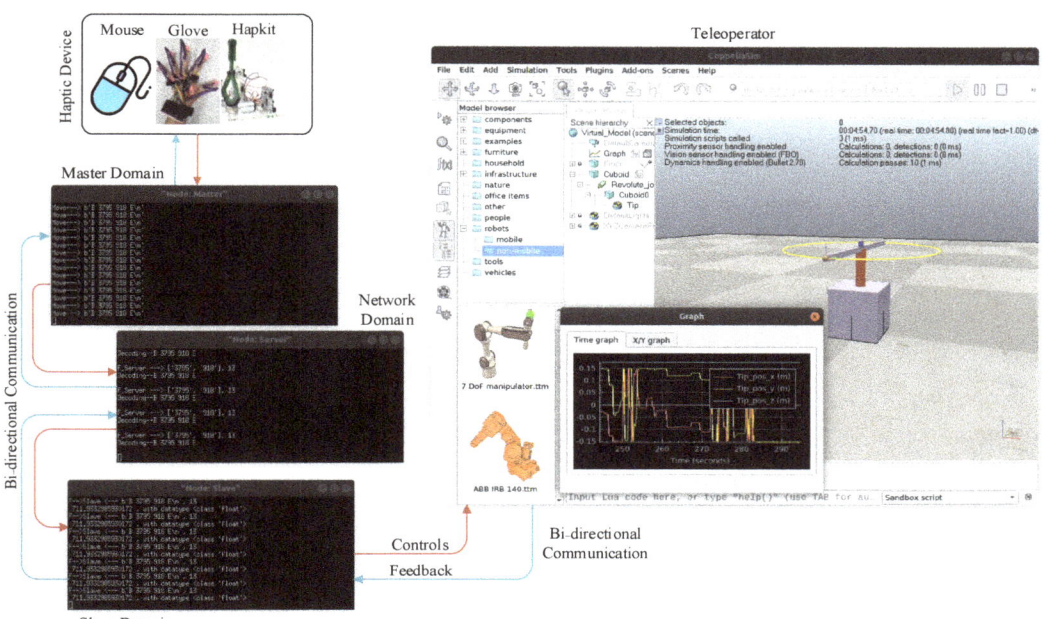

Figure 11. IoTactileSim interface for scenario II experiment.

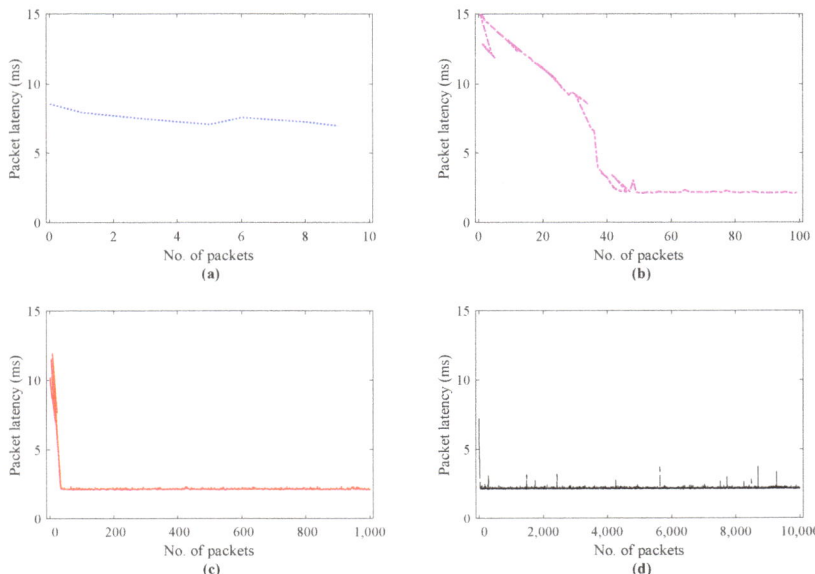

Figure 12. Packet latency investigation for scenario II real time haptic-driven remote operation; (**a**) data packets = 10, (**b**) data packets = 100, (**c**) data packets = 1000, (**d**) data packets = 10,000.

To elaborate this packet latency convergence in a better way, Figure 13 illustrates the histogram of the frequent data packets regarding packet latencies. The results in Figure 13c,d indicate that the packet latency reduces for an increase in the number of packets compared to results in Figure 10c,d. In addition, these results depict that the latency is more concentrated between 2 to 2.3 ms. It is also interesting to observe that, for the higher number of the data packet with a higher sampling rate, the proposed IoTactileSim is capable of reducing congestion and maintaining the application latency requirement. In addition Table 3 presents the in-depth reliability analysis for the scenario II experiment. As it can be seen clearly from Table 3 for scenario II the percentage of delayed or dropped packets decrease from 100% to 0.40% (10 to 39 data packets) with 100% to 0.30% out-of-order sending packets from 10 to 10,000 number of data packets.

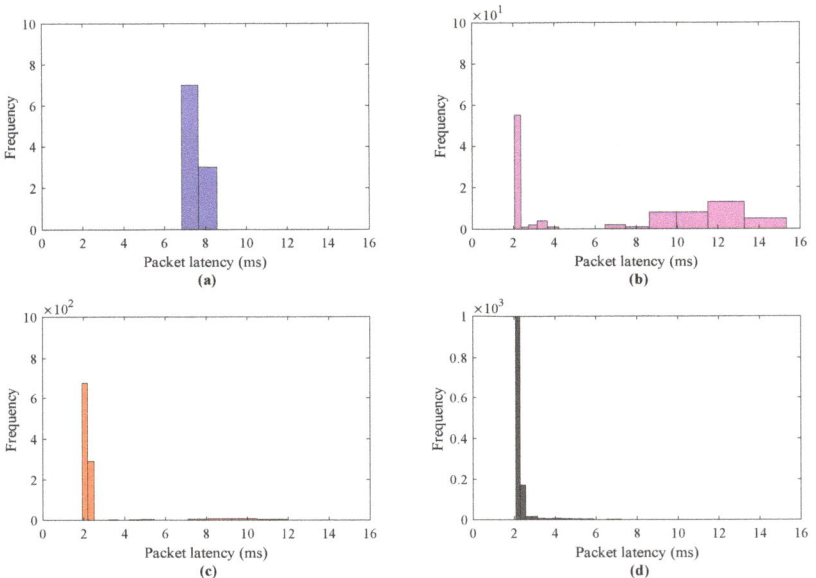

Figure 13. Packet latency histogram for scenario II real time haptic-driven remote operation; (**a**) data packets = 10, (**b**) data packets = 100, (**c**) data packets = 1000, (**d**) data packets = 10,000.

In summary, the packet latency convergence analysis in Figures 9 and 12 and periodic packet variation analysis Figures 10 and 13 for use case scenarios I and II indicates that the proposed virtual testbed IoTactileSim provides the facility to the users to implement complex tactile industrial use cases, evaluate their proposed strategy, and investigate the QoS and QoE requirement of the implemented tactile IoT services. Some of the complex tactile IIoT use cases are illustrated in Table 4. The proposed IoTactileSim concentrate on providing QoS and QoE provisioning by taking different network parameters into account. Based on the mentioned complex tactile IIoT use cases along with requirement specification (delay, packet size, packet rate, packet loss rate, etc.) in Table 4 it indicates that IoTactileSim can ensure strict QoS-based traffic. The main objective of this paper is to provide a tool to minimize the network development cost while realizing the stringent QoS/QoE requirements for tactile IIoT applications. Moreover, it also offers to implement edge intelligence to the designed tactile support engine, which can be leveraged to improve QoS and QoE provisioning in highly dynamic network environments. The users can deploy machine learning, specifically reinforcement learning models, to track the frequently dynamic network environment states and make online decisions to improve network conditions and support time-varying user demands.

Table 4. Tactile IIoT use cases requirements specifications and characteristics supporting by proposed IoTactileSim.

Applications	IIoT Use Cases and Requirements				
	Cycle Time	Message Size	Data Rate	Latency	Packet Loss Rate
	Control loop motion control				
Machine tools Packaging machines Printing machines	0.5~2 ms	20~50 Bytes	1~10 Mbps	0.25~1 ms	10^{-9}~10^{-8}
	Remote control				
Process automation Process monitoring Process maintenance Fault reporting	≤50 ms	≥10 Mbps	1~100 Mbps	≤50 ms	≤10^{-7}

4. Discussion and Future Work Directions

In our previous work [23], we analyzed the different haptic gloves and investigated how data processing increased the latency in the haptic communication loop and proposed a low-latency haptic open glove (LLHOG). Contrary to previous work, the focus of this paper is to provide network infrastructure to transmit haptic traffic between operator and teleoperator and simulate delay-sensitive and loss-intolerant tactile IIoT applications. However, there are various industry 4.0 applications under use cases class C, such as fleet management, tactile-driven logistics, cooperative robotics, and motion control, which demand higher QoS and QoE. To allow these real-time applications, the utilization of edge computing is required. Therefore, there is a need for edge-based network systems with native machine learning parts to provide the QoS and QoE requirement provisioning for these applications. In this regard, as a future, an edge-based ITE is developed as a tactile support engine to enable the ability for the user to train and deploy machine learning models at the edge to ensure QoS and QoE. The conceptual diagram to design and deploy the trained model on ITE is illustrated in Figure 14.

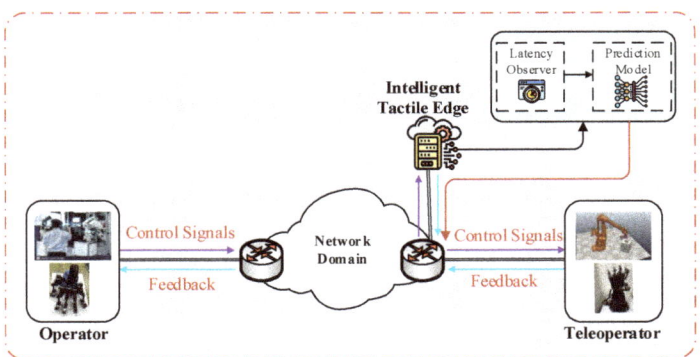

Figure 14. Conceptual architecture of ITE in IoTactileSim.

In future work, more practical challenges regarding tactile IIoT in the real-world scenario need to be considered. As discussed above, providing required QoS and QoE in the real-time complex industrial application is more challenging than simulation analysis. Therefore, we indented to test the proposed IoTactileSim in real-time physical IIoT scenarios and demonstrate the real-world experiment design overview in Figure 15. On the master side, we utilized the LLHOG, which consists of the rotary position sensors with a min-max scaling filter to send haptic data. The bionic robot hand, which consists of Arduino and servo controllers, is used at the slave side. The specification, sample code, and documentation are available at the official website (https://wiki.dfrobot.com/, accessed on 30 November 2021). The proposed IoTactileSim connects the LLHOG and bionic robot hand to develop a closed control loop. The ITE is also integrated with the proposed IoTactileSim to monitor the network dynamics and guarantee the QoS and QoE requirements for tactile IIoT applications.

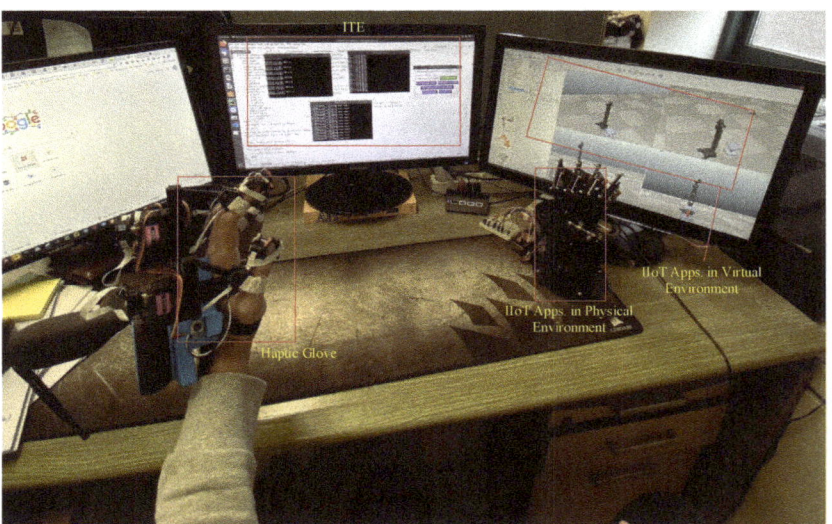

Figure 15. Modular representation of ITE in real-world scenario.

5. Conclusions

In this paper, we proposed a virtual testbed termed as IoTactileSim to investigate and provision QoS and QoE strict requirements for tactile industrial IoT applications. The proposed IoTactileSim is equipped with a network emulator Mininet and an industrial simulator CoppeliaSim to mimic the real-world communication network and industrial IoT environment. It provides the users to evaluate the efficacy of their designed strategies under possible settings, including advanced core network technologies (SND, NVF), edge intelligence, and application-agnostic parameters (packet size, sampling rate, etc.) for improving QoS and QoE. The proposed IoTactileSim is investigated for two different industrial use case scenarios with haptic data traces and real-time remote interaction. The simulation results indicate that the IoTactileSim is able to handle real-time data traffic then offline scenario by providing communication latency ranges from 6 to 2.2∼2.0 ms, and from 5.8 to 2.1 ms for 10 to 10,000 data packets, respectively. Moreover, the experimentation analysis indicates that the IoTactileSim allows the user to investigate network impairments (latency, jitter, reliability) and can support complex tactile industrial environments with a higher number of data packets. In the early future, we plan to extend the IoTactileSim with network coding and machine learning approaches like federated reinforcement learning at the tactile support engine to integrate it with the 6G network infrastructure.

Author Contributions: Methodology, R.A., A.H. and M.Z.I.; Software and coding, M.Z.I. and S.; Experimentation and formal analysis, M.Z.I., R.A. and A.H.; Writing—original draft preparation, M.Z.I. and A.H.; Writing—review and editing, M.Z.I., S., A.H. and R.A.; Supervision, H.K.; Funding acquisition, M.Z.I and H.K. All authors have read and agreed to the published version of the manuscript.

Funding: This work was supported by the National Research Foundation of Korea (NRF) grant funded by the Korean government (MSIT) (Nos. 2019R1A4A1023746, 2019R1F1A1060799) and the Strengthening R&D Capability Program of Sejong University.

Institutional Review Board Statement: Not applicable.

Informed Consent Statement: Not applicable.

Data Availability Statement: The 3DoF haptic datasets used in this work are available online with open access for academic research use. The static and dynamic interaction haptic dataset is available online at https://cloud.lmt.ei.tum.de/s/4FmHUCsoUvwRle3 (accessed on 1 November 2021), and the simulation experiment results files are availabe at https://github.com/zubair1811/IoTactileSimV1.git, (accessed on 30 November 2021).

Conflicts of Interest: The authors declare no conflict of interest.

Abbreviations

Abbreviations
TI	Tactile Internet
H2H	Human-to-Human
M2M	Machine-to-Machine
M2H	Machine-to-Human
IoT	Internet of Things
IIoT	Industrial Internet of Things
URLLC	Ultra Reliable and Low Latency Communication
QoS	Quality of Service
QoE	Quality of Experience
E2E	End-to-End
SDN	Software Define Network
NFV	Network Function Virtualization
MEC	Mobile Edge Computing
OVS	Open vSwitch

References

1. Atzori, L.; Iera, A.; Morabito, G. The internet of things: A survey. *Comput. Netw.* **2010**, *54*, 2787–2805. [CrossRef]
2. Maier, M.; Ebrahimzadeh, A. Towards immersive tactile Internet experiences: Low-latency FiWi enhanced mobile networks with edge intelligence. *J. Opt. Commun. Netw.* **2019**, *11*, B10–B25. [CrossRef]
3. Li, C.; Li, C.P.; Hosseini, K.; Lee, S.B.; Jiang, J.; Chen, W.; Horn, G.; Ji, T.; Smee, J.E.; Li, J. 5G-based systems design for tactile Internet. *Proc. IEEE* **2018**, *107*, 307–324. [CrossRef]
4. Fettweis, G.P. The tactile internet: Applications and challenges. *IEEE Veh. Technol. Mag.* **2014**, *9*, 64–70. [CrossRef]
5. Holland, O.; Steinbach, E.; Prasad, R.V.; Liu, Q.; Dawy, Z.; Aijaz, A.; Pappas, N.; Chandra, K.; Rao, V.S.; Oteafy, S.; et al. The IEEE 1918.1 "tactile internet" standards working group and its standards. *Proc. IEEE* **2019**, *107*, 256–279. [CrossRef]
6. Le, T.K.; Salim, U.; Kaltenberger, F. An overview of physical layer design for Ultra-Reliable Low-Latency Communications in 3GPP Releases 15, 16, and 17. *IEEE Access* **2020**, *9*, 433–444. [CrossRef]
7. Aijaz, A.; Sooriyabandara, M. The tactile internet for industries: A review. *Proc. IEEE* **2018**, *107*, 414–435. [CrossRef]
8. Samdanis, K.; Taleb, T. The road beyond 5G: A vision and insight of the key technologies. *IEEE Netw.* **2020**, *34*, 135–141. [CrossRef]
9. Zhang, C.; Ueng, Y.L.; Studer, C.; Burg, A. Artificial intelligence for 5G and beyond 5G: Implementations, algorithms, and optimizations. *IEEE J. Emerg. Sel. Top. Circuits Syst.* **2020**, *10*, 149–163. [CrossRef]
10. Ali, R.; Zikria, Y.B.; Bashir, A.K.; Garg, S.; Kim, H.S. URLLC for 5G and Beyond: Requirements, Enabling Incumbent Technologies and Network Intelligence. *IEEE Access* **2021**, *9*, 67064–67095. [CrossRef]
11. Saad, W.; Bennis, M.; Chen, M. A vision of 6G wireless systems: Applications, trends, technologies, and open research problems. *IEEE Netw.* **2019**, *34*, 134–142. [CrossRef]
12. Tataria, H.; Shafi, M.; Molisch, A.F.; Dohler, M.; Sjöland, H.; Tufvesson, F. 6G wireless systems: Vision, requirements, challenges, insights, and opportunities. *Proc. IEEE* **2021**, *109*, 1166–199. [CrossRef]
13. Van Den Berg, D.; Glans, R.; De Koning, D.; Kuipers, F.A.; Lugtenburg, J.; Polachan, K.; Venkata, P.T.; Singh, C.; Turkovic, B.; Van Wijk, B. Challenges in haptic communications over the tactile internet. *IEEE Access* **2017**, *5*, 23502–23518. [CrossRef]
14. Steinbach, E.; Strese, M.; Eid, M.; Liu, X.; Bhardwaj, A.; Liu, Q.; Al-Ja'afreh, M.; Mahmoodi, T.; Hassen, R.; El Saddik, A.; et al. Haptic codecs for the tactile internet. *Proc. IEEE* **2018**, *107*, 447–470. [CrossRef]

15. Polachan, K.; Prabhakar, T.; Singh, C.; Kuipers, F.A. Towards an Open Testbed for Tactile Cyber Physical Systems. In Proceedings of the 2019 11th International Conference on Communication Systems & Networks (COMSNETS), Bengaluru, India, 7–11 January 2019; pp. 375–382.
16. Engelhardt, F.; Behrens, J.; Güneş, M. The OVGU Haptic Communication Testbed (OVGU-HC). In Proceedings of the 2020 IEEE 31st Annual International Symposium on Personal, Indoor and Mobile Radio Communications, London, UK, 31 August–3 September 2020; pp. 1–6.
17. Gokhale, V.; Kroep, K.; Rao, V.S.; Verburg, J.; Yechangunja, R. TIXT: An Extensible Testbed for Tactile Internet Communication. *IEEE Internet Things Mag.* **2020**, *3*, 32–37. [CrossRef]
18. Günes, M.; Juraschek, F.; Blywis, B. An experiment description language for wireless network research. *J. Internet Technol.* **2010**, *11*, 465–471.
19. Parvez, I.; Rahmati, A.; Guvenc, I.; Sarwat, A.I.; Dai, H. A survey on low latency towards 5G: RAN, core network and caching solutions. *IEEE Commun. Surv. Tutor.* **2018**, *20*, 3098–3130. [CrossRef]
20. Lantz, B.; Heller, B.; McKeown, N. A network in a laptop: Rapid prototyping for software-defined networks. In Proceedings of the 9th ACM SIGCOMM Workshop on Hot Topics in Networks, Monterey, CA, USA, 20–21 October 2010; pp. 1–6.
21. Rohmer, E.; Singh, S.P.; Freese, M. V-REP: A versatile and scalable robot simulation framework. In Proceedings of the 2013 IEEE/RSJ International Conference on Intelligent Robots and Systems, Tokyo, Japan, 3–7 November 2013; pp. 1321–1326.
22. Bhardwaj, A.; Cizmeci, B.; Steinbach, E.; Liu, Q.; Eid, M.; AraUjo, J.; El Saddik, A.; Kundu, R.; Liu, X.; Holland, O.; et al. A candidate hardware and software reference setup for kinesthetic codec standardization. In Proceedings of the 2017 IEEE International Symposium on Haptic, Audio and Visual Environments and Games (HAVE), Abu Dhabi, United Arab Emirates, 22–23 October 2017; pp. 1–6.
23. Sim, D.; Baek, Y.; Cho, M.; Park, S.; Sagar, A.; Kim, H.S. Low-Latency Haptic Open Glove for Immersive Virtual Reality Interaction. *Sensors* **2021**, *21*, 3682. [CrossRef] [PubMed]
24. Stanford-University. Hapkit. Available online: https://hapkit.stanford.edu/ (accessed on 30 November 2021).

Article

SEMPANet: A Modified Path Aggregation Network with Squeeze-Excitation for Scene Text Detection

Shuangshuang Li and Wenming Cao *

Guangdong Key Laboratory of Intelligent Information Processing and Shenzhen Key Laboratory of Media Security, Shenzhen 518060, China; lishuangshuang2016@email.szu.edu.cn
* Correspondence: wmcao@szu.edu.cn

Abstract: Recently, various object detection frameworks have been applied to text detection tasks and have achieved good performance in the final detection. With the further expansion of text detection application scenarios, the research value of text detection topics has gradually increased. Text detection in natural scenes is more challenging for horizontal text based on a quadrilateral detection box and for curved text of any shape. Most networks have a good effect on the balancing of target samples in text detection, but it is challenging to deal with small targets and solve extremely unbalanced data. We continued to use PSENet to deal with such problems in this work. On the other hand, we studied the problem that most of the existing scene text detection methods use ResNet and FPN as the backbone of feature extraction, and improved the ResNet and FPN network parts of PSENet to make it more conducive to the combination of feature extraction in the early stage. A SEMPANet framework without an anchor and in one stage is proposed to implement a lightweight model, which is embodied in the training time of about 24 h. Finally, we selected the two most representative datasets for oriented text and curved text to conduct experiments. On ICDAR2015, the improved network's latest results further verify its effectiveness; it reached 1.01% in F-measure compared with PSENet-1s. On CTW1500, the improved network performed better than the original network on average.

Keywords: text detection; natural scene; feature fusion

Citation: Li, S.; Cao, W. SEMPANet: A Modified Path Aggregation Network with Squeeze-Excitation for Scene Text Detection. *Sensors* **2021**, *21*, 2657. https://doi.org/10.3390/s21082657

Academic Editors: Raffaele Bruno and Zihuai Lin

Received: 18 February 2021
Accepted: 30 March 2021
Published: 9 April 2021

Publisher's Note: MDPI stays neutral with regard to jurisdictional claims in published maps and institutional affiliations.

Copyright: © 2021 by the authors. Licensee MDPI, Basel, Switzerland. This article is an open access article distributed under the terms and conditions of the Creative Commons Attribution (CC BY) license (https://creativecommons.org/licenses/by/4.0/).

1. Introduction

The rapid development of deep learning has promoted the remarkable success of various visual tasks. Among them, the progress of text detection in natural scenes is increasing. Traditional CNN networks can effectively extract image features and train text classifiers. Other networks are gradually being derived from CNNs, such as segmentation, regression, and end-to-end methods. Deep learning brings algorithms that include more diverse structures, and the results are even more impressive [1,2].

Text detection in natural scenes is based on target detection, but it is different from target detection: it considers the diversity of text direction rotation and size ratio changes; the lighting of the scene, such as the actual streets and shopping mall scenes, (causing the image to be blurred); the inclined shooting angle; and the difficulty caused by the change of text language from horizontal text to curved text. The competition is still fierce. The disadvantage of most network structures is that the simple form cannot satisfy the improvement of the results. Generally speaking, models with high results have significant parameters and large models, while complex systems are time-consuming. Many algorithms are in the research stage, and it is difficult to enter the batch use stage, which still has a large unmet demand. Therefore, this type of application-based algorithm needs to produce state-of-the-art accuracy in theoretical research and consider the request for production in the application scenario and the lightweight model in the portable device.

A series of target detection algorithms [3,4] have been applied in the scene text detection field and promoted the research and development of natural scene text detection recently. The SSD algorithm [5] proposed by Liu et al. uses a pyramid structure and feature maps of different sizes to perform softmax classification and position regression on multiple feature maps simultaneously. The location box of the real target is obtained through classification and bounding box regression. Based on SSD, many researchers improve their methods for the detection of scene text. Shi et al. proposed the SegLink algorithm [6], which is enhanced based on the SSD target detection method. It detects partial fragments at first, and connects all fragments through rules to obtain the final text line, which can better detect text lines of any length. Ren et al. [7] proposed the Faster-RCNN target detection algorithm. Reference [2] proposed a hybrid framework that integrates Persian dependency-based rules and DNN models, including long short-term memory (LSTM) and convolutional neural networks (CNN). Tian et al. proposed the CTPN algorithm [8], which combines CNN and LSTM networks, and adds a two-way LSTM to learn the text-based sequence features via Faster-RCNN; this kind of approach is conducive to the prediction of text boxes. Ma et al. proposed the RRPN algorithm [9] based on Faster-RCNN, a rotation area suggestion network using text inclination angle information, which adjusts the angle information for border regression to fit the text area better.

It is worth noting that many new tasks based on ResNet [10,11] and FPN [12] have appeared and have attracted more attention in recent years. At the same time, ResNet and FPN have many improved methods. SENet [13] adds an SE module to the residual learning unit and integrates a learning mechanism to explicitly model the interdependence between channels so that the network can automatically obtain the importance of each feature channel. This importance enhances the valuable features and suppresses the features that are not useful for the current task. The SE module is also added to some target detection algorithms. Take M2Det [14] as an example: the SFAM structure in this paper uses an SE block to perform an attention operation on the channel to capture useful features better. PANet [15] uses the element addition operation by layer, different levels of information are fused, and a shortcut path is introduced. The bottom-up way is enhanced, making the low-level information more easily spread to the top, and the top-level can also obtain fine-grained local information. Each level is a richer feature map. It can be seen from the above that the latest improved methods also have apparent effects on the improvement of other tasks. Based on the above, this paper introduces a new basic network framework for scene text detection tasks, namely, SEMPANet.

Compared with the previous scene text detection systems, the proposed architecture has two different characteristics:

(1) Compared with the standard ResNet residual structure, the addition of SENet in this paper enables the network to enhance the beneficial feature channel selectively and suppress the useless feature channel by using the global information to realize the feature channel adaptive calibration, reflected in the improvement of the value in the experimental results.

(2) Considering the information flow between the network layers during the training period, the bottom-up path of MPANet is enhanced, making the bottom-up information more easily spread to the top. This paper verifies the influence of PANet on the detection method and modifies the process of PANet to make it more effective. Experimental results show that it can get a more accurate text detection effect than the model with FPN.

The paper is organized as follows:

Section 2 introduces the popular experimental framework in scene text detection in recent years, which describes related work from the following three aspects: whether the detector is based on anchoring, whether it is one stage or two-stage, and whether it is based on RESNET and FPN. Section 3 presents the overall network framework of this paper; the principle of the algorithm is introduced as well, including the SE module and MPANet module. Section 4 includes testing results and their evaluation by the proposed methods. Conclusions are given in Section 5.

2. Related Work

2.1. Anchor-Based and Anchor-Free

Both anchor-based detectors and anchor-free detectors have been used in recent natural scene text detection tasks.

Specifically, anchor-based methods traverse the feature maps calculated by convolutional layers, and place a large number of pre-defined anchors on each picture, the categories are predicted, and the coordinates of these anchors are optimized, which will be regarded as detection results. According to the text area's aspect ratio characteristics, TextBoxes [16] equips each point with six anchors with different aspect ratios as the initial text detection box. TextBoxes+ [17] can detect text in any direction, which uses text boxes with oblique angles to detect irregularly shaped text. DMPNet [18] retains the traditional horizontal sliding window and separately sets six candidate text boxes with different inclination angles according to the inherent shape characteristics of the text: add two 45-degree rectangular windows in the square window; add two long parallelogram windows in the long rectangular window; add two tall parallelogram windows inside the tall rectangular window. The four vertices coordinates of the quadrilateral are used to represent the text candidate frame.

Anchor-free detectors can find objects directly in two different ways without defining anchors in advance. One method is to locate several pre-defined or self-learning key points and limit the spatial scope of the target. Another method is to use the center point or area of the object to define the positive, and then predict the four distances from the positive to the object boundary. For example, in FCOS [19], the introduction of centerness can well inhibit these low-quality boxes' production. Simultaneously, it avoids the complex calculation of anchor frames, such as calculating the overlap in the training process, and saves the memory consumption in the training process. AF-RPN [20] solves the problem that the classic RPN algorithm cannot effectively predict text boxes in any direction. Instead of detecting fusion features from different levels, it detects the text size by the size of the multi-scale components extracted by the feature pyramid network. The RPN stage abandons the use of anchors and uses a point directly to return the coordinates of the four corners of the bounding box, and then shrinks the text area to generate the text core area.

PSENet [21] is slightly different from anchor-free methods. It segments the fusion features of different scales' outputs by the FPN network. Each text instance is reduced to multiple text segmentation maps of different scales through the shrinkage method. The segmentation maps of different scales are merged by the progressive expansion algorithm based on breadth-first-search, which focuses on reconstructing the text instance as a whole to get the final detected text. The progressive scale expansion algorithm can detect the scene text more accurately and distinguish the text that is close or stuck together, which is another method that can process text well without an anchor.

2.2. One-Stage and Two-Stage Algorithms

The representative one-stage and two-stage algorithms are YOLO and Faster-R-CNN, respectively.

The most significant advantage of the single-stage detection algorithm is that it is fast. It provides category and location information directly through the backbone network without using the RPN network to display the candidate area. The accuracy of this algorithm is slightly lower than that of the two-stage. With the development of target detection algorithms, the accuracy of single-stage target detection algorithms has also been improved. Gupta et al. proposed the FCRN model [22], which extracts features based on the full convolutional network, and then performs regression prediction on the feature map by convolution operation. Unlike the prediction of a category label in FCN [23], it predicts the bounding box parameters of each enclosing word, including the center coordinate offset, width, height, and angle information. EAST [24] directly indicates arbitrary quadrilateral text based on the full convolutional network (FCN). It uses NMS to process overlapping bounding boxes and generates multi-channel pixel-level text scoring maps and geometric figures with an end-to-end model. R-YOLO [25] proposed a real-time detector including a fourth-scale detection branch based on YOLOv4 [26], which improved the detection ability of small-scale text effectively.

The precision of the two-stage is higher, while the speed is slower than that of the one-stage. The two-stage network extracts deep features through a convolutional neural network, and then divides the detection into two stages: The first step is to generate candidate regions that may contain objects through the RPN network, and complete the classification of the regions to make a preliminary prediction of the position of the target; the second step is to further accurately classify and calibrate the candidate regions to obtain the final detection result. The entire network structure of RRPN [9] is the same as Faster-R-CNN, which is divided into two parts: one is used to predict the category, and the other one is used to regress the rotated rectangular box to detect text in any direction. Its two-stage is embodied in the use of RRPN to generate a candidate area with rotation angle information, and then adding an RROI pooling layer to generate a fixed-length feature vector, followed by two layers fully connected for the classification of the candidate area. Mask TextSpotter [27] is also a two-stage text detection network based on Mask R-CNN [28], it replaces the RoI pooling layer of Faster-R-CNN with the RoIAlign layer, and adds an FCN branch that predicts the segmentation mask. TextFuseNet [29] merged the ideas of masktextspotter and Mask R-CNN to extract multi-level features from different paths to obtain richer features.

2.3. ResNet and FPN

In addition to the design and improvement of various target detection algorithms that focus on different positions, a detector that can be applied currently in either one stage or two stages usually has the following two parts: the backbone network and the neck part.

It comprises a series of convolution layers, nonlinear layers, and downsampling layers for CNN. The features of images are captured from the global receptive field to describe the images. VGGNet [30] improves performance by continuously deepening the network structure. The increase in the number of network layers will not bring about an explosion in the number of parameters, and the ability to learn features is more vital. The BN layer in batch normalization [31] suppresses the problem that small changes in parameters are amplified as the characteristic network deepens and is more adaptable to parameter changes. Its superior performance makes it the standard configuration in current convolutional networks. ResNet establishes a direct correlation channel between input and output. The robust parameterized layer concentrates on learning the residual between input and output, and improves gradient explosion and gradient disappearance when the network develops deeper.

The backbone of target detection includes VGG, ResNet, etc. In CTPN [8], the VGG16 backbone is first used for feature extraction, SSD network [5] also uses VGG-16 as the primary network. ResNet-50 module was first used for feature extraction in the method proposed by Yang et al. [32], and most of the later networks adopt the ResNet series. The backbone part has also helped develop many excellent networks, such as DenseNet. DenseNet establishes the connection relationship between different layers through feature reuse and bypass settings to further reduce the problem of gradient disappearance and achieve a good training effect, instead of deepening the number of network layers in ResNet and widening network structure in Inception to improve network performance. Besides, the use of the bottleneck layer and translation layer makes the network narrower and reduces the parameters, suppressing overfitting effectively. Some detectors use DenseNet as a backbone for feature extraction.

With the popularity of multi-scale prediction methods such as FPN, many lightweight modules integrating different feature pyramids have been proposed. In FPN, the information from the adjacent layers of bottom-up and top-down data streams will be combined. The target texts of different sizes use the feature map at different levels and detect them separately, leading to repeated prediction results. It is not possible to use the information of the other level feature maps. The neck part of the network has also further developed PANet and other networks. In the target detection algorithm, Yolov4 [26] also uses the PANet method based on the FPN module of YOLOv3 [33] to gather parameters for the training phase to improve the performance of its detector, which proves the effectiveness of PANet. That multi-level fusion architecture has been widely used recently.

3. Principle of the Method

This paper is based on PSENet: without an anchor and in one stage, it explores common text detection frameworks such as ResNet and FPN in other directions. The proposed framework is mainly divided into two modules: the SENet module and the MPANet module. In the residual structure of ResNet, the original PANet processes adjacent layers through addition operations. The MPANet used in this paper is modified from original PANet and connects the characteristic graphs of adjacent layers together to improve the effect. Figure 1 clearly describes the proposed architecture of the scene text detection algorithm.

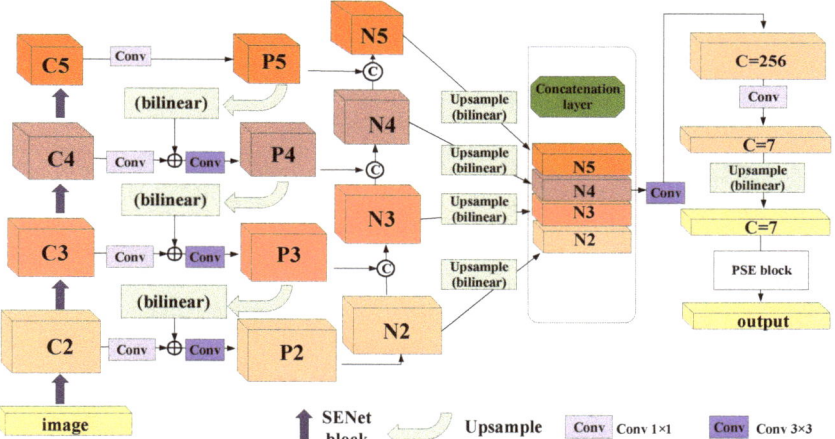

Figure 1. An illustration of our framework. It includes a basic structure with SE blocks; a backbone of feature pyramid networks; bottom-up path augmentation; the progressive scale expansion algorithm, which predicts text regions, kernels, and similarity vectors to describe the text instances. Note that we omit the channel dimensions of feature maps for brevity.

3.1. SENet Block

Convolution neural networks can only learn the dependence of local space according to the receptive field's size. A weight is introduced in the feature map layer considering the relationship between feature channels. In this way, different weights are added to each channel's features to improve the learning ability of features. It should be noted that the SE module adds weights in the dimension of channels. YOLOv4 uses the SE module to do target detection tasks, proving that the SE module can improve the network.

In terms of function, the framework shown in Figure 2 consists of three parts: firstly, a backbone network is constructed to generate the shared feature map, and then a squeeze and excitation network is inserted. This framework's key is adding three operations to the residual structure: squeeze feature compression, exception incentive, and weight recalibration.

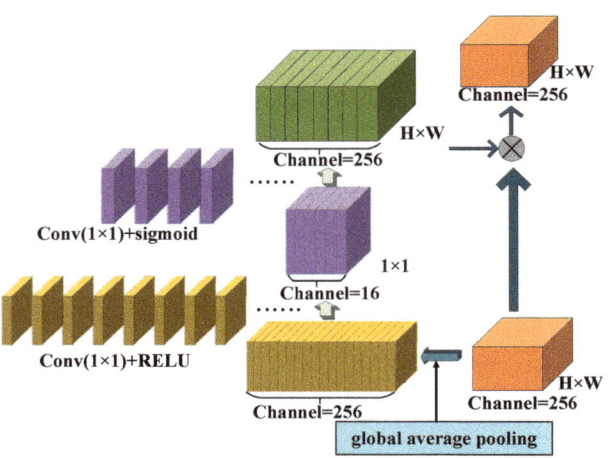

Figure 2. Illustration of an SE block in our model.

Main steps of SENet:

(1) The spatial dimensions of features are compressed, and global average pooling is used Capture the global context, compress all the spatial information to generate channel statistics, compress the size of the graph from H × W to 1 × 1, and the one-dimensional parameter 1 × 1 can obtain the global view of H × W, and the perception area is wider, that is, the statistical information z, z ∈ R ^{C}. The c-th element of z in the formula is calculated by the following formula:

$$z_c = F_{sq}(u_c) = \frac{1}{H \times W} \sum_{i=1}^{H} \sum_{j=1}^{W} u_c(i,j) \tag{1}$$

where $F_{sq}(\cdot)$ is the compression operation, and u_c is the c-th feature.

(2) A 1 × 1 convolution and Relu operation follow, reducing the dimension by 16 times from 256; that is, the channel is transformed to 16—Relu activation function $\delta(x) = \max(0,x)$, dimension reduction layer parameter, $W_1 \in R^{C \times \frac{C}{r}}$; then, the dimension increment layer of 1 × 1 convolution stimulates the number of channels to the original number of 256.

$$S = F_{ex}(Z, W) = \sigma(g(Z, W)) = \sigma(W_2 \delta(W_1 Z)) \tag{2}$$

where the sigmoid activation function $\sigma(x) = \frac{1}{(1+e^{-x})}$, and the dimension increase layer parameter $W_2 \in R^{C \times \frac{C}{r}}$, $F_{ex}(\cdot)$ is the excitation operation, S = $[s_1, s_2, s_3, ..., s_c]$, $s_k \in R^{H \times W}$ (k = 1, 2, 3, ..., c);

(3) The weight is generated for each feature channel's importance after feature selection is obtained, which are multiplied one by one with the previous features to complete the calibration of the original features in the channel dimension.

$$\tilde{X}_C = F_{scale}(u_C, s_C) = s_C \cdot u_C \tag{3}$$

where $\tilde{X} = [\tilde{x}_1, \tilde{x}_2, ..., \tilde{x}_C]$, $F_{scale}(u_C, s_C)$ refers to the corresponding channel product between the feature map $u_C \in R^{H \times W}$ and the scalar s_C.

3.2. Architecture of MPANet

Inspired by FPN, which obtains the semantic features of multi-scale targets, we propose a path aggregation network described in Figure 3; it can be added to the FPN to make the features of different scales more in-depth and more expressive. The emphasis is on fusing low-level elements and adaptive features at the top level.

Our framework improves the bottom-up path expansion. We follow FPN to define the layer that generates the feature map. The same space size is in the same network stage. Each functional level corresponds to a specific stage. We also need ResNet-50 as the basic structure; the output vector of Conv2-x, Conv3-x, Conv4-x, and Conv5-x in the ResNet network is C_2, C_3, C_4, C_5. P_5, P_4, P_3, and P_2 are used to represent the feature levels from top to bottom of FPN generation.

$$P_i = \begin{cases} f_1^{3\times3}(C_i) & i = 5. \\ f_2^{3\times3}\{C_i \oplus F_{upsample}^{\times 2}[f_1^{3\times3}(P_{i+1})]\} & i = 2,3,4. \end{cases} \quad (4)$$

where $f_1^{3\times3}$ means that each P_{i+1} first passes a 3×3 convolutional layer to reduce the number of channels; then the feature map is upsampled to the same size as C_i and adds to the C_i feature map elements; $f_2^{3\times3}$ means that the summed feature map undergoes another 3×3 convolution operation to generate P_i.

$$N_i = \begin{cases} P_i & i = 2. \\ f_2^{3\times3}\{f^{1\times1}[P_{i+1} \| f_1^{3\times3}(N_i)]\} & i = 3,4,5. \end{cases} \quad (5)$$

Our augmented path starts from the bottom P_2 and gradually approaches P_5. The spatial size is gradually sampled down by factor 2 from P_2 to P_5. We use N_2, N_3, N_4, N_5 to represent the newly generated feature graph. Note that N_2 is P_2, without any processing, and retains the original feature map's information.

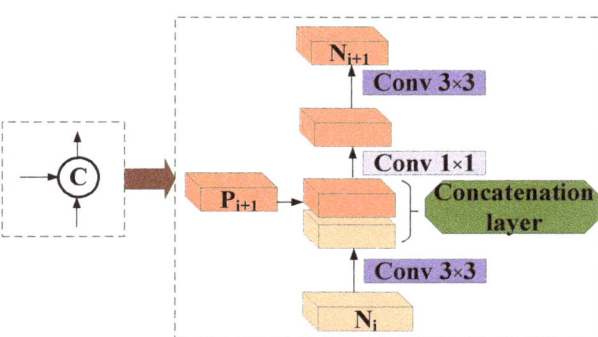

Figure 3. An illustration of our modification of the bottom-up path augmentation.

As shown in Figure 3, each building block needs a higher resolution feature map N_i and a coarser P_{i+1} to generate a new feature map N_{i+1}.

$f_1^{3\times3}$ means that each feature map N_i passes through a 3×3 convolution layer with a step size of 2 to reduce the space size firstly.

"$\|$" means that the feature map P_{i+1} of each layer is connected horizontally, not added, but concatenated with the downsampled map.

After this operation, $f^{1\times1}$ means that the number of channels in the concatenated feature map will be doubled, through 1×1 convolution layer, the step size is 1, and then the channel number is restored to 256.

$f_2^{3\times3}$ means that the fused feature map is then processed by 3×3 convolution fusion to generate N_{i+1} layer for the next step. This is an iterative process, which ends when it approaches P_5. In these building blocks, we mostly use each feature map with 256 channels.

$$N = N_2 \| F_{upsample}^{\times 2}(N_3) \| F_{upsample}^{\times 4}(N_4) \| F_{upsample}^{\times 8}(N_5) \quad (6)$$

Then, the suggestions of each function are collected from the new feature mapping, namely, N_2, N_3, N_4, N_5. The N_3, N_4 and N_5 are upsampled to the size of N_2, $F_{upsample}^{\times 2}$, $F_{upsample}^{\times 4}$, $F_{upsample}^{\times 8}$ refers to 2, 4, 8 times unsampling, and the four layers are concatenated into a feature map.

$$input_{PSE} = F_{upsample}^{\times 2}\{f^{1\times 1}[f^{3\times 3}(N)]\} \tag{7}$$

where $f^{3\times 3}$ refers to convolution operation for reducing the number of channels to 256, $f^{1\times 1}$ refers to the generation of 7 segmentation results. $F_{upsample}$ refers to upsampling to the size of the original image, and the output channel is 7, which is input into the PSE block.

4. Experiments

4.1. Experiment Configuration

The computer configuration shows in Table 1, the training details are as follows:

When training ICDAR2015 [34] and CTW1500 [35] datasets separately, we use a single dataset, note that there are no extra data available for pretraining, e.g., SynthText [22] and IC17-MLT [36]. Before loading them into the network for training, we preprocess images with data augmentation, images are rescaled and returned with random ratios of 0.5,1.0,2.0,3.0; the rotated images randomize in the range $[-10°, 10°]$. Samples are randomly selected from the transformed images, and the minimum output area of the bounding box is calculated for ICDAR2015, the final result is generated by PSE results for CTW1500. All the networks are using SGD. We train each independent dataset with a batch size of 10 on two GPUs for 600 iterations. The training time for each lightweight model is only 24 h. The initial learning rate is set to 1×10^{-3}, divided by 10 at 200 and 400 iterations. We ignore all the text areas labeled as "DO NOT CARE" in the dataset during the training stage, which are not shown as data. Other hyper-parameter settings of the loss function are consistent with PSENet, such as the number of λ is set to 0.7, the positive value of ohem is set to 3, etc. During the testing stage, the confidence threshold is set to 0.89.

Table 1. Computer configuration.

Software Platform	System	Code Edit	Framework
	Ubuntu 16.04 LTS	Python2.7	PyTorch1.2
Hardware Platform	**Memory**	**GPU**	**CPU**
	25 GB	GeForce RTX 2080Ti 11G memory	28 core

4.2. Benchmark Datasets

4.2.1. ICDAR2015

This is a standard dataset proposed for scene text detection in the Challenge4 of ICDAR2015 Robust Reading Competition, which is divided into two categories: the training part contains 1000 image-text pairs; the testing part contains 500 image-text pairs. Each picture is associated with one or more labels annotated with four vertices of the quadrangle. Unlike the previous datasets (such as ICDAR2013 [37]) that only contain horizontal text, the orientations of the reference text in this benchmark are arbitrary.

4.2.2. CTW1500

It is a challenging text detection dataset in long curve format, 1000 for training and 500 for testing form a total of 1500 images. Unlike traditional text datasets (such as ICDAR2017 MLT), the text instance in CTW1500 is marked by a 14-point polygon. The annotations in this dataset are labelled in textline level, which can describe the arbitrary curved form.

4.3. Performance Evaluation Criteria

In this detection algorithm, three evaluation indexes are involved, namely:

4.3.1. Recall

Recall rate(R) is the ratio of the number of positive classes predicted as positive classes to the number of positive real positive classes in the dataset, that is, how much of all the accurate text has been detected.

$$recall = \frac{TP}{TP + FN} \tag{8}$$

4.3.2. Precision

The precision rate(P) represents the ratio of all samples to the total number of samples predicted correctly, that is, how much text detected is accurate.

$$precision = \frac{TP}{TP + FP} \tag{9}$$

4.3.3. F-measure

We aim to have higher precision and recall in the evaluation results, but they are rarely in high results at the same time. Generally speaking, the former is higher while the latter is often inclined to the lower side; the latter is higher while the former is usually lower.

Therefore, when considering the performance of the algorithm, the precision rate and recall rate are not unique. We need to link the two to evaluate. Generally, the weighted average of the two is used to measure the quality of the algorithm and reflect the overall index, namely, F-measure(F). The formula is as follows:

$$\frac{2}{F} = \frac{1}{precision} + \frac{1}{recall} \tag{10}$$

the formula is transformed to:

$$F = \frac{2PR}{P + R} = \frac{2TP}{2TP + FP + FN} \tag{11}$$

Here, TP, FP, and FN are the numbers of True Postive(the instance is a positive class while the prediction is a positive class), False Postive(the instance is a negative class while the prediction is a positive class), and False Negative(the instance is a positive class while the prediction is a negative class), respectively.

4.4. Ablation Study

4.4.1. Effects of MPANet

We conduct several ablation studies on ICDAR2015 and CTW1500 datasets to verify the effectiveness of the proposed MPANet(see Table 2). Note that all the models are trained using only official training images. As shown in Table 2, MPANet obtains 1.01% and 1.21% improvement in F-measure on ICDAR2015 and CTW1500, respectively.

Table 2. The performance gain of MPANet. * and † are results from ICDAR2015 and CTW1500, respectively. FPN * and FPN † represent the results of using the FPN network model in PSE [21] on ICDAR2015 and CTW1500, respectively.

Method	Recall	Precision	F-Measure
FPN *	79.68	81.49	80.57
MPANet *	79.97	83.26	81.58
Gain *	0.29	1.77	1.01
FPN †	75.55	80.57	78.00
MPANet †	75.52	83.29	79.21
Gain †	−0.03	2.72	1.21

Figure 4 shows the train loss difference between modified PANet with SE block (SEMPANet) and MPANet without SE block (MPANet). It demonstrates that the loss function of SEMPANet drops faster on ICDAR2015. Figure 5 shows the loss comparison of two models with and without SE block, which proves that the loss function of MPANet model has a slightly faster convergence effect on average than the other one on CTW1500. The difference of the loss function on the two datasets is reflected in the last two rows of Table 4 and Table 5.

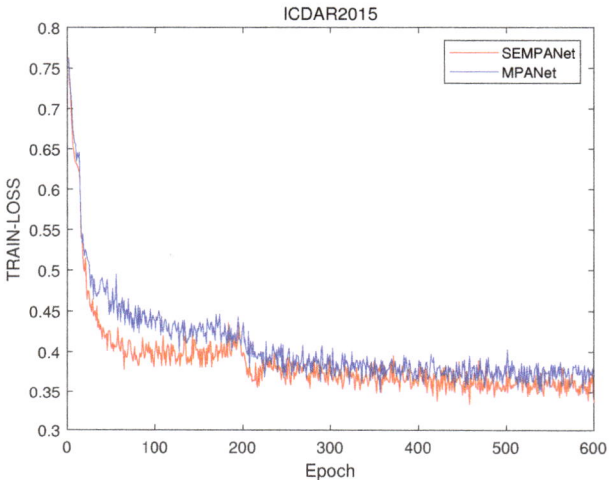

Figure 4. Ablation study of an SE block on ICDAR2015. These results are based on (ResNet 50 and SE block) and (ResNet 50 block) trained on MPANet.

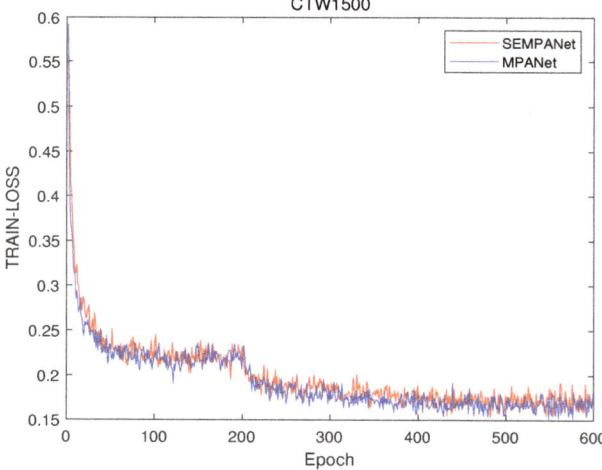

Figure 5. Ablation study of an SE block on CTW1500. These results are based on (ResNet 50 and SE block) and (ResNet 50 block) trained on MPANet.

4.4.2. Effects of the Threshold λ in the Testing Phase

The hyper-parameter λ in the final test balances the influence between the three evaluation indexes. Table 3 compares the prediction effects of MPANet and SEMPANet with different λ within a short fluctuation range on the dataset ICDAR2015. We see that when SEMPANet with a λ of 0.89 is used, even if the performance is robust to changes in λ, in the average performance of the three evaluation indexes, F-measure is higher than PSENet, and Recall also performs best.

Table 3. The performance comparison of λ.

λ in MPANet	Recall	Precision	F-Measure
0.93	78.77	85.92	82.19
0.91	79.82	84.25	81.98
0.89	79.97	83.26	81.58
λ in SEMPANet	Recall	Precision	F-Measure
0.93	78.57	84.74	81.54
0.91	79.83	83.57	81.65
0.89	80.45	82.80	81.61

4.5. Experimental Results

4.5.1. Evaluation on Oriented Text Benchmark

In order to verify the effectiveness of the bankbone proposed in this paper, we have carried out comparative experiments on ICDAR2015 with CTPN, Seglink, EAST, PSENet and other mainstream methods. The ICDAR2015 dataset mainly includes horizontal, vertical and slanted text. As shown in Table 4, the proposed method without external data achieves a state-of-the-art result of 80.45%, 82.80% and 81.61% in recall, precision and F-measure, respectively. Each paper in Table 4 has its representative detection method for natural scene text characteristics. Compared with EAST, our precision is reduced by 0.8%, while recall and F-measure are increased by 6.95% and 3.41%, respectively. Compared with WordSup, the recall, precision and F-measure are increased by 3.45%, 3.5% and 3.41%, respectively. Compared with PAN, our precision is slightly decreased by 0.1%, while recall is increased by 2.65%, the F-measure reflecting the comprehensive detection ability is increased by 1.31%. We have also compared with several lightweight networks in 2020. As shown in Table 3, we selected the results of three indicators that have been improved to above 80 when considering the overall performance. Compared with [38–40], our recall are increased by 3.75% 0.25% and 0.77%, respectively.

Table 4. The single-scale results on ICDAR2015. "Ext" indicates external data. MPANet is a model without an SE module.

Method	Year	Ext	Recall	Precision	F-Measure
CTPN [8]	2016	-	51.6	74.2	60.9
Seglink [6]	2017	√	73.1	76.8	75.0
SSTD [41]	2017	√	73.9	80.2	76.9
EAST [24]	2017	-	73.5	83.6	78.2
WordSup [42]	2017	√	77.0	79.3	78.2
DeepReg [43]	2017	-	80.0	82.0	81.0
RRPN [9]	2018	-	73.0	82.0	77.0
Lyu et al. [44]	2018	√	70.7	94.1	80.7
PAN [45]	2019	-	77.8	82.9	80.3
PSENet-1s [21]	2019	-	79.7	81.5	80.6
Pelee-Text++ [39]	2020	√	76.7	87.5	81.7
Qin et al. [40]	2020	-	80.20	82.86	81.56
Jiang et al. [38]	2020	-	79.68	85.79	82.62
MPANet		-	79.97	83.26	81.58
SEMPANet		-	80.45	82.80	81.61

Compared with PSENet-1s, we can find that this paper's method has improved recall, precision, and F-measure. The rates are increased by 0.75%, 1.3% and 1.01%, respectively. The comparison with the above methods on the ICDAR2015 dataset shows that the method proposed in this paper has a high level of detection results for regular text and slanted text. Overall, SEMPANet has a higher recall rate than MPANet on ICDAR2015, and its recall also achieves state-of-the-art result in Table 4. Some qualitative results are visualized in Figure 6.

Figure 6. Results on ICDAR2015. The green boxes in (**a**,**b**) and the red boxes in (**b**) represent the evaluation results of the text and the error detection boxes of them, respectively.

4.5.2. Evaluation on Curve Text Benchmark

We have verified the superiority of our method in the Curve text by conducting experiments on the public dataset CTW1500. The experimental results are shown in Table 5. The data for the comparison methods in the table are all from their corresponding papers. The CTW1500 dataset contains many curved letters. Methods such as CTPN and Seglink often fail to detect and label with rectangular boxes accurately. The bankbone proposed in this paper extracts richer features, combined with the post-processing part of PSENet, which is not limited by rectangular boxes and can detect any shape well. Compared with the benchmark method CTD+TLOC of the CTW1500 dataset, our accuracy rate has been improved by 3.02%, 6.68%, and 4.64% in recall, precision and F-measure, respectively. Compared with TextSnake, our recall is lower, while the precision is higher, which is 16.2% higher than TextSnake. The F-measure is lower by 2.16% compared with TextSnake. Compared with [38,40], our precision are increased by 2.28% and 3.48%, respectively.

Compared with PSENet-1s, the method proposed in this paper has a lower recall of 2.78%, however, the precision is greatly improved 3.48%. Due to the fact that many letters in the CTW1500 dataset are too close or even glued and overlapped, they are still difficult to separate. The F-measure of the method proposed in this paper reached 78.04%, indicating that it can detect curved text well. Figure 7 demonstrates some detection results of SEMPANet on CTW1500.

Table 5. The single-scale results from CTW1500. * indicates the results from [35]. Ext is short for external data used in the training stage. MPANet is a model without an SE module.

Method	Year	Ext	Recall	Precision	F-Measure
CTPN * [8]	2016	-	53.8	60.4	56.9
Seglink * [6]	2017	-	40.0	42.3	40.8
EAST * [24]	2017	-	49.1	78.7	60.4
CTD+TLOC [35]	2017	-	69.8	77.4	73.4
TextSnake [46]	2018	√	85.3	67.9	75.6
CSE [47]	2019	√	76.0	81.1	78.4
PSENet-1s [21]	2019	-	75.6	80.6	78.0
Jiang et al. [38]	2020	-	75.9	80.6	78.2
Qin et al. [40]	2020	-	76.8	81.8	79.4
MPANet		-	75.52	83.29	79.21
SEMPANet		-	72.82	84.08	78.04

Figure 7. Some visualization results from CTW1500. The green boxes in (**a**,**b**) and the red boxes in (**a**) represent the evaluation results of the text and the error detection boxes of them respectively.

4.6. Discussion of Results

Most of the text can be well detected: see the green text detection boxes in Figures 6 and 7. Invalid examples are shown in the red boxes in Figures 6b and 7b, some of which are missing. We have analyzed the failure results of the proposed method. The following briefly introduces several sets of test results and analyzes environmental factors. In Figure 6b, the first image shows multiple text targets on the billboard. The red target of overly large size cannot be detected correctly, which is mistakenly divided into three target boxes. Due to the influences of the text and the background environment, the characters on the building in the second picture are omitted; due to the impact of the surrounding colors and the dense arrangement, the characters "3" and "20" in the third picture were left out. In Figure 7b, the "HO" in the first image is omitted; the two small samples in the second image are omitted;

characters in the third image are close to the white background. In short, the test results in an austere environment are good. For example, for text with a complex environment, a small portion of the text with shallow definition can be detected. Since there are scenes with many lines and colorful spots in the image, the existing model will classify the text as clearly recognizable by the human eye but not detected as background.

The proposed method can achieve outstanding detection results. However, PSENet still has limitations in processing small-sized text. Compared with the previous methods, this paper uses SEMPANet to improve the overall structure and adjusts the network parameters. In ICDAR2015, the recall rate R has been improved; P and F perform well; there are still deficiencies in the curved text CTW1500.

5. Conclusions

In this paper, our network can be divided into two parts: feature extraction and post-processing. The post-processing part using the progressive expansion algorithm can guarantee the accuracy of text detection, but the experimental results prove that the simple use of FPN network in the feature extraction part has insufficient feature extraction, which leads to the decline of the text detection effect. This paper proposes a new scene text detection method based on feature fusion. This method uses SENet as the basic network and integrates the features of the MPANet to make up for the lack of features extracted from the original network. The fusion strategy proposed in this paper enables the text detection model to reach a detection level higher than that of the original network. Finally, the progressive expansion algorithm is used for post-processing so that the entire model can detect the text quickly and accurately. With the aim of improving the experimental results, the method in this paper avoids the introduction of end-to-end networks with too many parameters, and finally achieves the purpose of accurate and fast text detection, which is of great significance for the research of natural scene text detection technology oriented toward actual application scenarios. Furthermore, I hope to introduce new mathematical tools for research and discussion. In that regard, a recent approach based on geometric algebra [48] extracts features for multispectral images to be investigated. Finally, other multi-dimensional data processing such as L1-norm minimization [49] and hashing networks [50] remain primarily unexplored and can benefit from further research.

Author Contributions: Conceptualization, S.L. and W.C.; methodology, S.L.; software, S.L.; validation, S.L.; formal analysis, S.L. and W.C.; investigation, S.L. and W.C.; resources, S.L. and W.C.; data curation, S.L. and W.C.; writing—original draft preparation, S.L. and W.C.; writing—review and editing, S.L. and W.C.; visualization, S.L. and W.C.; supervision, W.C.; project administration, W.C.; funding acquisition, W.C. All authors have read and agreed to the published version of the manuscript.

Funding: This research was funded by the National Natural Science Foundation of China under grant 61771322 and grant 61871186, and funded by the Fundamental Research Foundation of Shenzhen under Grant JCYJ20190808160815125.

Institutional Review Board Statement: Not applicable.

Informed Consent Statement: Not applicable.

Data Availability Statement: Not applicable.

Conflicts of Interest: The authors declare no conflict of interest.

References

1. Aljuaid, H.; Iftikhar, R.; Ahmad, S.; Asif, M.; Afzal, M.T. Important citation identification using sentiment analysis of in-text citations. *Telemat. Inform.* **2021**, *56*, 101492. [CrossRef]
2. Dashtipour, K.; Gogate, M.; Li, J.; Jiang, F.; Kong, B.; Hussain, A. A hybrid Persian sentiment analysis framework: Integrating dependency grammar based rules and deep neural networks. *Neurocomputing* **2020**, *380*, 1–10. [CrossRef]
3. Cao, W.; Liu, Q.; He, Z. Review of pavement defect detection methods. *IEEE Access* **2020**, *8*, 14531–14544. [CrossRef]
4. Huang, G.; Liu, Z.; Maaten, L.V.D.; Weinberger, K.Q. Densely connected convolutional networks. In Proceedings of the IEEE Conference on Computer Vision and Pattern Recognition, Honolulu, HI, USA, 21–26 July 2017; pp. 4700–4708.

5. Liu, W.; Anguelov, D.; Erhan, D.; Szegedy, C.; Berg, A.C. Ssd: Single shot multibox detector. In Proceedings of the European Conference on Computer Vision, Amsterdam, The Netherlands, 11–14 October 2016.
6. Shi, B.; Bai, X.; Belongie, S. Detecting oriented text in natural images by linking segments. In Proceedings of the IEEE Conference on Computer Vision and Pattern Recognition, Honolulu, HI, USA, 21–26 July 2017; pp. 2550–2558.
7. Ren, S.; He, K.; Girshick, R.; Sun, J. Faster r-cnn: Towards real-time object detection with region proposal networks. *arXiv* **2015**, arXiv:1506.01497.
8. Tian, Z.; Huang, W.; He, T.; He, P.; Qiao, Y. Detecting text in natural image with connectionist text proposal network. In Proceedings of the European Conference on Computer Vision, Amsterdam, The Netherlands, 11–14 October 2016; Springer: New York, NY, USA, 2016; pp. 56–72 .
9. Ma, J.; Shao, W.; Ye, H.; Wang, L.; Wang, H.; Zheng, Y.; Xue, X. Arbitrary-oriented scene text detection via rotation proposals. *IEEE Trans. Multimed.* **2018**, *20*, 3111–3122. [CrossRef]
10. He, K.; Zhang, X.; Ren, S.; Sun, J. Identity mappings in deep residual networks. In Proceedings of the European Conference on Computer Vision, Amsterdam, The Netherlands, 11–14 October 2016; Springer: New York, NY, USA, 2016; pp. 630–645.
11. He, K.; Zhang, X.; Ren, S.; Sun, J. Deep residual learning for image recognition. In Proceedings of the IEEE Conference on Computer Vision and Pattern Recognition, Las Vegas, NV, USA, 27–30 June 2016; pp. 770–778.
12. Lin, T.-Y.; Dollár, P.; Girshick, R.; He, K.; Hariharan, B.; Belongie, S. Feature pyramid networks for object detection. In Proceedings of the IEEE Conference on Computer Vision and Pattern Recognition, Honolulu, HI, USA, 21–26 July 2017; pp. 2117–2125.
13. Hu, J.; Shen, L.; Sun, G. Squeeze-and-excitation networks. In Proceedings of the IEEE Conference on Computer Vision and Pattern Recognition, Salt Lake City, UT, USA, 18–23 June 2018; pp. 7132–7141.
14. Zhao, Q.; Sheng, T.; Wang, Y.; Tang, Z.; Chen, Y.; Cai, L.; Ling, H. M2det: A single-shot object detector based on multi-level feature pyramid network. In Proceedings of the AAAI Conference on Artificial Intelligence, Honolulu, HI, USA, 27 January–1 February 2019; Volume 33, pp. 9259–9266.
15. Liu, S.; Qi, L.; Qin, H.; Shi, J.; Jia, J. Path aggregation network for instance segmentation. In Proceedings of the IEEE Conference on Computer Vision and Pattern Recognition, Salt Lake City, UT, USA, 18–23 June 2018; pp. 8759–8768.
16. Liao, M.; Shi, B.; Bai, X.; Wang, X.; Liu, W. Textboxes: A fast text detector with a single deep neural network. *arXiv* **2016**, arXiv:1611.06779.
17. Liao, M.; Shi, B.; Bai, X. Textboxes++: A single-shot oriented scene text detector. *IEEE Trans. Image Process.* **2018**, *27*, 3676–3690. [CrossRef] [PubMed]
18. Liu, Y.; Jin, L. Deep matching prior network: Toward tighter multi-oriented text detection. In Proceedings of the IEEE Conference on Computer Vision and Pattern Recognition, Honolulu, HI, USA, 21–26 July 2017; pp. 1962–1969.
19. Tian, Z.; Shen, C.; Chen, H.; He, T. Fcos: Fully convolutional one-stage object detection. In Proceedings of the IEEE International Conference on Computer Vision, Seoul, South Korea, 27 October–3 November 2019; pp. 9627–9636.
20. Zhong, Z.; Sun, L.; Huo, Q. An anchor-free region proposal network for faster r-cnn-based text detection approaches. *Int. J. Doc. Anal. Recognit. (IJDAR)* **2019**, *22*, 315–327. [CrossRef]
21. Wang, W.; Xie, E.; Li, X.; Hou, W.; Lu, T.; Yu, G.; Shao, S. Shape robust text detection with progressive scale expansion network. In Proceedings of the IEEE Conference on Computer Vision and Pattern Recognition, Long Beach, CA, USA, 15–21 June 2019; pp. 9336–9345.
22. Gupta, A.; Vedaldi, A.; Zisserman, A. Synthetic data for text localisation in natural images. In Proceedings of the IEEE Conference on Computer Vision and Pattern Recognition, Las Vegas, NV, USA, 27–30 June 2016; pp. 2315–2324.
23. Long, J.; Shelhamer, E.; Darrell, T. Fully convolutional networks for semantic segmentation. In Proceedings of the IEEE Conference on Computer Vision and Pattern Recognition, Boston, MA, USA, 7–12 June 2015; pp. 3431–3440.
24. Zhou, X.; Yao, C.; Wen, H.; Wang, Y.; Zhou, S.; He, W.; Liang, J. East: An efficient and accurate scene text detector. In Proceedings of the IEEE Conference on Computer Vision and Pattern Recognition, Honolulu, HI, USA, 21–26 July 2017; pp. 5551–5560.
25. Wang, X.; Zheng, S.; Zhang, C.; Li, R.; Gui, L. R-YOLO: A Real-Time Text Detector for Natural Scenes with Arbitrary Rotation. *Sensors* **2021**, *21*, 888. [CrossRef] [PubMed]
26. Bochkovskiy, A.; Wang, C.Y.; Liao, H.Y.M. YOLOv4: Optimal Speed and Accuracy of Object Detection. *arXiv* **2020**, arXiv:2004.10934.
27. Lyu, P.; Liao, M.; Yao, C.; Wu, W.; Bai, X. Mask textspotter: An end-to-end trainable neural network for spotting text with arbitrary shapes. In Proceedings of the European Conference on Computer Vision (ECCV), Munich, Germany, 8–14 September 2018; pp. 67–83.
28. He, K.; Gkioxari, G.; Dollár, P.; Girshick, R. Mask r-cnn. In Proceedings of the IEEE International Conference on Computer Vision, Venice, Italy, 22–29 October 2017; pp. 2961–2969.
29. Ye, J.; Chen, Z.; Liu, J.; Du, B. TextFuseNet: Scene Text Detection with Richer Fused Features. In Proceedings of the 29th International Joint Conference on Artificial Intelligence (IJCAI-20), Yokohama, Japan, 7–15 January 2021.
30. Simonyan, K.; Zisserman, A. Very deep convolutional networks for large-scale image recognition. *arXiv* **2014**, arXiv:1409.1556.
31. Ioffe, S.; Szegedy, C. Batch normalization: Accelerating deep network training by reducing internal covariate shift. *arXiv* **2015**, arXiv:1502.03167.
32. Yang, Q.; Cheng, M.; Zhou, W.; Chen, Y.; Qiu, M.; Lin, W.; Chu, W. Inceptext: A new inception-text module with deformable psroi pooling for multi-oriented scene text detection. *arXiv* **2018**, arXiv:1805.01167.

33. Redmon, J.; Farhadi, A. Yolov3: An incremental improvement. *arXiv* **2018**, arXiv:1804.02767.
34. Karatzas, D.; Gomez-Bigorda, L.; Nicolaou, A.; Ghosh, S.; Valveny, E. Icdar 2015 competition on robust reading. In Proceedings of the International Conference on Document Analysis & Recognition, Tunis, Tunisia, 23–26 August 2015.
35. Yuliang, L.; Lianwen, J.; Shuaitao, Z.; Sheng, Z. Detecting curve text in the wild: New dataset and new solution. *arXiv* **2017**, arXiv:1712.02170.
36. Nayef, N.; Fei, Y.; Bizid, I.; Choi, H.; Ogier, J.M. Icdar2017 robust reading challenge on multi-lingual scene text detection and script identification—Rrc-mlt. In Proceedings of the 2017 14th IAPR International Conference on Document Analysis and Recognition (ICDAR), Kyoto, Japan, 9–15 November 2017.
37. Karatzas, D.; Shafait, F.; Uchida, S.; Iwamura, M.; Heras, L.P.D.L. Icdar 2013 robust reading competition. In Proceedings of the 2013 12th International Conference on Document Analysis and Recognition (ICDAR), Washington, DC, USA, 25–28 August 2013.
38. Jiang, X.; Xu, S.; Zhang, S.; Cao, S. Arbitrary-Shaped Text Detection with Adaptive Text Region Representation. *IEEE Access* **2020**, *8*, 102106–102118. [CrossRef]
39. Córdova, M.; Pinto, A.; Pedrini, H.; Torres, R.D.S. Pelee-Text++: A Tiny Neural Network for Scene Text Detection. *IEEE Access* **2020**, *8*, 223172–223188. [CrossRef]
40. Qin, X.; Jiang, J.; Yuan, C.A.; Qiao, S.; Fan, W. Arbitrary shape natural scene text detection method based on soft attention mechanism and dilated convolution. *IEEE Access* **2020**, *8*, 122685–122694. [CrossRef]
41. He, P.; Huang, W.; He, T.; Zhu, Q.; Li, X. Single shot text detector with regional attention. In Proceedings of the IEEE International Conference on Computer Vision, Venice, Italy, 22–29 October 2017.
42. Hu, H.; Zhang, C.; Luo, Y.; Wang, Y.; Han, J.; Ding, E. Wordsup: Exploiting word annotations for character based text detection. In Proceedings of the IEEE International Conference on Computer Vision, Venice, Italy, 22–29 October 2017; pp. 4940–4949.
43. He, W.; Zhang, X.Y.; Yin, F.; Liu, C.L. Deep direct regression for multi-oriented scene text detection. In Proceedings of the IEEE International Conference on Computer Vision, Venice, Italy, 22–29 October 2017.
44. Lyu, P.; Yao, C.; Wu, W.; Yan, S.; Bai, X. Multi-oriented scene text detection via corner localization and region segmentation. In Proceedings of the IEEE Conference on Computer Vision and Pattern Recognition, Salt Lake City, UT, USA, 18–23 June 2018; pp. 7553–7563.
45. Wang, W.; Xie, E.; Song, X.; Zang, Y.; Wang, W.; Lu, T.; Yu, G.; Shen, C. Efficient and accurate arbitrary-shaped text detection with pixel aggregation network. In Proceedings of the IEEE/CVF International Conference on Computer Vision, Seoul, Korea, 27 October–3 November 2019.
46. Long, S.; Ruan, J.; Zhang, W.; He, X.; Wu, W.; Yao, C. Textsnake: A flexible representation for detecting text of arbitrary shapes. In Proceedings of the European Conference on Computer Vision (ECCV), Munich, Germany, 8–14 September 2018; pp. 20–36.
47. Liu, Z.; Lin, G.; Yang, S.; Liu, F.; Lin, W.; Goh, W.L. Towards robust curve text detection with conditional spatial expansion. In Proceedings of the IEEE Conference on Computer Vision and Pattern Recognition (CVPR), Long Beach, CA, USA, 15–21 June 2019.
48. Wang, R.; Shen, M.; Wang, X.; Cao, W. RGA-CNNs: Convolutional neural networks based on reduced geometric algebra. *Sci. China Inf. Sci.* **2021**, *64*, 1–3. [CrossRef]
49. Wang, R.; Shen, M.; Wang, T.; Cao, W. L1-norm minimization for multi-dimensional signals based on geometric algebra. *Adv. Appl. Clifford Algebr.* **2019**, *29*, 1–18. [CrossRef]
50. Lin, Q.; Cao, W.; He, Z.; He, Z. Mask Cross-Modal Hashing Networks. *IEEE Trans. Multimed.* **2021**, *23*, 550–558. [CrossRef]

Article

Green IoT and Edge AI as Key Technological Enablers for a Sustainable Digital Transition towards a Smart Circular Economy: An Industry 5.0 Use Case

Paula Fraga-Lamas [1,2,*], Sérgio Ivan Lopes [3,4,] and Tiago M. Fernández-Caramés [1,2,]

1 Department of Computer Engineering, Faculty of Computer Science, Universidade da Coruña, 15071 A Coruña, Spain; tiago.fernandez@udc.es
2 Centro de Investigación CITIC, Universidade da Coruña, 15071 A Coruña, Spain
3 ADiT-Lab, Instituto Politécnico de Viana do Castelo, Rua da Escola Industrial e Comercial de Nun'Alvares, 4900-347 Viana do Castelo, Portugal; sil@estg.ipvc.pt
4 IT—Instituto de Telecomunicações, Campus Universitário de Santiago, 3810-193 Aveiro, Portugal
* Correspondence: paula.fraga@udc.es; Tel.: +34-981-167-000

Abstract: Internet of Things (IoT) can help to pave the way to the circular economy and to a more sustainable world by enabling the digitalization of many operations and processes, such as water distribution, preventive maintenance, or smart manufacturing. Paradoxically, IoT technologies and paradigms such as edge computing, although they have a huge potential for the digital transition towards sustainability, they are not yet contributing to the sustainable development of the IoT sector itself. In fact, such a sector has a significant carbon footprint due to the use of scarce raw materials and its energy consumption in manufacturing, operating, and recycling processes. To tackle these issues, the Green IoT (G-IoT) paradigm has emerged as a research area to reduce such carbon footprint; however, its sustainable vision collides directly with the advent of Edge Artificial Intelligence (Edge AI), which imposes the consumption of additional energy. This article deals with this problem by exploring the different aspects that impact the design and development of Edge-AI G-IoT systems. Moreover, it presents a practical Industry 5.0 use case that illustrates the different concepts analyzed throughout the article. Specifically, the proposed scenario consists in an Industry 5.0 smart workshop that looks for improving operator safety and operation tracking. Such an application case makes use of a mist computing architecture composed of AI-enabled IoT nodes. After describing the application case, it is evaluated its energy consumption and it is analyzed the impact on the carbon footprint that it may have on different countries. Overall, this article provides guidelines that will help future developers to face the challenges that will arise when creating the next generation of Edge-AI G-IoT systems.

Keywords: Green IoT; IIoT; edge computing; AI; edge AI; sustainability; digital transition; digital circular economy; Industry 5.0

1. Introduction

The current digital transformation offers substantial opportunities to industry for building competitive and innovative business models and complex circular supply chains; however, such a transformation also implies severe implications concerning sustainability, since the Information and Communications Technology (ICT) industry has a significant environmental footprint. In order to reach the milestones defined by the United Nations Agenda for Sustainable Development [1] and to implement the visions of circular economy, it is necessary to provide solutions in an efficient and sustainable way during their whole life cycle. Such a sustainable digital transition towards a smart circular economy is enabled by three key technologies: IoT, edge computing, and Artificial Intelligence (AI).

It is estimated that Internet of Things (IoT) and Industrial IoT (IIoT) technologies, which enable ubiquitous connectivity between physical devices, can add, only in industrial applications, USD 14 trillion of economic value to the global economy by 2030 [2]. In addition, the development of the classic view of the Internet of People (IoP) [3] and the Internet Protocol (IP) led to a convergence of IoT technologies over the last two decades, which paved the way for the so-called Internet of Everything (IoE) [4]. Such a concept is rooted in the union of people, things, processes, and data to enrich people's lives.

The explosion of IoT/IIoT technologies and their potential to pave the way to a more sustainable world (in terms of full control of the entire life cycle of products), can also lead to some pitfalls that represent a major risk in achieving the milestones defined by the UN Agenda for Sustainable Development [1]. As part of the IoT Guidelines for Sustainability that were addressed in 2018 by the World Economic Forum, a recommendation to adopt a framework based on the UN Sustainable Development Goals (SDGs) [1] to evaluate the potential impact and measure the results of the adoption of such recommendations was put forward [2]; however, in 2010–2019, and considering *Goal 12: Ensure sustainable consumption and production* [1], electronic waste grew by 38% and less than 20% has been recycled. Paradoxically, although these technologies have a huge potential for the digital transformation towards sustainability, they are not yet contributing to the sustainable development of the ICT sector. Specifically, such a contribution is expected for the IoT sector, which has been seen as the driving force for a sustainable digital transition. The need for policies that effectively promote the sustainable development of new products and services is crucial and can be seen as a societal challenge in the years to come.

The concept of Green IoT (G-IoT) [5,6] is defined in [7] as: *"energy-efficient procedures (hardware or software) adopted by IoT technologies either to facilitate the reduction in the greenhouse effect of existing applications and services or to reduce the impact of the greenhouse effect of the IoT ecosystem itself"*. In the former case, the use of IoT technologies may help to reduce the greenhouse effect, whereas the latter focuses on the optimization of IoT greenhouse footprints. Moreover, the entire life cycle of a G-IoT system should focus on green design, green production, green utilization, and finally, green disposal/recycling, to have a neutral or very small impact on the environment [7].

IoT devices have increasingly higher computational power, are more affordable and more energy-efficient, which helps to sustain the progress of Moore's law to bring a sustainable IoT revolution in the global economy [8]; however, this vision directly collides with the advent of the concept of Edge Intelligence (EI) or Edge Artificial Intelligence (Edge-AI), where the processing of the IoT collected data is performed at the edge of the network, which imposes additional challenges in terms of latency, cybersecurity, and more specially, energy efficiency.

This article summarizes the most relevant emerging trends and research priorities for the development of Edge-AI G-IoT systems in the context of sustainability and circular economy. In particular, the following are the main contributions of the article:

- The essential concepts and background knowledge necessary for the development of Edge-AI G-IoT systems are detailed.
- The most recent Edge-AI G-IoT communications architectures are described together with their main subsystems to allow future researchers to design their own systems.
- The latest trends on the convergence of AI and edge computing are detailed. Moreover, a cross-analysis is provided in order to determine the main issues that arise when combining G-IoT and Edge-AI.
- The energy consumption of a practical Industry 5.0 application case is analyzed to illustrate the theoretical concepts introduced in the article.
- The most relevant future challenges for the successful development of Edge-AI G-IoT systems are outlined to provide a roadmap for future researchers.

The remainder of this article is structured as follows. Section 2 introduces the essential concepts that will be used in the article. Section 3 analyzes the main aspects related to the development of G-IoT systems, including their communications architecture and their main subsystems. Section 4 analyzes the convergence of AI and edge computing to create Edge-AI systems. Section 5 provides a cross-analysis to determine the key issues that arise when combining G-IoT and Edge-AI systems. Section 6 presents a practical Industry 5.0 application case and evaluates the energy consumption of a mist computing Edge-AI G-IoT model. Section 7 outlines the main future challenges that stand in the way of leveraging Edge-AI G-IoT systems. Finally, Section 8 is devoted to the conclusions.

2. Background
2.1. Digital Circular Economy
2.1.1. Circular Economy

Circular Economy (CE) promotes an enhanced socio-economic paradigm for sustainable development. It aims to fulfill current needs without jeopardizing the needs of future generations under three dimensions: economic, social, and environmental. The European Green Deal [9], Europe's new agenda for sustainable growth, is an ambitious action plan to move to a clean circular economy, to restore biodiversity, to reduce emissions by at least 55% by 2030, and to become the world's first climate neutral continent by 2050. The EC strategy is well aligned with the United Nations (UN) 2030 Agenda for Sustainable Development [10]. The 17 Sustainable Development Goals (SDGs) are at the heart of the EU policymaking across all sectors.

CE reforms current linear "take-make-dispose" economic models based on unsustainable mass production and consumption and proposes a new model that is restorative by design (materials, components, platforms, resources, and products add as much value as possible throughout their life cycle). Such a model also aligns the needs of the different stakeholders through business models, government policies, and consumer preferences [11]. At the end of their lifetime, much of these products and components are regenerated and/or recycled.

The European Commission adopted a new Circular Economy Action Plan (CEAP) in March 2020, as one of the main key elements of the European Green Deal [12]. Such an action plan promotes initiatives along the entire life cycle of products, from design to the end of their lifetime, encouraging sustainable consumption and waste reduction. According to the World Economic Forum [13], achieving a CE transition will require unprecedented collaboration, given that, in 2019, only 8.6% of the world was circular, although CE can yield up to USD 4.5 trillion in economic benefits in 2030 [14].

2.1.2. Digital Circular Economy (DCE)

Data centers and digital infrastructures require substantial levels of energy. ICT accounts for 5 to 9% of the total electricity demand with a potential increase to 20% by 2030 [15]. In addition, materials (e.g., physical resources, raw materials) linked to the digital transformation are also a problem: the world produces over 50 million tons of electronic and electrical waste (e-waste) annually and just 20% is formally recycled. Such an amount of waste will reach 120 million tons annually by 2050 [16].

The challenge posed by the increase in digital technologies requires the application of circular economy principles to the digital infrastructure. While currently, the focus of the sector is mainly on meeting the needs in a sustainable way (e.g., energy efficiency and cybersecurity), the supply of critical raw materials will be an issue in the coming years. Moreover, the opportunities provided by the DCE to the digital transition should be also explored (e.g., new business models, new markets, and reduced information asymmetry).

2.1.3. G-IoT and Edge-AI for Digital Circular Economy (DCE)

Digital technologies are a key enabler for the upscaling of the circular economy, as they allow for creating and processing data required for new business models and complex circular supply chains. In addition, they can close the information and transparency gaps that currently slow down the scale-up of DCE.

There is a need for further integration of digital enabling technologies such as functional electronics (e.g., nanoelectronics, flexible, organic and printed electronics or electronic smart systems), blockchain [17], edge computing [18], UAVs [19], 5G/6G [20], big data, and

AI [21] into existing circular business approaches to provide information and additional services.

Specifically, G-IoT and Edge-AI have the potential to substantially leverage the adoption of DCE concepts by organizations and society in general in two main ways. First, by considering an open G-IoT architecture [11], where G-IoT devices have circularity enabling features (e.g., end-to-end cybersecurity, privacy, interoperability, energy harvesting capabilities). Second, by having a network of Edge-AI G-IoT connected devices that provide fast smart services and real-time valuable information to the different stakeholders (e.g., designers, end users, suppliers, manufacturers, and investors). Thus, supply chain visibility and transparency of the product, of the production system, and the whole business, are ensured. Moreover, stakeholders can rely on real-time accurate information to make the right decisions at the right time to use resources effectively, to improve the efficiency of the processes, and to reduce waste. Furthermore, asset monitoring and predictive maintenance can increase product lifetime. Figure 1 illustrates the previous concepts and provides an overall view of the main areas impacted by the combined use of G-IoT and Edge-AI.

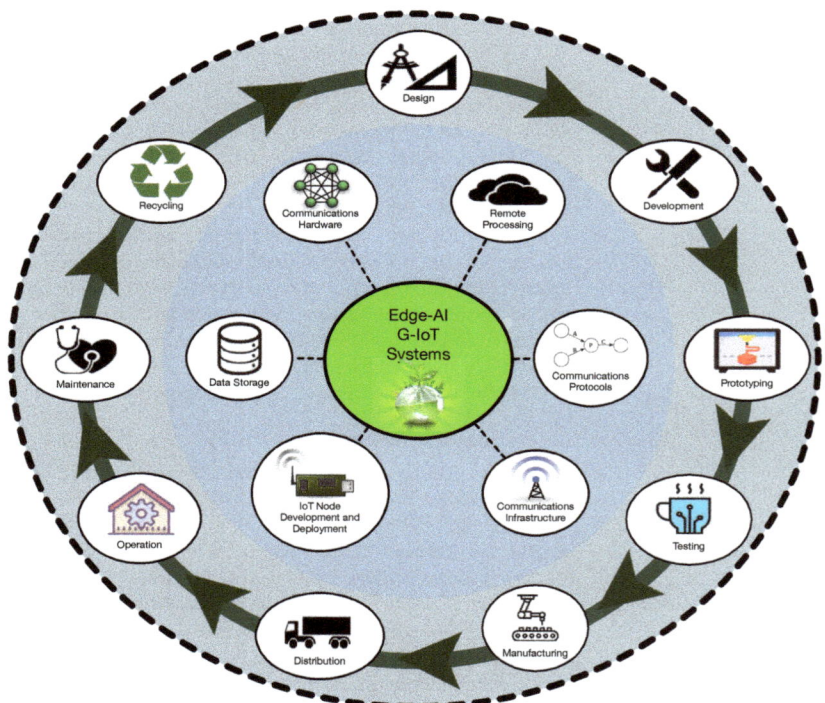

Figure 1. Edge-AI G-IoT main areas and their digital circular life cycle.

2.2. Industry 5.0 and Society 5.0

The Industry 5.0 paradigm is still being characterized by industry and academia, but the European Commission has already defined its foundations, due to the impact that such a concept will have in the coming years for the European industry [22]. The proposed concept seeks to correct some Industry 4.0 aspects that have not been properly addressed or that have become controversial due to forgetting essential values such as social fairness and sustainability. Thus, according to the European Commission, the foundations of Industry 5.0 have to be completely aligned with societal goals and to aim higher than just considering jobs and economic growth. As a consequence, Industry 5.0 is focused on sustainable manufacturing and industrial operator well-being [23].

It is important to note that Industry 5.0 has not been conceived as a complete industrial revolution, but as a complement to Industry 4.0 that contemplates aspects that link emerging societal trends to industrial development [24]; therefore, the Industry 5.0 paradigm looks for the improvement of smart factory efficiency through technology, while minimizing environmental and social impacts.

It is also worth pointing out that the vision of Industry 5.0 according to the European Commission seems to be clearly inspired by a previous concept: Society 5.0. Such a concept was first put forward by the Japanese government in 2015 [25] and later (in 2016) it was fostered by Keidanren, one of the most relevant business federations of Japan [26]. Society 5.0 goes beyond industrial company digitalization and proposes a collaborative strategy for the whole Japanese society, as it happened throughout history with the four previous society revolutions: Society 1.0 and Society 2.0 are related to hunters and gatherers; Society 3.0 is associated with the industrial revolution that occurred at the end of the 18th century; and Society 4.0 arose from the information-based economies related to the spread of the Internet and on industrial digitalization. As a continuation to Society 4.0, Society 5.0 still looks for expanding economic development, but, at the same time, it keeps in mind societal and environmental concerns.

2.3. Technology Enablers

In order to reach the UN Sustainable Development Goals and to implement the visions of the digital circular economy, Society 5.0, and Industry 5.0, it is necessary to provide solutions to integrate the physical and virtual worlds in an efficient and sustainable way. Thus, the next subsections describe the three key technology enablers that this article is focused on and that need to be optimized to make our daily lives and industrial processes greener.

2.3.1. IoT and IIoT

The term IoT refers to a network of physical devices (i.e., "things") that can be connected among themselves and with other services that are deployed over the Internet. Such devices are usually composed of sensors, actuators, communications transceivers, and computationally constrained processing units (e.g., microcontrollers). IoT devices have multiple applications in fields such as appliance remote monitoring [27], home automation [28], or precision agriculture [29]. The adaptation of the IoT principles to industrial environments is referred to as IIoT and allows for deploying many remotely monitored and controlled sensors, actuators, and smart machinery in industrial scenarios [30–32].

2.3.2. Cloud and Edge Computing

Most current IoT applications are already deployed on cloud computing based systems since they allow for centralizing data storage, processing, and remote monitoring/interaction; however, such centralized solutions have certain limitations. The cloud itself is considered a common point of failure, since attacks, vulnerabilities, or maintenance tasks can block it and, as a consequence, the whole system may stop working [33]. In addition, it is important to note that the number of connected IoT devices is expected to increase in the next years [34] and, consequently, the number of predicted communications with the cloud may overload it if it is not scaled properly.

Due to the previous constraints, in recent years, new architectures have been proposed. In the case of edge computing, it is aimed at offloading the cloud from tasks that can be performed by devices placed at the edge of an IoT network, close to the end IoT nodes. Thus, different variants of the edge computing paradigm have been put forward, such as fog computing [35], proposed by Cisco to make use of low-power devices on the edge, or cloudlets [36], which consist of high-end computers that perform heavy processing tasks on the edge [37,38].

2.3.3. AI

AI is a field that looks for adding intelligence to machines [39]. Such intelligence can be demonstrated in the form of recommendation systems, human-speech recognition solutions, or autonomous vehicles that are able to make decisions on their own. The mentioned examples are able to collect information from the real world and then process it in order to provide an output (i.e., a solution to a problem). In some cases, AI systems need to learn previously how to solve a specific problem, so they need to be trained.

In the case of IoT systems, AI systems receive data from the deployed IoT nodes, which usually collect them from their sensors. In traditional IoT architectures, such data are transmitted to a remote cloud where they are processed by the AI system and a result is generated, which usually involves making a decision that is communicated to the user or to certain devices of the IoT network.

The problem is that real-time IoT systems frequently cannot rely on cloud-based architectures, since latency prevents the system from responding timely. In such cases, the use of Edge-AI provides a solution: edge computing devices are deployed near the IoT end nodes, so lag can be decreased, and IoT node requests are offloaded from the cloud, thus avoiding potential communications bottlenecks when scaling the system.

Although Edge-AI is a really useful technology for IoT systems, their combination derives into systems that can consume a significant amount of energy, so Edge-AI IoT systems need to be optimized in terms of power consumption. The next sections deal with such a problem: first, the factors that impact the development of energy-efficient (i.e., green) IoT systems are studied and then the power consumption of Edge-AI systems is analyzed.

3. Energy Efficiency for IoT: Developing Green IoT Systems

3.1. Communications Architectures for G-IoT Systems

Before analyzing how G-IoT systems try to minimize energy consumption, it is first necessary to understand which components make up an IoT architecture. Thus, Figure 2 depicts a cloud-based architecture, currently the most popular IoT architecture, which is built around the cloud computing paradigm. Such a cloud collects data from remote IoT sensors and can send commands to IoT actuators. The cloud is also capable of interacting with third-party services (usually hosted in servers or other cloud computing systems) and with remote users, to whom it provides management software.

Cloud-based IoT systems have allowed the spread of IoT systems, but, since they are commonly centralized, they suffer from known bottlenecks (e.g., Denial of Service (DoS) attacks) and from relatively long response latency [33]. To tackle such issues, in recent years, new IoT paradigms have been explored, such as edge, fog, or mist computing [35,40], which offload the cloud from certain tasks to decrease the amount of node requests and to reduce latency response. In the case of edge computing, it adds a new layer between the cloud and the IoT devices (where the gateway is placed in Figure 2) to provide them with fast-response services through edge devices such as cloudlets or fog computing gateways [41]. Fog computing gateways are computationally constrained devices (e.g., routers and Single-Board Computers (SBCs)) that provide support for physically distributed, low-latency, and Quality of Service (QoS) aware applications [35,37]. Cloudlets allow for providing real-time rendering or compute-intensive services, which require deploying high-end PCs in the local network [36]. Regarding mist computing devices, they perform tasks locally at the IoT nodes and can collaborate with other IoT nodes to perform complex tasks without relying on a remote cloud [37,40,42–44]. Thus, mist nodes reduce the need for exchanging data to the higher layers of the architecture (thus saving battery power), but, in exchange, they are responsible for carrying out multiple tasks locally.

Figure 3 depicts an example of mist computing based architecture. In this figure, for the sake of clarity, no edge computing layer is included, but it is standard to make use of it in practical applications [42]. The two layers that are present are the cloud layer, which works similarly to the previously described architectures, and the mist computing layer, which is composed by mist nodes. Such nodes embed additional hardware to perform the necessary local processing tasks. In addition, it is worth noting that mist nodes often can communicate directly among themselves, thus avoiding the need for using intermediate gateways.

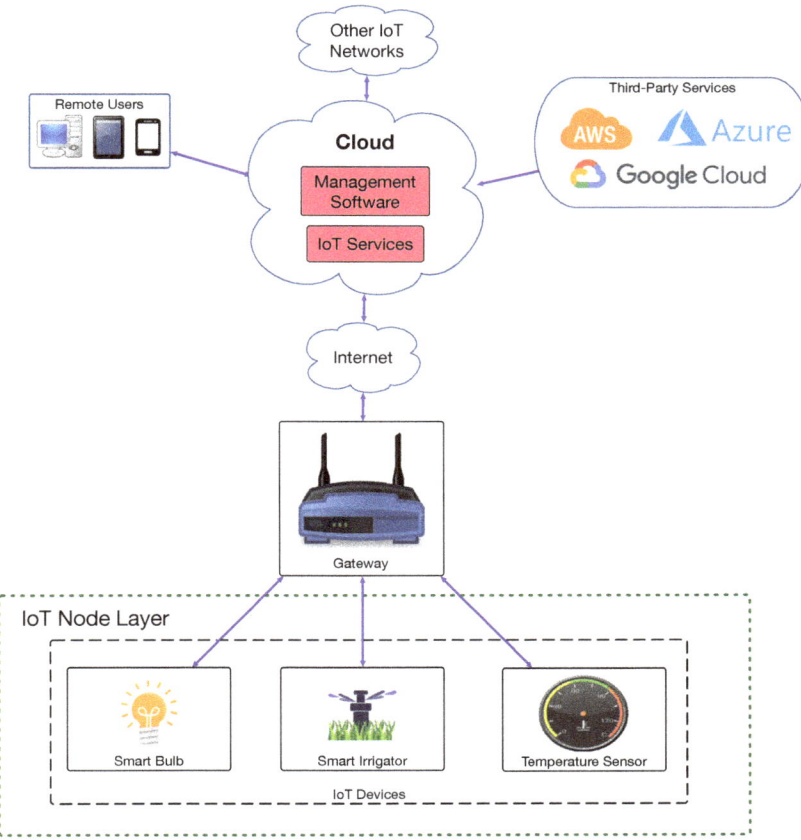

Figure 2. Cloud-based IoT architecture.

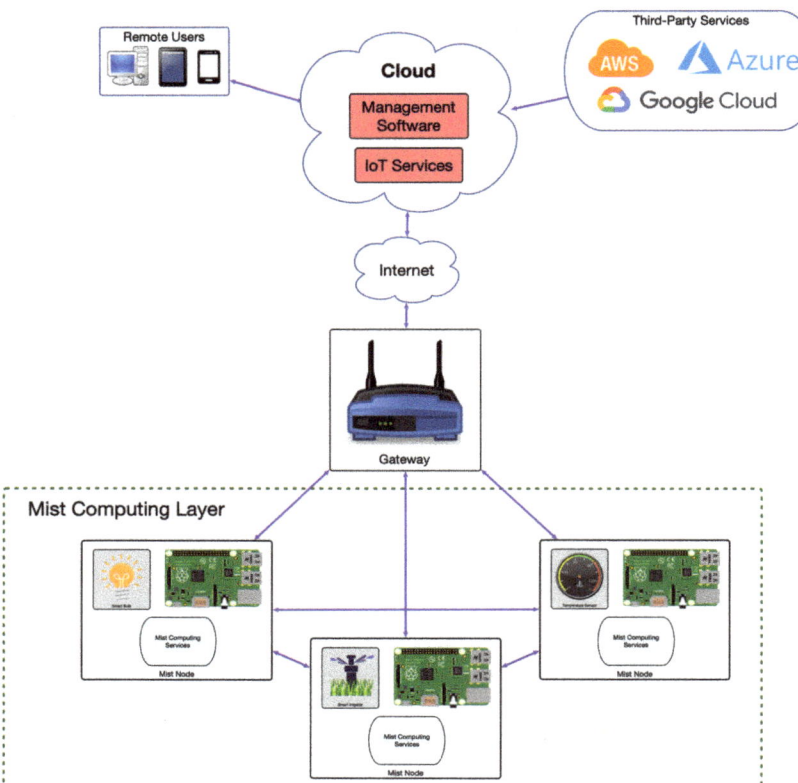

Figure 3. Example of mist computing architecture.

After analyzing the previous architectures, it can be stated that, to create G-IoT systems, it is necessary to consider the efficiency of the hardware and software of their main components: the IoT nodes, the edge computing devices, and the cloud. The next sections delve into such a topic, reviewing the most relevant contributions of the state of the art.

3.2. Types of G-IoT Devices

The development and deployment of efficient G-IoT devices is conditioned by their hardware and software. It is also important to note that the requirements of the G-IoT devices differ significantly: G-IoT nodes do not have the same energy consumption needs as edge devices (e.g., fog computing gateways, cloudlets, Mobile Edge Computing (MEC) hardware) or the cloud. Nonetheless, all the involved hardware have in common the fact that it is essential to select the main parts that allow for optimizing their energy efficiency (the control and power subsystems), and the communications interfaces. Regarding software, the control software, the implemented communications protocols, and the used security algorithms are essential when minimizing energy consumption. The next subsections analyze such hardware and software components in order to guide future G-IoT developers.

3.3. Hardware of the Control and Power Subsystems

There are different approaches to maximize the energy efficiency of IoT deployments. One of the most important is to find the right trade-off between the different capabilities of the control hardware and their energy consumption. Currently, the most popular IoT nodes are based on microcontrollers. Such devices are usually cheap, have enough processing power to perform control tasks, can be easily reprogrammed, and have low-energy consumption. There are other more sophisticated alternatives, such as Digital Signal Processors (DSPs), System-On-Chips (SOCs), Central Processing Units (CPUs), Field-Programmable Gate Arrays (FPGAs), Complex Programmable Logic Devices (CPLDs), Graphics Processing Units (GPUs), and Application-Specific Integrated Circuits (ASICs).

DSPs are usually power efficient, especially certain models designed specifically for low power consumption (e.g., Texas Instruments TMS320C5000). Central Processing Units (CPUs) (e.g., Intel Xeon) are general-purpose processing units that offer an adequate trade-off between performance and power consumption, but they are usually optimized for high-speed and parallel processing. With respect to SoCs, they integrate medium-to-high performance microcontrollers and peripherals, so they consume more power than traditional microcontrollers, but they are more appropriate for lightweight systems. In the case of FPGAs, they offer very good performance for executing deterministic tasks, but its programming is not as easy as with microcontrollers, and they require to power the used logic continuously. There are also hybrid solutions that combine the benefits of FPGAs and CPUs, known as Field-Programmable Systems-on-Chips (FPSoCs) [45]. In the case of CPLDs, they can execute tasks faster than FPGAs, but their maximum allowed design complexity is inferior to the one offered by FPGAs. GPUs were created to offload graphic computation from the CPUs, but current products can include several thousands of cores designed for the efficient execution of complex functions. Regarding ASICs, they offer even higher performance than FPGAs and other embedded devices, since they are optimized for power consumption, but their development cost is very high (usually in the order of millions of dollars).

Besides choosing the right control hardware, it is necessary to optimize the power subsystems. Most current IoT node deployments rely on batteries. Such batteries can store a finite amount of energy, and they need to be replaced or recharged frequently. Maintenance tasks are costly and cumbersome, especially in large deployments, industrial confined spaces, or remote areas. In addition, such tasks are critical when developing power-hungry applications. Battery replacement also leads to a heavy carbon footprint due to the use of scarce raw materials, the battery manufacturing process, and the involved recycling processes; therefore, there is a need for self-sustainable solutions such as environmental energy harvesting. Such solutions exploit ubiquitous energy sources in the deployment area without requiring external power sources and ease maintenance tasks. The most common harvesting techniques are related to solar and kinetic energy sources. Examples of different energy harvesting techniques are presented in [46–49].

3.4. Communications Subsystem

G-IoT devices can make use of different technologies for their communications interfaces. The communications with the cloud are usually through the Internet or a wired intranet, so this section focuses on the energy efficiency of the wireless communications technologies used by G-IoT nodes and edge devices. Table 1 compares the characteristics of some of the most relevant communications technologies according to their power consumption, operating band, maximum range, expected data rate, their relevant features, and main applications.

Table 1. Main characteristics of the most relevant communications technologies for G-IoT nodes.

Technology	Power Consumption	Frequency Band	Maximum Range	Data Rate	Main Features	Popular Applications
NFC	Tags require no batteries, no power	13.56 MHz	<20 cm	424 kbit/s	Low cost	Ticketing and payments
Bluetooth 5 LE	1–20 mW, Low power and rechargeable (days to weeks)	2.4 GHz	<400 m	1360 kbit/s	Trade-off among different PHY modes	Beacons, wireless headsets
EnOcean	Very low consumption or battery-less thanks to using energy harvesting	868–915 MHz	300 m	120 kbit/s	Up to 2^{32} nodes	Energy harvesting building automation applications
HF RFID	Tags require no batteries	3–30 MHz (13.56 MHz)	a few meters	<640 kbit/s	NLOS, low cost	Smart Industry, payments, asset tracking
LF RFID	Tags require no batteries	30–300 KHz (125 KHz)	<10 cm	<640 kbit/s	NLOS, durability, low cost	Smart Industry and security access
UHF RFID	Batteries last from days to years	30 MHz–3 GHz	tens of meters	<640 kbit/s	NLOS, durability, low cost	Smart Industry, asset tracking and toll payment
UWB/IEEE 802.15.3a	Low power, rechargeable (hours to days)	3.1 to 10.6 GHz	<10 m	>110 Mbit/s	Low interference	Fine location, short-distance streaming
Wi-Fi (IEEE 802.11b/g/n/ac)	High power consumption, rechargeable (hours)	2.4–5 GHz	<150 m	up to 433 Mbit/s (one stream)	High-speed, ubiquity, easy to deploy and access	Wireless LAN connectivity, Internet access
Wi-Fi HaLow/IEEE 802.11ah	Power consumption of 1 mW	868–915 MHz	<1 km	100 Kbit/s per channel	Low power, different QoS levels (8192 stations per AP)	IoT applications
ZigBee	Very low power consumption, 100–500 µW, batteries last months to years	868–915 MHz, 2.4 GHz	<100 m	Up to 250 kbit/s	Up to 65,536 nodes	Smart Home and industrial applications
LoRa	Long battery life, it lasts >10 years	2.4 GHz	kilometers	0.25–50 kbit/s	High range, resistant to interference	Smart cities, M2M applications
SigFox	Battery lasts 10 years sending 1 message, <10 years sending 6 messages	868–902 MHz	50 km	100 kbit/s	Global cellular network	M2M applications

G-IoT node communications need to provide a trade-off between features and energy consumption. For example, Near-field Communication (NFC) [50] is able to deliver a reading distance of up to 30 cm, but NFC tags usually do not need to make use of batteries since they are powered by the readers through inductive coupling. NFC is a technology derived from Radio Frequency Identification (RFID), which, despite certain security constraints [51], in recent years, has experienced significant growth in home and industrial scenarios [52,53] thanks to its very low power consumption. It must be noted that RFID and NFC are essentially aimed at identifying items, but they can be used for performing regular wireless communications among G-IoT nodes (e.g., for reading embedded sensors). Nonetheless, there are technologies that have been devised to provide more complex interactions. For instance, Bluetooth implementations such as Bluetooth Low Energy (BLE) can provide wireless communications distances between 10 and 100 m [54] and very low energy consumption thanks to the use of beacons [55], which are a sort of lightweight IoT devices able to transmit packets at periodic time intervals.

The widely popular Wi-Fi (i.e., IEEE 802.11 standards) can also provide indoor and outdoor coverage easily and inexpensively for IoT nodes; however, its energy consumption is usually relatively high and proportional to the speed rate. Nonetheless, new IEEE 802.11 standards have been proposed in recent years so as to reduce energy consumption. For instance, Wi-Fi Hallow offers low power consumption (comparable with Bluetooth) while maintaining high data rates, and a wider coverage range.

In terms of green communications, the following are currently the most popular and promising technologies:

- ZigBee [56]. It was conceived for deploying Wireless Sensor Networks (WSNs) that are able to provide overall low energy consumption by being asleep most of the time,

just waking up periodically. In addition, it is easy to scale ZigBee networks, since they can create mesh networks to extend the IoT node communications range.
- LoRA (Long-Range Wide Area Network) and LoRAWAN [57]. These technologies have been devised to deploy Wide Area IoT networks while providing low energy consumption.
- Ultrawideband (UWB). It is able to provide low-energy wide-bandwidth communications as well as centimeter-level positioning accuracy in short-range indoor applications. Mazhar et al. [58] evaluate different UWB positioning methods, algorithms, and implementations. The authors conclude that some techniques (e.g., hybrid techniques combining both Time-of-Arrival (TOA) and Angle-of-Arrival (AOA)), although more complex, are able to offer additional advantages in terms of power consumption and performance.
- Wi-Fi Hallow/IEEE 802.11ah. In contrast to Wi-Fi, it offers very low energy consumption by adopting novel power-saving strategies to ensure an efficient use of energy resources available in IoT nodes. It was specifically created to address the needs of Machine-to-Machine (M2M) communications based on many devices (e.g., hundreds or thousands), long range, sporadic traffic needs, and substantial energy constraints [59].

3.5. Green Control Software

There is a significant number of recent publications that propose different techniques and protocols for network control and power saving. For instance, there are G-IoT protocols for interference reduction, optimized scheduling (e.g., switching selectively inactive sensor nodes and put them into deep sleep mode), resource allocation and access control, temporal and spatial redundancy, cooperative techniques in the network, dynamic transmission power adjustment, or energy harvesting [6].

Power-efficient network routing is also a hot topic. For instance, Xie et al. [60] reviewed recent works on energy-efficient routing and propose a novel method for relay node placement. Other authors focused on solutions for service-aware clustering [61]. Another interesting work can be found in [62], where the authors present an energy-efficient IoT architecture able to predict the adequate sleep interval of sensors. The experimental results show significant energy savings for sensor nodes and improved resource utilization of cloud resources. Nonetheless, this solution is not valid for applications with real-time requirements or that require constant availability. Finally, recent approaches such as [63] proposed solutions that combine distributed energy harvesting-enabled mobile edge computing offloading systems with on-demand computing resource allocation and battery energy level management.

3.6. Energy Efficient Security Mechanisms

A number of attacks can be performed to break the confidentiality, integrity, and availability of IoT/IIoT networks (e.g., jamming, malicious code injection, Denial of Service (DoS) attacks, Man-in-the-Middle (MitM) attacks, and side-channel attacks) [64]. In order to have protection for such attacks, secure deployment of G-IoT networks should involve three main elements: architecture, hardware, and the security mechanisms across the different devices.

The resource-constrained nature of IoT devices, specially IoT nodes, imposes limitations on the inclusion of complex protocols to encrypt and secure communications [65]. This is particularly challenging when implementing cryptosystems that require substantial computational resources. Hash functions, symmetric cryptography, and public-key cryptosystems (i.e., asymmetric cryptographic systems such as Rivest–Shamir–Adleman (RSA) [66], Elliptic Curve Cryptography (ECC) [67,68], or Diffie–Hellman (DH) [69]) are among the most used cryptosystems.

Public-key cryptosystems are essential for authenticating transactions and are part of Internet standards such as Transport Layer Security (TLS) (TLS v1.3 [70]), currently the most-extended solution for securing TCP/IP communications. Regarding cipher suites recommended for TLS, Rivest–Shamir–Adleman (RSA) and Elliptic Curve Diffie–Hellman Ephemeral (ECDHE) are the most popular ones.

The execution of cryptographic algorithms must be fast and energy efficient, but still provide adequate security levels. Such a trade-off has attracted scientific attention, which is currently an active area of research [71], especially since recent advances in computation have made it easy to break certain schemes (e.g., 1024-bit RSA is broken [72]); however, there are few articles in the literature that address the impact of security mechanisms on energy consumption for G-IoT systems. For instance, in [42], the authors compare the energy consumption of different cryptographic schemes, showing that, at the same security level, some schemes are clearly more efficient in terms of energy and data throughput than others when executed on certain IoT devices.

Moreover, hardware acceleration can be used for keeping energy consumption and throughput values at a reasonable level when executing public-key cryptography algorithms [73]. Furthermore, the use of specific hardware can also speed up the execution of cryptographic algorithms such as hash algorithms [74] or block ciphers [75].

3.7. G-IoT Carbon Footprint

The concept of carbon footprint (or carbon dioxide emissions coefficient) measures the amount of greenhouse gases (including CO_2) caused by human or non-human activities. In the case of the development and use of a technology, it involves a carbon footprint related to its life cycle: from the design stage to the recycling of products. This is especially critical for IoT, since a large number of connected devices is expected in the coming years (up to 30.9 billion in 2025 [76]), which will consume a significant amount of electricity and, as a consequence, a high volume of carbon dioxide will be emitted into the environment. G-IoT has emerged as an attractive research area whose objective is to study how to minimize the environmental impact related to the deployment of IoT networks in smart homes, factories, or smart cities [77].

The following are some of the challenges that must be faced in order to reduce IoT network carbon footprint and environmental impact [78,79]:

- Hardware power consumption. The used IoT hardware is the basis for the IoT network, so its energy consumption should be as energy efficient as possible while preserving its functionality and required computing power.
- IoT node software energy consumption. Software needs to be optimized together with the hardware, so developers need to introduce energy-aware constraints during the development of G-IoT solutions. Such optimizations are especially critical for certain digital signal processing tasks such as compression, feature extraction, or machine learning training [80].
- IoT protocol energy efficiency. The IoT relies on protocols that enable communicating between the multiple nodes and routing devices involved in an IoT network. As a consequence, such protocols need to be energy efficient in terms of software implementation and should consider the minimization of the usage of communication interfaces. For instance, Peer-to-Peer (P2P) protocols are well-known for being intensive in terms of the number of communications they manage, although some research has been dedicated to reducing their energy consumption [81–85].
- RF spectrum management optimization. The increasing number of deployed IoT nodes will derive into the congestion of the RF spectrum, so its management will need to be further optimized to minimize node energy consumption [77].
- Datacenter sustainability. As the demand for IoT devices grows, ever-increasing amounts of energy are needed to power the datacenters where remote cloud services are provided. This issue is especially critical for corporations such as Google or Microsoft, which rely on huge data centers and, in fact, the U.S. Environmental Protection Agency (EPA) already warned about this problem in 2007 [86]. As a consequence of

such a warning, carbon footprint estimations were performed in order to determine the emissions related to the construction and operation of a datacenter [87].
- Data storage energy usage. In cloud-centric architectures, most of the data are stored in a server or in a farm of servers in a remote datacenter, but some of the latest architectures decentralize data storage to prevent single-point-of-failure issues and avoid high operation costs. Thus, for such decentralized architectures, G-IoT requires minimizing node energy consumption and communications. This is not so easy, since devices are physically scattered, and they usually make use of heterogeneous platforms whose energy optimization may differ significantly.
- Use of green power sources. IoT networks can become greener by making use of renewable power sources from wind, solar, or thermal energy. IoT nodes can also make use of energy-harvesting techniques to minimize their dependence on batteries or extend their battery life [46–49]. Moreover, battery manufacturing and end-of-life processes have their own carbon footprint and impact the environment with their toxicity. Furthermore, IoT architectures can be in part powered through decentralized green smart grids, which can collaborate among them to distribute the generated energy [78].
- Green task offloading. Traditional centralized architectures have tended to offload the computing and storage resources of IoT devices to a remote cloud, which requires additional power consumption and network communications that are proportional to the tasks to be performed and to the latency of the network. In contrast, architectures such as the ones described in Section 3.1, can selectively choose which tasks to offload to the cloud. Thus, most of the node requests are processed in the edge of the network, which reduces latency and network resource consumption due to the decrease in the number of involved gateways and routers [88]. Nonetheless, G-IoT designers must be aware of the energy implications of decentralized systems [89].

4. AI and Edge Computing Convergence

As previously mentioned in Section 2.3.3, AI can be broadly defined as a science capable of simulating human cognition to incorporate human intelligence into machines. Machine Learning (ML) can be seen as a specific subset of AI, as a technique for training algorithms that focuses on empowering computer systems with the ability to learn from data, perform accurate predictions, and therefore, make decisions. The training stage in ML involves the collection of huge amounts of data (train set) to train an algorithm that allows the machine to learn from the processed information. Then, after training, the algorithm is used for inference in new data [90]. Deep Learning (DL) is a subset of ML that can be seen as the natural evolution of ML. DL algorithms are inspired by the human brain cognitive processing patterns (i.e., by its ability for pattern identification and classification), using DL algorithms that are trained to perform the same tasks in computer systems. By analogy, the human brain typically attempts to interpret a new pattern by labeling it and performing subsequent categorization [91]. Once new information is received, the brain attempts to compare it to a known reference before reasoning, which is conceptually what DL algorithms perform (e.g., Artificial Neural Networks (ANNs) algorithms aim to emulate the way the human brain works). In [91], Samek et al. identified two major differences between ML and DL:

1. DL can automatically identify and select the features that will be used in the classification stage. In contrast, ML requires the features to be provided manually (i.e., unsupervised vs. supervised learning).
2. DL requires high-end hardware and large training data sets to deliver accurate results, as opposed to ML, which can operate in low-end hardware with smaller data sets in the training stage (i.e., ML is typically adopted in resource contained embedded hardware).

The use of such AI techniques is highly dependent, not only on the hardware specifications and the available computational power, but also on the adopted inference approach [92].

4.1. AI-Enabled IoT Hardware

AI-enabled IoT devices are paving the way to implement new and increasingly complex cyber–physical systems (CPS) in distinct application domains [93–95]. The increasing complexity of such devices is typically specified based on SWaP requirements (i.e., reduced Size, Weight, and Power) [96]. When considering the IoT/IIoT ecosystems, changes in SWaP requirements, and also in unit cost, may impact the overall performance and functionality of the end devices, since the number of devices tends to increase at a steady pace, the cost per unit becomes more and more relevant. Note that the number of devices deployed is expected to increase massively in the coming years, with many of these devices operating as sensors and/or actuators, which will demand increasing processing power enabling effective edge AI deployment. On the other hand, portability is also relevant, and therefore, power will often come from an external battery or an energy harvesting subsystem, which imposes several challenges in the design of AI-enabled IoT devices. For example, in [97], a study regarding low-power ML architectures has been put forward and results have shown that sub-mW power consumption can potentially be deployed in "always-ON" AI-enabled IoT nodes.

4.1.1. Common Edge-AI Device Architectures

The G-IoT hardware previously described in Section 3.2 has evolved in recent years as illustrated in Figure 4 in order to provide AI-enable functionality. Thus, basic IoT hardware (represented at the top of Figure 4), typically uses a traditional computing approach that combines an embedded processor (CPU) or a microcontroller (MCU) with on-board memory, sensor/actuator interfacing—digital (e.g., SPI, I2C, 1-Wire) and analog (ADCs, DACs) inputs/outputs—and basic connectivity (e.g., Wi-Fi, Bluetooth).

AI-enabled IoT device architectures (depicted in the middle of Figure 4), use a near-memory computing approach based on a multicore CPU or FPGA, and typically includes external sensors and actuators, and extended connectivity options such as NB-IoT, LoRaWAN, or 5G/6G support.

Lastly, an AI-specific IoT device also includes cognitive capabilities and typically uses an in-memory computing approach, which may be supported by a dedicated AI SoC, specifically included to execute learning algorithms (this architecture is depicted at the bottom of Figure 4). IoT devices are getting increasingly powerful and computationally efficient as new SoCs with integrated AI chips become available. For example, the usage of FPGAs in AI-enabled IoT devices allows high-speed inference, parallel execution, and the implementation of application-specific computational architectures without the need for expensive ASICs; however, the total power consumption may be a problem when using FPGAs in power-sensitive applications [96].

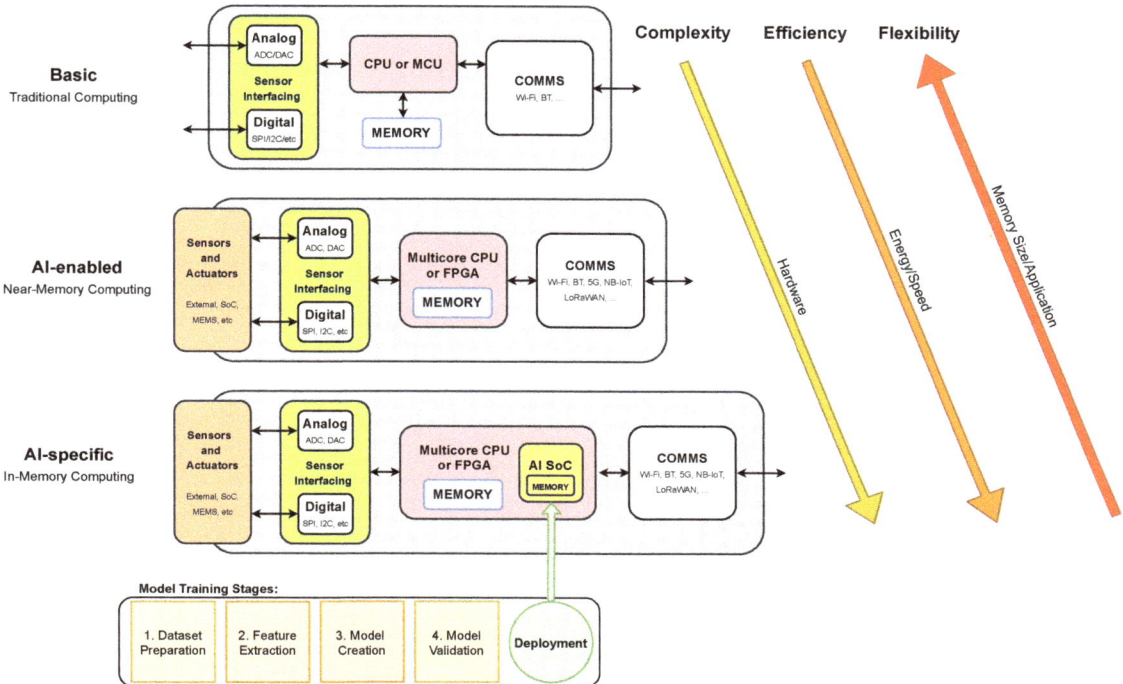

Figure 4. Basic, AI-enabled and AI-specific IoT device architectures.

4.1.2. Embedded AI SoC Architectures

Embedded AI SoCs are used in specific IoT architectures [98], allowing for the execution of ML algorithms directly on the end device, and therefore detecting patterns and trends in data, and enabling the transmission of low-bandwidth data streams with contextual information to enhance decision-making and empower prognosis throughout the use in-device prediction models and ML, as it is represented at the bottom in Figure 4. In [96], Mauro et al. achieved high performance in power saving for both logic and SRAM design, using Binary Neural Networks (BNNs). BNNs enable the deployment of deep models on resource-constrained devices [99], because they may be trained to produce outcomes comparable to full-precision alternatives while maintaining a smaller footprint, a more scalable structure, and better error resilience. Such characteristics enable the implementation of completely programmable SoC IoT end-devices capable of performing hardware-accelerated and software-defined algorithms at ultra-low power, reaching 22.8 Inference/s/mW while using 674 µW [98].

4.1.3. AI-Enabled IoT Hardware Selection Criteria

Running an AI model at an AI-enable IoT device presents four main advantages when compared with the classical cloud-based approach:

1. Reliable Connectivity: data can be gathered and processed on the same device instead of relying on a network connection to transmit data to the cloud, which reduces the probability of network connection problems.
2. Reduced Latency: when processing is performed locally, all communications-related latencies are avoided, resulting in an overall latency that converges to the inference latency.

3. Increased Security and Privacy: reducing the need for communicating between the IoT edge device and the cloud means reducing the risk that data will be compromised, lost, stolen, or leaked.
4. Bandwidth Efficiency: reducing the communications between IoT edge devices and the cloud, also reduces bandwidth needs and the overall communications cost.

Table 2 compiles several AI-enabled IoT hardware boards that are able to run ML libraries, such as Tensorflow Lite [100]. TensorFlow Lite is an open-source ML library specifically designed for resource-constrained IoT devices, that typically use MCU-based architectures.

Table 2. AI-enabled IoT hardware compatible with TensorFlow Lite.

Board	Processor	Power	Connectivity	Architecture Type	Cryptographic Engine	Cost
Arduino Nano 33 BLE Sense [101]	ARM Cortex-M0 32-bit@64 MHz	52 µA/MHz	BLE	AI-enabled	Yes	€27
SparkFun Edge [102]	ARM Cortex-M4F 32-bit@48/96 MHz	6 µA/MHz	BLE 5	AI-enabled	Yes	€15
Adafruit EdgeBadge [103]	ATSAMD51J19A 32-bit@120 MHz	65 µA/MHz	BLE/WiFi	AI-enabled	Yes	€35
ESP32-DevKitC [104]	Xtensa dual-core 32-bit@160/240 MHz	2 mA/MHz	BLE/WiFi	AI-enabled	Yes	€10
ESPEYE-DevKit [105]	Xtensa dual-core 32-bit@160/240 MHz	2 mA/MHz	BLE/WiFi	AI-enabled	Yes	€50
STM32 Nucleo-144 [106]	ARM Cortex-M4 Nucleo-L4R5ZI 32-bit@160/120 MHz	43 µA/MHz	Ethernet	AI-enabled	No	€100

4.2. Edge Intelligence or Edge-AI

Typically, in cloud-centric architectures, IoT devices can transfer data to the cloud using an Internet gateway. In this architecture, the raw data produced by IoT devices are pushed to a centralized server without processing; however, since IoT devices are becoming more efficient and powerful, new possibilities arise at the network edge, enabling real-time intelligent processing with minimal latency. Edge Intelligence (EI) or Edge-AI are the common names given to this approach, and its performance is often expressed in terms of model accuracy and overall latency [107].

A common IoT device (also known as a "dumb" device) tends to generate large quantities of raw and low-quality data, which may have no operational relevance. In most cases, data are noisy, intermittent, or change slowly, being useless in specific periods. Moreover, the management and transmission of these useless data streams consume vital power and tend to be bandwidth-intensive. On the other hand, the inclusion of in-device/edge intelligence results in the reduction in the data dimension by turning data into relevant information, lowering power consumption, latency, and the overall bandwidth needs. Intelligence at the edge of the network enables the distribution of the computational cost among edge devices. In this computational approach, data can be classified and aggregated before its transmission up to the cloud. By using this approach, only information with historical value is archived, which can be later used for tuning prediction models and optimizing the cloud-based processing.

4.2.1. Model Inference Architectures

The three major Edge-AI computing paradigms are [108]:

(i) On-device computation: it relies on AI techniques (e.g., Deep Neural Networks (DNNs)) that are executed on the end device.
(ii) Edge-based computation: it computes on edge devices the information collected from end devices.
(iii) Joint computation: it allows for processing data on the cloud during training and inference stages.

Given the limited resources that are typically available in most IoT devices, bringing AI to the edge can be challenging. Reducing model inference time has been implemented successfully at the cost of decreasing the overall model inference accuracy. According to Merenda et al. [109], to effectively run an AI model (after the compression stage) in an embedded IoT device, the hardware selection must be carefully performed.

4.2.2. Edge-AI Levels

Besides the well-known Cloud Intelligence (CI), which consists in training and inferencing the DNN models fully in the cloud, EI, as described in [110], can be classified into the six levels depicted in Figure 5. The quantity of data sent up to the cloud tends to decrease as the level of EI increases, resulting in lower communications bandwidth and lower transmission delay; however, this comes at the cost of increased computational latency and energy consumption at the network's edge (including IoT nodes), implying that the EI level is application-dependent and must be carefully chosen based on several criteria: latency, energy efficiency, and privacy and communications bandwidth cost.

Inference and training are the two main computing stages in an NN. Depending on the Edge-AI level (as illustrated in Figure 5), the computational power is typically distributed between the IoT node or the edge layer, which requires increased computational power. In recent years, AI-specific hardware accelerators have enhanced high-performance inference computation at the edge of the network, namely in embedded and resource-constrained devices. For example, in [111], Karras et al. present an FPGA-based SoC architecture to accelerate the execution of ML algorithms at the edge. The system presents a high degree of flexibility and supports the dynamic deployment of ML algorithms, which demonstrate an efficient and competitive performance of the proposed hardware to accelerate AI-based inference at the edge. Another example is presented in [112] by Kim et al., where they propose a co-scheduling method to accelerate the convolution layer operations of CNN inferences at the edge by exploiting parallelism in the CNN output channels. The developed FPGA-based prototype presented a global performance improvement of up to 200%, and an energy reduction between 14.9% and 49.7%. Finally, in [113], the authors introduce NeuroPipe, a hardware management method that enables energy-efficient acceleration of DNNs on edge devices. The system incorporates a dedicated hardware accelerator for neural processing. The proposed method enables the embedded CPU to operate at lower frequencies and voltages, and to execute faster inferences for the same energy consumption. The provided results show a reduction in energy consumption of 11.4% for the same performance.

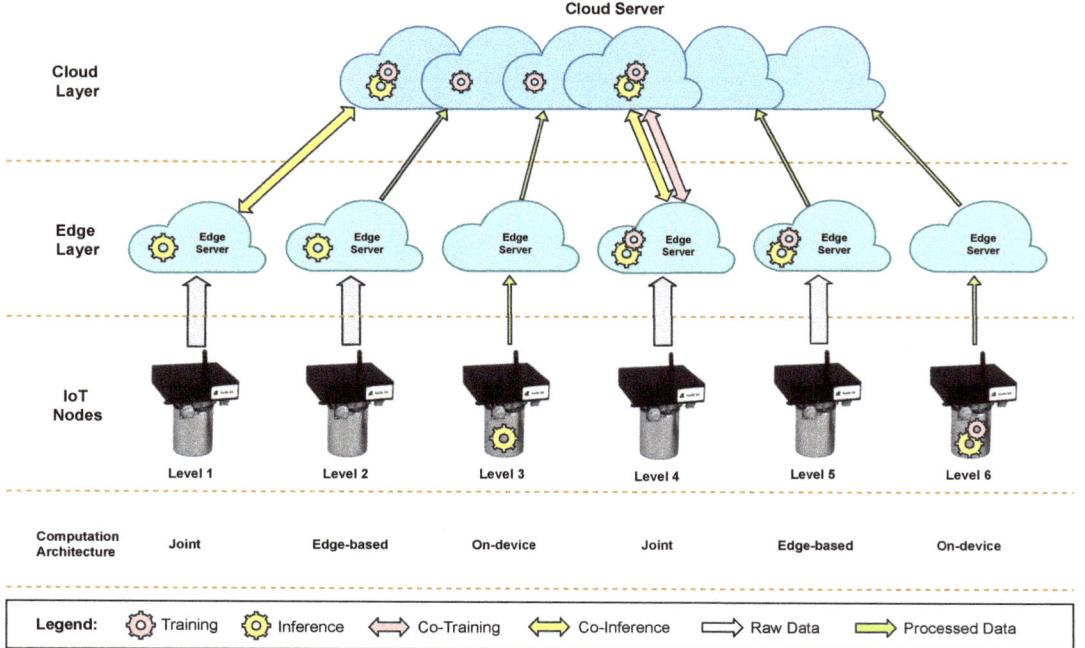

Figure 5. Edge-AI Levels and model inference computation architectures: on-device, edge-based, and joint.

4.2.3. Embedded ML

Conventional IoT devices are ubiquitous and low-cost, but natively resource-constrained, which limits their usage in ML tasks; however, data generated at the edge are increasingly being used to support applications that run ML models. Until now, edge ML has been predominantly focused on mobile inference, but recently several embedded ML solutions have been developed to operate in ultra-low-power devices, typically characterized by its hard resource constraints [97]. Recently, a new field of ML, known as Tiny ML, was put forward to enable inference at the edge endpoints. ML inference at the edge can optimize the overall computational resource needs, increases privacy within applications, and enhances system responsiveness. TinyML, which has been coined due to its ML inference power consumption of under a milliWatt, overcomes the power limitations of such devices, enabling low-power and low-cost distributed machine intelligence. TinyML is an open-source ML framework specifically designed for resource-constrained embedded devices. It is fully compatible with several low-cost, globally accessible hardware platforms and was designed to streamline the development of embedded ML applications [114].

TinyML technologies and applications target battery-operated devices, including hardware, algorithms, and software for on-device inference and data analytics at the edge. In [115], MLCommons, an open engineering consortium, presented a recent benchmark (MLPerf™ Tiny Inference v0.5). This inference benchmark suite targets ML use cases on embedded devices by measuring how rapidly a trained NN can process new data in ultra-low-power devices. Embedded ML is a new field in which AI-based sensor data analytics is carried out near to where the data are collected in real time. The benchmark presented in [115] focuses on a number of use cases that rely on tiny NNs (i.e., models lower than 100 kB) to analyze sensor data such as audio and video to provide intelligence at the edge of the network. The benchmark consists of four ML tasks that include the use of microphone and camera sensors in different embedded devices:

1. Visual Wake Words (**VWW**): classification task for binary images that detects the presence of a person. For instance, an application use case is in-home security monitoring.
2. Image Classification (**IC**): small image classification benchmark with 10 classes, with several use cases in smart video recognition applications.
3. Keyword Spotting (**KWS**): uses a neural network to detect keywords from a spectrogram, with several use cases in consumer end devices, such as virtual assistants.
4. Anomaly Detection (**AD**): uses a neural network to identify anomalies in machine operating sounds, and has several application cases in industrial manufacturing (e.g., predictive maintenance, asset tracking, and monitoring).

This benchmark aims to measure performance for ML in embedded systems, which operate at a microwatt level and include cameras, wearables, smart sensors, and other IoT devices that demand a certain level of intelligence. Thus, the objective of the benchmark is to measure the performance of such constrained systems in order to achieve higher efficiency over time. The results have been reported based on the embedded ML approach and its hardware and software. Table 3 compares the benchmark results for distinct embedded hardware when running a trained model by measuring the processing latency in milliseconds (i.e., how fast systems can process inputs to produce a valid result) and the respective consumed energy in µJ [116].

4.3. Edge-AI Computational Cost

Computation needs for AI are growing rapidly. Recent numbers show that large AI training runs are doubling every 3.5 month and, since 2012, the computational needs have increased by more than 300,000 times [117]. In recent years, a lot of effort has been put into increasing AI accuracy and, especially with DL, accuracy has increased at a steady pace. This increase in accuracy has been very important in making AI a reality in real-world applications; however, to run such high accuracy models, more and more computational resources need to be considered. In the short and medium term, AI will face major challenges that put its sustainability and ecological footprint into perspective. Due to the explosion of its use in several application domains, increased pressure on computational resources is already happening, not only to train but also to run these models, which are increasingly more accurate but also, computationally heavier.

Due to this novel and more sustainable practices regarding AI implementation and deployment are yet to come. In [118], Schwartz et al. introduced the concepts of Red and Green AI, as a way to clarify and distinguish the two major currents AI approaches.

Red AI is known for relying on large models and datasets, as its performance is typically evaluated through accuracy, which is usually obtained through the use of massive processing power. In this context, the relation between model performance and model complexity is known to be logarithmic (i.e., an exponentially bigger model is required for a linear improvement in performance [119]). Furthermore, the quantity of training data and the number of tuning experiments, present the same exponential growth [118]. In each of these cases, a small performance improvement comes at an increased computational cost.

Green AI, on the other hand, focuses on achieving results without increasing or, preferably, lowering computational costs. Unlike Red AI, which results in rapidly increasing computing costs and, as a result, a rising carbon footprint, Green AI has the opposite effect [118]. In Green AI, efficiency is usually prioritized over accuracy when evaluating performance. As a result, Green AI focuses on model efficiency, which includes the amount of effort necessary to create a given result using AI, the amount of work required to train a model, and, if appropriate, the total of all tuning experiments. Efficiency may be assessed using a variety of metrics, including carbon emissions, power consumption, real-time elapsed time, number of parameters, and so on.

Table 3. MLPerf™ Tiny Inference v0.5 benchmark results. Data from [115].

ID	Submitter	Device	Processor	Software	Task: #1 - VWW Data: Visual Wake Words Dataset Model: MobileNetV1 (0.25x) Accuracy: 80% (Top 1)		Task: #2 - IC Data: CIFAR-10 Model: ResNet-V1 Accuracy: 85% (Top 1)		Task: #3 - KS Data: Google Speech Commands Model: DSCNN Accuracy: 90% (Top 1)		Task: #4 - AD Data: ToyADMOS (ToyCar) Model: FC AutoEncoder Accuracy: 0.85 (AUC)	
					Latency (ms)	Energy (uJ)	Latency (ms)	Energy (uJ)	Latency (ms)	Energy (uJ)	Latency (ms)	Energy (uJ)
0.5-464	Harvard (Reference)	Nucleo-L4R5ZI	Arm Cortex M4 w/ FPU	Tensorflow Lite for Microcontrollers	603.14	24,320.84	704.23	29,207.01	181.92	7373.70	10.40	416.31
0.5-465	Peng Cheng Laboratory	PCL Scepu02	RV32IMAC with FPU	TensorFlowLite for Microcontrollers 2.3.1 (modified)	846.74	-	1239.16	-	325.63	-	13.65	-
0.5-466	Latent AI	RPi 4	Broadcom BCM2711	LEIP Framework	3.75	-	1.31	-	0.39	-	0.17	-
0.5-467	Latent AI	RPi 4	Broadcom BCM2711	LEIP Framework	2.60	-	1.07	-	0.42	-	0.19	-

4.4. Measuring Edge-AI Energy Consumption and Carbon Footprint

The overall cost of using AI can be obtained by considering the resources involved in all processing stages, which include energy consumption and CO_2 emissions.

4.4.1. Energy Consumption

In [120], Pinto et al. define energy consumption as an accumulation of power dissipation over time:

$$\text{Energy Consumption} = P \times t \tag{1}$$

Note that Energy Consumption is measured in joules and Power (P) is measured in watts. The relationship between these two quantities can be easily interpreted through an example: if a software program takes 5 s to execute and dissipates 5 watts, it consumes 25 joules of energy. In the case of software energy consumption, attention must be paid not only to the software under execution, but also to the hardware that executes the software, the environmental context of execution, and its duration.

4.4.2. CO_2 Emissions

In [121], Strubell et al. presented a study that focused on the estimation of the financial and environmental cost of training a variety of recently successful NN models. To estimate CO_2 emissions (CO_2e), they proposed a simple method based on the multiplication of the energy consumption with the average produced CO_2. After measuring the CO_2e for several models using different hardware, they concluded that the CO_2 required for training one model can range from 12 kg up to 284 t. Note that this CO_2e footprint is highly significant when compared with the world average CO_2 emissions per capita, whose estimate was 4.56 t in 2016 [122]. Moreover, they evaluated the cost of training these models in the cloud, which raised from USD 41 up to USD 3,201,722, respectively.

4.5. Measuring Edge-AI Performance

Although this article focuses on Edge-AI sustainability, there are other factors that should be considered during the evaluation of the performance of an Edge-AI system. Specifically, four main metrics are often used for the performance evaluation of AI algorithms [123]: accuracy, memory bandwidth, energy efficiency, and execution time.

4.5.1. Accuracy

Classification accuracy is the simplest performance metric and is commonly used with balanced datasets (i.e., the number of samples per class is balanced). Accuracy is defined as the number of correct predictions, divided by the total number of predictions, and is implemented by comparing the annotated ground truth data with the predicted results:

$$\text{Accuracy} = \frac{tp + tn}{tp + tn + fp + fn} \tag{2}$$

where tp represents the true positives, tn the true negatives, fp are the false positives, and fn the false negatives. Note that, if unbalanced data are considered (i.e., the number of samples per class is not balanced), a new accuracy metric, known as balanced accuracy, should be computed. The balanced accuracy is computed by normalizing tp and tn by the number of positive and negative samples, respectively, then perform their sum, and divide by two, as indicated in Equation (3):

$$\text{Balanced accuracy} = \frac{TP + TN}{2} \tag{3}$$

where TP represents the normalized true positives and TN the normalized true negatives; however, a fair performance evaluation between algorithms should not only rely on the accuracy, as Red AI tends to favor.

4.5.2. Memory Bandwidth

In [124], Jouppi et al. compare the performance of several processors used by Google cloud-based systems on inference tasks when running various types of NNs. The analysis uses a roofline model, where the performance of the algorithms is plotted based on the computational performance (operations per second) versus the operational intensity (number of operations per byte of data). Typically, in cloud-based architectures, the overall performance is limited by the memory bandwidth, and as the operational intensity tends to increase, the performance is limited by the computational capacity of the computer system architecture. Recent hardware architectures, notably SoC architectures, are focused on increasing the memory bandwidth to address the continuously growing demand of AI [98].

4.5.3. Energy Efficiency

A simple metric that can be used to measure the software energy efficiency is presented in [123] and is shown in Equation (4). In Edge-AI, the useful work performed can be defined as the number of model inferences. As a result, Energy Efficiency can be measured as the number of inferences per Joule.

$$\text{Energy Efficiency} = \frac{\text{Useful Work Performed}}{\text{Energy Consumption}} = \frac{\text{Number of Inferences}}{\text{Energy Consumption}} \quad (4)$$

4.5.4. Execution Time

This metric represents the execution time of a specific task in the ML process to obtain a valid result, which may include, model training or model inference [123], and are measured in seconds, being typically referred as the "training time" and "inference time", respectively.

5. Cross-Analysis of G-IoT and Edge-AI: Key Findings

Although Edge-AI G-IoT system deployment in real-world applications has already started, the research and development are still undergoing, and some issues compromise its wider acceptance, of which we highlight: trustworthiness (e.g., algorithm transparency, traceability, privacy, and data integrity); capacity (e.g., communications bandwidth and coverage, hardware constraints such as power and computational power, security in edge distributed architectures); heterogeneity (e.g., dealing with distinct data sources and formats as well as adapting with a variety of operational, technical, and human requirements); and scale (e.g., inadequate volume of publicly available data, high-quality data required to effectively simulate the physical world's complexity). In addition, the cross-analysis of the G-IoT and Edge-AI literature allows for obtaining the following key findings that can be useful for future developers and researchers:

- Communications between G-IoT nodes and Edge-AI devices are essential, so developers should consider the challenges related to the use of energy efficient transceivers and fast-response architectures. Thus, researchers need to contemplate aspects such as the use of low-power communications technologies (e.g., ZigBee, LoRa, UWB, and Wi-Fi Hallow), the management of the RF spectrum or the design of distributed AI training, learning algorithms, and architectures that achieve low-latency inference (either distributed or decentralized [107]).
- Although the most straightforward way to implement Edge-AI systems is to deploy the entire model on edge devices, which eliminates the need for any communications overhead, when the model size is large or the computational requirements are very high, this approach is unfeasible and it is necessary to include additional techniques that involve the cooperation among nodes to accomplish the different AI training and inference tasks (e.g., federated learning techniques [107]). Such techniques should minimize the network traffic load and communications overhead in resource-constrained devices.

- Edge-AI G-IoT systems should consider that the different nodes of the architecture (e.g., mist nodes, edge computing devices, and cloudlets) have heterogeneous capabilities in terms of communications, computation, storage, and power; therefore, the tasks to be performed should be distributed in a smart way among the available devices according to their capabilities.
- Besides heterogeneity, developers should take into account that G-IoT node hardware constrains the performance of the developed Edge-AI systems. Such hardware must be far more powerful than traditional IoT nodes and provide a suitable trade-off between performance and power consumption. In addition, such hardware should be customized to the selected Edge-AI G-IoT architecture and application.
- Currently, most G-IoT systems rely on traditional cloud computing architectures, which do not meet some of the needs of Edge-AI G-IoT applications in terms of high availability, low latency, high network bandwidth, and low power consumption. Moreover, current cloud-based approaches may be compromised by cyberattacks; therefore, new architectures such as the ones based on fog, mist, and edge computing should be considered to increase the robustness against cyberattacks and to allow for choosing which AI tasks to offload to the cloud, if any, while reducing network resource consumption.
- Green power sources and energy-harvesting capabilities for Edge-AI G-IoT systems still need to be studied further. Although batteries are typically used to meet power requirements, future developers should analyze the use of renewable power sources or energy-harvesting mechanisms to minimize energy consumption. In addition, the use of decentralized green smart grids for Edge-AI G-IoT architectures can be considered.
- High-security mechanisms are usually not efficient in terms of energy consumption, so it is important to analyze their performance and carry out practical energy measurements for the developed Edge-AI G-IoT systems.
- Developers should consider using energy efficiency metrics for the developed AI solutions. For instance, in [123] the authors propose four key indicators for an objective assessment of AI models (i.e., accuracy, memory bandwidth, energy efficiency, and execution time). The trade-off between such metrics will depend on the environment where the model will be employed (e.g., "increased safety" scenarios impose low execution time).

6. Application Case: Developing a Smart Workshop

6.1. Workshop Characterization and Edge-AI System Main Goals

To illustrate the concepts described in the previous sections, it was selected a practical Industry 5.0 use case in a real-world scenario. Specifically, the selected Industry 5.0 scenario consists in an industrial workshop that looks for improving operator safety through IIoT sensors/actuators and Edge-AI. The chosen scenario is based on the previous work of the authors [125–127], which participated in a Joint Research Unit together with one of the largest shipbuilders in the world (Navantia). The specific scenario is the pipe workshop that such a company owns in its shipyard in Ferrol (Spain). The workshop manufactures pipes as follows:

1. First, raw pipes are stored in the Reception Area (shown in Figure 6a). Thus, they are collected by the workers as they are needed. If the pipes are delivered with dirt or grease, then, before being stored in the Reception Area, they are cleaned in the Cleaning Area (in Figure 6b). Operators need to keep away from the Cleaning Area unless authorized because of the presence of dangerous chemical products (e.g., chloridric acid, caustic soda) and water that is pressurized and hot.

(a) Reception Area.   (b) Cleaning Area.

Figure 6. Relevant areas of the workshop.

2. Second, every pipe is first cut in the Cutting Area according to the required dimensions. Really powerful (and dangerous) mechanical and plasma saws (shown in Figure 7a,b) are used in the Cutting Area. It is important to note that pipes are moved from the Reception Area to the Cutting Area (or from one area to any other area) by stacking them on pallets, which are carried by big gantries installed in the ceiling of the workshop (several pallets can be seen on the foreground of Figure 7b).

(a) Mechanical saws in the Cutting Area.   (b) Plasma saw in the Cutting Area.

Figure 7. Saws of the Cutting Area.

3. Third, pipes are bent in the Bending Area. There are three large bending machines in such an area. Operators need to always keep a safe distance and safety glasses when operating a bending machine.
4. Fourth, pipes are cleaned and moved to the Manufacturing Area, where accessories are added. For instance, operators may need to weld a valve to a pipe. Welding requires taking specific safety measures and only the authorized operators can access the welding area when someone is working.
5. Finally, pipes are stacked into pallets, packed, and then stored in two different areas of the workshop (shown in Figure 8a,b).

(a) Outbound storage area.   (b) Another outbound storage area.

Figure 8. Main storage areas.

Figure 9 depicts the main areas of the workshop floor map and shows the position of the IIoT cameras that monitor the presence of the workers. In addition, the dashed semicircles indicate the estimation of the field of view of such cameras. Specifically, Figure 9 shows 18 distinct areas of the factory floor that are equipped with cameras for continuous monitoring (24 h a day, 7 days a week) of a complete manufacturing process. Note that, in this specific application case, images should be neither transmitted nor recorded in the cloud, not only due to bandwidth and connectivity limitations, but also due to the impositions of the General Regulation on Data Protection (GDPR) in force.

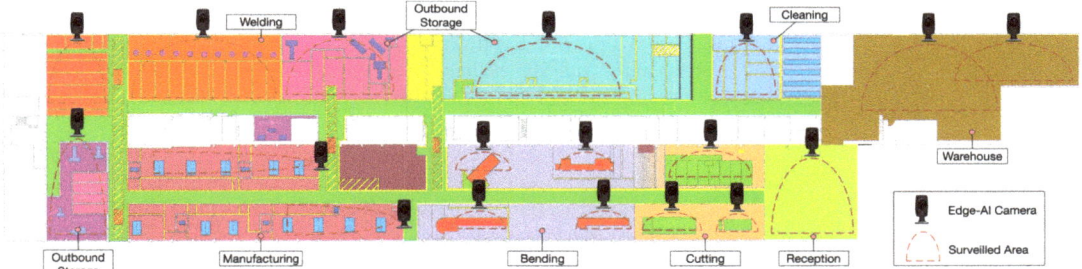

Figure 9. Floor map of the smart workshop.

The objective of the proposed solution is to harness "visual wake words" in order to detect the presence of the workers with the help of cameras and then lock or unlock the deployed industrial devices and machinery, and automate the available security mechanisms. For instance, industrial robot arms or cutting machines can harm a worker during their operation when safety distance is not respected. Thus, the system takes advantage of the proposed mist AI-enabled architecture (described next in Section 6.2) to achieve two specific application goals:

- **Increased Safety**: automatically detect humans in the proximity of machinery that is operating. After detection, a sound warning should be physically generated in the surrounding zone. After triggering the sound warning, if the detection persists and the estimated distance between the operating machine and the human does not increase, a shutdown command should be sent to the operating machine.
- **Operation Tracking**: automatically detect and track human operators and moving machinery. The tracking information is then used for the continuous improvement of manufacturing processes.

Besides the mentioned goals, it is important to note that the proposed system impacts different circular economy aspects:

- Smarter use of resources: the detection of the presence of operators allows for determining when machinery should be working and when it should be shut down.
- Reduction of total annual greenhouse gas emissions: the smarter use of resources decreases energy consumption and, as a consequence, carbon footprint.
- Enhanced process safety: human proximity detection allows for protecting against possible incidents or accidents with the deployed industrial devices and machinery.

6.2. System Architecture

The architecture proposed for the application case is shown in Figure 10. As it can be observed, there are two main layers:

- Mist Computing Layer: it is composed of AI-enabled IIoT nodes that run AI algorithms locally. Thus, after the AI training stage, nodes avoid exchanging image data through the network with edge computing devices or with the cloud, benefiting from:
    - Lower latency. Since most of the processing is carried out locally, the mist computing device can respond faster.

- Communications problems in complex environments can be decreased. Local processing avoids continuous communications with local edge devices or remote clouds. Thus, potential communications problems are reduced, which is really important in industrial scenarios that require wireless communications [126].
- Fewer privacy issues. Camera images do not need to be sent to other devices through the network, so potential attacks to such devices or man-in-the-middle attacks can be prevented and thus avoid image leakages.
- Improved local communications with other nodes. Mist devices can implement additional logic to communicate directly with other mist devices and machines, so responses and data exchanges are faster, and less traffic is generated due to not needing to make use of intermediate devices such as edge computing servers or the cloud.

Despite the benefits of using mist AI-enabled nodes, it is important to note that IIoT nodes, since they integrate cameras/sensors and the control hardware, are more expensive and complex (i.e., there are more hardware parts that can fail).

- Cloud: it behaves like in the edge computing based architecture. As a consequence, it deals with the requests of the mist devices that cannot be handled locally.

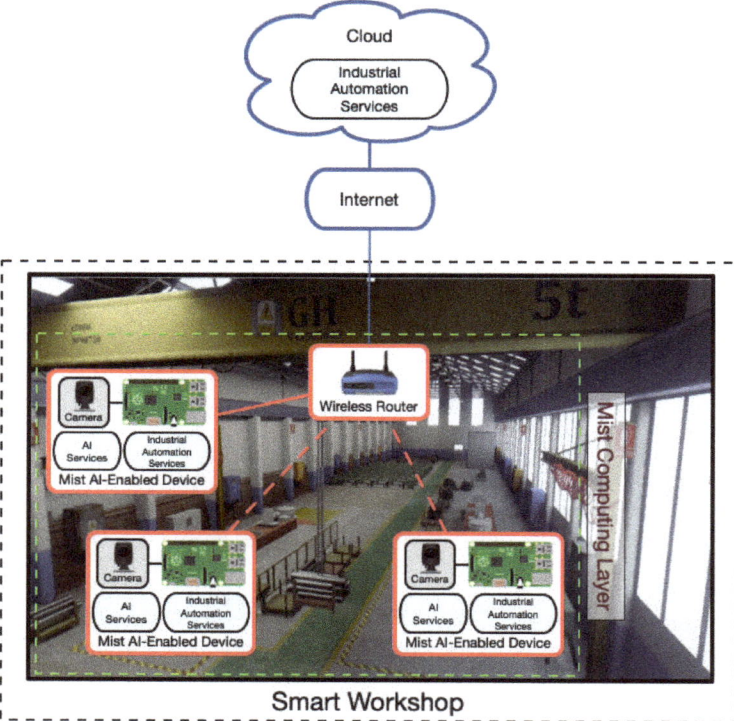

Figure 10. Mist-computing-based communications architecture.

6.3. Energy Consumption of the Mist AI-Enabled Model

In this application case, latency is a critical factor, and a low fault-tolerance policy needs to be implemented. To achieve the "Increase Safety" goal, the use of object detection models with low inference latencies is mandatory. In this case, the human movement dynamics are typically low, since, running on the factory floor is typically not allowed. Moreover, with respect to the "Operations Tracking" goal, the inference latency is not critical, since it does not affect the obtained results, due to the deterministic nature of the inference latency.

To estimate the energy cost of the overall system, it was considered the data presented in Table 3 for an STM32 Nucleo-L4R5ZI processor running TensorFlow Lite with a Mobinet-V1 model (Task #1-Visual Wake Words) to simulate the "Increase Safety" task and a Resnet-V1 model (Task #2-Image Classification) for simulating the "Operations Tracking" task. The former is a classification task for binary images that detect the presence of a person with an inference latency of 603.14 ms and energy consumption of 24,320.84 µJ per inference (1 joule = 2.77777778 × 10^{-7} kWh). The latter is an image classification benchmark with 10 classes for smart video recognition applications with an inference latency of 704.23 ms and energy consumption of 29,207.84 µJ per inference. At this stage, it is important to notice that only inference is being considered, since no information is available regarding the training stage, namely the consumed energy.

First, the number of inferences can be estimated for a year and one camera, and then the overall power consumption can be extrapolated to all cameras, based on the previous assumptions:

$$N_{VWW} = \frac{365 \times 24 \times 3600 \text{ s}}{603.14 \text{ ms}} = 52{,}286{,}368 \text{ inferences/year} \quad (5)$$

$$E_{VWW} = N_{VWW} \times 24{,}320.84 \text{ µJ} = 12{,}716{,}483.9 \text{ J} = 0.353 \text{ kWh/device} \quad (6)$$

$$N_{IC} = \frac{365 \times 24 \times 3600 \text{ s}}{704.23 \text{ ms}} = 44{,}780{,}824 \text{ inferences/year} \quad (7)$$

$$E_{IC} = N_{IC} \times 29{,}207.84 \text{ µJ} = 1{,}307{,}951.2 \text{ J} = 0.363 \text{ kWh/device} \quad (8)$$

where N_x represents the number of inferences per year for model x (VWW or IC) and E_x represents the total equivalent energy consumed in one year per device. In this particular case, the energy refers only to the one consumed by the inference task. Given that, in this study, we are only focused on the additional power consumption of the inference stage, the power consumed by all functional hardware blocks has not been included.

Equation (6) indicates that each camera, when running the VWW model, consumes approximately 0.353 kWh in a year. When running the IC model for the same period (Equation (8)), each camera consumes approximately 0.363 kWh; therefore, by extrapolating for the 18 cameras, we achieve a total consumption (in one year) of 6.354 kWh and 6.534 kWh, for the VWW and IC models, respectively. This power consumption is on the Green-AI magnitude scale, and the yearly inference cost of all the 18 cameras can easily be maintained by a conventional renewable energy source, such as a photovoltaic panel.

6.4. Carbon Footprint

Carbon footprint can be estimated by using the formula in Equation (9) [128]:

$$CO_2e(g) = E_x(KWh) \times I_N(g/KWh) \quad (9)$$

where CO_2e is the number of grams of emitted CO_2, E_x (x equal to VWW or IC) is the consumed energy (in KWh) and I_N is the carbon intensity (in grams of emitted CO_2 per KWh). This latter parameter can be obtained through the data published publicly by many countries or by organizations such as the European Union, but it is easier to obtain it from Electricity Maps [129], an open-source project that collects such data automatically and plots them through a user-friendly interface. Such a website also indicates the energy sources used by each country (an example of such sources for France, Portugal, Spain, California, and the province of Alberta is shown in Figure 11). The data were obtained for 25 July 2021 and, as it can be observed, energy sources differ significantly from one country to another:

- France (data source: Réseau de Transport d'Electricité (RTE)): it has almost got rid of CO_2-intensive energy sources thanks to generating most of its electricity through nuclear power. Nonetheless, on 25 July 2021, when the data in Figure 11 were collected, only roughly 31% of France's energy came from renewable sources.
- Portugal (data source: European Network of Transmission System Operators for Electricity (ENTSOE)): its most relevant energy source is natural gas, but, when the data were gathered, approximately 43% of its energy came from renewable sources and none from nuclear power.
- Spain (data source: ENTSOE): like Portugal, it has a dependency on natural gas, but, thanks to a powerful solar energy sector, it generates roughly 53% of its energy from renewable sources. In addition, almost 24% of the Spanish energy comes from nuclear power, so a total of 77% of the energy is generated from low-carbon technologies.
- California (data source: California Independent System Operator (CAISO)): in spite of being a state of the U.S., it was selected due to its crucial role in IT and cloud-based services. Nearly 42% of its energy on 25 July 2021 was generated through low-carbon technologies, but almost 58% came from natural gas.
- Alberta (data source: Alberta Electric System Operator (AESO)): it was included as an example of a rich area with a key role in the oil and natural gas production in North America. As it can be observed in Figure 11, most of its energy (almost 84%) is generated by natural gas and coal, which results in the generation of a large amount of CO_2 emissions.

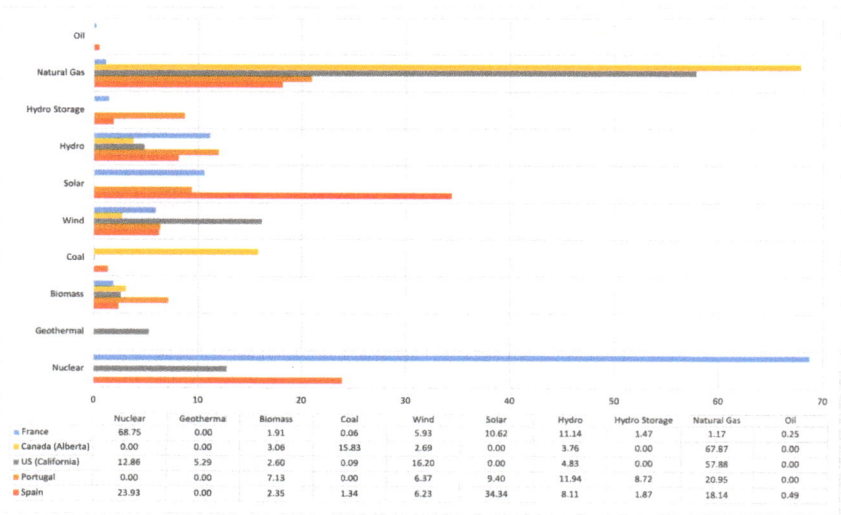

Figure 11. Energy sources for France, Portugal, Spain, California, and Alberta (25 July 2021).

Figure 12 shows the estimated CO_2 emissions for the energy consumption estimated in the previous section. As it can be easily guessed, emissions increase with the number of deployed mist AI-enabled devices; however, such growth changes dramatically from one country to another depending on the energy source: while near-zero emission countries like France are barely impacted by the increase in the number of deployed devices, a province like Alberta emits more than 17 times more CO_2 for 1000 deployed devices.

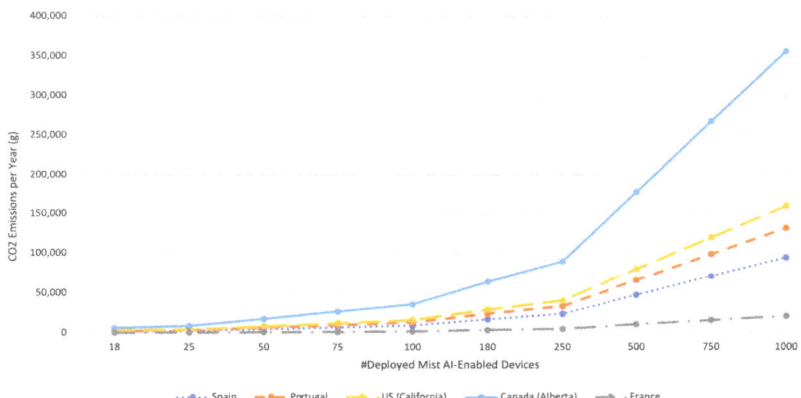

Figure 12. Estimated CO_2 emissions for different number of deployed devices for different countries.

It is also possible to obtain the monetary cost of running the mist AI-enabled devices (as an example, the average prices for April 2021 for each territory were considered), which is depicted in Figure 13. As it can be seen in the figure, the cost of running the system in Alberta would be cheaper but will result in more CO_2 emissions. In contrast, the countries with the largest shares of renewable energy sources (Spain and Portugal) are the ones with the most expensive electricity. Nonetheless, please note that such a link between the use of renewable energies and cost is impacted by other external factors (e.g., taxes, environmental policy, and energy trading).

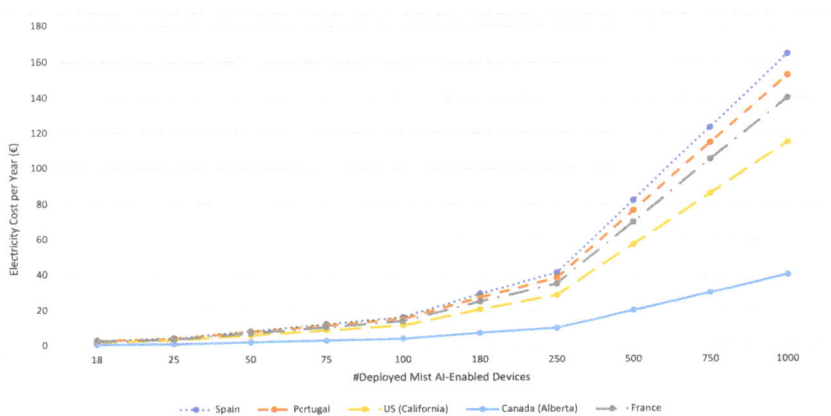

Figure 13. Electricity cost for different number of deployed devices and for different countries.

7. Future Challenges of Edge-AI G-IoT Systems

Despite the promising foreseen future of Edge-AI G-IoT systems, it is possible to highlight some open challenges that must be faced by future researchers:

- Additional mechanisms are needed to offer protection against network, physical, software, and encryption attacks. In addition, it is critical to have protection against adversarial attacks during on-device learning [130].
- Future communications networks. 5G/6G are intended to deliver low-latency communications and large capacity; therefore, moving the processing tasks to the network edge will demand higher edge computing power, which puts G-IoT and Edge-AI convergence as fundamental technology enablers for the next 6G mobile infrastructure.

Moreover, the rapid proliferation of new products and devices and their native connectivity (at a global level) will force the convergence of not only G-IoT and Edge-AI, but also 5G/6G communication technologies, the latter being a fundamental prerequisite for future deployments. Indeed, future communications services should also provide better dependability and increased flexibility to effectively cope with a continuously changing environment.

- Edge-AI G-IoT Infrastructure. The IoT market is currently fragmented, so it is necessary to provide a comprehensive standardized framework that can handle all the requirements of Edge-AI G-IoT systems.
- Decentralized storage. Cloud architectures store data in remote data centers and digital infrastructures that require substantial levels of energy. Luckily, recent architectures for Edge-AI G-IoT systems are able to decentralize data storage to prevent cyberattacks and avoid high operating costs; however, to achieve energy optimizations for such decentralized architectures, sophisticated P2P protocols are needed.
- G-IoT supply chain visibility and transparency. To increase the adoption of the DCE and limit the environmental impact of a huge number of connected devices, further integration of value chains and digital enabling technologies (e.g., functional electronics, UAVs, blockchain) is needed. End-to-end trustworthy G-IoT supply chains that produce, utilize, and recycle efficiently are required.
- Development of Edge-AI G-IoT applications for Industry 5.0. The applications to be developed should be first analyzed in terms of its critical requirements (e.g., latency, fault tolerance) together with the appropriate communications architecture, while considering its alignment with social fairness, sustainability, and environmental impact. In addition, hardware should be customized to the selected Edge-AI G-IoT architecture and the specific application.
- Complete energy consumption assessment. For the sake of fairness, researchers should consider the energy consumption of all the components and subsystems involved in an Edge-AI G-IoT system (e.g., communications hardware, remote processing, communications protocols, communications infrastructure, G-IoT nodes, and data storage), which may be difficult in some practical scenarios and when using global networks.
- Digital circular life cycle of Edge-AI G-IoT systems. In order to assess the impact of circular economy based applications, all the different stages of the digital circular life cycle (i.e., design, development, prototyping, testing, manufacturing, distribution, operation, maintenance, and recycling stages) should be contemplated.
- CO_2 emission minimization for large-scale deployments. Future developers will need to consider that CO_2 emissions increase with the number of deployed Edge-AI IoT devices. In addition, such growth changes dramatically from one country to another depending on the available energy source.
- Corporate governance, corporate strategy, and culture. Organization willingness to explore new business strategies and long-term investments will be critical in the adoption of Edge-AI G-IoT systems, as a collaborative approach is required to involve all the stakeholders and establish new ways for creating value while reducing the carbon footprint. New business models will emerge (e.g., Edge-AI as a service, such as NVIDIA Clara [131]).

8. Conclusions

This article reviewed the essential concepts related to the development of Edge-AI G-IoT systems and their carbon footprint. In particular, the most relevant Edge-AI G-IoT communications architectures were analyzed together with their main subsystems. In addition, the most recent trends on the convergence of AI and edge computing were analyzed and a cross-analysis on the fusion of Edge-AI and G-IoT was provided. Furthermore, an Industry 5.0 application case was described and evaluated in order to illustrate the theoretical concepts described throughout the article. The obtained results show how CO_2

emissions increase depending on the number of deployed Edge-AI G-IoT devices and on how greener is the energy generated by a country. Finally, the main open challenges for the development of the next generation of Edge-AI G-IoT systems were enumerated to guide future researchers.

Author Contributions: Conceptualization, P.F.-L., S.I.L., and T.M.F.-C.; methodology, P.F.-L., S.I.L., and T.M.F.-C.; investigation, P.F.-L., S.I.L., and T.M.F.-C.; writing—original draft preparation, P.F.-L., S.I.L., and T.M.F.-C.; writing—review and editing, P.F.-L., S.I.L., and T.M.F.-C.; supervision, T.M.F.-C.; project administration, T.M.F.-C.; funding acquisition, T.M.F.-C. All authors read and agreed to the published version of the manuscript.

Funding: This work has been supported through funds ED431G 2019/01 provided by Centro de Investigación de Galicia "CITIC" for a three-month research stay in Instituto Técnico de Viana do Castelo between 15 June and 15 September 2021. This work has also been funded by the Xunta de Galicia (by grant ED431C 2020/15), the Agencia Estatal de Investigación of Spain (by grants RED2018-102668-T and PID2019-104958RB-C42) and ERDF funds of the EU (FEDER Galicia 2014–2020 & AEI/FEDER Programs, UE).

Institutional Review Board Statement: Not applicable.

Informed Consent Statement: Not applicable.

Data Availability Statement: Not applicable.

Acknowledgments: The authors would like to thank CITIC for its support for the research stay that led to this article. CITIC, as Research Center accredited by Galician University System, is funded by "Consellería de Cultura, Educación e Universidades from Xunta de Galicia", supported in an 80% through ERDF Funds, ERDF Operational Programme Galicia 2014-2020, and the remaining 20% by "Secretaría Xeral de Universidades" (Grant ED431G 2019/01).

Conflicts of Interest: The authors declare no conflict of interest.

References

1. UN General Assembly, Transforming Our World: The 2030 Agenda for Sustainable Development, 21 October 2015, A/RES/70/1. Available online: https://www.refworld.org/docid/57b6e3e44.html (accessed on 28 July 2021).
2. World Economic Forum, Internet of Things Guidelines for Sustainability, Future of Digital Economy and Society System Initiative, 2018. Available online: http://www3.weforum.org/docs/IoTGuidelinesforSustainability.pdf (accessed on 28 July 2021).
3. Miranda, J.; Mäkitalo, N.; Garcia-Alonso, J.; Berrocal, J.; Mikkonen, T.; Canal, C.; Murillo, J. M. From the Internet of Things to the Internet of People. *IEEE Internet Comput.* **2015**, *19*, 40–47. [CrossRef]
4. Miraz, M. H.; Ali, M.;Excell, P. S.; Picking, R. A review on Internet of Things (IoT), Internet of Everything (IoE) and Internet of Nano Things (IoNT). In Proceedings of the 2015 Internet Technologies and Applications (ITA), Wrexham, UK, 8–11 September 2015; pp. 219–224.
5. Arshad, R.; Zahoor, S.; Shah, M.A.; Wahid, A.; Yu, H. Green IoT: An Investigation on Energy Saving Practices for 2020 and Beyond. *IEEE Access* **2017**, *5*, 15667–15681. [CrossRef]
6. Albreem, M.A.; Sheikh, A.M.; Alsharif, M.H.; Jusoh, M.; Mohd Yasin, M.N. Green Internet of Things (GIoT): Applications, Practices, Awareness, and Challenges. *IEEE Access* **2021**, *9*, 38833–38858. [CrossRef]
7. Zhu, C.; Leung, V.C.M.; Shu, L.; Ngai, E.C.-H. Green Internet of Things for Smart World. *IEEE Access* **2015**, *3*, 2151–2162. [CrossRef]
8. Mulay, A. Sustaining Moore's Law: Uncertainty Leading to a Certainty of IoT Revolution. *Morgan Claypool* **2015**, *1*, 1–109 [CrossRef]
9. Communication from the Commission to the European Parliament, the European Council, the Council, the European Economic and Social Committee and the Committee of the Regions. The European Green Deal. Bussels. 11 December 201 . Available online: https://ec.europa.eu/info/sites/info/files/european-green-deal-communication_en.pdf (accessed on 28 July 2021).
10. United Nation's 2030 Agenda and Sustainable Development Goals. Available online: https://sustainabledevelopment.un.org/ (accessed on 28 July 2021).
11. Askoxylakis, I. A Framework for Pairing Circular Economy and the Internet of Things. In Proceedings of the 2018 IEEE International Conference on Communications (ICC), Kansas City, MO, USA, 20–24 May 2018; pp. 1–6.
12. European Commission. Circular Economy Action Plan. Available online: https://ec.europa.eu/environment/strategy/circular-economy-action-plan_es (accessed on 28 July 2021).
13. World Economic Forum. Circular Economy and Material Value Chains. Available online: https://www.weforum.org/projects/circular-economy (accessed on 28 July 2021).

14. Explore the Circularity Gap Report 2021. Available online: https://www.circularity-gap.world (accessed on 28 July 2021).
15. Digital Circular Economy: A Cornerstone of a Sustainable European Industry Transformation. White Paper-ECERA European Circular Economy Research Alliance. 20 October 2020. Available online: https://www.era-min.eu/sites/default/files/publications/201023_ecera_white_paper_on_digital_circular_economy.pdf (accessed on 28 July 2021).
16. PACE (Platform for Accelerating the Circular Economy) and World Economic Forum. A New Cicular Vision for Electronics. Time for a Global Reboot. January 2019. Available online: http://www3.weforum.org/docs/WEF_A_New_Circular_Vision_for_Electronics.pdf (accessed on 28 July 2021).
17. Fernández-Caramés, T.M.; Fraga-Lamas, P. Design of a Fog Computing, Blockchain and IoT-Based Continuous Glucose Monitoring System for Crowdsourcing mHealth. *Proceedings* **2019**, *4*, 37. [CrossRef]
18. De Donno, M.; Tange, K.; Dragoni, N. Foundations and Evolution of Modern Computing Paradigms: Cloud, IoT, Edge, and Fog. *IEEE Access* **2019**, *7*, 150936–150948. [CrossRef]
19. Fraga-Lamas, P.; Ramos, L.; Mondéjar-Guerra, V.; Fernández-Caramés, T.M. A Review on IoT Deep Learning UAV Systems for Autonomous Obstacle Detection and Collision Avoidance. *Remote Sens.* **2019**, *11*, 2144. [CrossRef]
20. Alsamhi, S.H.; Afghah, F.; Sahal, R.; Hawbani, A.; Al-qaness, A.A.; Lee, B.; Guizani, M. Green internet of things using UAVs in B5G networks: A review of applications and strategies. *Ad Hoc Netw.* **2021**, *117*, 102505. [CrossRef]
21. Kibria, M.G.; Nguyen, K.; Villardi, G.P.; Zhao, O.; Ishizu, K.; Kojima, F. Big Data Analytics, Machine Learning, and Artificial Intelligence in Next-Generation Wireless Networks. *IEEE Access* **2018**, *6*, 32328–32338. [CrossRef]
22. European Commission, Industry 5.0: Towards a Sustainable, Human-centric and Resilient European Industry, January 2021. Available online: https://ec.europa.eu/info/news/industry-50-towards-more-sustainable-resilient-and-human-centric-industry-2021-jan-07_en (accessed on 28 July 2021).
23. Nahavandi, S. Industry 5.0—A Human-Centric Solution. *Sustainability* **2019**, *11*, 16. [CrossRef]
24. Paschek, D.; Mocan, A.; Draghici, A. Industry 5.0—The expected impact of next industrial revolution. In Proceedings of the MakeLearn & TIIM Conference, Piran, Slovenia, 15–17 May 2019; pp. 1–8.
25. Council for Science, Technology and Innovation, Government of Japan, Report on The 5th Science and Technology Basic Plan. 18 December 2015. Available online: https://www8.cao.go.jp/cstp/kihonkeikaku/5basicplan_en.pdf (accessed on 28 July 2021).
26. Keidanren (Japan Business Federation), Society 5.0: Co-Creating the Future. November 2018. Available online: https://www.keidanren.or.jp/en/policy/2018/095.html (accessed on 28 July 2021).
27. Blanco-Novoa, O.; Fernández-Caramés, T.M.; Fraga-Lamas, P.; Castedo, L. An Electricity-Price Aware Open-Source Smart Socket for the Internet of Energy. *Sensors* **2017**, *17*, 643. [CrossRef]
28. Suárez-Albela, M.; Fraga-Lamas, P.; Fernández-Caramés, T.M.; Dapena, A.; González-López, M. Home Automation System Based on Intelligent Transducer Enablers. *Sensors* **2016**, *16*, 1595. [CrossRef]
29. Pérez-Expósito, J.M.; Fernández-Caramés, T.M.; Fraga-Lamas, P.; Castedo, L. VineSens: An Eco-Smart Decision Support Viticulture System. *Sensors* **2017**, *3*, 465. [CrossRef]
30. Wang, H.; Osen, O.L.; Li, G.; Li, W.; Dai, H.-N.; Zeng, W. Big data and industrial Internet of Things for the maritime industry in Northwestern Norway. In Proceedings of the TENCON, Macao, China, 1–4 November 2015.
31. Shu, L.; Mukherjee, M.; Pecht, M.; Crespi, N.; Han, S.N. Challenges and Research Issues of Data Management in IoT for Large-Scale Petrochemical Plants. *IEEE Syst. J.* **2017**, *99*, 1–15. [CrossRef]
32. Fraga-Lamas, P.; Fernández-Caramés, T.M.; Suárez-Albela, M.; Castedo, L.; González-López, M. A Review on Internet of Things for Defense and Public Safety. *Sensors* **2016**, *16*, 1644. [CrossRef] [PubMed]
33. Kshetri, N. Can Blockchain Strengthen the Internet of Things? *IT Prof.* **2017**, *19*, 68–72. [CrossRef]
34. Gartner. Forecast: The Internet of Things, Worldwide, 2013. Available online: https://www.gartner.com/en/documents/2625419/forecast-the-internet-of-things-worldwide-2013 (accessed on 28 July 2021).
35. Bonomi, F.; Milito, R.; Zhu, J.; Addepalli, S. Fog Computing and its Role in the Internet of Things. In Proceedings of the First Edition of the MCC Workshop on Mobile Cloud Computing, Helsinki, Finlad, 17 August 2012; pp. 13–16.
36. Dolui, K.; Datta, S.K. Comparison of edge computing implementations: Fog computing, cloudlet and mobile edge computing. In Proceedings of the Global Internet of Things Summit (GIoTS), Geneva, Switzerland, 6–9 June 2017.
37. Suárez-Albela, M.; Fernández-Caramés, T.M.; Fraga-Lamas, P.; Castedo, L. A Practical Evaluation of a High-Security Energy-Efficient Gateway for IoT Fog Computing Applications. *Sensors* **2017**, *17*, 1978. [CrossRef] [PubMed]
38. Fraga-Lamas, P.; Fernández-Caramés, T.M.; Blanco-Novoa, Ó.; Vilar-Montesinos, M. A Review on Industrial Augmented Reality Systems for the Industry 4.0 Shipyard. *IEEE Access* **2018**, *6*, 13358–13375. [CrossRef]
39. Schuetz, S.; Venkatesh, V. The Rise of Human Machines: How Cognitive Computing Systems Challenge Assumptions of User-System Interaction. *J. Assoc. Inf. Syst.* **2020**, *21*, 460–482.
40. Preden, J.S.; Tammemäe, K.; Jantsch, A.; Leier, M.; Riid, A.; Calis, E. The Benefits of Self-Awareness and Attention in Fog and Mist Computing. *Computer* **2015**, *48*, 37–45. [CrossRef]
41. Fernández-Caramés, T.M.; Fraga-Lamas, P.; Suárez-Albela, M.; Vilar-Montesinos, M. A Fog Computing and Cloudlet Based Augmented Reality System for the Industry 4.0 Shipyard. *Sensors* **2018**, *18*, 1798. [CrossRef] [PubMed]
42. Suárez-Albela, M.; Fraga-Lamas, P.; Fernández-Caramés, T.M. A Practical Evaluation on RSA and ECC-Based Cipher Suites for IoT High-Security Energy-Efficient Fog and Mist Computing Devices. *Sensors* **2018**, *18*, 3868. [CrossRef]

43. Markakis, E.K.; Karras, K.; Zotos, N.; Sideris, A.; Moysiadis, T.; Corsaro, A.; Alexiou, G.; Skianis, C.; Mastorakis, G.; Mavromoustakis, C.X.; et al. EXEGESIS: Extreme Edge Resource Harvesting for a Virtualized Fog Environment. *IEEE Commun. Mag.* **2017**, *55*, 173–179. [CrossRef]
44. Yeow, K.; Gani, A.; Ahmad, R.W.; Rodrigues, J.J.P.C.; Ko, K. Decentralized Consensus for Edge-Centric Internet of Things: A Review, Taxonomy, and Research Issues. *IEEE Access* **2018**, *6*, 1513–1524. [CrossRef]
45. Radner, H.; Stange, J.; Büttner, L.; Czarske, J. Field-Programmable System-on-Chip-Based Control System for Real-Time Distortion Correction in Optical Imaging. *IEEE Trans. Ind. Electron.* **2021**, *68*, 3370–3379. [CrossRef]
46. De Mil, P.; Jooris, B.; Tytgat, L.; Catteeuw, R.; Moerman, I.; Demeester, P.; Kamerman, A. Design and implementation of a generic energy- harvesting framework applied to the evaluation of a large-scale electronic shelf-labeling wireless sensor network. *J. Wireless Com. Netw.* **2010**, *2020*, 343690. [CrossRef]
47. Ercan, A.Ö.; Sunay, M.O.; Akyildiz, I.F. RF Energy Harvesting and Transfer for Spectrum Sharing Cellular IoT Communications in 5G Systems. *IEEE Commun. Mag.* **2018**, *17*, 1680–1694. [CrossRef]
48. Ejaz, W.; Naeem, M.; Shahid, A.; Anpalagan, A.; Jo, M. Efficient Energy Management for the Internet of Things in Smart Cities. *IEEE Commun. Mag.* **2017**, *55*, 84–91. [CrossRef]
49. Lazaro, A.; Villarino, R.; Girbau, D. A Survey of NFC Sensors Based on Energy Harvesting for IoT Applications. *Sensors* **2018**, *18*, 3746. [CrossRef]
50. Coksun, V.; Ok, K.; Ozdenizci, B. *Near Field Communications: From Theory to Practice*, 1st ed.; Wiley: Chichester, UK, 2012.
51. Fraga-Lamas, P.; Fernández-Caramés, T.M. Reverse engineering the communications protocol of an RFID public transportation card. In Proceedings of the IEEE International Conference RFID, Phoenix, AZ, USA, 9–11 May 2017; pp. 30–35.
52. Murofushi, R.H.; Tavares, J.J.P.Z.S. Towards fourth industrial revolution impact: Smart product based on RFID technology. *IEEE Instrum. Meas. Mag.* **2017**, *20*, 51–56. [CrossRef]
53. Fernández-Caramés, T.M. An Intelligent Power Outlet System for the Smart Home of the Internet-of-Things. *Int. J. Distrib. Sens. Netw.* **2015**, *11*, 214805. [CrossRef]
54. Held, I.; Chen, A. Channel Estimation and Equalization Algorithms for Long Range Bluetooth Signal Reception. In Proceedings of the IEEE Vehicular Technology Conference, Taipei, Taiwan, 16–19 May 2010.
55. Hernández-Rojas, D.L.; Fernández-Caramés, T.M.; Fraga-Lamas, P.; Escudero, C.J. Design and Practical Evaluation of a Family of Lightweight Protocols for Heterogeneous Sensing through BLE Beacons in IoT Telemetry Applications. *Sensors* **2017**, *18*, 57. [CrossRef] [PubMed]
56. ZigBee Alliance. Available online: http://www.zigbee.org (accessed on 28 July 2021).
57. Khutsoane, O.; Isong, B.; Abu-Mahfouz, A.M. IoT devices and applications based on LoRa/LoRaWAN. In Proceedings of the Annual Conference of the IEEE Industrial Electronics Society, Beijing, China, 29 October–1 November 2017.
58. Mazhar, F.; Khan, M.G.; Sällberg, B. Precise indoor positioning using UWB: A review of methods, algorithms and implementations. *Wirel. Pers. Commun.* **2017**, *97*, 4467–4491. [CrossRef]
59. Adame, T.; Bel, A.; Bellatta, B.; Barcelo, J.; Oliver, M. IEEE 802.11AH: The WiFi approach for M2M communications. *IEEE Wirel. Commun.* **2014**, *21*, 144–152. [CrossRef]
60. Xie, J.; Zhang, B.; Zhang, C. A Novel Relay Node Placement and Energy Efficient Routing Method for Heterogeneous Wireless Sensor Networks. *IEEE Access* **2020**, *8*, 202439–202444. [CrossRef]
61. Bagula, A.; Abidoye, A.P.; Zodi, G.-A.L. Service-Aware Clustering: An Energy-Efficient Model for the Internet-of-Things. *Sensors* **2016**, *16*, 9. [CrossRef] [PubMed]
62. Kaur, N.; Sood, S.K. An Energy-Efficient Architecture for the Internet of Things (IoT). *IEEE Syst. J.* **2017**, *11*, 796–805. [CrossRef]
63. Xia, S.; Yao, Z.; Li, Y.; Mao, S. Online Distributed Offloading and Computing Resource Management with Energy Harvesting for Heterogeneous MEC-enabled IoT. *IEEE Trans. Wirel. Commun.* **2021**. [CrossRef]
64. Xiao, L.; Wan, X.; Lu, X.; Zhang, Y.; Wu, D. IoT Security Techniques Based on Machine Learning: How Do IoT Devices Use AI to Enhance Security? *IEEE Signal Process. Mag.* **2018**, *35*, 41–49. [CrossRef]
65. Suárez-Albela, M.; Fernández-Caramés, T.M.; Fraga-Lamas, P.; Castedo, L. A Practical Performance Comparison of ECC and RSA for Resource-Constrained IoT Devices. In Proceedings of the 2018 Global Internet of Things Summit (GIoTS), Bilbao, Spain, 4–7 June 2018; pp. 1–6.
66. Rivest, R.L.; Shamir, A.; Adleman, L.M. A method for obtaining digital signatures and public-key cryptosystems. *Commun. ACM* **1978**, *21*, 120–126. [CrossRef]
67. Koblitz, N. Elliptic curve cryptosystems. *Math. Comput.* **1987**, *48*, 203–209. [CrossRef]
68. Miller, V.S. Use of elliptic curves in cryptography. *Proc. Adv. Cryptol.* **1985**, *218*, 417–426.
69. Diffie, W.; Hellman, M.E. New directions in cryptography. *IEEE Trans. Inf. Theory* **1976**, *IT-22*, 644–654. [CrossRef]
70. The Transport Layer Security (TLS) Protocol Version 1.3. Available online: https://datatracker.ietf.org/doc/html/rfc8446 (accessed on 28 July 2021).
71. Burg, A.; Chattopadhyay, A.; Lam, K. Wireless Communication and Security Issues for Cyber-Physical Systems and the Internet-of-Things. *Proc. IEEE* **2018**, *106*, 38–60. [CrossRef]
72. Scharfglass, K.; Weng, D.; White, J.; Lupo, C. Breaking Weak 1024-bit RSA Keys with CUDA. In Proceedings of the 2012 13th International Conference on Parallel and Distributed Computing, Applications and Technologies, Beijing, China, 14–16 December 2012; pp. 207–212.

73. Liu, Z.; Großschädl, J.; Hu, Z.; Järvinen, K.; Wang, H.; Verbauwhede, I. Elliptic Curve Cryptography with Efficiently Computable Endomorphisms and Its Hardware Implementations for the Internet of Things. *IEEE Trans. Comput.* **2017**, *66*, 773–785. [CrossRef]
74. Chaves, R.; Kuzmanov, G.; Sousa, L.; Vassiliadis, S. Cost-Efficient SHA Hardware Accelerators. *IEEE Trans. Very Large Scale Integr. Syst.* **2008**, *16*, 999–1008. [CrossRef]
75. Gielata, A.; Russek, P.; Wiatr, K. AES hardware implementation in FPGA for algorithm acceleration purpose. In Proceedings of the 2008 International Conference on Signals and Electronic Systems, Krakow, Poland, 14–17 September 2008; pp. 137–140.
76. IoT Analytics, State of IoT Q4/2020 & Outlook 2021. November 2020. Available online: https://iot-analytics.com/product/state-of-iot-q4-2020-outlook-2021/ (accessed on 28 July 2021).
77. Salameh, H.B.; Irshaid, M.B.; Ajlouni, A.A.; Aloqaily, M. Energy-Efficient Cross-layer Spectrum Sharing in CR Green IoT Networks. *IEEE Trans. Green Commun. Netw.* **2021**, *5*, 1091–1100. [CrossRef]
78. Sharma, P.K.; Kumar, N.; Park, J.H. Blockchain Technology Toward Green IoT: Opportunities and Challenges. *IEEE Netw.* **2020**, *34*, 4. [CrossRef]
79. Bol, D.; de Streel, G.; Flandre, D. Can we connect trillions of IoT sensors in a sustainable way? A technology/circuit perspective. In Proceedings of the IEEE SOI-3D-Subthreshold Microelectronics Technology Unified Conference (S3S), Rohnert Park, CA, USA, 5–8 October 2015.
80. Fafoutis, X.; Marchegiani, L. Rethinking IoT Network Reliability in the Era of Machine Learning. In Proceedings of the 2019 International Conference on Internet of Things (iThings) and IEEE Green Computing and Communications (GreenCom) and IEEE Cyber, Physical and Social Computing (CPSCom) and IEEE Smart Data (SmartData), Atlanta, GA, USA, 14–17 July 2019; pp. 1112–1119.
81. Zhou, Z.; Xie, M.; Zhu, T.; Xu, W.; Yi, P.; Huang, Z.; Zhang, Q.; Xiao, S. EEP2P: An energy-efficient and economy-efficient P2P network protocol. In Proceedings of the International Green Computing Conference, Dallas, TX, USA, 3–5 November 2014.
82. Sharifi, L.; Rameshan, N.; Freitag, F.; Veiga, L. Energy Efficiency Dilemma: P2P-cloud vs. Datacenter. In Proceedings of the IEEE 6th International Conference on Cloud Computing Technology and Science, Singapore, 15–18 December 2014.
83. Zhang, P.; Helvik, B.E. Towards green P2P: Analysis of energy consumption in P2P and approaches to control. In Proceedings of the International Conference on High Performance Computing & Simulation (HPCS), Madrid, Spain, 2–6 July 2012.
84. Miyake, S.; Bandai, M. Energy-Efficient Mobile P2P Communications Based on Context Awareness. In Proceedings of the IEEE 27th International Conference on Advanced Information Networking and Applications (AINA), Barcelona, Spain, 25–28 March 2013.
85. Liao, C.C.; Cheng, S.M.; Domb, M. On Designing Energy Efficient Wi-Fi P2P Connections for Internet of Things. In Proceedings of the IEEE 85th Vehicular Technology Conference (VTC Spring), Sydney, Australia, 4–7 June 2017.
86. EPA, Report on Server and Data Center Energy Efficiency. August 2007. Available online: https://www.energystar.gov/ia/partners/prod_development/downloads/EPA_Report_Exec_Summary_Final.pdf (accessed on 28 July 2021).
87. Schneider Electric. *White Paper: Estimating a Data Center's Electrical Carbon Footprint*; Schneider Electric: Le Creusot, Franc , 2010.
88. Fernández-Caramés, T.M.; Fraga-Lamas, P.; Suárez-Albela, M.; Díaz-Bouza, M.A. A Fog Computing Based Cyber-Physical System for the Automation of Pipe-Related Tasks in the Industry 4.0 Shipyard. *Sensors* **2018**, *18*, 1691. [CrossRef]
89. Yan, M.; Chen, B.; Feng, G.; Qin, S. Federated Cooperation and Augmentation for Power Allocation in Decentralized Wireless Networks. *IEEE Access* **2020**, *8*, 48088–48100. [CrossRef]
90. Ray, S. A Quick Review of Machine Learning Algorithms. In Proceedings of the 2019 International Conference on Machine Learning, Big Data, Cloud and Parallel Computing (COMITCon), Faridabad, India, 14–16 February 2019; pp. 35–39.
91. Samek, W.; Montavon, G.; Vedaldi, A.; Hansen, L.K.; Muller, K.-R. *Explainable AI: Interpreting, Explaining and Visualizing Deep Learning*; Lecture Notes in Computer Science Book Series (LNCS, Volume 11700); Springer Nature Switzerland AG: Cham, Switzerland, 2019.
92. Rausch, T.; Dustdar, S. Edge Intelligence: The Convergence of Humans, Things, and AI. In Proceedings of the 2019 IEEE International Conference on Cloud Engineering (IC2E), Prague, Czech Republic, 24–27 June 2019; pp. 86–96.
93. Bhardwaj, K.; Suda, N.; Marculescu, R. EdgeAI: A Vision for Deep Learning in IoT Era. *IEEE Des. Test* **2021**, *38*, 37–43. [CrossRef]
94. Martins, P.; Lopes, S.I.; Rosado da Cruz, A.M.; Curado, A. Towards a Smart & Sustainable Campus: An Application-Oriented Architecture to Streamline Digitization and Strengthen Sustainability in Academia. *Sustainability* **2021**, *13*, 3189.
95. Martins, P.; Lopes, S.I.; Curado, A.; Rocha, Á.; Adeli, H.; Dzemyda, G.; Moreira, F.; Ramalho Correia, A.M. (Eds.) Designing a FIWARE-Based Smart Campus with IoT Edge-Enabled Intelligence. In *Trends and Applications in Information Systems and Technologies*; WorldCIST 2021; Advances in Intelligent Systems and Computing; Springer Nature Switzerland AG: Cham, Switzerlan , 2021; Volume 1367.
96. Alvarado, I. AI-enabled IoT, Network Complexity and 5G. In Proceedings of the 2019 IEEE Green Energy and Smart Systems Conference (IGESSC), Long Beach, CA, USA, 4–5 November 2019; pp. 1–6.
97. Zou, Z.; Jin, Y.; Nevalainen, P.; Huan, Y.; Heikkonen, J.; Westerlund, T. Edge and Fog Computing Enabled AI for IoT-An Overview. In Proceedings of the 2019 IEEE International Conference on Artificial Intelligence Circuits and Systems (AICAS), Hsinchu, Taiwan, 18–20 March 2019; pp. 51–56.
98. Mauro, A.D.; Conti, F.; Schiavone, P.D.; Rossi, D.; Benini, L. Always-On 674μW@4GOP/s Error Resilient Binary Neural Networks with Aggressive SRAM Voltage Scaling on a 22-nm IoT End-Node. *IEEE Trans. Circuits Syst. I Regul. Pap.* **2020**, *67*, 3905–3918. [CrossRef]

99. Qin, H.; Gong, R.; Liu, X.; Bai, X.; Song, J.; Sebe, N. Binary neural networks: A survey, Pattern Recognition. *Pattern Recognit.* **2020**, *105*, 107281. [CrossRef]
100. Tensor Flow Light Guide. Available online: https://www.tensorflow.org/lite/guide (accessed on 28 July 2021).
101. Arduino Nano 33 BLE Sense. Available online: https://store.arduino.cc/arduino-nano-33-ble-sense (accessed on 28 July 2021).
102. SparkFun Edge Development Board-Apollo3. Blue. Available online: https://www.sparkfun.com/products/15170 (accessed on 28 July 2021).
103. Adafruit EdgeBadge-Tensorflow Lite for Microcontrollers. Available online: https://www.adafruit.com/product/4400 (accessed on 28 July 2021).
104. Espressif Systems. ESP32 Overview. Available online: https://www.espressif.com/en/products/devkits/esp32-devkitc/overview (accessed on 28 July 2021).
105. Espressif Systems. ESP-EYE-Enabling a Smarter Future with Audio and Visual AIoT. Available online: https://www.espressif.com/en/products/devkits/esp-eye/overview (accessed on 28 July 2021).
106. STM32, Nucleo-L4R5ZI Product Overview. Available online: https://www.st.com/en/evaluation-tools/nucleo-l4r5zi.html (accessed on 28 July 2021).
107. Shi, Y.; Yang, K.; Jiang, T.; Zhang, J.; Letaief, K.B. Communication-Efficient Edge AI: Algorithms and Systems. *IEEE Commun. Surv. Tutorials* **2020**, *22*, 2167–2191. [CrossRef]
108. Rocha, A.; Adeli, H.; Dzemyda, G.; Moreira, F.; Ramalho Correia, A.M. Trends and Applications in Information Systems and Technologies Features. In Proceedings of the 9th World Conference on Information Systems and Technologies (WorldCIST'21), Terceira Island, Azores, Portugal, 30 March–2 April 2021; Volume 3.
109. Merenda, M.; Porcaro, C.; Iero, D. Edge Machine Learning for AI-Enabled IoT Devices: A Review. *Sensors* **2020**, *9*, 2533. [CrossRef]
110. Zhou, Z.; Chen, X.; Li, E.; Zeng, L.; Luo, K.; Zhang, J. Edge Intelligence: Paving the Last Mile of Artificial Intelligence With Edge Computing. *Proc. IEEE* **2019**, *107*, 1738–1762. [CrossRef]
111. Karras, K.; Pallis, E.; Mastorakis, G.; Nikoloudakis, Y.; Batalla, J.M.; Mavromoustakis, C.X.; Markakis, E. A Hardware Acceleration Platform for AI-Based Inference at the Edge. *Circuits Syst Signal Process.* **2020**, *39*, 1059–1070. [CrossRef]
112. Kim, Y.; Kong, J.; Munir, A. CPU-Accelerator Co-Scheduling for CNN Acceleration at the Edge. *IEEE Access* **2020**, *8*, 211422–211433. [CrossRef]
113. Kim, B.; Lee, S.; Trivedi, A.R.; Song, W.J. Energy-Efficient Acceleration of Deep Neural Networks on Realtime-Constrained Embedded Edge Devices. *IEEE Access* **2020**, *8*, 216259–216270. [CrossRef]
114. Reddi, V.J.; Plancher, B.; Kennedy, S.; Moroney, L.; Warden, P.; Agarwal, A.; Banbury, C.; Banzi, M.; Bennett, M.; Brown, B.; et al. Widening Access to Applied Machine Learning with TinyML. *arXiv* **2021**, arXiv:2106.04008.
115. MLCommons™, MLPerf™ Tiny Inference Benchmark v0.5 Benchmark, San Francisco, CA. 16 June 2021. Available online: https://mlcommons.org/en/inference-tiny-05/ (accessed on 28 July 2021).
116. Banbury, C.; Reddi, V.J.; Torelli, P.; Holleman, J.; Jeffries, N.; Kiraly, C.; Montino, P.; Kanter, D.; Ahmed, S.; Pau, D.; et al. MLPerf Tiny Benchmark. *arXiv* **2021**, arXiv:2106.07597.
117. Amodei, D.; Hernandez, D. AI and Compute. May 2018. Available online: https://openai.com/blog/ai-and-compute/ (accessed on 28 July 2021).
118. Schwartz, R.; Dodge, J.; Smith, N.A.; Etzioni, O. Green AI. *arXiv* **2019**, arXiv:1907.10597.
119. Huang, J.; Rathod, V.; Sun, C.; Zhu, M.; Korattikara, A.; Fathi, A.; Fischer, I.; Wojna, Z.; Song, Y.; Guadarrama, S.; et al. Speed/Accuracy Trade-Offs for Modern Convolutional Object Detectors. In Proceedings of the 2017 IEEE Conference on Computer Vision and Pattern Recognition (CVPR), Honolulu, HI, USA, 21–26 July 2017; pp. 3296–3297.
120. Pinto, G.; Castor, F. Energy efficiency: A new concern for application software developers. *Commun. ACM* **2017**, *60*, 68–75. [CrossRef]
121. Strubell, E.; Ganesh, A.; McCallum, A. Energy and Policy Considerations for Deep Learning in NLP. *arXiv* **2019**, arXiv:1906.02243.
122. World Data Bank, CO_2 Emissions (Metric Tons per Capita). Available online: https://data.worldbank.org/indicator/EN.ATM.CO2E.PC (accessed on 28 July 2021).
123. Lundegård, A.D.Y. GreenML—A Methodology for Fair Evaluation of Machine Learning Algorithms with Respect to Resource Consumption. Master's Thesis, Department of Computer and Information Science, Linköping University, Linköping, Sweden, 2019.
124. Jouppi, N.P.; Young, C.; Patil, N.; Patterson, D.; Agrawal, G.; Bajwa, R.; Bates, S.; Bhatia, S.; Boden, N.; Borchers, A.; et al. In-Datacenter Performance Analysis of a Tensor Processing Unit. In Proceedings of the 44th Annual International Symposium on Computer Architecture (ISCA '17), Toronto, ON, Canada, 24–28 June 2017; pp. 1–12.
125. Fraga-Lamas, P.; Fernández-Caramés, T.M.; Noceda-Davila, D.; Díaz-Bouza, M.A.; Pena-Agras, J.D.; Castedo, L. Enabling automatic event detection for the pipe workshop of the shipyard 4.0. In Proceedings of the 2017 56th FITCE Congress, Madrid, Spain, 14–15 September 2017; pp. 20–27.
126. Fraga-Lamas, P.; Fernández-Caramés, T.M.; Noceda-Davila, D.; Vilar-Montesinos, M. RSS stabilization techniques for a real-time passive UHF RFID pipe monitoring system for smart shipyards. In Proceedings of the 2017 IEEE International Conference on RFID (RFID), Phoenix, AZ, USA, 9–11 May 2017; pp. 161–166.
127. Fraga-Lamas, P.; Noceda-Davila, D.; Fernández-Caramés, T.M.; Díaz-Bouza, M.A.; Vilar-Montesinos, M. Smart Pipe System for a Shipyard 4.0. *Sensors* **2016**, *16*, 2186. [CrossRef]

128. Stoll, C.; Klaaßen, L.; Gallersdörfer, U. The Carbon Footprint of Bitcoin. *Joule* **2019**, *3*, 1647–1661. [CrossRef]
129. Electricity Maps Public Website. Available online: https://www.electricitymap.org (accessed on 28 July 2021).
130. Dong, Y.; Cheng, J.; Hossain, M.; Leung, V.C.M. Secure distributed on-device learning networks with byzantine adversaries. *IEEE Netw.* **2019**, *33*, 180–187. [CrossRef]
131. NVIDIA, NVIDIA Clara: An Application Framework Optimized for Healthcare and Life Sciences Developers. Available online: https://developer.nvidia.com/clara (accessed on 28 July 2021).

Article

Fast and Accurate Approach to RF-DC Conversion Efficiency Estimation for Multi-Tone Signals

Janis Eidaks, Romans Kusnins, Ruslans Babajans *, Darja Cirjulina, Janis Semenjako and Anna Litvinenko

Institute of Radioelectronics, Riga Technical University, Kalku St. 1, LV-1050 Riga, Latvia; janis.eidaks@rtu.lv (J.E.); romans.kusnins@rtu.lv (R.K.); darja.cirjulina@rtu.lv (D.C.); janis.semenako@rtu.lv (J.S.); anna.litvinenko@rtu.lv (A.L.)
* Correspondence: ruslans.babajans@rtu.lv

Citation: Eidaks, J.; Kusnins, R.; Babajans, R.; Cirjulina, D.; Semenjako, J.; Litvinenko, A. Fast and Accurate Approach to RF-DC Conversion Efficiency Estimation for Multi-Tone Signals. *Sensors* **2022**, *22*, 787. https://doi.org/10.3390/s22030787

Academic Editor: Zihuai Lin

Received: 6 December 2021
Accepted: 17 January 2022
Published: 20 January 2022

Publisher's Note: MDPI stays neutral with regard to jurisdictional claims in published maps and institutional affiliations.

Copyright: © 2022 by the authors. Licensee MDPI, Basel, Switzerland. This article is an open access article distributed under the terms and conditions of the Creative Commons Attribution (CC BY) license (https:// creativecommons.org/licenses/by/ 4.0/).

Abstract: The paper presents a computationally efficient and accurate numerical approach to evaluating RF–DC power conversion efficiency (PCE) for energy harvesting circuits in the case of multi-tone power-carrying signal with periodic envelopes. This type of signal has recently received considerable attention in the literature. It has been shown that their use may result in a higher PCE than the conventional sine wave signal for low to medium input power levels. This reason motivated the authors to develop a fast and accurate two-frequency harmonic balance method (2F-HB), as fast PCE calculation might appreciably expedite the converter circuit optimization process. In order to demonstrate the computational efficiency of the 2F-HB, a comparative study is performed. The results of this study show that the 2F-HB significantly outperforms such extensively used methods as the transient analysis (TA), the harmonic balance method (HB), and the multidimensional harmonic balance method (MHB). The method also outperforms the commercially available non-linear circuit simulator Keysight ADS employing both HB and MHB. Furthermore, the proposed method can be readily integrated into commonly used commercially available non-linear circuit simulation software, including the Keysight ADS, Ansys HFSS, just to name a few—minor modifications are required. In addition, to increase the correctness and reliability of the proposed method, the influence of PCB is considered by calculating Y parameters of its 3D model. The widely employed voltage doubler-based RF–DC converter for energy harvesting and wireless power transfer (WPT) in sub-GHz diapason is chosen to validate the proposed method experimentally. This RF–DC converter is chosen for its simplicity and capability to provide sufficiently high PCE. The measurements of the PCE for a voltage doubler prototype employing different multi-tone waveform signals were performed in laboratory conditions. Various combinations of the matching circuit element values were considered to find the optimal one in both—theoretical model and experimental prototype. The measured PCE is in very good agreement with the PCE calculated numerically, which attests to the validity of the proposed approach. The proposed PCE estimation method is not limited to one selected RF–DC conversion circuit and can also be applied to other circuits and frequency bands. The comparison of the PCE obtained by means of the proposed approach and the measured one shows very good agreement between them. The PCE estimation error reaches as low as 0.37%, and the maximal estimation error is 32.65%.

Keywords: wireless power transfer; energy harvesting; power conversion efficiency; single diode rectifier; voltage doubler; harmonic balance method; autonomous sensor node; wireless sensor network; multi-tone signal; full-wave simulations of PCB

1. Introduction

The current decade has witnessed a rapid increase in the number of smart wireless devices, their influence over social and economic development has also been growing. Wireless devices have become increasingly compact, it has become much easier to integrate them into various environments, which in turn promotes development of the Internet of Things (IoT) and the underlying wireless sensor networks (WSNs). Smart cities [1], agriculture [2], and medicine [3] are just some areas where WSNs are employed to control smart environments via the IoT. The increasing use of WSNs has caused exponential growth in the number of autonomous individual sensor nodes (SN), which in turn poses powering-related challenges for the sensor networks. Battery power is the most common source for powering autonomous devices. Along with the increase of the number of autonomous devices used in the network, more time and attention are required to monitor the power level of every single device; the batteries should also be changed when necessary. However, devices situated in confined areas cannot be easily maintained, which may compromise the integrity of the WSN. Radio frequency (RF) wireless power transfer (WPT) offers a solution for preserving the integrity of the WSN during operation, providing control over the amount of energy each SN receives to perform its duties. The key benefits of using WPT for powering autonomous devices consist of a reduced need for batteries, which in its turn mitigates inconveniences related to powering of these devices, and the opportunity to maintain closer control over device energy levels. The use of RF allows transferring power to secluded SNs from a sufficient distance, it also allows for ambient energy harvesting.

The rectenna (receiving antenna paired with an RF–DC converter) with high power conversion efficiency (PCE) is the most important element of an efficient RF WPT. High PCE increases the amount of useful energy the autonomous device receives, which is particularly relevant in case of relatively long distances between the power transmitter and receiver, which cause reduction of the amount of received RF power. Over the years, many studies proposed various rectennas for WPT. Table 1 lists the key properties of the proposed rectennas ordered by frequencies and input powers. The table also includes the results of this study for comparison. The results will be further elaborated upon in this manuscript.

Table 1. Comparison of the experimentally studied rectennas.

Ref.	Substrate	RF–DC Topology	Frequency, GHz	RF Input Power, dBm	Waveform	PCE, %
[4]	-	1 diode	24	27.0	Single-tone [1]	43.6
				16.0		42.9
[5]	Custom [3]	4 diodes	5.8	30	Single-tone	92.8
[6]	FR4	2 diodes	5.76	20	Single-tone	84.0
[7]	RT/Duroid 5870	1 diode	5.80	16.9	Single-tone	82.7
			2.45	19.5		84.4
[8]	Custom [4]	1 diode	2.45	37	Single-tone	91.0
[9]	FR4	2 diodes	2.45	24.7	Single-tone	78.0
[10]	RO4003C	1 diode	2.45	3	Multi-tone [2]	54.5
[11]	FR4	4 diodes	2.4	27	Multi-tone	75.0
[12]	PTFE	4 diodes	2.4	26.2	Single-tone	80.0
[13]	FR4	2 diodes	2.4	22	Single-tone	82.3
[14]	RO4003C	1 diode	2.4	10	Single-tone	60.0
[15]	-	1 diode	2.4	−10	Multi-tone	42.0
[16]	FR4	4 diodes	2.15	0	Single-tone	70.0
[17]	Arlon A25N	1 diode	0.915	0	Multi-tone	67.8
This work	FR4	2 diodes	0.865	−2	Single-tone	64.8
					Multi-tone	63.2
[18]	RT/Duroid 5880	2 diodes	0.860	−4	Single-tone	60.0
[19]	-	1 diode	0.433	−10	Multi-tone	55.0

[1] All instances of "single-tone" refer to an unmodulated carrier. [2] All instances of "multi-tone" refer to a sum of several subcarriers. [3] Relative permittivity $\varepsilon_r = 3.4$, the dielectric loss tangent $\tan\delta = 0.0015$. [4] Relative permittivity $\varepsilon_r = 2.55$, the dielectric loss tangent $\tan\delta = 0.0018$.

As seen from Table 1, different rectenna configurations have been proposed and studied. Rectennas that show PCE above 70% [5–9,11–13] use high RF input power (>15 dBm), which greatly limits the range of effective distances between the power transmitter and the secluded SN if this received input power is to be achieved. Increase of the distance calls for increase of transmission power to maintain the required input power and PCE, which potentially exceeds power restrictions for the given frequencies. Studies [5,8] reached efficiencies over 90%. This can only be achieved with the receiver optimized for such high input powers (the RF–DC conversion is done using GaAs diodes), which is not optimal for practical applications in powering secluded SN using the given frequencies. The use of input power in the range around 0 dBm implies application of both SN and low-power technologies, such as RFID and E-ink [20,21]. This range of input RF power was less frequently addressed in literature than high and low (<-15 dBm) power ranges. Comparing rectennas in terms of frequencies, Table 1 demonstrates that rectennas were mainly developed for 2.45 GHz ISM frequency band. The use of high frequency also limits the effective distance between the transmitter and the SN. Sub-GHz ranges, such as 433 MHz (ISM) and 860 MHz (GSM-850), allow transferring of power to greater distances. Regarding the waveform of the power-carrying signal, rectennas listed in Table 1 mainly use a single-tone signal (an unmodulated carrier). However, studies, such as [11,15,19,22] and [23–25], reported an increase in PCE when multi-tone (formed by a sum of several subcarriers) power-carrying signals are used. The topology of the RF–DC circuit is another crucial parameter of rectenna design. The most common RF–DC topologies are presented in Figure 1: one-diode-based (half-wave rectifier), two-diode-based (voltage doubler), and four-diode-based (diode bridge rectifier) topologies. These topologies with slight variations were used in the studies listed in Table 1. Analyzing information in Table 1, it may be concluded that rectenna based on a voltage doubler RF–DC converter working at a sub-GHz frequency and multi-tone power-carrying signals proved to be the most well-balanced solution in terms of cost and efficiency for RF WPT applications targeted at powering SN and low power electronics.

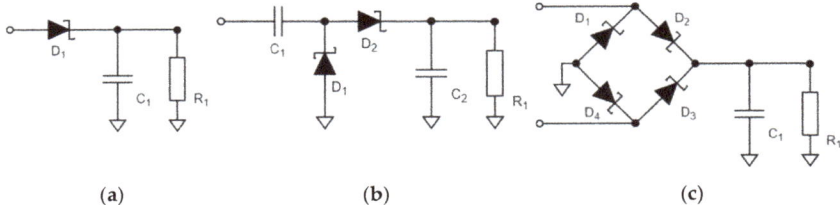

Figure 1. One-diode-based rectifier (**a**), two-diode-based rectifier (**b**), diode bridge rectifier (**c**).

The considered studies mainly focused on enhancing performance of rectennas with experimental validation of results, aiming at development of reliable theoretical models for the WPT and RF–DC converters. Numerous theoretical models exist in the field of AC–DC [26] and DC–DC [27,28] converters, several modeling approaches have also been proposed over the years for RF–DC circuits. Development of an accurate computer model and its use in simulations is a feasible alternative to experimental studies of RF–DC power converters. In contrast to experiments, simulation is a more convenient and cost-effective solution, as it does not require fabrication of prototypes, especially when circuit design optimization is needed.

Despite recent advances in the field, the analysis of non-linear circuits not amenable to linearization is usually very time-consuming. This issue becomes even more pronounced when complex input waveforms are employed. Although transient analysis (TA) is a robust circuit analysis method [29], it is not suitable for analyzing RF–DC converters because long simulation times are required due to the presence of transients [30]. Furthermore, in case of narrow-band signals with periodic envelopes, the time step must be much smaller compared to the period of the carrier wave that leads to a very large number of iterations. Though some attempts have been made to speed-up the TA [31], the aforementioned restriction on the time step size considerably limits the performance of the method, as will be shown in this paper (see Section 2.5). Another widely used non-linear circuit analysis method is the Volterra series method [32]. However, this method is mainly applied to weakly non-linear circuits, since for circuits with highly pronounced non-linearity the convergence is very slow. The harmonic balance (HB) method was initially proposed in [33] to solve problems in mechanical engineering, it has subsequently been adapted to treat non-linear circuits under sinusoidal excitation [34]. The issue of transients does not pose problems within HB, as this method allows computing the steady state response directly, involving the solving of a system of non-linear equations [35]. The system of equations can be reduced by partitioning the original circuit into linear and non-linear parts [36]. The resulting non-linear equations can be solved by means of Newton's method (NM) [37], or iteration relaxation method (IRM) [38,39], among others. The evaluation of the Jacobian matrix can be significantly accelerated using FFT algorithms [40] and the continuation method was developed to ensure convergence at high input powers [41]. The HB has also been extended to handle multi-tone input signals [42,43]. However, in such cases the Jacobian matrix is significantly larger, resulting in the high computational burden. This issue can be mitigated by exploiting useful properties of multidimensional FFT algorithms [44]. Over the last several decades, the method has found use in a number of applications, including the analysis of the behavior of both autonomous and non-autonomous oscillators [45–47]. Additionally, in an effort to reduce the simulation time, several extensions and modifications of the HB, as well as its multidimensional extensions, have been proposed, such as the hierarchical harmonic balance method [48], several parallel versions of the HB [49,50], the multi-level frequency decomposition-based HB [51], and the HB using the graph sparsification [52].

Although the methods mentioned above are accurate, they are highly computationally intensive. As a result, a number of approximate closed-form expression-based models have been proposed to analyze rectennas sharing a common load [53], single diode rectifiers [54,55], and Class-F rectifiers converters [56]. In [8], PCE up to 90% has been achieved for the input power range of 30–35 dBm at 2.4 GHz, using the SPICE model with the parameters obtained from experimental data by means of curve fitting. Similar results were obtained in [57] for a single shunt diode rectifier using an analytical model that also considers the effect of the transmission line. In [58], an approximate model was used to find PCE for multi-tone excitation with equally spaced frequencies. Unfortunately, the analytical models give only approximate results that may not be sufficient for the precise evaluation and circuit optimization, like in the case of [59], where the nonlinearity of the diodes and the possible influence of the PCB are not taken into account, resulting in a highly idealized theoretical model.

The method proposed in this paper allows for more computationally efficient treatment of RF–DC converters in the case of input signals with evenly spaced subcarriers. The method has been successfully validated experimentally, as it will be shown in Section 3. In contrast to the multidimensional HB method (MHB) that treats each subcarrier frequency as a fundamental frequency, the proposed approach requires only two fundamental frequencies. Thus, fewer harmonics are needed to approximate the voltages and currents, thereby significantly reducing CPU time.

The aforementioned studies of the rectennas and RF–DC converters focused largely on experimental research and design-specific modeling, paying limited attention to development of reliable and computationally effective models considering the influence of the PCB material for estimating the PCE, whose great importance has been comprehensively demonstrated in [18].

In the current paper, a novel theoretical approach to evaluating the PCE of a rectenna is introduced. The proposed approach offers the following advantages:

(1) Employment of the two-frequency harmonic balance (2F-HB) method is less computationally demanding than other methods, while it still ensures adequate accuracy.
(2) It allows for investigating the impact of different multi-tone power-carrying signal waveforms on the PCE, especially in the sub-GHz band.

(3) It offers an effective approach to considering various effects of the PCB and their impact on the PCE.
(4) It offers an opportunity to examine the influence of variation in the nominal values of several RF–DC circuit elements on the PCE, including the matching circuit.

The validity and accuracy of the proposed approach were verified by measuring the PCE of a prototype RF–DC converter. A voltage doubler circuit with a sub-GHz carrier frequency was selected as a test case and a comprehensive analysis of the effect of multi-tone power-carrying signals with different peak-to-average power ratio (PAPR) levels on its PCE was conducted. To the best of the authors' knowledge, no exhaustive study of such combination of the circuit and signals has been reported in the literature thus far.

The paper is structured as follows: Section 2 describes the novel theoretical approach to PCE estimation and presents a comparative analysis of its performance against conventional methods with a voltage doubler circuit employed as a test object. Discussion and comparison of the results obtained by means of the proposed theoretical estimation approach and its experimental verification are presented in Section 3. Section 4 presents conclusions of the research.

2. Development of a Realistic Model of RF–DC Conversion

This section describes a computationally efficient theoretical approach (model) developed to estimate the PCE for RF–DC converter circuits. For the sake of completeness, the general case of the circuit containing an arbitrary number of diodes is considered. A voltage doubler-based RF–DC converter circuit illustrated in Figure 2 used to validate the approach (see Section 3) can be viewed as a special case. The approach is adapted to power-carrying signals with periodic envelopes. The spectra of such signals comprise harmonics whose frequencies can be expressed as linear combinations of two fundamental frequencies only. This property allows for the employment of a two-dimensional FFT algorithm, which accelerates computation. Performing PCE estimation in shorter times is particularly important, since converter optimization involving PCE calculation for various circuit configurations is tremendously time-consuming, especially for a large number of carriers.

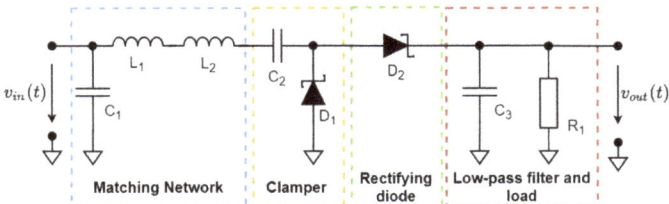

Figure 2. Voltage doubler circuit topology.

Unfortunately, due to the sufficiently high complexity of circuit PCB layout and high operating frequency, some of the existing and extensively used non-linear circuit methods fall short of expectations. For instance, despite numerous advantages, the TA is not suitable for the analysis of converters driven by a multi-carrier signal for several reasons. First, quite a large ratio of the period of the envelope to that of the carrier wave (in the present study, it is in the order of 1000) leads to a large number of iterations needed to calculate at least one period of the output voltage. Second, the presence of a filtering capacitor causes transients; therefore, many periods have to be computed until the steady state is reached. Third, for the equivalent circuit of the PCB to be valid over a frequency range encompassing at least 7–10 harmonics of the carrier wave, it must possess quite a complicated topology that is difficult to handle [60]. Therefore, TAs have been abandoned in favor of their frequency (or time-frequency) domain counterparts, such as the HB.

The HB relies upon Fourier series representation of circuit voltages (currents) and leverages some useful properties of well-established FFT algorithms leading to reduced consumption of computational resources. However, the method is not well-suited for multi-tone excitation. To tackle this issue, the authors propose to employ a two-frequency harmonic balance (2F-HB) method described in this paper. In contrast to its conventional counterpart, the 2F-HB exploits the fact that the spectra of the circuit currents and voltages consist of a number of sub-bands centered at integer multiples of the carrier frequency. Furthermore, each sub-band contains harmonics that are equally spaced. This property of the spectrum enables one to leverage the power of 2D versions of FFT [61] to achieve a substantial reduction in CPU time.

Experimental studies and simulations using RF–DC converter circuit models that do not consider the effect of the PCB show large discrepancies between the experimental and theoretical results [62]. Discrepancies are generally caused by the fact that the contribution of the PCB is either completely neglected, or its effect is only partially accounted for via some approximations. The proposed approach, in contrast, considers the contribution of the PCB through the calculation of the Y parameters obtained by means of full-wave numerical analysis. More precisely, the PCB is treated as a multi-port network formed from the original circuit by disconnecting discrete circuit components, as illustrated in Figure 3. The main advantage of this approach is that the accuracy of the PCE estimation depends solely on the accuracy of the 3D model. It should be noted that the approach is by no means perfect—3D models are typically idealized, neglecting some imperfections of real-world circuits. Nevertheless, it provides more accurate PCE estimation for PCBs having a complex layout, such as the one studied herein. Regarding the nonlinearity of the circuit, the proposed approach utilizes the standard SPICE diode model [63], as it describes the behavior of Schottky diodes with reasonable accuracy. Furthermore, the model closely approximates the diode breakdown behavior, which is particularly important, since the diodes under study possess quite low breakdown voltages (in the order of 2–4 V).

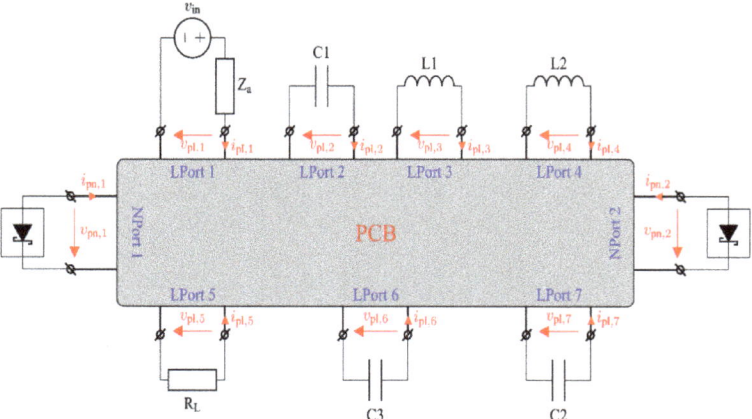

Figure 3. An equivalent circuit of PCB with equivalent two-port networks of linear devices connected to its linear ports.

It is noteworthy that the approach can be integrated into existing non-linear circuit simulators employing the HB or its extended multi-tone version, namely, the MHB. In the case of simulators using the MHB, only the subroutines responsible for the evaluation of the Jacobian entries have to be replaced or modified. Specifically, the approach proposed in this work requires the use of 2D-FFT and its inverse algorithms to perform time-to-frequency and reverse transformations of the non-linear element voltages (diode voltages). Regarding solvers capable of handling multi-tone signals driven non-linear circuits, only minor modifications in the existing codes are required. In fact, the proposed method can be viewed as a two-dimensional MHB where one of the fundamental frequencies is that of the carrier wave, whereas the other is the subcarrier separation frequency.

2.1. Two-Frequency Harmonic Balance Method

As mentioned previously, the conventional HB is not a good candidate for handling multi-tone excitation, since FFT algorithms require uniform spectra, thus, a large number of harmonics should be considered. More specifically, all harmonics up to a specific order must be used for the approximation. In contrast, in the case of multi-tone excitation, the spectrum is not uniform—it consists of a number of sub-bands formed from the nonlinear conversion products. Therefore, it would not be wise to consider the harmonics between the sub-bands with negligibly small amplitudes. On the other hand, neglecting these harmonics prohibits the use of FFT, thereby reducing the computational efficiency. To overcome this issue, a multidimensional extension of the harmonic balance method (MHB) has been proposed [42–44]. The method approximates voltages (currents) with the truncated multidimensional Fourier series [64], enabling the use of multidimensional FFT algorithms (NFFT) to speed up calculations. While the MHB can be used to analyze multi-tone signal-powered RF–DC converters, the CPU time grows rapidly with the number of subcarriers. To mitigate this problem, the 2F-HB was developed and validated on a voltage doubler circuit.

The 2F-HB handles multi-tone signals in a more time-efficient way, since it requires fewer voltage (current) phasors than the HB and MHB and thus outperforms them. The proposed method relies upon the approximation of the voltage across each circuit element by a truncated two-dimensional extension of the Fourier series of the form:

$$v_m(t) \Rightarrow v_m(t_1, t_2) = \sum_{n_1=-N_1/2}^{N_1/2} \sum_{n_2=-N_2/2}^{N_2/2} \widetilde{V}_{n_1,n_2}^{(m)} e^{jn_1\omega_1 t_1} e^{jn_2\omega_2 t_2}, \quad (1)$$

where $\widetilde{V}_{n_1,n_2}^{(m)}$ are the phasors of voltage $v_m(t)$, ω_1 denotes the carrier frequency (CF) and ω_2— the subcarrier separation frequency (CSF), N_1 and N_2 determine the numbers of harmonics of ω_1 and ω_2, respectively, used to approximate the voltage.

The circuit currents are approximated in the same way. The main benefit of using Equation (1) is that it yields a compact equation system, owing to derivative-free relations between the linear element voltage and current phasors. The introduction of time variables t_1 and t_2 associated with ω_1 and ω_2, respectively, allows evaluating the Jacobian matrix, which will be discussed further, in a considerably more time-efficient manner via the use of 2D-FFT [65].

The voltage (current) phasors can be found by solving a system of circuit equations derived by applying nodal analysis to the equivalent circuit (EC) obtained by reducing the linear sub-circuit to a mesh network. The equation for the n-th node of the EC is:

$$i_{\Sigma,n} = i_n(v_n) + Y_n v_n + \sum_{m=1, m \neq n}^{M} Y_{nm} v_m + i_{\text{eq},n} = 0, \quad (2)$$

where Y_n denotes an operator transforming the phasors according to the self-admittance of the n-th node, Y_{nm} is the mutual admittance operator for the n-th and m-th nodes, $i_n(v_n)$ is the current through the n-th diode, $i_{\text{eq},n}$ is an equivalent current representing the effect of independent sources contained in the circuit, $i_{\Sigma,n}$ is the total current at the n-th node, and M is the number of circuit nodes.

NM is used to solve the system of equations obtained by collecting Equation (2) for all nodes, iteratively constructing and solving the following systems of linear equations:

$$\hat{\mathbf{J}}^{(l)} \Delta \mathbf{v}^{(l)} = -\hat{\mathbf{r}}^{(l)}, \quad (3)$$

where $\hat{\mathbf{J}}^{(l)}$ is the Jacobian matrix, $\hat{\mathbf{r}}^{(l)}$ is the residual vector calculated at the l-th iteration of the NM, and $\Delta \mathbf{v}^{(l)}$ is the phasor correction vector.

The Jacobian matrix entries are 2D-FFT transformed partial derivatives of each $i_{\Sigma,n}$ with respect to the real and imaginary parts of each $\widetilde{V}_{n_1,n_2}^{(m)(l)}$. The residual vector contains 2D-FFT transformed $i_{\Sigma,n}^{(l)}$. It is worth noting that the Jacobian matrix can be transformed column-wise using 2D-FFT algorithms. Alternatively, the 2D-FFT algorithm needs to be run only once to evaluate the first column of the Jacobian matrix, while the other columns can be obtained by applying cyclic shifts to the entries of the first column. Equation system (3) can be solved by using either a plain linear equation solver [66] or various iterative methods, e.g., Krylov subspace methods [67].

2.2. Diode Equations

As follows from (2), relations I–V for diodes are required to evaluate $\hat{\mathbf{j}}^{(l)}$ and $\hat{\mathbf{r}}^{(l)}$. While diodes play a crucial role in the RF–DC converters, their inherent nonlinearity renders circuit analysis considerably more complex. As in the present study, a sub-GHz range is concerned, the choice of diode model becomes even more critical with regard to the PCE estimation reliability. This is due to a number of effects that may be neglected at low frequencies, while they start to manifest themselves at high frequencies dramatically affecting the overall efficiency of the power conversion.

In the proposed approach, the standard SPICE model was selected to describe the behavior of the Schottky diodes. The model has the following advantages: ease of implementation, high stability when used in conjunction with 2F-HB, as well as accurate modeling of breakdown current and junction capacitance. The parameters of the SPICE model used in the theoretical analysis of voltage doubler PCE are taken from the datasheet for the SMS7630 Schottky diode [68]. The main part of the diode equivalent circuit (DEC) is shown in Figure 4. Throughout the paper, the voltage across the junction of the m-th diode is denoted as v_m.

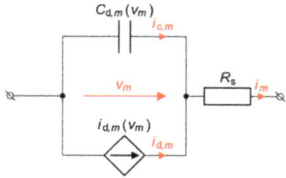

Figure 4. Low frequency diode SPICE model.

As indicated in Figure 4, the current flowing through the diode is comprised of two components: the junction current and the current determined by the junction capacitance. The former depends non-linearly on the voltage across it, and in the framework of the SPICE model it can be calculated as:

$$i_{d,m} = i_s \left(e^{v_m/(Nv_t)} - 1 \right) - i_{bv} e^{-\frac{v_m + v_{bv}}{N_{bv} v_t}}, \qquad (4)$$

where $i_{d,m}$ denotes the junction current of the m-th diode, i_s—the saturation current, v_t—the thermal voltage of the diode junction, N—the ideality factor, v_{bv} denotes the breakdown voltage, and N_{bv} and i_{bv}, are the ideality factor and the knee current of the breakdown current, respectively.

The contribution of the non-linear diode capacitance to the total diode current plays an important role in the behavior of diodes at high frequencies, therefore, it has to be considered as well. The total capacitance of the diode is given by:

$$C_{d,m} = C_{t,m} + C_{j,m} = t_t \frac{d(i_s(e^{v_m/(Nv_t)} - 1))}{dv_m} + C_{j,m}, \qquad (5)$$

where $C_{t,m}$ is the transit time capacitance of the m-th diode and t_t is the transit time. Since for the Schottky diodes this quantity is typically negligibly small and therefore does not have a substantial effect on diode performance, it is assumed that $C_{t,m} = 0$. The other component is the junction capacitance given by $C_{j,m} = C_{j0}(1 - FC)^{-(M+1)}(K + Mv_m/v_j)$, if $v_m > FC \cdot v_j$ and $C_{j,m}(v_m) = C_{j0}(1 - v_m/v_j)^{-M}$, otherwise, where C_{j0} is zero bias voltage capacitance, M is the grading coefficient, v_j is the junction built-in voltage, $K = 1 - FC(M+1)$, and FC represents the forward-bias depletion capacitance coefficient. Using (4) and (5), the expressions for the Jacobian matrix and residual vector entries can be derived in a straightforward manner, however, for the sake of brevity they are not presented here. Parameters of the SMS7630 Schottky diode are compiled in Table 2.

Table 2. SPICE model parameters of the SMS7630 Schottky diode [68].

Parameter	Value	Unit
i_{bv}	1×10^{-4}	A
R_S	20	Ω
C_{j0}	0.14	pF
v_{bv}	2	V
i_s	5×10^{-6}	A
t_t	1×10^{-11}	s
M	0.4	-
N	1.05	-
v_j	0.51	V

2.3. Evaluating Y Parameters for the Linear Sub-Network

In addition to the diode I–V relation, Equation (3) also requires the knowledge of the behavior of the linear sub-network composed of all linear elements, including the PCB. As it was mentioned previously, within the proposed approach the PCB is treated as a separate circuit element—multi-port network. In the frequency domain, the behavior of the PCB can be fully described in terms of Y parameters. Similar to diodes, a proper model of the PCB is essential, since the impact of the PCB upon the converter plays a crucial role and therefore should not be neglected.

Conventional lumped element equivalent circuits (LEEC) are not suitable for the excitation and the working frequency at hand due to highly pronounced non-linear distortions. More specifically, the equivalent circuit must be usable for a frequency range encompassing at least 6–8 harmonics of the CW, which is quite challenging to meet owing to the frequency-dependent nature of different parasitic effects. As it is rather difficult to evaluate the values of the LEEC constituents, the authors decided to perform a full-wave analysis (FWA) of the PCB for the RF–DC circuit under study. The main advantage of the FWA is that it allows capturing of the effects that other methods cannot because of their approximate nature. Thus, the FWA is the most reliable method for characterizing non-linear high frequency circuits.

The discrete circuit components are modeled as lumped elements (LE), or equivalent circuits composed of LE. Since the PCB of the circuit under study has a complex layout and it may be complicated to construct an LEEC that would be valid over a relatively wide band, the Y parameters of the circuit are obtained using an FWA.

For this purpose, commercially available software Ansys HFSS is employed [69], which solves Maxwell's equations using the well-established finite element method [70]. Each discrete element is replaced with a lumped port. The PCB model of the voltage doubler circuit can be seen in Figure 5a. The model is enclosed by a fictitious absorbing surface that truncates the solution domain [52]. The dimensions of the PCB model itself and its conducting parts are the same as for the prototype circuit used for the experimental validation. As the main objective is to eliminate all linear equations, the Y matrix for the PCB model can be partitioned as follows:

$$\begin{pmatrix} \mathbf{i}_L \\ \mathbf{i}_N \end{pmatrix} = \begin{pmatrix} \mathbf{Y}_{LL} & \mathbf{Y}_{LN} \\ \mathbf{Y}_{NL} & \mathbf{Y}_{NN} \end{pmatrix} \begin{pmatrix} \mathbf{v}_L \\ \mathbf{v}_N \end{pmatrix}, \quad (6)$$

where vectors $\mathbf{v}_L$ and $\mathbf{v}_N$ contain voltages at linear and non-linear ports, respectively, whereas $\mathbf{i}_L$ and $\mathbf{i}_N$ contain the vectors of current at linear and non-linear ports, respectively.

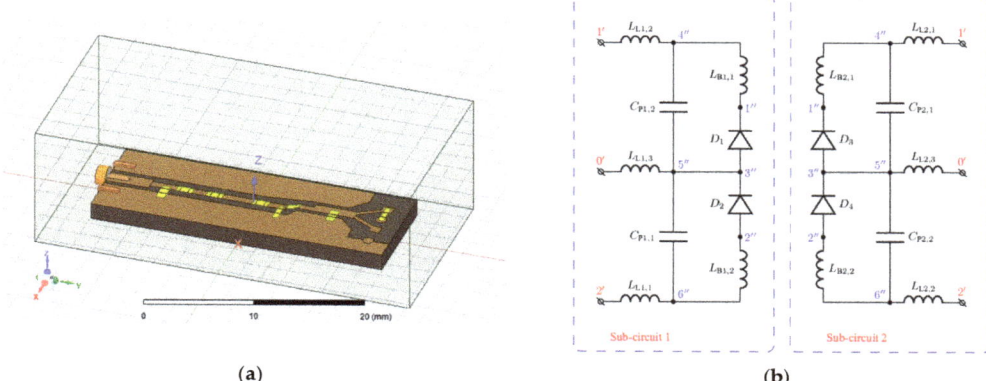

Figure 5. HFSS model of the voltage doubler-based RF–DC converter PCB (**a**), an equivalent network for 4 diodes in a single package (**b**).

In order to make the model even more reliable, various parasitic effects associated with diodes should also be considered by introducing a number of lumped elements, each modeling the corresponding effect, such as bond wire inductance, lead inductance, package capacitance, etc. An extended diode equivalent circuit (EDEC) incorporating parasitic inductances and capacitances of four diodes within a single package is illustrated in Figure 5b. A port composed of the reference terminal (indicated by 0' in Figure 5b) and a non-referenced one is termed an internal port (IP), whereas a port obtained by eliminating the non-linear part of the DEC depicted in Figure 4 is termed an external port (EP). Similar to the Y matrix of PCB-EC, the Y matrix for the EDEC can be partitioned as follows:

$$\begin{pmatrix} \mathbf{i}^{(i)} \\ \mathbf{i}^{(e)} \end{pmatrix} = \begin{pmatrix} \mathbf{Y}_r^{(ii)} & \mathbf{Y}_r^{(ie)} \\ \mathbf{Y}_r^{(ei)} & \mathbf{Y}_r^{(ee)} \end{pmatrix} \begin{pmatrix} \mathbf{v}^{(i)} \\ \mathbf{v}^{(e)} \end{pmatrix}, \quad (7)$$

where vectors $\mathbf{v}^{(o)}$ and $\mathbf{i}^{(o)}$ contain voltages and currents at the EPs, respectively, while vectors $\mathbf{v}^{(i)}$ and $\mathbf{i}^{(i)}$ correspond to the IPs.

Finally, combining (6) and (7), as well as using the Norton equivalent circuit parameters for all elements other than diodes connected to PCB-EC, yields the relation:

$$\mathbf{i}_d = \mathbf{i}_{d,eq} + \mathbf{Y}_d \mathbf{v}_d, \quad (8)$$

where $\mathbf{i}_d$ is a vector of total diode currents, $\mathbf{v}_d$ is the voltages across the non-linear part of the DEC, $\mathbf{Y}_d$ is the admittance matrix for the linear subcircuit of the RF–DC converter, and $\mathbf{i}_{d,eq}$ contains equivalent currents that represent the effect of the voltage source.

2.4. Estimation of the PCE for a Voltage Doubler Circuit

A voltage doubler circuit shown in Figure 2 was considered as an example. Following the methodology described in the previous subsection, the circuit can be regarded as a multi-port network representing the effect of the PCB on the circuit behavior. The voltage doubler circuit can then be represented as the multi-port network with other circuit elements, or their equivalent circuits connected to its ports. Since only 2, not 4, diodes in a single package are used for the prototype circuit, only one of the two subcircuits shown in Figure 6 must be considered. The entire circuit of the voltage doubler is represented as a multi-port network corresponding to the PCB, to which lumped circuit elements are connected, including the generator, as illustrated in Figure 3. The Y matrix of the PCB is computed using Ansys HFSS as described in the previous subsections. The impedance of the generator is assumed to be 50 Ω. The values of the elements of the DEC are taken from the relevant datasheet.

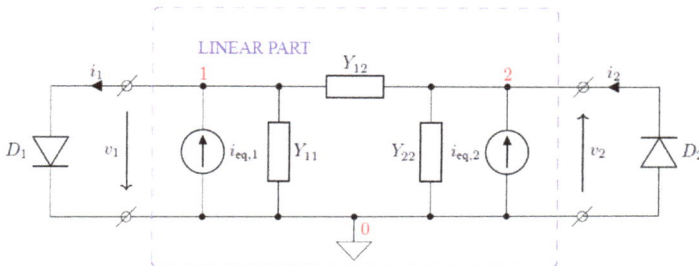

Figure 6. The equivalent circuit of the voltage doubler with the PCB replaced by the equivalent mesh network.

The equivalent circuit of the entire linear part of the voltage doubler circuit is depicted in Figure 6. It should be noted that the diode symbol in the equivalent circuits represents the non-linear part of the low frequency DEC, while the effect of R_S (see Figure 4) is incorporated in the equivalent circuit for the linear part of the original one. The current sources $i_{eq,1}$ and $i_{eq,2}$ are the equivalent current sources representing the contribution of the voltage source to the total currents at nodes 1 and 2. Thus, the behavior of the circuit can be described using two non-linear equations:

$$\begin{cases} i_{\Sigma,1} = i_1 + v_1(Y_{11} + Y_{12}) + v_2 Y_{12} - i_{eq,1} = 0 \\ i_{\Sigma,2} = -i_2 - v_2(Y_{22} + Y_{12}) - v_1 Y_{12} - i_{eq,2} = 0 \end{cases} \quad (9)$$

To determine the phasors of v_1 and v_2, system (9) is solved using the NM. The NM is employed as it has proven itself as a rapidly converging method, provided the initial guess is close enough to the actual solution. If it is not the case, the continuation method [41] can be utilized to take advantage of the fact that the convergence of the NM is more stable for small amplitudes. The convergence is ensured by gradually increasing the input excitation amplitude, starting with the smallest one. Each time the NM fails, the values of the equivalent current sources are reduced. The phasors of both the initial guess and input currents are multiplied by a scaling factor F. The NM is then applied to the altered (scaled) input data. If the algorithm still fails to converge, the scaling is applied repeatedly until the convergence is achieved. In addition, upon each failure, the scaling factor is reduced, thus making the procedure more adaptive. In the case of successful convergence, the algorithm does the opposite—it increases the scaling coefficient until its value reaches the desired one (the one before the scaling). The last successfully calculated set of phasors is used as an initial guess for the next iteration of the continuation method.

The flowchart of the algorithm employed to find the PCE of the circuit under study is depicted in Figure 7, where the scaling coefficients are denoted by S_i, and $i = 0$ corresponds to the smallest magnitude. At the very first iteration of the algorithm, the spectral coefficients of diode voltages are initialized using some a priori knowledge about them. The optimal value of NM damping factor (β) is found to be in the range from 0.9 to 1.1. Values of β beyond this range result in an increase in the number of iterations. The value of the DC voltage is utilized as a convergence criterion—the execution of the algorithm is terminated once the DC voltage falls below the prescribed threshold.

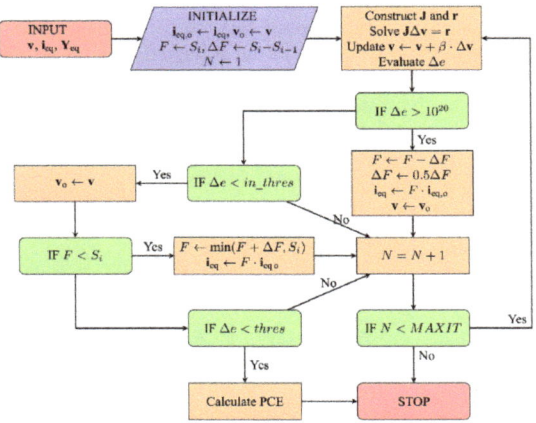

Figure 7. The flowchart of the algorithm to compute diode voltages.

2.5. Comparison of the Proposed Method with Other Methods

In order to demonstrate the efficiency of 2F-HB, a comparative study of the most commonly used non-nonlinear circuit analysis methods was undertaken. The methods were applied to an idealized voltage doubler circuit shown in Figure 2. The circuit element values are $C_1 = 2.4$ pF, $C_2 = 8.5$ nF, $C_3 = 1$ µF, $R_1 = 7.5$ kΩ, and $L_1 = L_2 = 17$ µH. The diode SPICE model parameters used in the analysis are summarized in Table 2 that correspond to the SMS7630 Schottky diode. The effect of the PCB, as well as parasitic inductances and capacitances of diodes and other circuit elements, were not taken into account in this study due to the lack of the appropriate PCB model for the TA (LTSpice [71]). The voltage doubler PCE obtained using the TA, 2F-HB, MHB, and HB is shown in Figure 8. Since both the HB and MHB are implemented in the commercially available Keysight ADS software [72] that has proven itself as a reliable and powerful non-linear circuit simulator, we employ it to calculate the PCE in place of custom programs. In order to compute the PCE using the TA, the well-established circuit simulator LTSpice is employed. The time required to compute the output voltage at 100 values of the input power level taken uniformly in the range of −20-0 dBm using each method is summarized in Table 3. The circuit was excited by a multi-carrier with 8 subcarriers occupying a 4.5 MHz band centered at 865.5 MHz (the CSF is 0.5 MHz). The Y of the PCB is computed for the two frequency ranges separately: 0.1-100 MHz and 0.1-10 GHz and exported into two MATLAB script files. The entire frequency range is divided into two subranges is to improve the calculation accuracy at low frequencies. More specifically, when applied to a wide frequency range, the interpolative sweep may result in a poor accuracy at the lower end. Once the computations are done, the exported MATLAB files are used by the program written in C++ to evaluate the entries of both the Jacobian matrix and the right-hand side vector, as well as to solve the resulting non-linear equations with Newton's method. Additionally, it should be noted that although the HB method can yield accurate results while solving the problem under consideration, it requires considering a large number of harmonics, which in turn would call for a considerable amount of computational resources. However, in this study, the issue is mitigated by considering signals whose CF is an integer multiple of the CSF. Unfortunately, such an approach imposes serious restrictions on the shape of the input signals.

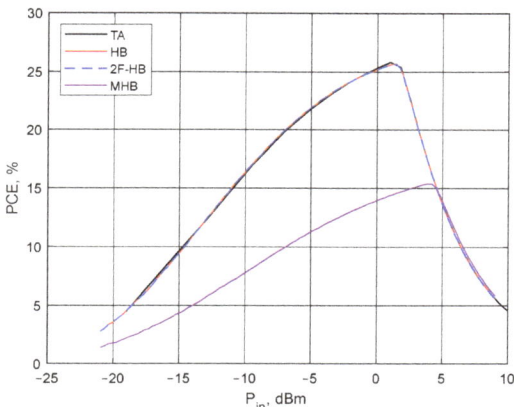

Figure 8. The PCE of the ideal voltage doubler obtained using four different methods as a function of the input power level.

Table 3. Comparison of different analysis methods.

Method	CPU Time, s	Number of Harmonics
HB (Keysight ADS)	6833	20,000 (fund. freq. 0.5 MHz)
MHB (Keysight ADS)	26,441	425 (8 fundamental freq. with the max. mixing order of 3)
2F-HB (proposed method)	71	683 (2 fund. freq.: 0.5 and 865.5 MHz)
TA with SM	3227	No harmonics. Time step: 0.01 ns Max. num. of SM iterations: 20

As can be seen in Figure 8, the HB, 2F-HB, and TA show sufficiently high accuracy, while the accuracy of the PCE obtained using the MHB method is much lower. The low accuracy is conditioned by a small number of harmonics used to approximate voltages (currents) in the circuit. However, as can be seen in Table 3, even with the small number of harmonics (425 harmonics), the CPU time required by the MHB is larger than that of other methods. The fundamental frequencies for the MHB were set to be equal to those of the subcarriers, i.e., 8 frequencies.

It should be noted that in this particular case the conventional HB method solves the task faster than the MHB, since the fundamental frequency was chosen to be equal to the CSF, and CF can be expressed as an integer multiple of CSF. In a more general case, however, the MHB considerably outperforms its conventional counterpart.

Although the computational time of the TA scales linearly with the number of harmonics provided the bandwidth is kept fixed, the main drawback of the TA is the lack of simple and reliable PCB-EC. In order to expedite simulation time, the TA was accelerated through the shooting method (SM) with the maximum number of iterations set to 20 and time step of 0.05 ns. The first period of the input signal envelope was skipped to avoid transients due to energy storage elements other than the filtering capacitor. The SM has been implemented as a MATLAB script that modifies the circuit netlist, runs the LTSpice simulations in the batch mode, and processes the results of the intermediate simulations, as well as performs postprocessing.

The proposed method (2F-HB) demonstrates good accuracy, allowing performing of computations considerably faster than other harmonic balance methods and TA. The reason why the 2F-HB outperforms the MBH when applied to multi-tone signals is the spectral redundancy of the latter. More specifically, because subcarriers are evenly spaced, a great deal of non-linear conversion products may have the same frequency, which is not considered by the MBH. Therefore, to ensure the same accuracy, the MBH requires much larger matrices than 2F-HB, and that explains the huge difference in the computational time. However, the MBH is more general. In contrast, the 2F-HB can handle multi-tone signals with unevenly distributed tone frequencies.

3. Comparison of Theoretical and Experimental Results

This section discusses experimental verification of the validity of the proposed theoretical PCE evaluation method, especially in the case of employment of the multi-tone power-carrying signals. For this purpose, the voltage doubler circuit discussed in the previous section was chosen as a test object. The set of non-linear equations describing the circuit was derived in the preceding section. The PCE can be calculated by applying the approach presented in the previous section to the set of equations. From the calculated current spectrum of the second diode it is then possible to retrieve the output DC voltage in a straightforward way. To obtain a full picture of the performance of the voltage doubler circuit under different conditions, including different types of excitations, the calculations were carried out for different values of the inductance and capacitance of the matching circuit in order to find an optimal combination for achieving the highest PCE.

The power-carrying signals considered in the present study are a classical sine wave (SW). The three types of the considered multi-tone periodic envelope signals are listed below:

- Signals formed by adding a certain number of sine waves (subcarriers) with different frequencies arranged to form a uniform spectrum with equal amplitudes and phases. These signals have high peak-to-average power ratio (PAPR) values and thus for notational simplicity will be referred to throughout this paper as HPAPR signals. The HPAPR signals considered in the present study have 4, 8, 16, 32, 64, 128, and 256 subcarriers with PAPR levels of 9.03 dB, 12.04 dB, 15.05 dB, 18.06 dB, 21.07 dB, 24.08 dB, and 27.09 dB, respectively.
- Signals formed by adding a certain number of sine waves with different frequencies (forming a uniform spectrum) and with amplitudes and phases generated using Zadoff–Chu sequences [73] and an inverse fast Fourier transform (IFFT). These signals have low PAPR values and will be referred to as LPAPR signals. The numbers of carriers of the LPAPR signals under study are 4, 8, 16, 32, 64, 128, and 256 subcarriers with PAPR levels of 6.6 dB, 6.06 dB, 6.0 dB, 7.47 dB, 7.43 dB, 6.78 dB, and 7.44 dB, respectively.
- Signals formed by adding a certain number (4–256) of sine waves with different frequencies (forming a uniform spectrum) and with random amplitudes and phases following a uniform distribution. Regarding, the PAPR level for these kinds of signals, it can take arbitrary values, depending on a random combination of amplitude and phase values and are referred to as RPAPR signals.

3.1. Calculation of the PCE by Means of the Theoretical Model

The doubler circuit was selected for being one of the most widespread RF–DC converter topologies. It has been used in a wide variety of applications and demonstrates sufficiently high efficiency [74]. The converter employs an SMS7630-005LF Schottky diode [68] that possesses a low forward voltage, small junction capacitance, and is capable of operating in the desired license free sub-GHz ISM band around 865.5 MHz.

The results of the theoretical analysis are displayed using a color plot shown in Figure 9. The plot is composed of colored squares, different colors correspond to different values of the PCE. Darker colors correspond to lower PCE values, while brighter colors are used for higher PCE values. The squares are arranged into a two-dimensional array. Each row corresponds to a particular value of the matching network capacitance, and each column corresponds to a specific value of the matching network inductance according to the topology illustrated in Figure 2. Each array element is also a two-dimensional array, whose rows correspond to different values of the input power level in dBm. The columns of subarrays correspond to different waveforms of the input signals in the following order: SW, HPAPR with 4, 8, and 16 subcarriers, LPAPR with 4, 8, and 16 subcarriers, and RPAPR with 4, 8, and 16 subcarriers. The results obtained for signals with the number of subcarriers greater than 16 are omitted in this example, since for HPAPR signals the highest achieved PCE does not exceed 25% and thus they are of little practical interest in WPT. Additionally, the obtained results demonstrate that consideration of the LPAPR and RPAPR signals with the number of subcarriers greater than 16 is completely irrelevant, since for these types of signals the PCE does not exhibit any dependence on the number of subcarriers.

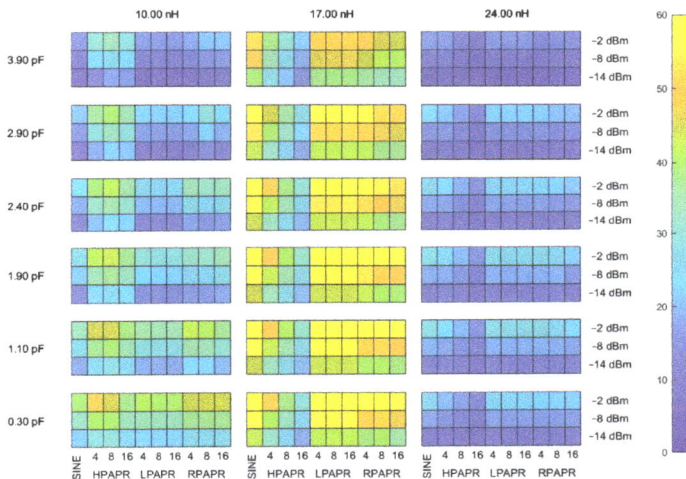

Figure 9. Color plot showing the PCE value of the voltage doubler for different values of the matching circuit parameters, waveforms, and average input power levels.

The power levels considered are -2 dBm, -8 dBm, and -14 dBm. The frequency of the carrier SW in all cases was 865.5 MHz. The reason why the results are given for the range of -14–2 dBm is due to a relatively low breakdown voltage of the diode employed in the experimental studies, namely, SMS7630. The breakdown voltage for this diode is just 2 V, resulting in considerable degradation of the PCE as the input power level exceeds approximately 0 dBm. Another reason is the nonlinearity of the generator that manifests itself at power levels close to -2 dBm when producing HPAPR signals with a large number of subcarriers, as they exhibit high peak voltages. The primary factor determining the lower limit of the input power level range being considered is the total noise level due to the generator, and both diodes. More specifically, the noise power measured by the oscilloscope when the generator power level was set to -30 dBm was in the vicinity of 4 µW that corresponds to about -23.9 dBm. This noise has not been considered during the theoretical modeling, which might result in huge discrepancies between the calculated data and the experimentally obtained data for input power levels below -14 dBm.

Figure 9 shows that the optimal value of the inductances is $L = L_1 = L_2 = 17$ nH, while the optimal value of C_1 is 2.4 pF. The SW and LPAPR signals are the waveforms with the highest achieved PCE (approx. 70%). The PCE obtained for the HPAPR signals is lower than that of the SW and LPAPR signals. Furthermore, it deteriorates as the number of subcarriers increases, attaining the maximum and minimum values for 4 and 16 subcarriers, respectively. The PCE obtained for the RPAPR signal with different subcarriers is slightly lower than that of the SW and LPAPR signals.

3.2. Experimental Validation of the Theoretical Model

In order to validate the proposed theoretical approach, experimental verification is performed with a specially designed prototype (see Figure 10) of the voltage doubler with SMS7630-005LF Schottky diode [68] capable of effectively operating at the required frequency of 865.5 MHz. The circuit components of voltage doubler are mounted on the top layer of PCB made of FR-4 with a dielectric constant of 4.2 and the thickness of the substrate of 1.6 mm. The SMA type connector is used to feed the power-carrying signal via a coaxial cable with characteristic impedance of 50 Ω during the current experimental study, or via antennas during wireless power transfer or harvesting in the real employment scenario. The matching network component values are selected by enumeration, obtaining the input impedance of the matching network closest to 50 Ω resistive load at 865.5 MHz and 0 dBm. Table 4 shows the matching process, where the optimal values of L_1, L_2, and C_1 (matching network elements) are examined. The initial values are $L_1 = L_2 = 20$ nH, $C_1 = 2.4$ pF. The values of other circuit elements are: $C_2 = 8.2$ pF, $C_3 = 1$ μF, and $R_1 = 7.5$ kΩ.

Figure 10. The fabricated prototype of the voltage doubler circuit for 865.5 MHz carrier frequency.

Table 4. Determining nominal values of the matching network.

C_1, pF	L_1, nH	L_2, nH	Input Impedance at 865.5 MHz	$|S11|$ at 865.5 MHz, dB	Frequency for $|S11|$ Minimum, MHz	$|S11|$ Minimum Value, dB
2.4	20	20	72.76 − 67.50j [1]	−5.87	790.63	−22.862
2.4	10	10	7.35 − 15.06j	−2.35	1126.90	−20.270
2.4	16	16	37.59 − 1.45j	−16.91	888.13	−47.431
2.4	18	18	73.26 − 13.70j	−13.24	839.38	−29.442
2.4	17	17	48.03 − 2.55j	−29.67	866.25	−29.759
2.9	17	17	44.99 − 5.76j	−21.97	870.01	−22.550
1.9	17	17	53.91 + 6.15j	−23.08	858.13	−26.166

[1] All instances of "j" mean the imaginary unit.

Measurements are made for different multi-tone signals with a different number of subcarriers and at different average input signal power levels. The SW is considered as the reference signal for comparison of the obtained PCE. The measurement setup is shown in Figure 11, it demonstrates the average input power level measurement (a) and converted power level measurement (b), and PCE estimation as the ratio of the average input and output powers.

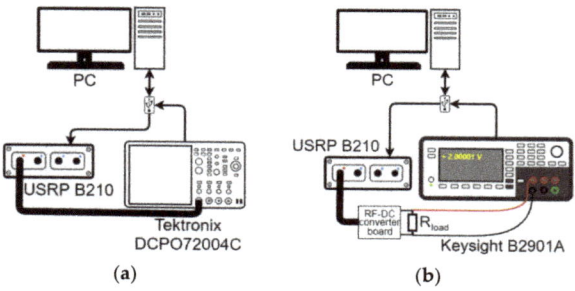

Figure 11. Measurement setup for evaluating the RF–DC conversion efficiency: (**a**) setup used for measuring the average power level of the input signal using a digital oscilloscope with the embedded average power estimation function, (**b**) setup used for measuring the RF–DC converted power level.

3.3. Evaluation of the Effect of the Matching Network Parameters on the PCE

In order to evaluate how the values of the matching network elements, namely, L and C_1, affect the performance of the voltage doubler circuit in terms of PCE, the following two different case study scenarios are considered:

a. dependence of the circuit PCE on the inductance of both inductors contained in circuit (L) for the fixed value of capacitance C_1 is calculated.
b. dependence of the circuit PCE on capacitance C_1 for the fixed value of L is found.

In order to validate the theoretical model, the aforementioned dependences are obtained experimentally as well.

As can be observed in Figures 12 and 13, the results of the theoretical analysis are in good agreement with those achieved experimentally, which means that the proposed methodology allows predicting of the behavior of diode-based RF–DC converters with a reasonably small discrepancy between the measurements and simulations. It is particularly apparent in the case of HPAPR signal, i.e., the shapes of the curves corresponding to different number of subcarriers match the calculated ones well. In the case of the dependence of the PCE on C_1 the largest discrepancy between the results is observed for small values of C_1. Similar to L sweep, in this case, the largest difference is also observed at the input power level of −14 dBm. The highest PCE of 64.8% was achieved for the SW. As for the multi-tone signals, the LPAPR signals exhibit the highest PCE of 63.15%. Furthermore, the PCE of LPAPR signals varies only slightly with the number of carriers. In the case of the HPAPR signals, the highest PCE reaches 51.64% for the signal with 4-subcarriers.

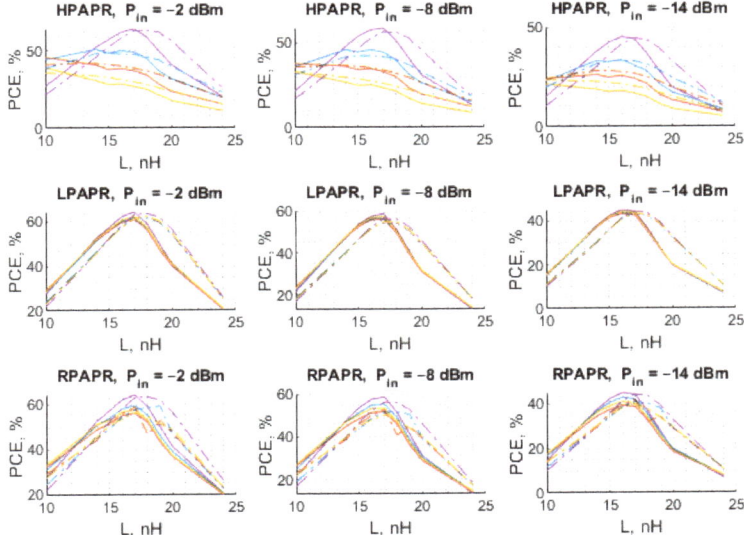

Figure 12. The calculated (dashed line) and measured (solid line) PCE of the voltage doubler RF–DC converter as a function of L when $C_1 = 2.4$ pF for different numbers of subcarriers: 4 (blue), 8 (red), and 16 (yellow) with the SW (purple) used as a reference.

The highest PCE of 60% that is very close to the one obtained in this work for a sine wave-driven single diode rectifier operating at 10 GHz was achieved in [75]. Though the working frequency is about an order of magnitude higher than the one considered in the present study, the input power level at which such a high efficiency has been attained is much higher. To compute the PCE the authors employed both the closed form expressions and LIBRE software employing the harmonic balance.

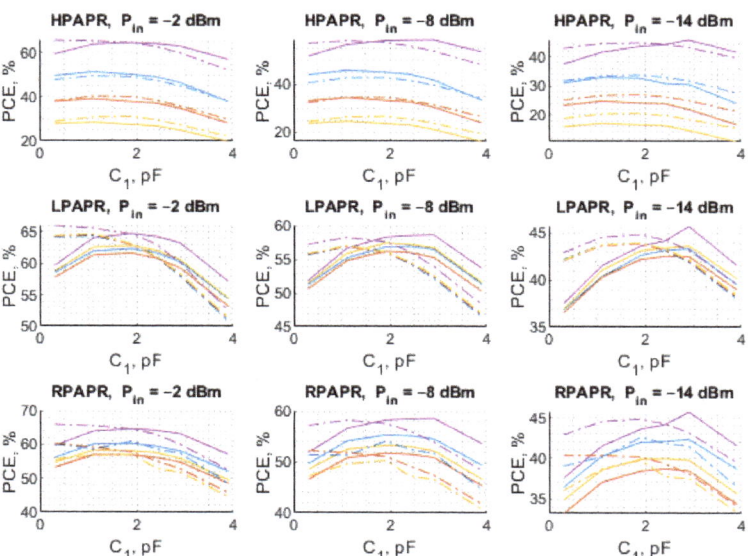

Figure 13. The calculated (dashed line) and measured (solid line) PCE of the voltage doubler RF–DC converter as a function of C_1 when $L = 17$ nH for different numbers of subcarriers: 4 (blue), 8 (red), and 16 (yellow) with the SW (purple) used as a reference.

A comprehensive comparative analysis of the efficiencies attainable by means of various RF–DC converters, including diode converters and CMOS technology-based converters, is presented in [76,77]. From this analysis it follows that using a pure sine wave, i.e., a single tone signal, the maximum achievable PCE does not exceed 60% when input power levels below 1 mW (0 dBm) are considered. Nevertheless, the same analysis also demonstrates that it is possible to achieve a PCE of up to 90% for sufficiently high power levels (around 1 W). However, to obtain such an amount of received power for medium distances (few tens of meters), that are typical distances in IoT sensor networks, according to the well-known Friis transmission equation one needs to maintain a high transmitted power that, in turn, necessitates more expensive equipment. This makes the deployment and wireless charging process costly, while the goal of the present study is to develop an affordable medium power alternative with sufficiently high PCE not the highest possible.

Although the voltage doubler circuit studied in this work has a limited range of the input power level (<0 dBm) due to a relatively low breakdown voltage of the diodes, as well as exhibiting the highest PCE that is just about 65%, the proposed approach has no limitation with respect to the circuit topology, PCB layout, working frequency, and power levels of input signals as it relies on the full-wave analysis. Alternatively, the only limitation of the full-wave analysis is the CPU time that increases with the frequency, and the complexity of the layout.

3.4. Simulation and Experimental Results for HPAPR Signals

The results discussed in the previous subsection show that the notable difference of PCE for different carrier number is observed only in the case of the high PAPR level. Thus, they deserve more detailed consideration. The PCE for the circuit under study is obtained for a larger number of subcarriers to obtain a more in-depth insight into the circuit behavior driven by such signals. The considered signals are HPAPR signals with 4, 8, 16, 32, 64, 128, and 256 subcarriers. Both the calculated and experimental PCE are graphically represented with scatterplots. For the graphs shown in Figure 14, the horizontal axis represents values of the matching network inductance, while the vertical one—the input power level. Each circle corresponds to a specific number of subcarriers. The size of circles increases with the number of subcarriers, i.e., the innermost circle corresponds to 4 subcarriers, while the outermost—to 256 subcarriers. The color of each circle represents different values of the PCE (both calculated and measured), where darker colors show the lower values of the PCE, while brighter ones—the higher values of the PCE. Regarding the graphs shown in Figure 15, the same format is used, but the horizontal axis represents different values of C_1.

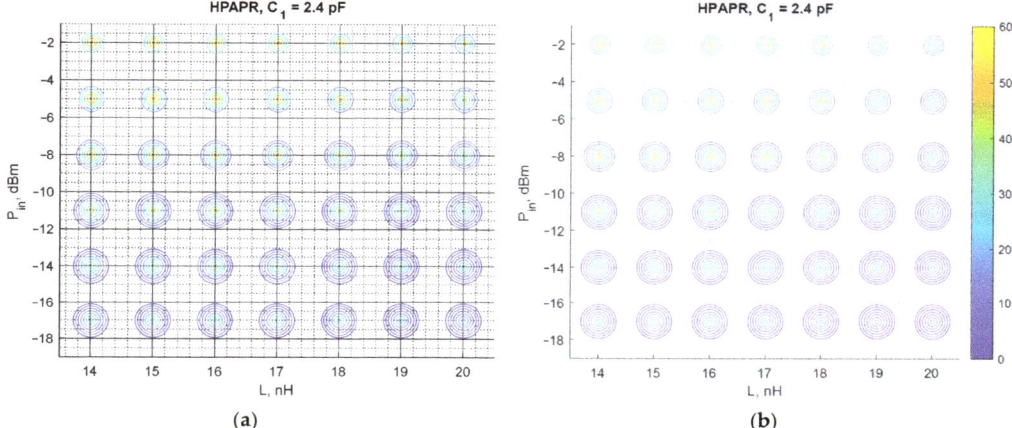

Figure 14. The calculated (**a**) and measured (**b**) PCE of the voltage doubler RF–DC converter as a function of both L and input signal power for $C_1 = 2.4$ pF. PCE is represented by color. Size of the circle represents the number of subcarriers (4, 8, 16, 32, 64, 128, 256), the smallest is for 4 carriers and the largest—for 256 carriers.

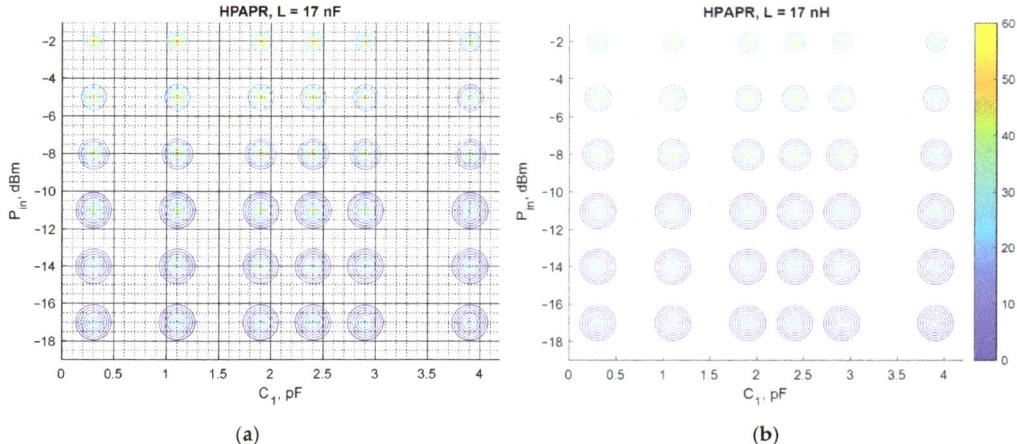

Figure 15. The calculated (**a**) and measured (**b**) PCE of the voltage doubler RF–DC converter as a function of C_1 values and input signal power for $L = 17$ nH. PCE is represented by color. Size of the circle represents the number of subcarriers (4, 8, 16, 32, 64, 128, 256), the smallest is for 4 carriers and the largest—for 256 carriers.

Again, Figures 14 and 15 show that the results of the measurements are consistent with the results obtained with the theoretical model, proving the validity of the proposed method. Both theoretical and experimental results show that in the case of input signal formed, to have the maximum possible PAPR level among signals with the same number of subcarriers, the PCE diminishes progressively with the number of subcarriers. In most of the cases considered, the highest PCE is attained by the signals with 4 subcarriers. The PCE for signals with 8 subcarriers is typically 10% lower than that of the signal with 4 subcarriers. For some combinations of the matching network element values (L and C_1), the opposite behavior is observed, i.e., a 4-subcarrier signal shows lower PCE than its 8-subcarrier counterpart. However, those combinations are not the optimal ones and the PCE of the sine wave in these cases is lower or comparable with that of the signals with 4 and 8 subcarriers. Another finding of this study concerns the sensitivity of the PCE to variations in the values of L, and C_1. Despite being the most optimal waveform in terms of the PCE, it was found that SW exhibits the highest sensitivity to variations in the matching network element values.

In some cases, the difference between the calculated and experimentally obtained results is quite small, e.g., for HPAPR signals with a small number of tones (<16). Although even in this case the error is large at large deviation from the optimal values of L and C (matching circuit elements), it occurs due to the shift between the theoretical and measured curves. A possible source of such a shift is likely the difference between the actual values of the discrete inductor used in the experimental studies and the one used in the theoretical model calculated from the data provided in the relevant datasheets.

The PCE is unacceptably low as far as signals with the number of subcarriers greater than 16. For this reason, such signals cannot be used for powering isolated sensor network nodes. This finding agrees with the results of a recent study undertaken by another group of researchers who also examined a voltage doubler circuit, but operating at lower frequencies [78]. The researchers also found that the use of signals with a high peak-to-average power ratio does not improve the PCE of RF–DC converters.

For numerical comparison of the theoretical and measured results from Figures 14 and 15, the estimation error is presented in Tables 5 and 6 corresponding to each figure. The error is taken as relative to the measured PCE. Since Figures 14 and 15 contain a substantial amount of data, tables show estimation errors for the input power of −2 dBm and in the range of 4–32 carriers. The tables show that the estimation error reaches as low as 0.37%, and the maximal estimation error is 32.65%. In Table 5 the estimation error notably increases when L is greater than 18 nH, which is visible in Figure 12 for the HPAPR. In Figure 12 the difference between the theoretical and measured curves increases with subcarrier number and L value. The source of such shift between the theoretical and measured curves is explained with the nominal mismatch of the two L elements (L_1 and L_2) and the SDR signal nonlinearity in the case of a large number of subcarriers.

Table 5. Relative error for theoretical and measured PCE results in % (C = 2.4 pF).

Pin, dBm	Subcar. No.	L = 10 nH	L = 14 nH	L = 15 nH	L = 16 nH	L = 17 nH	L = 18 nH	L = 19 nH	L = 20 nH	L = 24 nH
−2	4	10.29	4.34	0.37	3.49	3.45	3.10	16.45	20.75	18.58
	8	8.47	3.47	8.08	4.06	3.89	7.15	18.78	25.40	23.64
	16	8.53	9.00	13.72	9.06	10.36	11.60	19.44	27.38	28.32
	32	5.74	14.80	19.33	15.33	17.43	17.77	23.51	30.33	32.65

Table 6. Relative error for theoretical and measured PCE results in % (L = 17 nH).

Pin, dBm	Subcar. No.	C = 0.3 pF	C = 1.1 pF	C = 1.9 pF	C = 2.4 pF	C = 2.9 pF	C = 3.9 pF
−2	4	3.60	3.45	2.27	3.45	2.88	1.40
	8	0.83	3.63	6.26	3.89	3.71	6.81
	16	4.49	8.36	12.73	10.36	9.98	11.55
	32	8.97	14.48	19.63	17.43	17.52	17.95

4. Conclusions

The current paper proposes a novel theoretical approach to estimating the power conversion efficiency (PCE) of RF–DC converters for WPT applications. The approach relies on using the two-frequency harmonic balance (2F-HB) method in conjunction with full-wave simulations of the circuit PCB. A comparative numerical study showed that when applied to multi-tone signals, the 2F-HB appreciably outperforms the multi-dimensional harmonic balance method (MHB), conventional harmonic balance method (HB), and transient analysis (TA) in terms of required CPU time. The results of the HB and the MHB have been obtained using the commercially available Keysight ADS circuit simulator, whereas those of TA were computed by means of the LTSpice in conjunction with the shooting method (SM) implemented as a MATLAB script. To evaluate the accuracy of the theoretical model, the authors performed experimental measurements for the RF–DC converter prototype based on the voltage doubler rectifier topology. The PCE of the voltage doubler circuit was calculated and measured for different RF–DC converter matching network elements and different average input signal power levels and waveforms in the sub-GHz band.

The numerical results obtained using the proposed theoretical model have been found to be in good agreement with the results measured experimentally, which firmly attests the consistency between the simulations and experiments. The calculation accuracy reaches 0.37%. Furthermore, the results obtained for different values of the matching network elements exhibit the existence of optimal combinations for achieving the highest PCE, thus demonstrating the potential of the proposed estimation method in the design of highly efficient RF–DC converters.

Although only a voltage doubler was considered in this work, the applicability range of this approach is not limited to such a simple circuit, as it is capable of handling a wide range of RF–DC converter topologies involving an arbitrary number of diodes. The new method allows for editing and fine-tuning the design of an RF–DC converter much quicker than previous methods due to accelerated PCE estimation, which is 96 times faster than the broadly used harmonic balance method.

Author Contributions: Conceptualization, J.S. and A.L.; funding acquisition, A.L.; investigation, J.E., R.K. and D.C.; methodology, J.E., J.S. and A.L.; project administration, A.L.; software, J.E., R.K., R.B. and D.C.; supervision, J.S.; validation, R.B.; visualization, J.E. and D.C.; writing—original draft, J.E., R.B. and D.C.; writing—review and editing, R.K., R.B., J.S. and A.L. All authors have read and agreed to the published version of the manuscript.

Funding: This research was funded by the Latvian Council of Science, grant No. lzp-2020/2-0344, "Radio Frequency Wireless Power Transfer for Wireless Sensor Network Applications".

Institutional Review Board Statement: Not applicable.

Informed Consent Statement: Not applicable.

Data Availability Statement: Not applicable.

Conflicts of Interest: The authors declare no conflict of interest.

References

1. Zabasta, A.; Selmanovs-Pless, V.; Kunicina, N.; Ribickis, L. Wireless Sensor Networks for Optimumisation of District Heating. In Proceedings of the 15th International Power Electronics and Motion Control Conference (EPE/PEMC), Novi Sad, Serbia, 4–6 September 2012. [CrossRef]
2. Gulec, O.; Haytaoglu, E.; Tokat, S. A Novel Distributed CDS Algorithm for Extending Lifetime of WSNs with Solar Energy Harvester Nodes for Smart Agriculture Applications. *IEEE Access* **2020**, *8*, 58859–58873. [CrossRef]
3. Bharadwaj, R.; Koul, S.K. Assessment of Limb Movement Activities Using Wearable Ultra-Wideband Technology. *IEEE Trans. Antennas Propag.* **2021**, *69*, 2316–2325. [CrossRef]
4. Shinohara, N.; Nishikawa, K.; Seki, T.; Hiraga, K. Development of 24 GHz Rectennas for Fixed Wireless Access. In Proceedings of the 2011 30th URSI General Assembly and Scientific Symposium, Istanbul, Turkey, 13–20 August 2011. [CrossRef]
5. Sakai, N.; Noguchi, K.; Itoh, K. A 5.8-GHz Band Highly Efficient 1-W Rectenna with Short-Stub-Connected High-Impedance Dipole Antenna. *IEEE Trans. Microw. Theory Tech.* **2021**, *69*, 3558–3566. [CrossRef]

6. Singh, N.; Kanaujia, B.K.; Beg, M.T.; Kumar, S.; Khan, T.; Sachin, K. A Dual Polarized Multiband Rectenna for RF Energy Harvesting. *Int. J. Electron. Commun.* **2018**, *93*, 123–131. [CrossRef]
7. Suh, Y.H.; Chang, K. A high-efficiency dual-frequency rectenna for 2.45- and 5.8-GHz wireless power transmission. *IEEE Trans. Microw. Theory Tech.* **2002**, *50*, 1784–1789. [CrossRef]
8. Wang, C.; Yang, B.; Shinohara, N. Study and Design of a 2.45-GHz Rectifier Achieving 91% Efficiency at 5-W Input Power. *IEEE Microw. Wirel. Components Lett.* **2021**, *31*, 76–79. [CrossRef]
9. Yo, T.-C.; Lee, C.-M.; Hsu, C.-M.; Luo, C.-H. Compact Circularly Polarized Rectenna With Unbalanced Circular Slots. *IEEE Trans. Antennas Propag.* **2008**, *56*, 882–886. [CrossRef]
10. Chang, Y.T.; Claessens, S.; Pollin, S.; Schreurs, D. A Wideband Efficient Rectifier Design for SWIPT. 2019. Available online: https://ieeexplore-ieee-org.resursi.rtu.lv/document/9055666/ (accessed on 5 December 2021).
11. Rotenberg, S.A.; Podilchak, S.K.; Re, P.D.H.; Mateo-Segura, C.; Goussetis, G.; Lee, J. Efficient Rectifier for Wireless Power Transmission Systems. *IEEE Trans. Microw. Theory Tech.* **2020**, *68*, 1921–1932. [CrossRef]
12. Ito, M.; Hosodani, K.; Itoh, K.; Betsudan, S.-I.; Makino, S.; Hirota, T.; Noguchi, K.; Taniguchi, E. High Efficient Bridge Rectifiers in 100MHz and 2.4GHz Bands. In Proceedings of the 2014 IEEE Wireless Power Transfer Conference, Jeju, Korea, 8–9 May 2014; pp. 64–67. [CrossRef]
13. Chou, J.-H.; Lin, D.-B.; Weng, K.-L.; Li, H.-J. All Polarization Receiving Rectenna With Harmonic Rejection Property for Wireless Power Transmission. *IEEE Trans. Antennas Propag.* **2014**, *62*, 5242–5249. [CrossRef]
14. Wang, D.; Negra, R. Design of a Dual-Band Rectifier for Wireless Power Transmission. In Proceedings of the 2013 IEEE Wireless Power Transfer (WPT), Perugia, Italy, 15–16 May 2013; pp. 127–130. [CrossRef]
15. Ouda, M.H.; Mitcheson, P.; Clerckx, B. Robust Wireless Power Receiver for Multi-Tone Waveforms. In Proceedings of the 2019 49th European Microwave Conference (EuMC), Paris, France, 1–3 October 2019; pp. 101–104. [CrossRef]
16. Song, C.; Huang, Y.; Zhou, J.; Zhang, J.; Yuan, S.; Carter, P. A high-efficiency broadband rectenna for ambient wireless energy harvesting. *IEEE Trans. Antennas Propag.* **2015**, *63*, 3486–3495. [CrossRef]
17. Bolos, F.; Blanco, J.; Collado, A.; Georgiadis, A. RF Energy Harvesting from Multi-Tone and Digitally Modulated Signals. *IEEE Trans. Microw. Theory Tech.* **2016**, *64*, 1918–1927. [CrossRef]
18. Quddious, A.; Antoniades, M.A.; Vryonides, P.; Nikolaou, S. Voltage-Doubler RF-to-DC Rectifiers for Ambient RF Energy Harvesting and Wireless Power Transfer Systems. In *Recent Wireless Power Transfer Technologies*; IntechOpen: London, UK, 2019. [CrossRef]
19. Collado, A.; Georgiadis, A. Optimal Waveforms for Efficient Wireless Power Transmission. *IEEE Microw. Wirel. Compon. Lett.* **2014**, *24*, 354–356. [CrossRef]
20. Trotter, M.S.; Griffin, J.D.; Durgin, G.D. Power-Optimized Waveforms for Improving the Range and Reliability of RFID Systems. In Proceedings of the 2009 IEEE International Conference on RFID, Orlando, FL, USA, 27–28 April 2009; pp. 80–87. [CrossRef]
21. Terauds, M.; Malbranque, L.; Smolaninovs, V. Application of LoRaWAN for Interactive E-ink Based Schedule Board. In Proceedings of the 2020 IEEE Microwave Theory and Techniques in Wireless Communications (MTTW), Riga, Latvia, 1–2 October 2020; Volume 1, pp. 222–226. [CrossRef]
22. Collado, A.; Georgiadis, A. Improving Wireless Power Transmission Efficiency Using Chaotic Waveforms. In Proceedings of the 2012 IEEE/MTT-S International Microwave Symposium Digest, Montreal, QC, Canada, 17–22 June 2012. [CrossRef]
23. Eidaks, J.; Litvinenko, A.; Chiriyankandath, J.P.; Varghese, M.A.; Shah, D.D.; Prathakota, Y.K.T. Impact of Signal Waveform on RF-Harvesting Device Performance in Wireless Sensor Network. In Proceedings of the 2019 IEEE 60th International Scientific Conference on Power and Electrical Engineering of Riga Technical University (RTUCON), Riga, Latvia, 7–9 October 2019. [CrossRef]
24. Kim, J.; Clerckx, B.; Mitcheson, P.D. Signal and System Design for Wireless Power Transfer: Prototype, Experiment and Validation. *IEEE Trans. Wirel. Commun.* **2020**, *19*, 7453–7469. [CrossRef]
25. Litvinenko, A.; Eidaks, J.; Tjukovs, S.; Pikulins, D.; Aboltins, A. Experimental Study of the Impact of Waveforms on the Efficiency of RF-to-DC Conversion Using a Classical Voltage Doubler Circuit. In Proceedings of the 2018 Advances in Wireless and Optical Communications (RTUWO), Riga, Latvia, 15–16 November 2018; pp. 257–262. [CrossRef]
26. Mitani, K.; Kawamura, Y.; Kitagawa, W.; Takeshita, T. Circuit Modeling for Common Mode Noise on AC/DC Converter Using Sic Device. In Proceedings of the 2019 21st European Conference on Power Electronics and Applications (EPE '19 ECCE Europe), Genova, Italy, 3–5 September 2019. [CrossRef]
27. Pikulin, D. Complete bifurcation analysis of DC-DC converters under current mode control. *J. Phys. Conf. Ser.* **2014**, *482*, 012034. [CrossRef]
28. Pikulins, D. Exploring Types of Instabilities in Switching Power Converters: The Complete Bifurcation Analysis. *Elektron. ir Elektrotech.* **2014**, *20*, 76–79. [CrossRef]
29. Pederson, D. A historical review of circuit simulation. *IEEE Trans. Circuits Syst.* **1984**, *31*, 103–111. [CrossRef]
30. Liu, H.; Wong, N. Autonomous Volterra Algorithm for Steady-State Analysis of Nonlinear Circuits. *IEEE Trans. Comput. Des. Integr. Circuits Syst.* **2013**, *32*, 858–868. [CrossRef]
31. de Luca, G.; Bolcato, P.; Schilders, W.H.A. Proper Initial Solution to Start Periodic Steady-State-Based Methods. *IEEE Trans. Circuits Syst. I Regul. Pap.* **2019**, *66*, 1104–1115. [CrossRef]

32. Xiong, X.Y.Z.; Jiang, L.J.; Schutt-Ainé, J.E.; Chew, W.C. Volterra Series-Based Time-Domain Macromodeling of Nonlinear Circuits. *IEEE Trans. Components, Packag. Manuf. Technol.* **2017**, *7*, 39–49. [CrossRef]
33. Krylov, N.N.B.; Krylov, N.N.; Bogoliubov, N.N. *Introduction to Nonlinear Mechanics*; Princeton University Press: Princeton, NJ, USA, 1995; p. 147.
34. Baily, E.M. *Steady-State Harmonic Analysis of Non-Linear Networks*; Stanford University: Stanford, CA, USA, 1968.
35. Zhang, Q.-J.; Gad, E.; Nouri, B.; Na, W.; Nakhla, M. Simulation and Automated Modeling of Microwave Circuits: State-of-the-Art and Emerging Trends. *IEEE J. Microw.* **2021**, *1*, 494–507. [CrossRef]
36. Nakhla, M.; Vlach, J. A piecewise harmonic balance technique for determination of periodic response of nonlinear systems. *IEEE Trans. Circuits Syst.* **1976**, *23*, 85–91. [CrossRef]
37. Egami, S. Nonlinear, Linear Analysis and Computer-Aided Design of Resistive Mixers. *IEEE Trans. Microw. Theory Tech.* **1974**, *22*, 270–275. [CrossRef]
38. Kerr, A.R. A Technique for Determining the Local Oscillator Waveforms in a Microwave Mixer (Short Papers). *IEEE Trans. Microw. Theory Tech.* **1975**, *23*, 828–831. [CrossRef]
39. Hicks, R.G.; Khan, P.J. Numerical Analysis of Nonlinear Solid-State Device Excitation in Microwave Circuits. *IEEE Trans. Microw. Theory Tech.* **1982**, *30*, 251–259. [CrossRef]
40. Filicori, F.; Scalas, M.R.; Naldi, C. Nonlinear circuit analysis through periodic spline approximation. *Electron. Lett.* **1979**, *15*, 597–599. [CrossRef]
41. Cooley, J.W.; Lewis, P.A.W.; Welch, P.D. The Fast Fourier Transform and Its Applications. *IEEE Trans. Educ.* **1969**, *12*, 27–34. [CrossRef]
42. Gilmore', R.J.; Steer2, M.B. Nonlinear Circuit Analysis Using the Method of Harmonic Balance-A Review of the Art. Part I. Introductory Concepts. *Int. J. Microw. Millim. Wave Comput. Aided Eng.* **1991**, *1*, 22–37. [CrossRef]
43. Rizzoli, V.; Lipparini, A.; Marazzi, E. A General-Purpose Program for Nonlinear Microwave Circuit Design. *IEEE Trans. Microw. Theory Tech.* **1983**, *31*, 762–770. [CrossRef]
44. Rizzoli, V.; Cecchetti, C.; Lipparini, A. A General-Purpose Program for the Analysis of Nonlinear Microwave Circuits Under Multitone Excitation by Multidimensional Fourier Transform. In Proceedings of the 1987 17th European Microwave Conference, Rome, Italy, 7–11 September 1987; pp. 635–640. [CrossRef]
45. Ngoya, E.; Suárez, A.; Sommet, R.; Quéré, R. Steady state analysis of free or forced oscillators by harmonic balance and stability investigation of periodic and quasi-periodic regimes. *Int. J. Microw. Millim. Wave Comput. Eng.* **1995**, *5*, 210–223. [CrossRef]
46. Gourary, M.; Ulyanov, S.; Zharov, M.; Rusakov, S.; Gullapalli, K.K.; Mulvaney, B.J. A robust and efficient oscillator analysis technique using harmonic balance. *Comput. Methods Appl. Mech. Eng.* **2000**, *181*, 451–466. [CrossRef]
47. Gourary, M.M.; Rusakov, S.G.; Ulyanov, S.L.; Zharov, M.M. Improved Harmonic Balance Technique for Analysis of Ring Oscillators. In Proceedings of the 2009 European Conference on Circuit Theory and Design, Antalya, Turkey, 23–27 August 2009; pp. 327–330. [CrossRef]
48. Li, P.; Pileggi, L.T. Efficient Harmonic Balance Simulation Using Multi-Level Frequency Decomposition. In Proceedings of the IEEE/ACM International Conference on Computer Aided Design, 2004. ICCAD-2004., San Jose, CA, USA, 7–11 November 2004; pp. 677–682. [CrossRef]
49. Dong, W.; Li, P. Accelerating Harmonic Balance Simulation Using Efficient Parallelizable Hierarchical Preconditioning. In Proceedings of the 2007 44th ACM/IEEE Design Automation Conference, San Diego, CA, USA, 4–8 June 2007; pp. 436–439.
50. Dong, W.; Li, P. Hierarchical Harmonic-Balance Methods for Frequency-Domain Analog-Circuit Analysis. *IEEE Trans. Comput. Des. Integr. Circuits Syst.* **2007**, *26*, 2089–2101. [CrossRef]
51. Dong, W.; Li, P. A Parallel Harmonic-Balance Approach to Steady-State and Envelope-Following Simulation of Driven and Autonomous Circuits. *IEEE Trans. Comput. Des. Integr. Circuits Syst.* **2009**, *28*, 490–501. [CrossRef]
52. Han, L.; Zhao, X.; Feng, Z. An Efficient Graph Sparsification Approach to Scalable Harmonic Balance (HB) Analysis of Strongly Nonlinear RF Circuits. In Proceedings of the 2013 IEEE/ACM International Conference on Computer-Aided Design (ICCAD), San Jose, CA, USA, 18–21 November 2013; pp. 494–499. [CrossRef]
53. Gutmann, R.J.; Borrego, J.M. Power Combining in an Array of Microwave Power Rectifiers. In Proceedings of the 1979 IEEE MTT-S International Microwave Symposium Digest, Orlando, FL, USA, 30 April–2 May 1979; pp. 453–455. [CrossRef]
54. Guo, J.; Zhu, X. An Improved Analytical Model for RF-DC Conversion Efficiency in Microwave Rectifiers. In Proceedings of the 2012 IEEE/MTT-S International Microwave Symposium Digest, Montreal, QC, Canada, 17–22 June 2012. [CrossRef]
55. Guo, J.; Zhang, H.; Zhu, X. Theoretical analysis of RF-DC conversion efficiency for class-F rectifiers. *IEEE Trans. Microw. Theory Tech.* **2014**, *62*, 977–985. [CrossRef]
56. Gao, S.P.; Zhang, H.; Guo, Y.X. Closed-Form Expressions Based Automated Rectifier Synthesis. In Proceedings of the 2019 IEEE MTT-S International Wireless Symposium (IWS), Guangzhou, China, 19–22 May 2019. [CrossRef]
57. Hirakawa, T.; Shinohara, N. Theoretical Analysis and Novel Simulation for Single Shunt Rectifiers. *IEEE Access* **2021**, *9*, 16615–16622. [CrossRef]
58. Pan, N.; Belo, D.; Rajabi, M.; Schreurs, D.; Carvalho, N.B.; Pollin, S. Bandwidth Analysis of RF-DC Converters Under Multisine Excitation. *IEEE Trans. Microw. Theory Tech.* **2017**, *66*, 791–802. [CrossRef]
59. Ohira, T. Power efficiency and optimum load formulas on RF rectifiers featuring flow-angle equations. *IEICE Electron. Express* **2013**, *10*, 20130230. [CrossRef]

60. Lam, K.C.A.; Zwolinski, M. Circuit Transient Analysis Using State Space Equations. *Commun. Comput. Inf. Sci.* **2013**, *382*, 330–336. [CrossRef]
61. Guskov, M.; Sinou, J.J.; Thouverez, F. Multi-dimensional harmonic balance applied to rotor dynamics. *Mech. Res. Commun.* **2008**, *35*, 537–545. [CrossRef]
62. Valenta, C.R.; Morys, M.M.; Durgin, G.D. Theoretical Energy-Conversion Efficiency for Energy-Harvesting Circuits Under Power-Optimized Waveform Excitation. *IEEE Trans. Microw. Theory Tech.* **2015**, *63*, 1758–1767. [CrossRef]
63. PSpice How to Use This Online Manual Reference Guide. 1985. Available online: https://www.orcad.com/technical/technical.asp (accessed on 5 December 2021).
64. Osborne, A.R. Multidimensional Fourier series. *Int. Geophys.* **2010**, *97*, 115–145. [CrossRef]
65. Tutatchikov, V.S. Two-Dimensional Fast Fourier Transform: Batterfly in Analog of Cooley-Tukey Algorithm. In Proceedings of the 2016 11th International Forum on Strategic Technology (IFOST), Novosibirsk, Russia, 1–3 June 2016; pp. 495–498. [CrossRef]
66. Foster, L.V. The growth factor and efficiency of Gaussian elimination with rook pivoting. *J. Comput. Apphed Math. Elsevier J. Comput. Appl. Math.* **1997**, *86*, 177–194. [CrossRef]
67. Dembo, R.S.; Eisenstat, S.C.; Steihaug, T. Inexact Newton Methods. *SIAM J. Numer. Anal.* **2006**, *19*, 400–408. [CrossRef]
68. Skyworks Solutions. Surface Mount Mixer and Detector Schottky Diodes. Datasheet. 2018, pp. 1–9. Available online: https://www.skyworksinc.com/Products/Diodes/SMS7630-Series (accessed on 5 December 2021).
69. Ansys HFSS | 3D High Frequency Simulation Software. Available online: https://www.ansys.com/products/electronics/ansys-hfss (accessed on 21 September 2021).
70. Jin, J.-M. *The Finite Element Method in Electromagnetics*; John Wiley & Sons: Hoboken, NJ, USA, 2015; p. 846.
71. LTspice Simulator | Analog Devices. Available online: https://www.analog.com/en/design-center/design-tools-and-calculators/ltspice-simulator.html (accessed on 5 December 2021).
72. PathWave Advanced Design System (ADS) | Keysight. Available online: https://www.keysight.com/zz/en/products/software/pathwave-design-software/pathwave-advanced-design-system.html (accessed on 5 December 2021).
73. Popovic, B.M. Generalized chirp-like polyphase sequences with optimum correlation properties. *IEEE Trans. Inf. Theory* **1992**, *38*, 1406–1409. [CrossRef]
74. Tjukovs, S.; Eidaks, J.; Pikulins, D. Experimental Verification of Wireless Power Transfer Ability to Sustain the Operation of LoRaWAN Based Wireless Sensor Node. In Proceedings of the 2018 Advances in Wireless and Optical Communications (RTUWO), Riga, Latvia, 15–16 November 2018; pp. 83–88. [CrossRef]
75. Yoo, T.W.; Chang, K. Theoretical and Experimental Development of 10 and 35 GHz Rectennas. *IEEE Trans. Microw. Theory Tech.* **1992**, *40*, 1259–1266. [CrossRef]
76. Hemour, S.; Wu, K. Radio-frequency rectifier for electromagnetic energy harvesting: Development path and future outlook. *IEEE RFID Virtual J.* **2014**, *102*, 1667–1691. [CrossRef]
77. Hemour, S.; Zhao, Y.; Lorenz, C.H.P.; Houssameddine, D.; Yongsheng, G.; Can-Ming, H.; Ke, W. Towards low-power high-efficiency RF and microwave energy harvesting. *IEEE Trans. Microw. Theory Tech.* **2014**, *62*, 965–976. [CrossRef]
78. Shariati, N.; Scott, J.R.; Schreurs, D.; Ghorbani, K. Multitone excitation analysis in RF energy harvesters—Considerations and limitations. *IEEE Internet Things J.* **2018**, *5*, 2804–2816. [CrossRef]

Article

Closed-Form UAV LoS Blockage Probability in Mixed Ground- and Rooftop-Mounted Urban mmWave NR Deployments

Vyacheslav Begishev [1,*], Dmitri Moltchanov [2], Anna Gaidamaka [2,3] and Konstantin Samouylov [1,4]

1. Department of Applied Probability and Informatics, Peoples' Friendship University of Russia (RUDN University), 117198 Moscow, Russia; samuylov-ke@rudn.ru
2. Unit of Electronics and Communications Engineering, Tampere University, 33100 Tampere, Finland; dmitri.moltchanov@tuni.fi (D.M.); anna.gaydamaka@tuni.fi (A.G.)
3. Laboratory of the Internet of Things and Cyber-Physical Systems, HSE University, 101000 Moscow, Russia
4. Institute of Informatics Problems, Federal Research Center Computer Science and Control of Russian Academy of Sciences, 119333 Moscow, Russia
* Correspondence: begishev-vo@rudn.ru; Tel.: +7-903-559-87-64

Abstract: Unmanned aerial vehicles (UAV) are envisioned to become one of the new types of fifth/sixth generation (5G/6G) network users. To support advanced services for UAVs such as video monitoring, one of the prospective options is to utilize recently standardized New Radio (NR) technology operating in the millimeter-wave (mmWave) frequency band. However, blockage of propagation paths between NR base stations (BS) and UAV by buildings may lead to frequent outage situations. In our study, we use the tools of integral geometry to characterize connectivity properties of UAVs in terrestrial urban deployments of mmWave NR systems using UAV line-of-sight (LoS) blockage probability as the main metric of interest. As opposed to other studies, the use of the proposed approach allows us to get closed-form approximation for LoS blockage probability as a function of city and network deployment parameters. As one of the options to improve connectivity we also consider rooftop-mounted mmWave BSs. Our results illustrate that the proposed model provides an upper bound on UAV LoS blockage probability, and this bound becomes more accurate as the density of mmWave BS in the area increases. The closed-form structure allows for identifying of the street width, building block and BS heights, and UAV altitude as the parameters providing the most impact on the considered metric. We show that rooftop-mounted mmWave BSs allow for the drastic improvement of LoS blockage probability, i.e., depending on the system parameters the use of one rooftop-mounted mmWave BS is equivalent to 6–12 ground-mounted mmWave BSs. Out of all considered deployment parameters the street width is the one most heavily affecting the UAV LoS blockage probability. Specifically, the deployment with street width of 20 m is characterized by 50% lower UAV LoS blockage probability as compared to the one with 10 m street width.

Keywords: millimeter wave; new radio; unmanned aerial vehicles; LoS blockage; closed-from approximation; rooftop deployments

1. Introduction

The opportunities offered by unmanned aerial vehicles (UAVs) in a wide variety of fields have led to a dramatic increase in their production and deployment. Initially utilized in the military, UAVs today are applied in many fields including communications networks [1]. Specifically, UAVs can be used in wireless communications systems to reliably support connectivity in disaster management, public safety, and rescue operations [2–4].

The support of UAVs as a new network user in fifth generation (5G) systems opens up new opportunities related to the organization of services such as delivery, security surveillance, mapping navigation, and many others [5–7]. Furthermore, UAVs can be utilized by the network operator as repeaters and mobile base stations (BS). However, UAVs are characterized by the new unique properties compared to classic users (higher speed, higher position relative to the ground, etc.) and thus require new mechanisms to support them in 5G systems.

Many factors such as selected frequency range, line-of-sight (LoS) range, and signal attenuation play an important role in UAV communications. The work within 3GPP related to integration of UAV to 5G systems started with 3GPP SP-180909 (see Section 2 for a detailed overview of 3GPP standardization efforts) outlining requirements for communication delay, rate, and reliability as key performance indicators (KPI), for UAV applications. For example, security surveillance requires very high data rates in the downlink for air-to-ground communications [8]. Services such as private property monitoring, flying BSs, and mobile integrated access and backhaul (IAB) nodes also require high bandwidth at the air interface [9]. It is worth noting that some missions cannot be completed by one UAV. In such cases, a swarm of UAVs are needed, resulting in additional communications overheads [10]. Specifically, as a result of the movement of UAVs, the structure of the swarm may change dynamically, requiring regular updates.

Based on the abovementioned application requirements, UAVs need to be supported by all radio access technologies (RAT) within 5G systems. Within the range of technologies, the most challenging is the support of UAVs in the millimeter wave (mmWave) bands [11]. The rationale is that this band is highly susceptible to a blockage by buildings. A feasible solution to this problem would be to support the multi-connectivity functionality standardized for 5G NR systems [12]. According to it, when blockage occurs, it is possible to switch to another BS that is currently non-blocked. This technique has been shown to drastically improve performance of conventional terrestrial users, see, e.g., [13,14]. To assess the coverage of 5G mmWave NR deployments with multiconectivity functionality for UAV users, simple and accurate line-of-sight (LoS) blockage models are thus required [15].

The conventional approach to analyzing the coverage/outage phenomenon in the presence of blockage is to utilize the tools of stochastic geometry, see, e.g., [16–18] among others for human body blockage models. The core of the analysis is to estimate the probability that a LoS path between user equipment (UE) and BS is not blocked by obstacles having a certain shape. The major step is thus to determine the number of blockers falling to the so-called LoS blockage zone, see [19] for details. The approach has proved itself as a versatile tool for analysis of human blockage in mmWave systems with purely random deployments of blockers, where the dimensions of obstacles are negligible compared to the length of the path between communicating entities.

Analyzing regular deployments, where dimensions of the obstacles are not negligible as compared to the LoS path between the communicating entities, the described approach results in a number of inherent limitations. In particular, the probability that a LoS path is blocked by an obstacle depending on relative positions of obstacles with respect to each other leading to complex expressions for coverage/outage probabilities that cannot be provided in closed-form. When dealing with such deployments the results are often provided in product form either having infinite sums [20] or involving integration [21]. An alternative approach is to utilize field measurements of LoS blockage, see, e.g., [22]. The latter approach is mainly dictated by the simplicity of the final expression but is limited to those conditions where the measurements data have been gathered. Thus, there is a need for a model providing simple closed-form approximation for UAV LoS blockage probability accounting for both system and environment characteristics.

In this paper, we will target the abovementioned two challenges. Specifically, we first provide closed-form approximation for LoS blockage probability of UAVs in urban terrestrial deployments of mmWave systems. To this end, we utilize the tools of integral geometry rather than stochastic geometry. Then, we proceed to apply the proposed methodology to estimate the UAV LoS blockage probability in the rooftop deployment of BSs. The proposed approach allows for the providing of UAV LoS probability in closed-form in grounded, rooftop, and mixed grounded-rooftop deployments as a function of environmental characteristics.

Our main contributions can be summarized as follows:

- closed-form approximation for UAV LoS blockage probability in urban deployments of mmWave NR technology showing excellent agreement with complex models;
- numerical results showing that the most impact on UAV LoS blockage probability in ground-mounted mmWave deployment is produced by UAV altitude, BS height, street width, and mean building block height while the effect of other parameters is of secondary importance;
- numerical results for mixed ground-rooftop deployments of mmWave BSs showing that it allows for the drastic increase of UAV LoS blockage probability and, depending on system parameters, adding one rooftop-mounted mmWave BS is equivalent to adding 6–12 ground-mounted mmWave BSs.

The rest of the paper is organized as follows. First, we overview recent efforts in the analysis of UAV blockage probability in Section 2. The system model utilized in our study is introduced in Section 3. UAV LoS blockage probability for grounded and rooftop deployments is derived in Section 4. Numerical results are provided in Section 5. Conclusions are provided in the last section.

2. Related Work

In this section, we first review recent vendors' and standardization bodies' activities related to UAV integration into cellular 5G systems. We then proceed by providing an outlook of UAV LoS blockage models proposed over the last few years.

2.1. UAV Integration into 5G

In recent years, UAVs support in modern wireless networks has attracted attention from network operators and standardization organizations. The 3GPP TR 36.777 summarizes the research done on LTE support for UAVs. In particular, it considers several cellular network improvements for efficient service of UAV users, quantifies the impact of UAVs on the network, and evaluates characteristics of UAV-based service in urban and rural environments. Computer simulations of such systems, augmented with measurement data, show that the use of UAVs may lead to increased interference in both uplink and downlink directions. TR 36.777 also suggests methods to eliminate interference. Another issue identified in TR.36.777 is related to UAV mobility. The standard defines methods for providing additional information about the deployed ground network that can be used for decision-making during flight.

Since 3GPP Release 16, UAV support has been seen as a critical feature of the 5G cellular network infrastructure. In this context, TR 22.829 summarizes the use cases and analyzes UAV functions that may require enhanced support from access networks. It includes video broadcast applications, command and control services, and the use of UAVs as aerial BSs. The latter UAV application is covered in detail in TR 38.811.

3GPP is currently continuing research in this area. In particular, some of the tasks are to reduce the negative effects caused by the mobility of UAVs and to adapt to the needs of business, security, and the remote identification of UAVs. Specifically, TR 22.125 defines operational requirements for 3GPP systems. The 3GPP is expected to improve UAV integration methods in 5G communication networks in future revisions of TR 23.754 and TR 23.755. Nevertheless, it is already clear that UAVs will soon provide a wide range of services in 5G access networks.

2.2. LoS blockage Probability

The question of LoS occlusion by large static objects such as buildings has been significantly investigated in the context of terrestrial users. One of the fundamental studies dating back to 1984 [23] uses a methodology based on the combination of mathematical modeling and field measurements. Specifically, the study proposed a mathematical model describing a statistical method for predicting LoS propagation paths for a receiver-transmitter pair in densely populated areas based on a statistical building distribution model. The core of the model is based on an analysis of the mean free path of moving particles in randomly distributed obstacles. The resulting LoS blockage probability was calculated for a scenario where buildings are located along a certain axis between the receiver and transmitter, with building heights distributed exponentially.

The work in [24] includes a description of a model for calculating the LoS blockage probability for a pair UE-BS in the Fresnel zone of a certain radius, applicable to typical European cities with dense and regular streets. This empirical model is based on empirical data from the city center of Bristol, UK. The model takes into account the height of the buildings, their dimensions, the width of the streets, and the distribution of street corners. The carried out numerical analysis demonstrated that the distribution and variance of building height has little impact on the LoS blockage probability. Furthermore, in [16], a random shape theory for modeling random blockage effects in urban cellular networks is utilized. A fundamental method has been established to determine the LoS blockage probability from irregularly placed buildings. Although no direct comparison with empirical measurements has been performed, the main finding was that the LoS blockage probability decreases exponentially fast with the link length. Another example of a similar model for terrestrial users is reported in [25], where cube-shaped structures with uniformly distributed height are utilized as a model for buildings. The authors report the LoS blockage probability in integral form.

Recently, a number of models for UAV LoS blockage probability have been reported. In [26], the authors carried out a large-scale simulation campaign based on real data taken from the city of Ghent for collecting UAV coverage data with both LTE and mmWave BS terrestrial deployments. The reported data highlights that mmWave NR coverage of UAV is insufficient even for the highly dense deployment of these BSs. In [27], a method to estimate LoS blockage probability based on a scanning laser is proposed. This methodology is applied to open parking situations to collect data and use them to form an exponentially decaying probabilistic LoS blockage model.

Both ITU-R and 3GPP have also defined their LoS blockage models for UAV. In particular, the ITU-R model, reported in [20], considers the frequency range from 20 to 50 GHz. The LoS blockage probability is calculated assuming that the terrain is flat and has a certain constant slope over the area of interest. The model also accounts for different heights of UE and BS and uniform distribution of the building height. The LoS blockage probability is produced in product-form. Contrarily, 3GPP models of LoS blockage defined in TR 38.901 are purely empirical, obtained by fitting the measurement data to the exponentially decaying function starting from a certain breaking point. The model specifically tailored to UAV and proposed in TR.36.777 [28] utilizes only two parameters: BS height and UAV altitude. Parameters such as the height of buildings, building density and others are not taken into account. Thus, the model can only be used for certain BS heights, significantly reducing the application scenarios.

Recently, the authors in [21] proposed a detailed and versatile UAV LoS blockage probability that accounts for most critical parameters including different UAV and BS heights, different building height distribution, and various widths of streets and building blocks. The standard city deployment is however limited to the regular one and captured by the Manhattan Poisson line process (MPLP). Owing to the model complexity, closed-form expressions have been provided for specific building height distributions only. The authors demonstrated that the LoS blockage probability is highly sensitive to the type of deployment, the distribution of building heights, and the flight altitude of the UAV. Also, according to the authors, the existing standardized models developed by 3GPP and ITU-R provides an overly optimistic approximation of the UAV LoS blockage probability.

2.3. Summary

In summarizing, we note that the accuracy of empirical models proposed so far for UAV LoS blockage analysis heavily depends on the similarities of the analyzed deployment and the one where measurements have been taken. Specifically, measurement-based models require large-scale measurement campaigns for each specific environment. Purely analytical models are either too simple to account for critical details or do not provide the solution in closed-form.

In this paper, we will fill the abovementioned gap by proposing an accurate analytical model accounting for all the major specifics of the environment. The distinguishing feature of the proposed model is that, as opposed to other models, it provides the result in closed-form and is capable of capturing the specifics of both ground- and rooftop-mounted BS simultaneously.

3. System Model

In this section, we first introduce the considered system model by defining the system and environmental input parameters. We then define the metrics of interest and outline the proposed methodology.

3.1. Deployment Model and Metrics of Interest

We assume deterministic Manhattan grid deployment with street width l, see Figure 1. The widths and lengths of building blocks are assumed to be b_w and b_l. The height of building blocks is assumed to be a random variable (RV), H_B, with probability density function (pdf) $f_{H_B}(x)$. We consider a certain zone of interest having M_V and M_H vertical and horizontal streets, respectively. We further assume that there are N ground-mounted mmWave BSs located on the streets leading to the spatial density of $N/[M_V * (l + b_l) * M_H(l + b_w)]$ mmWave BSs per squared meter. On top of this, we assume that there are M rooftop-mounted mmWave BS randomly located on the building roofs.

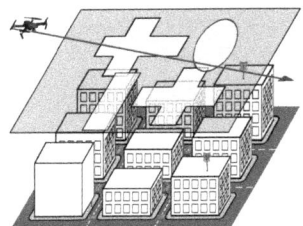

Figure 1. Illustration of the considered deployment.

MmWave BSs are assigned to streets randomly, i.e., first a discrete uniformly distributed RV between 0 and $M_H + M_V$ is used to determine the street index, and then the position of mmWave BS is determined by choosing x or y coordinate uniformly along the street width, excluding parts occupied by crossroads. Similarly, N_C, $N_C < N$, crossroad-installed mmWave BSs are assigned to crossroads randomly using discrete RV uniformly distributed between 0 and $M_V M_H$. A particular location of mmWave BS on a crossroad is defined with respect to the left upper corner and is fully determined by the distances $l_{A,1}$ and $w_{A,1}$. Similarly, we choose a particular position for BSs installed along the street at the distance $l_{A,1}$, from the building. Note that in practice these BSs can be installed on lampposts, for example, and distances $l_{A,1}$, and $w_{A,1}$ may coincide with the sidewalk width.

The UAV attitude is assumed to be constant, h_R. We assume that UAV is in coverage of BS if there is a LoS path between UAV and BS and this path is less than a certain r. UAV is assumed to cross this region following a random line at the constant speed v_U. We are interested in the UAV coverage probability–the probability that UAV is in coverage of at least one BS.

3.2. Methodology at a Glance

Instead of accounting for inherent dependencies between building positions and their shapes in regular urban deployments, we characterize LoS visibility regions in $\Re^2$ located at the UAV flying altitude, h_R, see Figure 1. Using these regions we then proceed by utilizing the tools of integral geometry to determine the probability that a random point in this plane is covered by at least one LoS visibility region immediately delivering the sought metrics of interest in a simple closed-form.

4. UAV Blockage Analysis

In this section, we develop our framework. We start by defining the so-called LoS visibility zones at the flying altitude of the UAV. Next, we utilize the integral geometry to specify the LoS probability for the ground deployment of mmWave BSs. Finally, we extend the methodology to account for rooftop-mounted mmWave BSs.

4.1. Geometric Structure of LoS Zones

We start by characterizing the LoS visibility zone induced by BR BS located along the street, see Figure 2a. As one may observe, this zone is of rectangular shape with sides that depend on (i) heights of buildings, $H_{B,1}$ and $H_{B,2}$, (ii) maximum coverage of BS, r, and (iii) UAV altitude h_R.

Observing Figure 3a, the length of the LoS visibility zone is

$$D = 2\sqrt{r^2 - (h_R - h_T)^2}, \qquad (1)$$

where r is the maximum communications distance,

$$r = \sqrt{10^{\frac{P_A+G_R+G_T-N_0-S_T-32.4-20\log_{10}F_C}{21}} - [h_R - h_T]^2}, \qquad (2)$$

where S_T is the SNR threshold, G_T and G_R are the transmit and receive antenna gains, P_A is the emitted power at mmWave BS, N_0 is the thermal noise, F_C is the carrier frequency.

The width of the LoS visibility zone, L, is an RV that is determined by building heights, $H_{B,1}$ and $H_{B,2}$, where both have the same pdf $f_{H_B}(x)$, see Figure 3b. Observe that angles α_1 and α_2 are given by

$$\alpha_i = \tan^{-1}\left(\frac{H_{B,i} - h_T}{l_{A,i}}\right), i = 1, 2, \qquad (3)$$

where l_1, l_2 are the distances to the buildings, see Figure 3b.

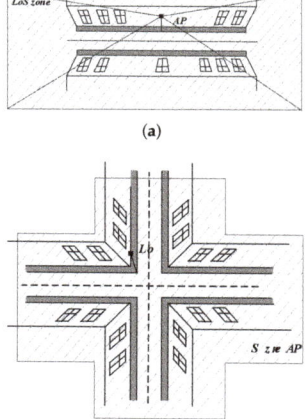

Figure 2. Two types of feasible LoS visibility zones in the considered scenario. (**a**) BS located along the street; (**b**) BS located at the crossroad.

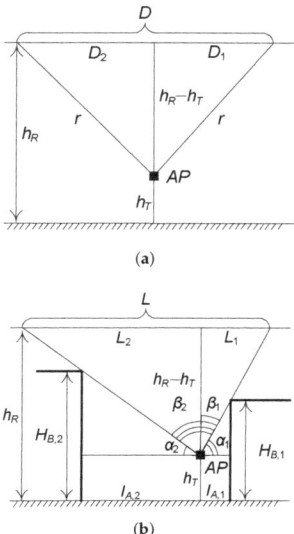

Figure 3. Geometrical illustration of the sides of LoS visibility zone. (**a**) Length of the LoS visibility zone; (**b**) Width of the LoS visibility zone.

Further, using $\tan \beta_i = L_i/(h_R - h_T)$, $i = 1, 2$ and observing that angles β_i are related to α_i as $\beta_i = \pi/2 - \alpha_i$ we arrive at the following expressions for RVs L_1 and L_2

$$L_i = (h_R - h_T) \tan\left(\frac{\pi}{2} - \tan^{-1}\left[\frac{H_{B,i} - h_T}{l_{A,i}}\right]\right) =$$
$$= \frac{l_{A,i}(h_R - h_T)}{H_{B,i} - h_T}, i = 1, 2. \qquad (4)$$

One may now determine the mean area of the LoS visibility zone as

$$E[S_B] = D \int_0^\infty \int_0^\infty f_{H_B}(x) f_{H_B}(y) [L_1(x) + L_2(y)] dx dy =$$
$$= 2\sqrt{r^2 - (h_R - h_T)^2} E[L_B], \qquad (5)$$

where the mean length of the LoS visibility zone, $E[L_B]$, is provided by

$$E[L_B] = \int_0^\infty \int_0^\infty \left(\frac{l_{A,i}(h_R - h_T)}{x - h_T}\right) \times$$
$$\times \left(\frac{l_{A,i}(h_R - h_T)}{y - h_T}\right) f_{H_B}(x) f_{H_B}(y) dx dy, \qquad (6)$$

that can be evaluated in closed-form for a given distribution of the building height.

A simple yet reliable approximation for (5) can be obtained by assuming the same random height of both buildings on the street, as it is usually the case in practice. In this case, the width of the blockage zone becomes

$$L_B = L_1 + L_2 = \frac{(h_T - h_R)(l_{A,1} + l_{A,2})}{x - h_T}, \qquad (7)$$

implying that (6) can be written as

$$E[L_B] = \int_0^\infty f_{H_B}(x) \frac{(h_T - h_R)(l_{A,1} + l_{A,2})}{x - hT} dx. \tag{8}$$

For example, for H_B having uniform distribution in (A, B) we have

$$E[L_B] = \frac{(h_T - h_R)(l_{A,1} + l_{A,2})(\log[1 - \frac{A}{h_T}] - \log[1 - \frac{B}{h_T}])}{B - A}. \tag{9}$$

Similarly, the mean perimeter of the LoS visibility zone B is $E[L_B] = 2D + 2E[L]$. The LoS visibility zones induced by BS deployments on the crossroads can be found similarly. Indeed, as one may observe in Figure 2b, they consist of two overlapping LoS visibility zones forming a "cross". Individually, parameters of these two zones can be estimated as shown above.

4.2. Blockage Probability with Grounded Infrastructure

We are now in a position to evaluate blockage probability, p_B, with ground-mounted BSs. The input parameters are the number of LoS visibility zones characterized by their mean areas and perimeters, $E[S_B]$ and $E[L_B]$, in $\Re^2$ plane positioned at the UAV flying altitude h_R.

To provide simple yet accurate expression for blockage probability, we will rely upon the tools of integral geometry. Further, we need two fundamental notions of integral geometry. A curious reader is referred to [29] for a basic account of information and to [30] for modern developments in the field.

Definition 1 (Kinematic density, [29]). *Let K denote the group of motions of a set A in the plane. The kinematic density dA for the group of motions K in the plane for the set A is*

$$dA = dx \wedge dy \wedge d\phi, \tag{10}$$

where $\wedge$ is the exterior product [31], x and y are Cartesian coordinates, ϕ is the rotation angle of A with respect to OX.

Definition 2 (Kinematic measure, [29]). *The kinematic measure m of a set of group motions K on the plane is defined as the integral of the kinematic density dA over K, that is,*

$$m_A = \int_K dA = \int_K dx \wedge dy \wedge d\phi. \tag{11}$$

Consider first a single mmWave BS in the area of interest A and let B define a LoS visibility zone. We are first interested in the probability p_C that UAV, located at a randomly chosen point P in A, is in coverage of this BS, that is, it is located in B. Using conditional probability we may write

$$p_C = \frac{Pr\{P \in A \cap B\}}{Pr\{A \cap B \neq 0\}}, \tag{12}$$

where the probability that UAV location P belongs to the intersection area of two sets, A and B, is in the nominator, while the probability that these sets do intersect is in the denominator.

Using the notion of kinematic measure, we get [29]

$$\begin{aligned} Pr\{P \in A \cap B\} &= m(A : P \in A \cap B\}), \\ Pr\{A \cap B \neq 0\} &= m(A : A \cap A \neq 0), \end{aligned} \tag{13}$$

where the first expression is the kinematic measure of the set of motions of A such that $P \in A$, while the second one provides the measure of all motions of A, for which the intersection between A and B is non-zero.

Following [29], the first measure is

$$m_j(P \in A \cap B\}) = \int_{P \in B} f(x,y) dx \wedge dy \wedge d\phi, \tag{14}$$

where $f(x,y)$ is the density of LoS visibility zone positions in A.

The measure of all motions of A, such that $A \cap B$, is [29]

$$m_j(A \cap B \neq 0) = \int_{A \cap B \neq 0} f(x,y) dx \wedge dy \wedge d\phi. \tag{15}$$

Finally, the sought probability is given by

$$p_C = \frac{\int_{P \in A \cap B} f(x,y) dx \wedge dy \wedge d\phi}{\int_{A \cap B \neq 0} f(x,y) dx \wedge dy \wedge d\phi}, \tag{16}$$

and can be computed for a particular form of A, B, and $f(x,y)$.

The numerator in (16) is computed as [29]

$$m_j(P \in A \cap B\}) = \int_{P \in B} dx \wedge dy \wedge d\phi =$$
$$= \int_{P \in B} dx \wedge dy \int_0^{2\pi} d\phi = 2\pi E[S_B], \tag{17}$$

where $E[S_B]$ is the mean area of LoS visibility zone provided in (5).

The measure of motions of A is such that $A \cap B \neq 0$ is [29]

$$m_j(A \cap B \neq 0) = \int_{A \cap B \neq 0} dx \wedge dy \wedge d\phi =$$
$$= 2\pi(S_A + E[S_B]) + L_A E[L_B], \tag{18}$$

where $E[L_B]$ is the perimeter of LoS visibility zone, S_A and L_A are the area and the perimeter of A, given by

$$S_A = [M_V * (l + b_l) + b_l] * [M_H(l + b_w) + b_l],$$
$$L_A = 2[M_V * (l + b_l) + b_l] + 2[M_H(l + b_w) + b_l]. \tag{19}$$

Substituting (17), (18) into (12) we obtain

$$p_C = \frac{2\pi E[S_B]}{2\pi(S_A + E[S_B]) + L_A E[L_B]}. \tag{20}$$

Recall that mmWave BSs are deployed randomly along the streets. When mmWave BS is deployed on the crossroad it creates two LoS visibility zones as illustrated in Figure 3b. Let u be the probability that mmWave BS is at the crossroad. This probability is found as the ratio of crossroad area to the overall area of streets as

$$u = \frac{M_V M_H l^2}{M_H(l[(b_w + l)M_V + b_w]) + M_V(l[(b_l + l)M_H + b_l]) - M_V M_H l^2}. \tag{21}$$

The mean number of LoS visibility zones of rectangular shape is then given by the mean of Binomial distribution with parameters N and u shifted by N, i.e., $N(1+u)$. Thus, the blockage probability can now be approximated as

$$p_B = 1 - (1 - p_C)^{N(1+u)}, \tag{22}$$

where p_C is provided in (20).

Substituting intermediate results and simplifying, we arrive at the closed-form expression for blockage probability in the presence of N ground-mounted mmWave BS as

$$p_B = 1 - \left(1 - \frac{2\pi E[S_B]}{2\pi(S_A + E[S_B]) + L_A E[L_B]}\right)^N \left(1 + \frac{M_V M_H l^2}{M_H(l[(b_w+l)M_V + b_w]) + M_V(l[(b_l+l)M_H + b_l]) - M_V M_H l^2}\right). \tag{23}$$

where S_A and L_A are provided in (19), $E[S_B]$ and $E[L_B]$ are calculated using (5) and (6) for a given $f_{H_B}(x)$.

4.3. Blockage Probability with Rooftop-Mounted BSs

The blockage probability heavily depends on the density of mmWave BSs, as well as on the heights of buildings. For some values of these input parameters, the blockage probability might be unacceptably high. In practical deployments, network operators may want to add additional dedicated mmWave BSs. Mounting these BSs on rooftops would allow for an unobstructed LoS of circular shape, drastically reducing blockage probability.

To assess joint deployment, one may apply the methodology developed in the previous section to rooftop mmWave BSs. The principal difference is that the LoS visibility zone is of circular form with radius $D/2$ as in (1) with H_B replacing h_T. However, as these mmWave BSs are now deployed on the roofs, D is a RV. Thus, we have

$$E[D] = \int_0^\infty f_{H_B}(x) 2\sqrt{r^2 - (h_R - x)^2} dx. \tag{24}$$

that can be evaluated for a given $f_{H_B}(x)$, $x < h_R$.

The blockage probability by M rooftop mmWave BSs is obtained similarly to (22). Finally, in the presence of N ground-mounted and M rooftop-mounted mmWave BSs the blockage probability is the product of individual blockage probabilities.

5. Numerical Results

In this section, we first assess the accuracy of the model identifying its application range, and then proceed to report on the impact of system parameters on the UAV blockage probability. Finally, we evaluate the effect of rooftop-mounted BSs. The values of input system parameters are provided in Table 1.

Table 1. Summary of notation and parameters.

Parameter	Value
mmWave BS height, h_T	5 m
UAV height, h_R	150 m
Carrier frequency, F_C	28 GHz
Emitted power, P_T	0.02 W
mmWave BS and UAV antenna gains, G_T, G_R	15 dB, 5 dB
Number of vertical and horizontal streets, M_H, M_V	10, 10
Length and width of building blocks, b_l, b_w	100 m, 100 m
Street width, l	20 m
SNR threshold, S_T	0 dB
Thermal noise, N_0	-174 dBm

5.1. Accuracy Assessment

To identify the application range of the developed closed-form approximation, we start assessing the accuracy of the model by comparing its results to those obtained using the computer simulations. To this end, Figure 4 shows the UAV LoS blockage probability obtained using the proposed model and computer simulations for various UAV altitudes and BS heights h_U and h_A, respectively, street width $l = 20$ m, mean building height and standard deviation $E[H_B] = 30$ m and $\sigma[H_B] = 10$ m, block width and length of

$b_w = b_l = 100$ m. The considered region of interest is formed by considering 10 horizontal and vertical building blocks interchanged with streets.

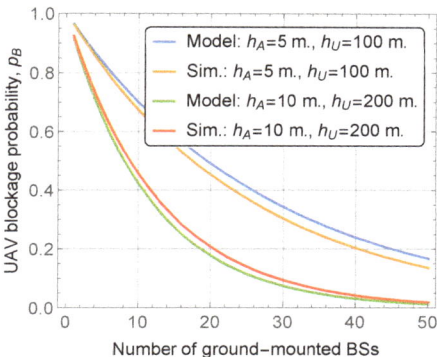

Figure 4. Comparison of the developed model and computer simulations.

By analyzing the results shown in Figure 4, one may deduce that the proposed model allows for the approximation of the results obtained via computer simulations quite closely. Similar observations have been made for rooftop-mounted mmWave BSs. Notably, the developed model slightly overestimates the actual value of the probability. This is explained by the inherent structure of the model that assumes that all the LoS visibility regions are completely independent. This observation allows us to identify the applicability regions of the model. First of all, observe that due to the abovementioned property the model always provides the upper bound on the UAV LoS blockage probability. Secondly, the results become more accurate as of the area of the zone and/or the density of the mmWave BSs increase. Based on these results, when discussing the response of the UAV blockage probability to system parameters and assessing the effect of rooftop-mounted mmWave BSs, we thus utilize the developed model.

5.2. Effects of System Parameters

We now proceed to evaluating the effect of system parameters on the UAV blockage probability including the BS height and altitude of UAV, the mean and variance of building height and, finally, the street width and building block's width and length.

We start with an assessment of the effects of mmWave height and UAV flying altitude. To this aim, Figure 5 shows UAV LoS blockage probability as a function of these parameters for street width $l = 20$ m, mean building height and standard deviation $E[H_B] = 30$ m and $\sigma[H_B] = 10$ m, block width and length of $b_w = b_l = 100$ m. By analyzing the presented results, we see that higher BS heights result in lower UAV LoS blockage probability, see Figure 5a. Particularly, the gain of changing mmWave BS height from 5 m to just 15 m leads to the decrease of UAV LoS blockage probability by approximately 0.15 for 20 mmWave BS deployed in the area. The rationale for these improvements is that higher mmWave BS heights make the visible regions at the UAV flying altitude larger, see Figure 2. Furthermore, this effect is non-linear as the area increases faster when the mmWave BS height increases. We also note that these gains depend heavily on BS deployment density and are minimal highly dense deployments.

Analyzing the effect of UAV flying altitude in Figure 5b, qualitatively similar conclusions can be made. More specifically, the higher the altitude the smaller the UAV LoS blockage probability. Specifically, for the density of 20 mmWave BS in the considered area, the gain of changing the altitude from 100 to 200 m is approximately 0.15 and is comparable to that of the change in BS height from 5 to 15 m. We also note that in practice this parameter should be tuned with care. The reason is that higher altitudes may lead to much lower received power, especially for ground-mounted mmWave BS that is usually downtilted to provide better coverage for terrestrial users, e.g., pedestrians.

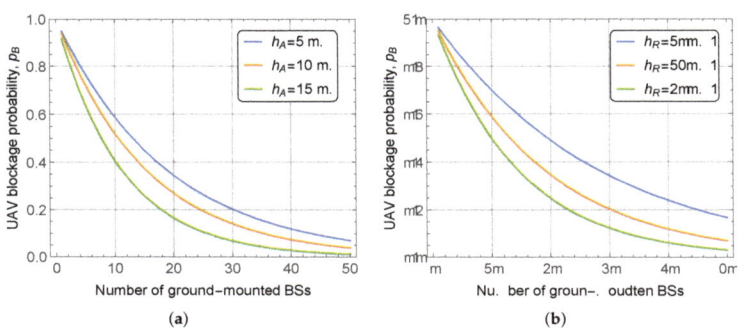

Figure 5. UAV blockage probability as a function of BS height and UAV altitude. (**a**) Various BS heights; (**b**) Various UAV altitudes.

In dense city deployments of mmWave BS, the characteristics of building block height may produce a significant impact on UAV LoS blockage probability. We investigate this hypothesis in Figure 6, where we illustrate the UAV LoS blockage probability as a function of the number of deployed mmWave BS for UAV altitude $h_U = 150$ m, BS height $h_A = 5$ m, street width $l = 20$ m, block width and length of $b_w = b_l = 100$ m. Here, in Figure 6a we show the effect of different mean values by keeping the standard deviation constant at $\sigma[H_B] = 10$, while in Figure 6b we vary standard deviation and keep the mean constant at $E[H_B] = 30$ m.

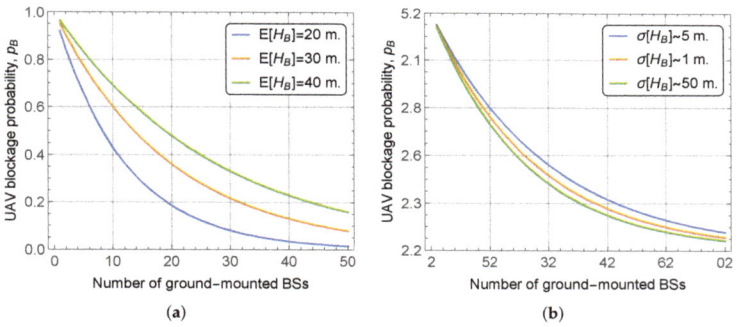

Figure 6. UAV blockage probability as a function of building block height parameters. (**a**) Various mean heights; (**b**) Various standard deviations.

By analyzing the presented data, we may conclude that the mean building height logically produces a significant effect on the UAV LoS blockage probability. The magnitude of this effect is comparable to that of BS height or UAV altitude. Particularly, when considering districts with high building heights, e.g., city centers, one needs to utilize additional ways to improve UAV LoS blockage probability. However, at the same time, the effect of standard deviation is rather limited, leading to differences in the range of 0.05–0.1 for the considered range of the number of deployed mmWave BS.

Finally, we consider the effect of street and building block widths on UAV LoS blockage probability illustrated in Figure 7 for UAV altitude $h_U = 150$ m, BS height $h_A = 5$ m, mean building height and standard deviation $E[H_B] = 30$ m and $\sigma[H_B] = 10$ m, respectively, block width and length of $b_w = b_l = 100$ m. As one may observe, both parameters drastically affect the considered metric of interest. However, the effects are different. Specifically, by increasing the street width the UAV LoS blockage probability drastically increases, see Figure 7a. The rationale is that this leads to much larger areas of LoS visibility zones, see Figure 2. At the same time, one may observe that by increasing the street and building block widths, the considered area increases as the number of streets and building blocks in both horizontal and vertical directions are kept constant. Thus, logically, larger building blocks dimensions lead to higher UAV LoS blockage probability, see Figure 7b. Nevertheless, this effect is attributed to the increase of the considered area.

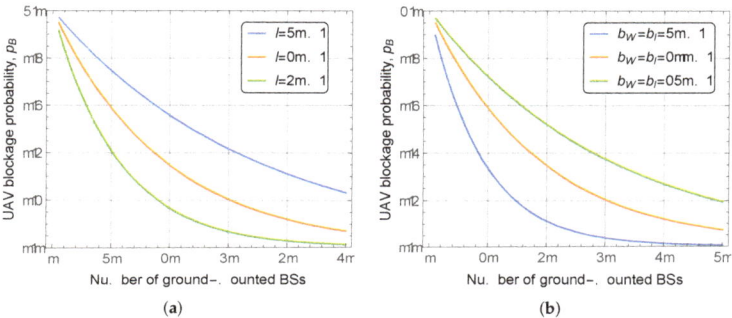

Figure 7. UAV blockage probability as a function of street and building block widths. (**a**) Various street widths; (**b**) Various building block width.

5.3. The Effect of Rooftop-Mounted BSs

Finally, we highlight the effect of rooftop-mounted BS on the UAV blockage probability. To this aim, Figure 8 shows the effect of rooftop-mounted mmWave BSs on the UAV blockage probability for UAV altitude $h_U = 150$ m, BS height $h_A = 5$ m, street width $l = 20$ m, mean building height and standard deviation $E[H_B] = 30$ m and $\sigma[H_B] = 10$ m, block width and length of $b_w = b_l = 100$ m. By analyzing the presented data, one may observe that mounting BSs on rooftops allows us to greatly reduce the BS blockage probability. More specifically, adding just three rooftop-mounted mmWave BSs to the considered area allows for the reduction of the UAV LoS blockage probability by multiple times. Recall that in the considered deployment the deployment area is $(b_l + l)M_V \times (b_w + l) * M_H \approx 1.44 \times 10^6$ m^2, implying that the density of rooftop BS is just $\approx 2 \times 10^{-6}$ BS/km^2. Specifically, by comparing the horizontal and vertical distances between lines in Figure 8, we observe that in terms of UAV LoS blockage probability, adding additional BS at the rooftop is equivalent to deploying 10 more ground-mounted mmWave BSs. This value is affected by system parameters and environmental characteristics of the deployment and may vary between six and twelve.

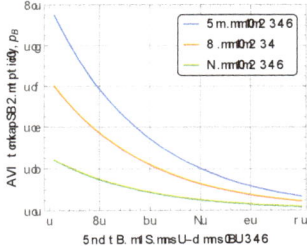

Figure 8. The effect of the rooftop-mounted BSs.

5.4. Discussion, Limitations and Applications

The presented results illustrate that out of all considered deployment parameters, street width and building block length are the ones impacting the UAV LoS blockage probability the most. The impact of BS and UAV heights as well as the mean building block height is also noticeable. These parameters all need to be accounted for when estimating the required density of BSs to support UAVs in mmWave 5G systems. Note that in real deployments, these parameters are not independent as specified in [20]. Thus, in general, in city centers, where the mean building heights and width are larger, much higher BS deployment density will be required for the same target UAV LoS blockage probability as compared to the suburbs.

We specifically emphasize the importance of rooftop-mounted BS. As we have observed, qualitatively, the density of ground-mounted BS deployment has to be extremely high, especially in city center deployment conditions. Here, to support the uninterrupted connectivity, it is much more economically sustainable for network operators to deploy dedicated BSs having almost unobstructed coverage for UAV. Our results demonstrate that one rooftop-mounted BS is equivalent to six to twelve ground-mounted ones in terms of UAV LoS blockage probability.

Although the proposed model by design can capture the specifics of different deployments, it also has its limitations. Specifically, as the model assumes that visibility areas are all convex, the visibility areas created by BSs deployed on the crossroads need to be treated as two independent rectangular visibility areas. This implies that the accuracy of the model increases as the size of the analyzed regions with homogeneous building deployments increases. Furthermore, the independence of all visibility areas also implies that the BS locations should be close to the Poisson point process (PPP, [32]). Note that due to restrictions of BS locations in the city center and also due to the need for high densification to satisfy the growing customer needs, BS deployment locations are far from regular cellular structures. Specifically, many studies assume PPP as the deployment process for 4G/5G systems.

The proposed model is especially usable in system-level simulations of mmWave NR deployments supporting UAVs. As noticed in [33], the handling of dynamic blockage events is one of the most time-consuming operations. Associating UAVs with the blockage process having the fraction of time in blockage coinciding with the UAV LoS blockage probability may efficiently address this challenge. Furthermore, the proposed model can be utilized by network operators at the network deployment phase to assess the density of mmWave BSs providing the required level of UAV coverage.

6. Conclusions

UAVs are expected to soon become a vital part of 5G deployments, acting as both users and aerial BSs. Motivated by the use of UAVs in future 5G deployments, in this paper, we utilize the tools of integral geometry to provide closed-form approximations for UAV blockage probability. In addition to LoS blockage with ground-mounted mmWave BSs, we also considered the case of the operator utilizing rooftop-mounted mmWave BSs.

The numerical results illustrate that the model can closely match the actual UAV LoS blockage probability. Furthermore, the accuracy of approximation increases as either the density of mmWave BSs or the area of interest increases. In analyzing the effect of the rooftop-mounted mmWave BSs, we have shown that one additional rooftop-mounted BS improves the UAV LoS blockage probability as six to twelve ground-mounted mmWave BS. Finally, the most impact on UAV blockage probability is produced by the mmWave BS height, UAV altitude, street width, and mean building block height. The developed model allows for the mathematical assessment of the sought metric for a given deployment condition and density of ground- and rooftop-mounted mmWave BSs.

We foresee two application areas of the proposed model. The first is with regard to system-level simulations, where one needs to utilize simple models for UAV LoS blockage probability. Additionally, the model can be utilized for assessment of the required density of mmWave NR BS to ensure a certain UAV LoS blockage probability. We also note that the accuracy of the model increases as the deployment area with the homogeneous building deployments increases. Thus, the proposed model needs to be applied to large city districts.

Author Contributions: Conceptualization, D.M., V.B. and K.S.; methodology, A.G.; software, V.B.; validation, D.M., A.G., V.B. and K.S.; formal analysis, V.B.; investigation, V.B.; resources, A.G.; data curation, A.G.; writing—original draft preparation, D.M., V.B. and K.S.; writing—review and editing, A.G.; visualization, A.G.; supervision, D.M.; project administration, D.M. and K.S.; funding acquisition, D.M. All authors have read and agreed to the published version of the manuscript.

Funding: Sections 3–5 were written by Vyacheslav Begishev under the support of the Russian Science Foundation, project no. 21-79-10139. This paper has been supported by the RUDN University Strategic Academic Leadership Program (recipients Konstantin Samouylov, Sections 1, 2 and 6).

Data Availability Statement: Not applicable.

Conflicts of Interest: The authors declare that they have no conflicts of interest.

References

1. Gupta, L.; Jain, R.; Vaszkun, G. Survey of Important Issues in UAV Communication Networks. *IEEE Commun. Surv. Tutorials* **2016**, *18*, 1123–1152. [CrossRef]
2. Wang, H.; Wang, J.; Chen, J.; Gong, Y.; Ding, G. Network-connected UAV communications: Potentials and challenges. *China Commun.* **2018**, *15*, 111–121.
3. Gapeyenko, M.; Petrov, V.; Moltchanov, D.; Andreev, S.; Himayat, N.; Koucheryavy, Y. Flexible and reliable UAV-assisted backhaul operation in 5G mmWave cellular networks. *IEEE J. Sel. Areas Commun.* **2018**, *36*, 2486–2496. [CrossRef]
4. Petrov, V.; Gapeyenko, M.; Moltchanov, D.; Andreev, S.; Heath, R.W. Hover or Perch: Comparing Capacity of Airborne and Landed Millimeter-Wave UAV Cells. *IEEE Wirel. Commun. Lett.* **2020**, *9*, 2059–2063. [CrossRef]
5. Pandey, S.R.; Kim, K.; Alsenwi, M.; Tun, Y.K.; Han, Z.; Hong, C.S. Latency-Sensitive Service Delivery With UAV-Assisted 5G Networks. *IEEE Wirel. Commun. Lett.* **2021**, *10*, 1518–1522. [CrossRef]
6. Matthew, U.O.; Kazaure, J.S.; Onyebuchi, A.; Daniel, O.O.; Muhammed, I.H.; Okafor, N.U. Artificial Intelligence Autonomous Unmanned Aerial Vehicle (UAV) System for Remote Sensing in Security Surveillance. In Proceedings of the 2020 IEEE 2nd International Conference on Cyberspac (CYBER NIGERIA), Abuja, Nigeria, 23–25 February 2021.
7. Lee, J.W.; Lee, W.; Kim, K.D. An Algorithm for Local Dynamic Map Generation for Safe UAV Navigation. *Drones* **2021**, *5*, 88. [CrossRef]
8. He, C.; Xie, Z.; Tian, C. A QoE-Oriented Uplink Allocation for Multi-UAV Video Streaming. *Sensors* **2019**, *19*, 3394. [CrossRef] [PubMed]
9. Bertizzolo, L.; Tran, T.X.; Buczek, J.; Balasubramanian, B.; Jana, R.; Zhou, Y.; Melodia, T. Streaming from the Air: Enabling Drone-sourced Video Streaming Applications on 5G Open-RAN Architectures. *IEEE Trans. Mob. Comput.* **2021**, accepted. [CrossRef]
10. Chen, M.; Wang, H.; Chang, C.Y.; Wei, X. SIDR: A Swarm Intelligence-Based Damage-Resilient Mechanism for UAV Swarm Networks. *IEEE Access* **2020**, *8*, 77089–77105. [CrossRef]
11. Xia, W.; Polese, M.; Mezzavilla, M.; Loianno, G.; Rangan, S.; Zorzi, M. Millimeter Wave Remote UAV Control and Communications for Public Safety Scenarios. In Proceedings of the 2019 16th Annual IEEE International Conference on Sensing, Communication, and Networking (SECON), Boston, MA, USA, 10–13 June 2019.
12. 3GPP. *NR; Multi-Connectivity; Stage 2 (Release 16)*; 3GPP TS 37.340 V16.0.0; 3GPP: Sophia Antipolis, France, 2019.
13. Moltchanov, D.; Samuylov, A.; Lisovskaya, E.; Kovalchukov, R.; Begishev, V.; Sopin, E.; Gaidamaka, Y.; Koucheryavy, Y. Performance Characterization and Traffic Protection in Street Multi-Band Millimeter-Wave and Microwave Deployments. *IEEE Trans. Wirel. Commun.* **2022**, *21*, 163–178. [CrossRef]
14. Begishev, V.; Sopin, E.; Moltchanov, D.; Kovalchukov, R.; Samuylov, A.; Andreev, S.; Koucheryavy, Y.; Samouylov, K. Joint Use of Guard Capacity and Multiconnectivity for Improved Session Continuity in Millimeter-Wave 5G NR Systems. *IEEE Trans. Veh. Technol.* **2021**, *70*, 2657–2672. [CrossRef]
15. Shahbazi, A.; Di Renzo, M. Analysis of Optimal Altitude for UAV Cellular Communication in Presence of Blockage. In Proceedings of the 2021 IEEE 4th 5G World Forum (5GWF), Montreal, QC, Canada, 13–15 October 2021.
16. Bai, T.; Vaze, R.; Heath, R.W., Jr. Analysis of Blockage Effects on Urban Cellular Networks. *IEEE Trans. Wirel. Commun.* **2014**, *13*, 5070–5083. [CrossRef]
17. Samuylov, A.; Gapeyenko, M.; Moltchanov, D.; Gerasimenko, M.; Singh, S.; Himayat, N.; Andreev, S.; Koucheryavy, Y. Characterizing Spatial Correlation of Blockage Statistics in Urban mmWave Systems. In Proceedings of the IEEE GLOBECOM Workshops, Washington, DC, USA, 4–8 December 2016.
18. Jain, I.K.; Kumar, R.; Panwar, S. Driven by capacity or blockage? A millimeter wave blockage analysis. In Proceedings of the 2018 30th International Teletraffic Congress (ITC 30), Vienna, Austria, 3–7 September 2018.

19. Gapeyenko, M.; Samuylov, A.; Gerasimenko, M.; Moltchanov, D.; Singh, S.; Aryafar, E.; Yeh, S.; Himayat, N.; Andreev, S.; Koucheryavy, Y. Analysis of human-body blockage in urban millimeter-wave cellular communications. In Proceedings of the IEEE International Conference on Communications (ICC), Kuala Lumpur, Malaysia, 23–27 May 2016.
20. ITU-R. *Propagation Data and Prediction Methods Required for the Design of Terrestrial Broadband Radio Access Systems Operating in a Frequency Range from 3 to 60 GHz*; Recommendation ITU-R P Series: P.1410; International Telecommunication Union: Geneva, Switzerland, 2012.
21. Gapeyenko, M.; Moltchanov, D.; Andreev, S.; Heath, R.W. Line-of-Sight Probability for mmWave-based UAV Communications in 3D Urban Grid Deployments. *IEEE Trans. Wirel. Commun.* **2021**, *20*, 6566–6579. [CrossRef]
22. 3GPP. *Study on Channel Model for Frequencies from 0.5 to 100 GHz (Release 16)*; 3GPP TR 38.901 V16.1.0; 3GPP: Sophia Antipolis, France, 2020.
23. Ogawa, E.; Satoh, A. Propagation Path Visibility Estimation for Radio Local Distribution Systems in Built-Up Areas. *IEEE Trans. Commun.* **1986**, *34*, 721–724. [CrossRef]
24. Feng, Q.; Tameh, E.K.; Nix, A.R.; McGeehan, J. WLCp2-06: Modelling the Likelihood of Line-of-Sight for Air-to-Ground Radio Propagation in Urban Environments. In Proceedings of the IEEE Globecom 2006, San Francisco, CA, USA, 27 November–1 December 2006.
25. Liu, X.; Xu, J.; Tang, H. Analysis of Frequency-Dependent Line-of-Sight Probability in 3-D Environment. *IEEE Commun. Lett.* **2018**, *22*, 1732–1735. [CrossRef]
26. Colpaert, A.; Vinogradov, E.; Pollin, S. Aerial Coverage Analysis of Cellular Systems at LTE and mmWave Frequencies Using 3D City Models. *Sensors* **2018**, *18*, 4311. [CrossRef]
27. Järveläinen, J.; Nguyen, S.L.H.; Haneda, K.; Naderpour, R.; Virk, U.T. Evaluation of Millimeter-Wave Line-of-Sight Probability With Point Cloud Data. *IEEE Wirel. Commun. Lett.* **2016**, *5*, 228–231. [CrossRef]
28. 3GPP. *Study on Enhanced LTE Support for Aerial Vehicles, (Release 15)*; TR 36.777 V15.0.0; 3GPP: Sophia Antipolis, France, 2018.
29. Santalo, L. *Integral Geometry and Geometric Probability*, 1st ed.; Addison-Wesley: Boston, MA, USA, 1976; pp. 23–45.
30. Schneider, R.; Weil, W. *Stochastic and Integral Geometry*; Springer: Berlin, Germany, 2008; pp. 220–248.
31. Flanders, H. *Differential Forms with Applications to the Physical Sciences*, 2nd ed.; Dover Publications: New York, NY, USA, 1989; pp. 32–64.
32. Moltchanov, D. Distance distributions in random networks. *Ad Hoc Netw.* **2012**, *10*, 1146–1166. [CrossRef]
33. Gapeyenko, M.; Samuylov, A.; Gerasimenko, M.; Moltchanov, D.; Singh, S.; Akdeniz, M.R.; Aryafar, E.; Himayat, N.; Andreev, S.; Koucheryavy, Y. On the Temporal Effects of Mobile Blockers in Urban Millimeter-Wave Cellular Scenarios. *IEEE Trans. Veh. Technol.* **2017**, *66*, 10124–10138. [CrossRef]

Review

Coverage Path Planning Methods Focusing on Energy Efficient and Cooperative Strategies for Unmanned Aerial Vehicles

Georgios Fevgas [1], Thomas Lagkas [1], Vasileios Argyriou [2],* and Panagiotis Sarigiannidis [3]

[1] Department of Computer Science, International Hellenic University, 65404 Kavala, Greece; gefevga@cs.ihu.gr (G.F.); tlagkas@cs.ihu.gr (T.L.)
[2] Department of Networks and Digital Media, Kingston University, Surrey KT1 2EE, UK
[3] Department of Informatics and Telecommunication Engineering, University of Western Macedonia, 50100 Kozani, Greece; psarigiannidis@uowm.gr
* Correspondence: vasileios.argyriou@kingston.ac.uk

Abstract: The coverage path planning (CPP) algorithms aim to cover the total area of interest with minimum overlapping. The goal of the CPP algorithms is to minimize the total covering path and execution time. Significant research has been done in robotics, particularly for multi-unmanned unmanned aerial vehicles (UAVs) cooperation and energy efficiency in CPP problems. This paper presents a review of the early-stage CPP methods in the robotics field. Furthermore, we discuss multi-UAV CPP strategies and focus on energy-saving CPP algorithms. Likewise, we aim to present a comparison between energy efficient CPP algorithms and directions for future research.

Keywords: coverage path planning; unmanned aerial vehicle; cell decomposition; decomposition methods; energy-aware approaches; energy optimal path; multi-robot systems; multi-UAV

Citation: Fevgas, G.; Lagkas, T.; Argyriou, V.; Sarigiannidis, P. Coverage Path Planning Methods Focusing on Energy Efficient and Cooperative Strategies for Unmanned Aerial Vehicles. *Sensors* 2022, 22, 1235. https://doi.org/10.3390/s22031235

Academic Editors: Zihuai Lin and Wei Xiang

Received: 3 December 2021
Accepted: 2 February 2022
Published: 6 February 2022

Publisher's Note: MDPI stays neutral with regard to jurisdictional claims in published maps and institutional affiliations.

Copyright: © 2022 by the authors. Licensee MDPI, Basel, Switzerland. This article is an open access article distributed under the terms and conditions of the Creative Commons Attribution (CC BY) license (https://creativecommons.org/licenses/by/4.0/).

1. Introduction

In recent years, due to rapid technological development, UAVs and sensors they can carry have been developed to the extent that they can cover a wide range of applications [1] that cannot be satisfied by other types of robots [2]. Some of the applications are precision agriculture [3,4], search and rescue [5], firefighting [6], law enforcement [7], powerline inspection [8], oil and gas [9], disaster management [10], and cell network expansion [11]. However, a fundamental problem is the optimal use of autonomous aircraft in terms of time and space [2].

Recently, CPP algorithms have been developed, considering the parameters required for more efficient data retrieval from remote sensing sensors [12,13]. In addition, algorithms have been developed that use multi-UAV to cover the area, thus reducing the coverage time of the area of interest [14,15]. The way to cover an area with autonomous robots differs depending on the algorithm [2,16,17]. In the literature, CPP algorithms use different methods (e.g., grids, graphs, and neural networks) with calculations performed online or offline for known or unknown areas [18].

The CPP algorithms can be classified into two main categories: offline and online [19]. Offline algorithms need to know the environment and the information included, such as obstacles and the geometry of the area of interest. Of course, in a real-life environment, many dynamic parameters cannot be known in advance. Offline algorithms have prior knowledge of the coverage area environment [20]. They also provide more efficient and convenient route plans and use less central processing unit (CPU) power than online algorithms [21].

The online algorithms are based on real-time environment data retrieved from onboard sensors to cover the area of interest. Online algorithms do not fully understand the coverage area environment, and the coverage path is executed in real-time by the UAV after processing the data using the sensors it carries. The benefits of online algorithms are the design of the in-flight route to complete the mission regardless of unforeseen situations and the unnecessary prior detailed knowledge of the coverage area [20,22].

Furthermore, there are two categories of problems in area coverage: single coverage and repeat coverage. The goal of single coverage is to cover the entire area of interest and, at the same time, minimize the time and distance traveled by the coverage route [23]. On the other hand, repetitive coverage aims to repeatedly cover all points of interest in the area, maximize the frequency of visits to points of interest, and minimize time and total coverage [24].

This paper aims to present the CPP methods and approaches used by UAVs, focusing on energy-saving CPP methods, such as using the direction of the wind in the cover area [25]. The CPP problem is the optimal motion of the robot in a specific area that includes obstacles to cover this area with minimum overlapping and the shortest path. In the case of a UAV in a three-dimensional area, the shortest path is related to the sensor's footprint. Of course, as the altitude of flight is higher, the footprint is more extensive, which means the shortest path. On the other hand, the higher the flight altitude of the UAV, the bigger the ground sample distance (GSD) and the lower the image quality. GSD is the distance between pixel centers measured in the ground. However, there are a lot of other limitations, such as no-flight zones, that must be computed during path planning to avoid obstacles [26].

Many surveys present studies related to UAV trajectory planning in an environment with obstacles [27], UAV autonomous guidance [28], and in specific applications, such as remote sensing with UAVs in precision agriculture [29]. A survey on CPP methods for mobile robots was presented by Choset, who classified the approaches in two classes [19]. The robots follow simple rules, but the success of area coverage is not guaranteed to be classified as a heuristic approach. On the other hand, the complete methods using cellular decomposition guarantee coverage. Moreover, the author mentions the flight time, which can be minimized by using multiple robots and reducing the number of turns.

The most recent surveys regarding the CPP methods for robotics or UAVs are presented in Table 1. Cabreira et al. [30] present a survey of the decomposition methods, UAV and Multi-UAV CPP methods, and energy-saving algorithms. Galceran and Carreras [20] present a survey of the decomposition methods and ground multi-robot strategies. Additionally, Almandhoun et al. [31] present Multi-UAV CPP methods in their survey. Chen et al. [32] present a survey of CPP methods using UAV or Multi-UAV. The existing surveys of CPP methods considering unmanned ground vehicles (UGV) and the surveys of CPP methods using UAVs extend the UGV's CPP methods. However, many factors, such as the sensors' weight, the flight endurance, direction, and intensity of the wind, must be considered when using UAVs on CPP methods developed for UGVs.

Table 1. Related surveys.

Related Work	Decomposition Methods	Multi-Robot Strategies	UAV CPP Methods	Multi-UAV CPP Methods	Energy-Saving Algorithms	Comparison of Energy-Saving CPP Methods
Cabreira et al. [30]	✓	✗	✓	✓	✓	✗
Galceran and Carreras [20]	✓	✓	✗	✗	✗	✗
Almandhoun et al. [31]	✓	✓	✗	✓	✗	✗
Chen et al. [32]	✗	✗	✓	✓	✗	✗
Our work	✓	✓	✓	✓	✓	✓

Table 1 compares the present work to already existing surveys of CPP methods for robotics or UAVs. The present paper is focused not only on surveying the CPP methods for UAVs, but also on: (a) examining all the decomposition methods, (b) reviewing the multi-robot strategies, (c) the multi-UAV's and standalone UAV's CPP methods, (d) UAVs' energy-saving CPP algorithms, and (e) the comparison of the energy-saving CPP methods. Our approach proves to be the most complete regarding the variables considered for the survey comparison.

The key contributions of this work can be summarized as follows:

- A review of the decomposition methods in different shapes of the area of interest, such as rectangular, concave, irregular, and convex polygons, has been presented.
- A presentation of multi-robot and multi-UAV CPP strategies based on single robot approaches, methods that guarantee the mission's completeness, and bio-inspired methods that perform coverage under uncertainty.
- A review of energy-saving algorithms and the limitations of them, according to the UAV's constraints and environmental conditions.
- A discussion of the CPP methods' limitations, how to overcome them, and directions for future research on energy-saving CPP algorithms.

This paper is organized considering the CPP methods, multi-UAV strategies, and energy-saving algorithms. Section 2 focuses on a detailed analysis of the systematic review research methodology. Section 3 reviews all decomposition algorithms, multi-robot CPP strategies, multi-UAV CPP methods, and presents UAVs' energy-saving CPP algorithms and a comparative table. Finally, directions for future research on energy-saving CPP algorithms are given in Section 4.

Our review considers the research gap concerning the differences between UGV CPP methods and the UAV CPP methods. Furthermore, our review presents the limitations of the UAVs considering environmental conditions, such as the intensity and direction of the wind. A detailed discussion about the main aspects of multi-robot and multi-UAV CPP methods is also provided. Our review focuses on approaches related to UAV energy-saving algorithms and a discussion of the combination of these algorithms considered for future research.

This paper aims to inform the reader of the coverage path planning approaches in different shapes of the area of interest, including rectangular, concave, and polygons, according to the decomposition method employed. Furthermore, we explore the limitations of the CPP methods between UGVs and UAVs, the latest multi-robot and multi-UAV CPP strategies, and the energy-efficient algorithms for UAVs. Finally, our review considers the performance metrics and the limitations of these methods.

2. Methods

For the present work, a systematic review research methodology was adopted. In that context, a range of platforms was sourced for information. Most of the sources cited in this survey were found in (a) the IEEE Xplore digital library, (b) the Google scholar platform, (c) the online Elsevier platform, and d) the online MDPI platform.

Keywords utilized were: "Coverage Path Planning", "Decomposition methods", "CPP methods", "Multi-robot CPP methods", "Cell decomposition", "Unmanned Aerial Vehicles", "Energy optimal path", "Energy-aware approaches", "Multi-robot systems", "Robot coverage", "Robot kinematics", and "UAV Remote Sensing". Initially, the resulting papers (approximately 170) were filtered by choosing the ones referring to CPP algorithms, decomposition methods, multi-robots and multi-UAV coverage path strategies, and energy-awareness CPP algorithms.

The 170 aforementioned publications reviewed for the decomposition methods, single or multi-robot CPP strategies, multi-UAV CPP methods, and UAV energy-saving algorithms. From the 170 papers, 128 were classified according to the relevance of the survey's scope and their overlapping information. In the end, 128 papers were analyzed for their approaches and their correlation to categorize in sub-sections of decomposition methods, CPP methods, and energy-saving algorithm, of which 88 made it into the refined version of the present survey.

3. Results

Choset [19] classified the CPP algorithms according to the decomposition used. Most CPP algorithms decompose the area of interest in cells. This method is preferable for irregular areas. On the other hand, when the area of interest is a regular shape, it does not require any decomposition for single coverage of UAV. Table 2 at the end of this section summarizes the decomposition methods and presents the CPP approach, the decomposition method, the algorithm processing, the shape of the area of interest, and the corresponding reference.

3.1. No Decomposition

There is no need for decomposition in areas with regular shapes and without complexity, such as rectangular areas. Patterns with simple path planning, such as boustrophedon or square, are adequate for total coverage of a non-complex area without overlapping. The boustrophedon method, which means "the way of the ox," is a pattern of simple back and forth motion along the longest side of the polygon, as shown in Figure 1 [33–35].

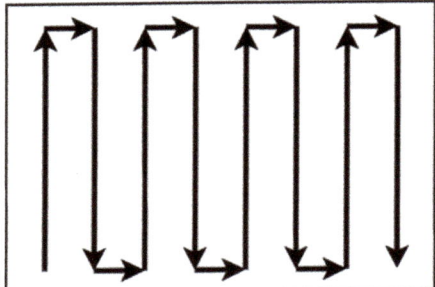

Figure 1. Boustrophedon pattern.

The literature assumes that the actual path is closely true to the plan when this method is executed from a ground vehicle. On the other hand, UAVs are aerodynamically directly affected by the direction and intensity of the wind, which means that the actual trajectory of the flight in most cases is not close to that planned.

The square method is represented by Andersen [36], and it is a pattern for a search and rescue mission. The flight path is straight lines with right 90 degrees turns. The pattern starts from the center of the area of interest and expands until the borders, as shown in Figure 2.

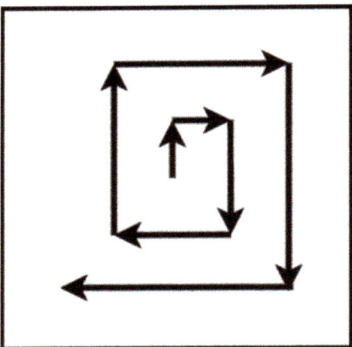

Figure 2. Square pattern.

3.2. Exact Cellular Decomposition

The cellular decomposition methods are based on dividing an irregular space into cells. One class of these methods is the exact one. The exact cellular decomposition method decomposes the irregular space into cells, and their connections produce an accurate free space composition. Accurate methods are complete because they guarantee the finding of an accessible path, if any [37]. The sub-areas that arise from the decomposition can be covered from a single UAV or multiple UAVs. There are patterns for a single UAV, such as boustrophedon and spiral for polygon and concave areas. Nevertheless, there are strategies for the cooperation of multiple UAVs in order to minimize the coverage time [38].

3.3. Trapezoidal Decomposition

One exact cellular decomposition technique for irregular spaces that can give a complete coverage path is trapezoidal decomposition. This method is classified in the offline category of algorithms because it does not use remote-sensing information [39,40]. Each cell is a trapezoid in this method, and simple methods such as back and forth can be used to cover every cell. The coverage can be achieved by an exhaustive walk that generates a path to cover each cell to execute the path using back and forth motions such as boustrophedon, as shown in Figure 3. Often, this method is used for agriculture applications where the fields are polygonal and clear from obstacles. Oksanen and Visala [41] introduced an algorithm for CPP in agricultural fields and used the path cost function to optimize the final path.

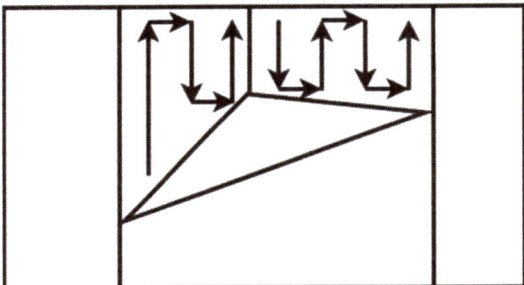

Figure 3. Trapezoidal decomposition.

3.4. Boustrophedon Decomposition

Trapezoidal decomposition produces many cells, some of which can be merged. This characteristic is a disadvantage because as many cells exist, the coverage path will be longer. To overcome this limitation, a method that creates nonconvex cells is needed. The boustrophedon cellular decomposition is similar to trapezoidal decomposition but considers vertices in the area called critical points [33,35]. The boustrophedon decomposition reduces the number of cells compared with trapezoidal decomposition, which means shorter path planning, as shown in Figure 4. As the trapezoidal decomposition, this method is for polygonal areas, and the environment of the coverage area should be known. For this reason, it can be classified as an offline method.

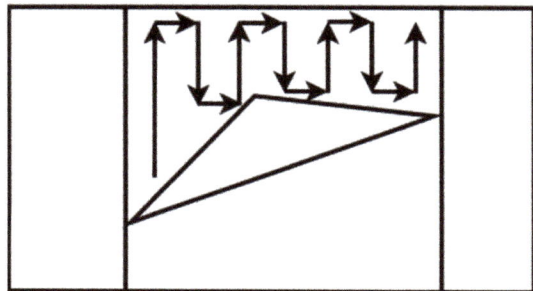

Figure 4. Boustrophedon decomposition.

3.5. Morse-Based Decomposition

Another cellular decomposition method proposed by Acar et al. [42] is based on Morse functions [43]. The Morse-based decomposition method has the advantage of different cells shapes such as circular and can be applied in any dimensional space, such as concave, polygon, and irregular space. The cell decomposition is succeeded with a slice that sweeps through the area of interest. A slice is discontinued at the critical point of the Morse function, which is restricted from the obstacle boundaries, as shown in Figure 5. This method uses information concerning the area during motion planning. For this reason, the method can be classified as online [44,45].

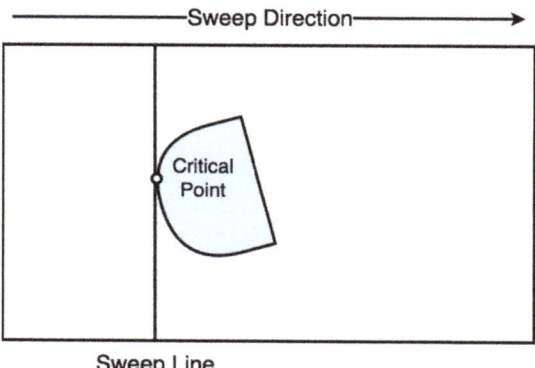

Figure 5. Morse-based decomposition.

3.6. Online Topological Coverage Algorithm

Wong [46] presented an algorithm that finds the cell boundaries online using slice decomposition. Slice decomposition is a method for determining the cell boundaries using a sweeping line over the area of interest. As the line sweeps over the area, it separates the obstacles and free space in two regions or more, as shown in Figure 6. The algorithm constructs a topological map using the slice decomposition on the area of interest [47].

Figure 6. Slice decomposition.

3.7. Contact Sensor-Based Coverage of Rectilinear Environments

Butler et al. [48] present an exact cell decomposition algorithm for contact sensor-based robots for online coverage of the rectilinear environment. In contact sensor-based coverage, the robot's path is cycling with retracing, while at the same time it repeatedly constructs a cellular decomposition of the area of interest. When a robot's full-cycle path is unsuccessful, it chooses a new path based on its position and environment. The robot's motion depends on the area's cell decomposition state, updated as the CPP progresses.

3.8. Grid-Based Methods

Grid-based methods are classified as approximate cellular decomposition due to the restriction of the grid's shape, which is uniform in space. It is impossible to represent precisely the shape of the target space and its obstacles [19]. The grid-based methods decomposed the space into uniform grid cells, which can be squares or other shapes, as shown in Figure 7. Moravec and Elfes [49] proposed a grid map presentation based on a sonar mounted on a mobile robot mapping an indoor environment.

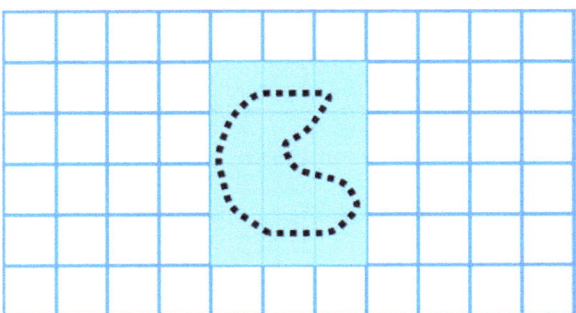

Figure 7. Grid-based decomposition.

3.8.1. Wavefront Algorithm

The first CPP's grid-based method was proposed by Zelinsky et al. [50]. Their method has a start cell and a goal cell. A grid represents the coverage area, and a wavefront algorithm is used from the goal cell to the start cell. Its operation is based on propagating a "wavefront" from the target cell passing through the free cells and bypassing all obstacles to the starting cell.

More specifically, the transmission of the "wavefront" from the target cell to the starting cell is used to assign specific numbers to each cell of the grid, as shown in Figure 8. Firstly,

0 is assigned to the target cell and then 1 to all adjacent cells. Then, all the other adjacent cells of 1 to which no number has been assigned are assigned 2. The process repeats incrementally until the wavefront reaches the starting point [13,20]. The environment should be known in this method, so the method can be classified offline.

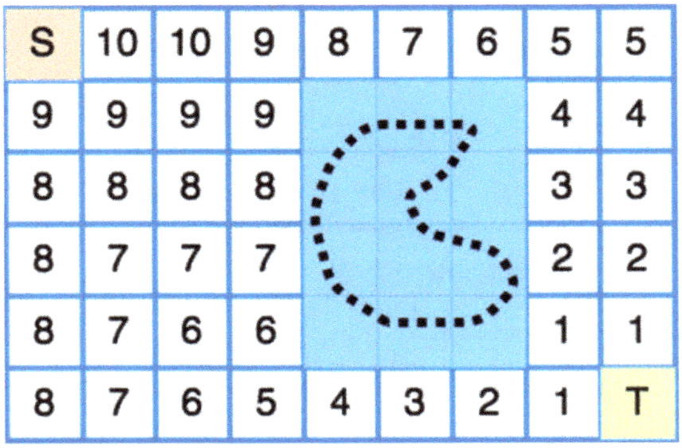

S	10	10	9	8	7	6	5	5
9	9	9	9				4	4
8	8	8	8				3	3
8	7	7	7				2	2
8	7	6	6				1	1
8	7	6	5	4	3	2	1	T

Figure 8. Wavefront Transmission from starting cell (S) to target cell (T).

Nevertheless, Shivashankar et al. [51] proposed a wavefront algorithm to accomplish an online CPP with a mobile robot in an unknown spatial environment.

3.8.2. Spanning Tree Coverage

The spanning tree coverage (STC) algorithm solves the problem of covering an area using a robot [38]. The method used by the STC algorithm is first to decompose the region into cells and calculate a connecting tree of the resulting graph. Finally, the robot's path starts near the "connecting tree" and follows its perimeter, as shown in Figure 9 [37]. A Spiral-STC algorithm was proposed by Gabriely and Rimon [52]. This online method converts the space into a grid map. The mobile robots execute a spanning tree-generated spiral path using onboard sensors.

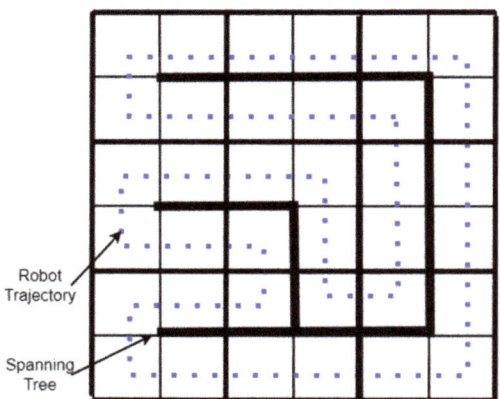

Figure 9. Spanning Tree-based coverage.

3.9. *Neural Network-Based Coverage on Grid Maps*

The CPP using a neural network is an online coverage method. First, in a 2D coverage area, a grid map is constructed where the length of the diagonal of each cell is equal to the coverage radius of the robot (e.g., the coverage radius of a robotic broom), and then a neuron is associated with each cell in the grid. Each neuron is connected to the eight primary neighboring neurons, as shown in Figure 10. Finally, the robot's path to the coverage area is executed by knowing each output value of each neuron at a given time, so that the robot is attracted to cells it has not visited while at the same time being rejected by cells it has visited [53,54].

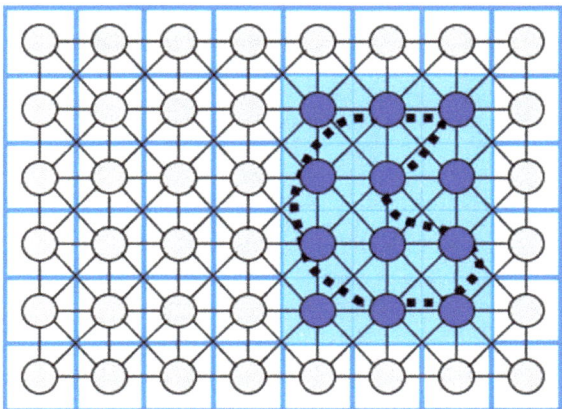

Figure 10. Neural Network-based coverage.

Table 2. CPP and decomposition methods.

CPP Approach	Decomposition Method	Algorithm Processing	Shape of Area	Reference
Boustrophedon	None	Offline	Rectangular	[33–35]
Square	None	Offline	Square	[36]
Boustrophedon, Spiral	Exact cellular	Offline	Polygon, Concave	[37]
Back and Forth	Trapezoidal	Offline	Polygon	[39,40]
Boustrophedon	Boustrophedon	Offline	Polygon	[33,35]
Boustrophedon	Morse-based	Online	Any dimensional	[42]
Online Topological	Slice	Online	Polygon	[46]
Contact Sensor-based	Exact cellular	Online	Rectilinear	[48]
Wavefront	Approximate cellular	Offline	Polygon, Concave	[50]
Wavefront	Approximate cellular	Online	Polygon, Concave	[51]
STC	Approximate cellular	Offline	Polygon, Concave	[37]
Spiral-STC	Approximate cellular	Online	Polygon, Concave	[52]
Neural Network-based	Approximate cellular	Online	Polygon, Concave	[53,54]

3.10. Multi-Robot CPP Strategies

Multiple robots have an advantage over single robotic systems [24]. The use of multiple robots accelerates coverage of an area of interest. The problem of covering an area with multiple robots lies in the calculation of optimal routes in order to minimize the coverage time [38]. Using multiple robots in a CPP work reduces the completion time due to workload division [20]. This section discusses multi-robot coverage methods based on single robot approaches, multi-robot strategies, and multiple UAVs to cover an area of interest. Some drawbacks of multiple UAV strategies are spatial orientation and communication difficulties. Table 3 at the end of this section summarizes the multi-robot CPP strategies and presents the CPP approach, the decomposition method, the algorithm processing, and the corresponding reference.

3.11. Multi-Robot Boustrophedon Decomposition

Rekleitis et al. [16] presented a set of online algorithms for solving the CPP using a group of mobile robots in an unknown environment. The algorithms employ the same planar cellular decomposition as the Boustrophedon single robot coverage algorithm, with additions to manage how robots cover a single cell and distribute among cells. Their solution takes into account the team members' communication limitations. The robots serve two roles to accomplish coverage where some members, known as explorers, cover the boundaries of the actual target cell, while others, known as coverers, conduct basic back-and-forth motions to cover the cell.

3.12. Multi-Robot Spanning Tree Coverage

Their experimental data reveal that their technique outperforms multi-robot spanning tree coverage (MSTC) by a significant margin. Nevertheless, the coverage time of an area with the multi-robot forest coverage (MFC) algorithm is shorter than the MSTC algorithm [38]. Moreover, an online, robust version of MSTC was provided by Hazon et al. [55]. They show that the approach is robust analytically, providing as much coverage as a single robot can.

3.13. Multi-Robot Neural Network-Based Coverage

A neural network approach for multi-robot coverage where each robot sees all the others as obstacles and the avoidance ability of stalemate situations was proposed by Luo and Yang [54,56,57]. The multi-robot neural-network based coverage is inspired by single robot neural-network coverage. During the coverage of the irregular-shaped area of interest, the robots see each other as moving obstacles.

3.14. Multi-Robot Graph-Based and Boundary Coverage

Easton and Burdick [58] presented a two-dimensional boundary coverage method for multiple robots. A team of robots must inspect all points on the boundary of the two-dimensional target environment, and each robot's inspection routes are planned to use a heuristic search. The planned paths cover the entire boundary. Moreover, the algorithm has been validated by simulations. The multi-robot boundary coverage is inspired by the need to inspect the blade surfaces inside a turbine.

Table 3. Multi-robot CPP strategies.

CPP Approach	Decomposition Method	Algorithm Processing	Reference
Boustrophedon	Exact cellular	Online	[16]
Spanning Tree Coverage	Approximate cellular	Online	[55]
Neural network-based	Approximate cellular	Online	[54,56,57]
Graph-based and Boundary	Approximate cellular	Offline	[58]

3.15. Multi-UAV CPP Methods

The number of applications where UAVs can be used is increasing as remote-sensing technology is developed. In the literature, there are a lot of multi-UAV CPP methods using different coverage algorithms with heterogeneous or homogeneous UAVs that were used in a variety of applications, such as agriculture [59], surveillance [60], mapping [61], and search and rescue missions [62]. Table 4 at the end of this section summarizes the multi-UAV CPP strategies and presents the CPP approach, the type of UAVs, the algorithm processing, the evaluation metrics, and the corresponding reference.

3.16. Multi-UAV Coverage

In the agricultural sector, Barrientos et al. [13] proposed a method for area coverage using a fleet of mini aerial robots. Their method divides the area of interest in k non-overlapping subtasks and assigns them in k UAVs. A decentralized method for surveillance missions using homogeneous UAVs was proposed by Acevedo et al. [63]. This method's primary goal is to minimize latency, which means a short sharing time of information between the UAVs. In a later work, Acevedo et al. [64] developed a method for surveillance in urban environments with heterogeneous UAVs that fly at low altitudes and avoid obstacles. Finally, in their most recent work, Acevedo et al. [65] developed a method based on grid-shape area partition, which can readjust the area shape and UAVs' capacity.

A terrain coverage method using a fleet of heterogeneous UAVs was presented by Maza and Ollero [61]. Their method divides the irregular-shaped area of interest per each UAV capability, such as total flight time. Each partition is assigned to a UAV that plans a zig-zag covering pattern according to the area's characteristics to minimize the number of turns. The method was validated in simulation.

A coverage algorithm for fixed-wing UAVs with the ability for obstacle and previously scanned regions avoidance was presented by Xu et al. [23,66]. Their method uses boustrophedon cellular decomposition [33], an exact cellular decomposition, and presents better accuracy than trapezoidal decomposition. The method can be classified as online in the phase of region scanning and offline in the coverage phase.

3.17. Back-and-Forth

Maza and Ollero [61] present a cooperative technique using heterogeneous UAVs in a convex polygonal area. A ground control station divides the area into sub-regions and assigns them to every UAV by the capability and starting position. Every UAV calculates back-and-forth patterns according to the camera footprint to reduce the number of turns.

3.18. Spiral

Balampanis et al. [67,68] present a spiral CPP algorithm using multiple heterogeneous UAVs. The area of interest is divided according to UAVs sensing capabilities using a constrained Delaunay triangulation (CDT) [69]. The CDT generates triangle cells that match almost exactly the shape of the area of interest. To make the triangles more uniform, they applied Lloyd optimization [70]. Then, a spiral algorithm generates the coverage pattern for each sub-area. This method can generate smoother trajectories considering avoiding no-fly zones and the shape of the coverage area. However, it generates more extensive coverage paths and a higher number of turns than classical grid decomposition and motion methods [71,72].

3.19. Multi-Objective Path Planning (MOPP) with Genetic Algorithm (GA)

Hayat et al. [73] propose multi-objective path planning (MOPP) with a genetic algorithm (GA) for search and rescue missions using multiple UAVs. The mission is divided into two phases: search and response. The search phase monitors an event to guarantee the total coverage in a given area, and the response phase spreads detection updates on the network. The MOPP algorithm performs the planning task during the search, while the GA minimizes the mission completion time. As a result, the method can be classified as offline in the search phase and online in the response phase.

3.20. Genetic Algorithm (GA) with Flood Fill Algorithm

Based on the Trujillo et al. [74] approach, Darrah et al. [75] present a CPP method for missions over more extensive areas using multi-UAVs. The method produces equitable sub-areas of the area of interest to cover by multi-UAVs or several flights performed by a single UAV. The flood fill algorithm integrated with game theory was applied to partition the area of interest. Each UAV is a player and has a starting position. According to a predefined pattern in a diamond shape, the UAVs take turns flooding the neighbor cells. The UAVs cannot fly over building cells or cells previously occupied by other UAVs. The partitioning method guarantees an approximate amount of work for each assigned UAV by balancing the tasks. An improved version of the approach proposed by Trujillo et al. [74] was used for each sub-area's coverage trajectories. The method can be classified initially as offline and then as online.

Table 4. Multi-UAV CPP strategies.

CPP Approach	Type of UAVs	Algorithm Processing	Evaluation Metrics	Reference
Sub-perimeter method	Homogeneous	Online	Minimize latency	[63]
Back-and-Forth	Homogeneous	Online/Offline	Total path length Time coverage	[23,66]
Back-and-Forth	Heterogeneous	Offline	Number of turns	[61]
Spiral	Heterogeneous	Offline	Coverage path, Number of turns	[67,68]
Multi-Objective Path Planning with GA	Homogeneous	Offline/Online	Mission Completion Time	[73]
GA with flood fill algorithm	Homogeneous	Offline/Online	Path length	[74,75]

3.21. Energy-Saving CPP Algorithms

In the literature, there are a lot of CPP strategies for energy saving. One method for energy saving proposed by Lawrance and Sukkarieh [25] is the energy exploitation of the wind using a small gliding UAV. The authors present an algorithm that generates energy gain paths according to the UAV's constraints, the field's wind conditions, and static and dynamic soaring. One of the limitations of this method is the requirement for prior knowledge of the field's wind conditions. In future research, an online stochastic wind estimation and planning method using current wind conditions of the field should be developed.

Another method for minimizing the power consumption of a UAV is reducing the number of turns of the CPP. Torres et al. [76] present an algorithm that reduces the number of turns and the total flying path to minimize battery consumption.

The effect of wind direction and intensity on the time of mission completion was presented by Coombes et al. [77]. The authors used a fixed-wing UAV and the boustrophedon method to cover the area of interest. Their simulated experiments used a constant direction of the wind and six different speeds, and for the coverage path used different directions of the fixed-wing UAV motion from 0 to 360 degrees in increments of 10 degrees. The results showed that the direction of the coverage path should be 90 degrees to the wind direction to minimize the coverage time. Furthermore, the direction of the turns is directly affected by the vertical component of the wind. In a later work, Coombes et al. [78] presented the flight time in wind (FTIW) function, which computes the total flight time for a total coverage of the area of interest. The flight time needed for the total coverage of the area is less than the previous methods. Their approach was validated after simulations and real flights.

An energy-efficient back and forth CPP algorithm proposed by Di Franco and Buttazzo [18] computes the best motion trajectory and the maximum altitude according to the ground sample distance (image resolution) to minimize the number of turns. Another approach for energy efficiency is to find an optimal constant speed according to the coverage path. An energy-aware spiral CPP algorithm uses wider angle turns to minimize the acceleration and deceleration to maintain an optimal constant speed Cabreira et al. [79]. After simulated and real flights, the most energy-efficient CPP method between energy-efficient back and forth CPP [18] and the energy-aware spiral CPP approach proposed by Cabreira et al. [79] for a convex area was the energy-aware spiral CPP method which adopted the energy model proposed by Di Franco and Buttazo [80].

Another energy-aware CPP algorithm for UAVs was proposed by Li et al. [81], where the algorithm has three stages. In the first stage, the algorithm builds a 3D terrain model. In the second stage, constant power consumption is computed by total take-off weight, flight speed, and air friction. In the third stage, a genetic algorithm generates an energy-optimal coverage path, which represents the amount of energy consumption in every part of the path.

Another problem concerning UAV energy consumption is the deceleration and acceleration at every turn of a conventional trajectory such as boustrophedon. Artemenko et al. [82] present an algorithm that modifies conventional trajectories using Bézier curves, smoothing the turns on a given path to minimize deceleration and acceleration before and after the turning point. The authors concluded that their algorithm could reduce energy spending compared to conventional algorithms. Restrictions, such as the UAV motion and camera's location, can be overcome using integer linear programming. Ahmadzadeh et al. [83] present a cooperative coverage technique with critical time for rectangular areas utilizing several fixed-wing heterogeneous UAVs, some carrying a frontal camera, flying circular paths and some of them carrying a camera on the left side, flying straight lines with left turn paths. Their proposed method uses four fixed-wing UAVs covering 100% of the area of interest instead of the simple methods covering 80%. The proposal was validated in simulation tests and real flights.

Araujo et al. [84] propose an algorithm where the workspace is divided into sub-areas assigned to each UAV according to its relative capability. According to the kinematics constraints of the UAVs, the algorithm generates an optimal number of stripes to minimize the number of stripes and eventually the number of turns, which means less energy consumption.

Majeed and Lee [85] present a CPP method for UAV low-altitude navigation in three-dimensional urban areas with fixed convex obstacles based on footprint sweeps fitting and a sparse waypoint graph. The primary goals of the proposed approach are to reduce computational time, the number of turns, and path overlapping while minimizing the total coverage path of the area of interest. The suggested method outperforms the similarly related CPP approaches according to simulation findings.

In a later work, Majeed and Hwang [86] present a CPP algorithm for UAV navigation to cover areas of interest (AOIs) surrounded by obstacles in three-dimensional urban areas with fixed obstacles. The proposed method is applicable in a wide range of practical applications that involve computing a low-cost coverage for spatially distributed AOIs in an urban environment. However, the proposed algorithm has not incorporated and tested for constraints and limitations, such as image resolution and UAV battery.

Cheng et al. [87] present a bio-inspired method for cooperative coverage. This method represents the trajectory of each UAV as the B-spline curve containing control points. This optimization problem aims to maximize the desirability of a path by combining four variables: path distance, minimum turning angle, maximum pitch rate, and superposition of the actual trajectory over different UAV trajectories. According to the authors, the beginning and last control points are at the area's borders because the UAV always travels from left to right. The ant colony optimization (ACO) algorithm was adapted for coverage with multiple UAVs by Kuiper and Nadjm-Tehrani [88]. The y-axis in the intermediate control points is optimized using the ACO algorithm to maximize the coverage. Several ants are launched during the algorithm repetitions, passing through the starting, intermediate, and endpoints.

Table 5 summarizes the energy-saving CPP methods reviewed in this paper according to the method used for energy saving. The table presents the CPP method, the energy-saving factor, the type of UAV, and the corresponding reference.

Table 5. CPP energy-aware methods.

CPP Method	Energy-Saving Approach	Type of UAV	Reference
Energy gain path	Energy exploitation of the wind	Fixed-wing	[25]
Back and Forth	Reducing the number of turns and the total flying path	Rotorcraft	[76]
Boustrophedon	The direction of the UAV path and the turns according to the wind direction	Fixed-wing	[77]
Back and Forth	Altitude maximization according to the Ground Sample Distance to reduce the number of turns	Rotorcraft	[18]
Spiral	Wider angle turns to minimize the acceleration and deceleration	Rotorcraft	[79]
Three stages energy optimal path	An energy-aware algorithm computes the take-off weight, flight speed, and air friction to generate an energy-optimal path	Rotorcraft	[81]
Smoothing turns	Smoothing the turns on a given path to minimize deceleration and acceleration before and after the turning point	Rotorcraft/Fixed-wing	[82]
Circular and straight lines with left turns paths	Cooperative coverage algorithm with critical time	Multiple Fixed-wing	[83]
Back and Forth	Minimizing the number of stripes and eventually the number of turns	Multiple Fixed-wing	[84]
Back and Forth	Reduce computational time, the number of turns, and path overlapping while minimizing the total coverage path	Rotorcraft	[85]
Back and Forth	Reducing the computational time and path length for the inter-regional path, the number of turning maneuvers, and path overlapping	Rotorcraft	[86]
ACO with Gaussian distribution functions	Path length, rotation angle and area overlapping rate	Rotorcraft/Fixed-wing	[87]

4. Discussion

The CPP problem using UAVs in areas of interest with different shapes and environmental conditions has been studied by several authors. Standard-shaped areas of interest, such as polygons and rectangles, do not require decomposition and can be covered by boustrophedon and spiral patterns. Generally, no decomposition methods, such as back-and-forth, require low computational cost to find the path trajectory. The main issue of these patterns is not considering that the UAVs are directly aerodynamically affected by the environmental conditions, which means the actual trajectory of the flight in most cases is not close to that planned.

In more complex and irregular areas of interest, a cellular decomposition method may be applied to split the area of interest into subregions. The subregions can be covered by different CPP methods to obtain the optimum path to minimize the total path and the total coverage flight time. Multi-UAV cooperative strategies are also being studied using the decomposition method according to the capabilities of the UAVs.

When the vehicle used for the proposed CPP algorithms is a UAV, there is the limitation of the motion constraints, such as the feasible trajectory of fixed-wing UAVs. However, the CPP methods plan the coverage path according to a performance metric. These approaches do not consider the UAVs' environmental factors and aerodynamic and flight limitations.

A further study is necessary for the area of CPP methods using UAVs. The coverage algorithms should consider the constraints of the aerial vehicles, such as the actual path trajectory rather than that planned. Moreover, the environmental factors in the area of interest that affect the path, the time, and the actual flight path should also be considered. According to all these mutable factors, an offline CPP method will not achieve optimal path planning, but an online CPP method considering all these factors and re-planning the trajectory will achieve the optimal coverage path within minimum time.

In recent years, many new CPP algorithms have been developed for energy-efficiency and awareness. The approach using a glider UAV for soaring limits early knowledge of the field's wind conditions. Otherwise, the method is less effective in a situation where the knowledge of wind conditions is limited [25]. In approaches where engine-driven UAVs are used, there are some methods or combinations for energy saving. A method for power saving in non-complex areas is reducing the number of turns in the UAV's trajectory to minimize the total path and the acceleration's power consumption after every turn, and eventually the total coverage time of the area of interest [20,76]. In approaches for energy saving, considering the direction and intensity of the wind was validated as the UAV's path should be vertical in the wind direction, and the turning maneuvers against the wind direction [77,78]. This approach can be combined with the previous method for greater energy saving.

Two more approaches that can be used in combination with the previous methods for further energy saving include minimizing the UAV's turns according to the GSD [18]. A spiral CPP algorithm uses wider angle turns to maintain a constant speed [67] or an algorithm for a conventional trajectory that modifies the turns for smoother motion [70] to minimize the deceleration and acceleration before and after the turning point. Another energy-aware algorithm computes the take-off weight, flight speed, and air friction to generate an energy-optimal path [81].

In convex areas, there are approaches using multiple UAVs to divide into sub-areas and assign each sub-area according to the UAV's capability, such as motion, sensors onboard, and total endurance flight time [83,84].

The proposed energy-efficient UAV CPP methods aim to minimize the total flight time and the coverage path length to save energy. However, the performance metrics are based on the path trajectory without considering other constraints, such as UAV aerodynamics and environmental conditions. For example, in a convex area, a CPP method with a performance metric for minimum path trajectory may produce very sharp turns. Meanwhile, it is infeasible for a fixed-wing UAV to obtain the planned trajectory due to its aerodynamics constraints. Another variable affecting the UAV's actual trajectory is the wind's direction and intensity. The UAV will consume more energy than a more extensive path length with smoother turns considering all these limitations.

A further study is necessary to combine all of the above constraints to develop new energy-efficient UAV CPP methods that consider variables, such as the vehicle kinematics and environmental conditions offline and online. A research direction to develop UAV CPP methods to maximize energy-saving should combine machine learning or deep learning and IoT onboard sensors in order to develop a CPP approach that will plan offline and adapt online the coverage path trajectory according to the main performance metrics, such as UAV kinematics constraints, and the information retrieved from onboard sensors such as wind conditions.

5. Conclusions

This paper presented a survey of coverage path planning according to the decomposition methods, such as no decomposition, exact, and approximate decomposition methods.

Different shapes of the area of interest, such as concave, rectangular, and polygon, are considered in this survey. We focused on simple path planning patterns, such as boustrophedon and spiral, and more complex approaches such as grid-based methods. We also presented multi-robot and multi-UAV CPP strategies that aim to accelerate the coverage area by focusing on optimal routes.

Some authors in more complex missions and areas use multiple UAVs to overcome their endurance limitations. However, this approach demands computational complexity to solve communication issues and coordinate the UAVs. The coordination of the UAVs requires a ground control station, which presents many communication failures in real-world scenarios.

CPP methods with simple path planning, such as boustrophedon [33] and square [36], are preferred over cellular decomposition methods for regular shapes without complexity. These CPP methods need less computational time, but they have limitations when UAVs use them. Exact cellular decomposition CPP methods are preferable in more complex area shapes, such as a polygon or concave. The boustrophedon cellular decomposition [37] is similar but better than trapezoidal decomposition [39,40] when the shape of the area has many vertices. The boustrophedon overcomes the trapezoidal decomposition by reducing the number of cells, which means shorter path planning. The morse-based decomposition [42] has the advantage over the other decomposition approaches in that it can produce different cell shapes such as circular and can be applied in any dimensional space. The contact sensor-based coverage is preferable in a rectilinear environment and for online coverage of the area because the coverage trajectory is updated as the CPP progresses.

Furthermore, we present UAVs' energy-saving CPP algorithms, which enhance the energy efficiency using optimal coverage methods and approaches, such as the sub-area assignment of the area of interest according to the capability of the UAV in a multi-UAV CPP strategy.

Finally, several kinds of research have been performed for UAV energy-aware methods in the literature. However, a remaining issue for further research is the combination of these techniques with machine learning, deep learning, and IoT sensors to develop a new, dynamic CPP method that will maximize energy-saving compared to the proposed energy-efficient CPP methods.

Author Contributions: Conceptualization, G.F. and T.L.; methodology, G.F. and T.L.; validation, T.L., V.A. and P.S.; investigation, G.F. and T.L.; resources, V.A. and P.S.; data curation, G.F.; writing—original draft preparation, G.F.; writing—review and editing, G.F. and T.L.; visualization, G.F.; supervision, T.L., V.A. and P.S.; project administration, V.A. and P.S.; funding acquisition, V.A. and P.S. All authors have read and agreed to the published version of the manuscript.

Funding: This research received no external funding.

Institutional Review Board Statement: Not applicable.

Informed Consent Statement: Not applicable.

Conflicts of Interest: The authors declare no conflict of interest.

References

1. Lagkas, T.; Argyriou, V.; Bibi, S.; Sarigiannidis, P. UAV IoT Framework Views and Challenges: Towards Protecting Drones as "Things". *Sensors* **2018**, *18*, 4015. [CrossRef] [PubMed]
2. Azpúrua, H.; Freitas, G.M.; Macharet, D.G.; Campos, M.F.M. Multi-Robot Coverage Path Planning Using Hexagonal Segmentation for Geophysical Surveys. *Robotica* **2018**, *36*, 1144–1166. [CrossRef]
3. Maes, W.H.; Steppe, K. Perspectives for Remote Sensing with Unmanned Aerial Vehicles in Precision Agriculture. *Trends Plant Sci.* **2019**, *24*, 152–164. [CrossRef] [PubMed]
4. Radoglou-Grammatikis, P.; Sarigiannidis, P.; Lagkas, T.; Moscholios, I. A Compilation of UAV Applications for Precision Agriculture. *Comput. Netw.* **2020**, *172*, 107148. [CrossRef]
5. Silvagni, M.; Tonoli, A.; Zenerino, E.; Chiaberge, M. Multipurpose UAV for Search and Rescue Operations in Mountain Avalanche Events. *Geomat. Nat. Hazards Risk* **2017**, *8*, 18–33. [CrossRef]

6. Yuan, C.; Liu, Z.; Zhang, Y. Aerial Images-Based Forest Fire Detection for Firefighting Using Optical Remote Sensing Techniques and Unmanned Aerial Vehicles. *J. Intell. Robot. Syst.* **2017**, *88*, 635–654. [CrossRef]
7. Straub, J. Unmanned Aerial Systems: Consideration of the Use of Force for Law Enforcement Applications. *Technol. Soc.* **2014**, *39*, 100–109. [CrossRef]
8. Deng, C.; Wang, S.; Huang, Z.; Tan, Z.; Liu, J. Unmanned Aerial Vehicles for Power Line Inspection: A Cooperative Way in Platforms and Communications. *J. Commun.* **2014**, *9*, 687–692. [CrossRef]
9. Cho, J.; Lim, G.; Biobaku, T.; Kim, S.; Parsaei, H. Safety and Security Management with Unmanned Aerial Vehicle (UAV) in Oil and Gas Industry. *Procedia Manuf.* **2015**, *3*, 1343–1349. [CrossRef]
10. Erdelj, M.; Natalizio, E. UAV-assisted disaster management: Applications and open issues. In Proceedings of the 2016 International Conference on Computing, Networking and Communications (ICNC), Kauai, HI, USA, 15–18 February 2016; pp. 1–5. [CrossRef]
11. Pliatsios, D.; Goudos, S.K.; Lagkas, T.; Argyriou, V.; Boulogeorgos, A.A.A.; Sarigiannidis, P. Drone-Base-Station for Next-Generation Internet-of-Things: A Comparison of Swarm Intelligence Approaches. *IEEE Open J. Antennas Propag.* **2021**, *3*, 32–47. [CrossRef]
12. Ballesteros, R.; Ortega, J.F.; Hernández, D.; Moreno, M.A. Applications of Georeferenced High-Resolution Images Obtained with Unmanned Aerial Vehicles. Part I: Description of Image Acquisition and Processing. *Precis. Agric.* **2014**, *15*, 579–592. [CrossRef]
13. Barrientos, A.; Colorado, J.; del Cerro, J.; Martinez, A.; Rossi, C.; Sanz, D.; Valente, J. Aerial Remote Sensing in Agriculture: A Practical Approach to Area Coverage and Path Planning for Fleets of Mini Aerial Robots. *J. Field Robot.* **2011**, *28*, 667–689. [CrossRef]
14. Maza, I.; Capitán, J.; Merino, L.; Ollero, A. Multi-UAV cooperation. In *Encyclopedia of Aerospace Engineering*; John Wiley & Sons, Ltd.: Hoboken, NJ, USA, 2015; pp. 1–10. [CrossRef]
15. Spyridis, Y.; Lagkas, T.; Sarigiannidis, P.; Zhang, J. Modelling and Simulation of a New Cooperative Algorithm for UAV Swarm Coordination in Mobile RF Target Tracking. *Simul. Model. Pract. Theory* **2021**, *107*, 102232. [CrossRef]
16. Rekleitis, I.; New, A.P.; Rankin, E.S.; Choset, H. Efficient Boustrophedon Multi-Robot Coverage: An Algorithmic Approach. *Ann. Math. Artif. Intell.* **2008**, *52*, 109–142. [CrossRef]
17. Nolan, P.; Paley, D.A.; Kroeger, K. Multi-UAS path planning for non-uniform data collection in precision agriculture. In Proceedings of the 2017 IEEE Aerospace Conference, Big Sky, MT, USA, 4–11 March 2017; pp. 1–12. [CrossRef]
18. Di Franco, C.; Buttazzo, G. Coverage Path Planning for UAVs Photogrammetry with Energy and Resolution Constraints. *J. Intell. Robot. Syst.* **2016**, *83*, 445–462. [CrossRef]
19. Choset, H. Coverage for Robotics—A Survey of Recent Results. *Ann. Math. Artif. Intell.* **2001**, *31*, 113–126. [CrossRef]
20. Galceran, E.; Carreras, M. A Survey on Coverage Path Planning for Robotics. *Robot. Auton. Syst.* **2013**, *61*, 1258–1276. [CrossRef]
21. Valente, J.; Del Cerro, J.; Barrientos, A.; Sanz, D. Aerial Coverage Optimization in Precision Agriculture Management: A Musical Harmony Inspired Approach. *Comput. Electron. Agric.* **2013**, *99*, 153–159. [CrossRef]
22. Paull, L.; Thibault, C.; Nagaty, A.; Seto, M.; Li, H. Sensor-Driven Area Coverage for an Autonomous Fixed-Wing Unmanned Aerial Vehicle. *IEEE Trans. Cybern.* **2014**, *44*, 1605–1618. [CrossRef]
23. Xu, A.; Viriyasuthee, C.; Rekleitis, I. Optimal complete terrain coverage using an unmanned aerial vehicle. In Proceedings of the 2011 IEEE International Conference on Robotics and Automation, Shanghai, China, 9–13 May 2011; pp. 2513–2519. [CrossRef]
24. Fazli, P.; Davoodi, A.; Mackworth, A.K. Multi-Robot Repeated Area Coverage. *Auton. Robots* **2013**, *34*, 251–276. [CrossRef]
25. Lawrance, N.; Sukkarieh, S. Wind Energy Based Path Planning for a Small Gliding Unmanned Aerial Vehicle. In Proceedings of the AIAA Guidance, Navigation, and Control Conference, American Institute of Aeronautics and Astronautics, Chicago, IL, USA, 10–13 August 2009. [CrossRef]
26. Felipe-García, B.; Hernández-López, D.; Lerma, J.L. Analysis of the Ground Sample Distance on Large Photogrammetric Surveys. *Appl. Geomat.* **2012**, *4*, 231–244. [CrossRef]
27. Goerzen, C.; Kong, Z.; Mettler, B. A Survey of Motion Planning Algorithms from the Perspective of Autonomous UAV Guidance. *J. Intell. Robot. Syst.* **2009**, *57*, 65. [CrossRef]
28. Dadkhah, N.; Mettler, B. Survey of Motion Planning Literature in the Presence of Uncertainty: Considerations for UAV Guidance. *J. Intell. Robot. Syst.* **2012**, *65*, 233–246. [CrossRef]
29. Colomina, I.; Molina, P. Unmanned Aerial Systems for Photogrammetry and Remote Sensing: A Review. *ISPRS J. Photogramm. Remote Sens.* **2014**, *92*, 79–97. [CrossRef]
30. Cabreira, T.; Brisolara, L.; Ferreira, P.R., Jr. Survey on Coverage Path Planning with Unmanned Aerial Vehicles. *Drones* **2019**, *3*, 4. [CrossRef]
31. Almadhoun, R.; Taha, T.; Seneviratne, L.; Zweiri, Y. A Survey on Multi-Robot Coverage Path Planning for Model Reconstruction and Mapping. *SN Appl. Sci.* **2019**, *1*, 847. [CrossRef]
32. Chen, Y.; Zhang, H.; Xu, M. The coverage problem in UAV network: A surve. In Proceedings of the Fifth International Conference on Computing, Communications and Networking Technologies (ICCCNT), Hefei, China, 11–13 July 2014; pp. 1–5. [CrossRef]
33. Choset, H.; Pignon, P. Path Planning: The Boustrophedon Cellular Decomposition. In Proceedings of the International Conference on Field and Service Robotics, Canberra, Australia, 12 October 1997; pp. 1311–1320.
34. Choset, H.; Pignon, P. Coverage path planning: The boustrophedon cellular decompositio. In *Field and Service Robotics*; Springer: London, UK, 1998; pp. 203–209. [CrossRef]

35. Choset, H.; Acar, E.; Rizzi, A.A.; Luntz, J. Exact cellular decompositions in terms of critical points of Morse functions. In Proceedings of the 2000 ICRA. Millennium Conference, IEEE International Conference on Robotics and Automation, Symposia Proceedings (Cat. No.00CH37065), San Francisco, CA, USA, 24–28 April 2000; Volume 3, pp. 2270–2277. [CrossRef]
36. Andersen, H.L. Path Planning for Search and Rescue Mission Using Multicopters. 2014. 135. Available online: https://ntnuopen.ntnu.no/ntnu-xmlui/handle/11250/261317 (accessed on 12 December 2021).
37. LaValle, S.M. *Planning Algorithms*; Cambridge University Press: Cambridge, UK, 2006.
38. Zheng, X.; Jain, S.; Koenig, S.; Kempe, D. Multi-robot forest coverage. In Proceedings of the 2005 IEEE/RSJ International Conference on Intelligent Robots and Systems, Edmonton, AB, Canada, 2–6 August 2005; pp. 3852–3857. [CrossRef]
39. Choset, H.; Lynch, K.M.; Hutchinson, S.; Kantor, G.A.; Burgard, W. *Principles of Robot Motion: Theory, Algorithms, and Implementations*; MIT Press: Cambridge, MA, USA, 2005.
40. Latombe, J.-C. *Robot Motion Planning*; Stanford University, Kluwer Academic Publishers: New York, NY, USA, 1991.
41. Oksanen, T.; Visala, A. Coverage Path Planning Algorithms for Agricultural Field Machines. *J. Field Robot.* **2009**, *26*, 651–668. [CrossRef]
42. Acar, E.U.; Choset, H.; Rizzi, A.A.; Atkar, P.N.; Hull, D. Morse Decompositions for Coverage Tasks. *Int. J. Robot. Res.* **2002**, *21*, 331–344. [CrossRef]
43. SStein, E.; Milnor, J.W.; Spivak, M.; Wells, R.; Wells, R.; Mather, J.N. *Morse Theory*; Princeton University Press: Princeton, NJ, USA, 1963; ISBN 978-0-691-08008-6.
44. Acar, E.U.; Choset, H.; Atkar, P.N. Complete sensor-based coverage with extended-range detectors: A hierarchical decomposition in terms of critical points and Voronoi diagrams. In Proceedings of the 2001 IEEE/RSJ International Conference on Intelligent Robots and Systems, Expanding the Societal Role of Robotics in the the Next Millennium (Cat. No.01CH37180), Maui, HI, USA, 29 October–3 November 2001; Volume 3, pp. 1305–1311. [CrossRef]
45. Acar, E.U.; Choset, H. Sensor-Based Coverage of Unknown Environments: Incremental Construction of Morse Decompositions. *Int. J. Robot. Res.* **2002**, *21*, 345–366. [CrossRef]
46. Wong, S. *Qualitative Topological Coverage of Unknown Environments by Mobile Robots*; The University of Auckland: Auckland, New Zealand, 2006.
47. Wong, S.C.; MacDonald, B.A. Complete coverage by mobile robots using slice decomposition based on natural landmarks. In *PRICAI 2004: Trends in Artificial Intelligence*; Springer: Berlin/Heidelberg, Germany, 2004; pp. 683–692. [CrossRef]
48. Butler, Z.J.; Rizzi, A.A.; Hollis, R.L. Contact sensor-based coverage of rectilinear environments. In Proceedings of the 1999 IEEE International Symposium on Intelligent Control Intelligent Systems and Semiotics (Cat. No.99CH37014), Cambridge, MA, USA, 17 September 1999; pp. 266–271. [CrossRef]
49. Moravec, H.; Elfes, A. High resolution maps from wide angle sonar. In Proceedings of the 1985 IEEE International Conference on Robotics and Automation Proceedings, St. Louis, MO, USA, 25–28 March 1985; Volume 2, pp. 116–121. [CrossRef]
50. Zelinsky, A.; Jarvis, R.A.; Byrne, J.C.; Yuta, S. Planning Paths of Complete Coverage of an Unstructured Environment by a Mobile Robot. In Proceedings of the International Conference on Advanced Robotics, Tsukuba, Japan, 11 September 1993; Volume 13, pp. 533–538.
51. Shivashankar, V.; Jain, R.; Kuter, U.; Nau, D. Real-Time Planning for Covering an Initially-Unknown Spatial Environment. In Proceedings of the Twenty-Fourth International FLAIRS Conference, Palm Beach, FL, USA, 20 March 2011.
52. Gabriely, Y.; Rimon, E. Spiral-STC: An on-Line Coverage Algorithm of Grid Environments by a Mobile Robot. In Proceedings of the 2002 IEEE International Conference on Robotics and Automation (Cat. No.02CH37292), Washington, DC, USA, 11–15 May 2002; Volume 1, pp. 954–960. [CrossRef]
53. Luo, C.; Yang, S.X.; Stacey, D.A.; Jofriet, J.C. A Solution to Vicinity Problem of Obstacles in Complete Coverage Path Planning. In Proceedings of the 2002 IEEE International Conference on Robotics and Automation (Cat. No.02CH37292), Washington, DC, USA, 11–15 May 2002; Volume 1, pp. 612–617. [CrossRef]
54. Yang, S.X.; Luo, C. A Neural Network Approach to Complete Coverage Path Planning. *IEEE Trans. Syst. Man Cybern. Part B Cybern.* **2004**, *34*, 718–724. [CrossRef]
55. Hazon, N.; Mieli, F.; Kaminka, G.A. Towards Robust On-Line Multi-Robot Coverage. In Proceedings of the 2006 IEEE International Conference on Robotics and Automation, ICRA 2006, Orlando, FL, USA, 15–19 May 2006; pp. 1710–1715. [CrossRef]
56. Luo, C.; Yang, S.X. A real-time cooperative sweeping strategy for multiple cleaning robots. In Proceedings of the of the IEEE Internatinal Symposium on Intelligent Control, Vancouver, BC, Canada, 30 October 2002; pp. 660–665. [CrossRef]
57. Luo, C.; Yang, S.X.; Stacey, D.A. Real-time path planning with deadlock avoidance of multiple cleaning robots. In Proceedings of the 2003 IEEE International Conference on Robotics and Automation (Cat. No.03CH37422), Taipei, Taiwan, 14–19 September 2003; Volume 3, pp. 4080–4085. [CrossRef]
58. Easton, K.; Burdick, J. Inspection'. In Proceedings of the Proceedings of the 2005 IEEE International Conference on Robotics and Automation, Barcelona, Spain, 18–22 April 2005; pp. 727–734. [CrossRef]
59. Ju, C.; Son, H. Multiple UAV Systems for Agricultural Applications: Control, Implementation, and Evaluation. *Electronics* **2018**, *7*, 162. [CrossRef]
60. Vincent, P.; Rubin, I. A framework and analysis for cooperative search using UAV swarms. In Proceedings of the 2004 ACM symposium on Applied Computing, New York, NY, USA, 14 March 2004; pp. 79–86. [CrossRef]

61. Maza, I.; Ollero, A. Multiple UAV cooperative searching operation using polygon area decomposition and efficient coverage algorithms. In *Distributed Autonomous Robotic Systems 6*; Springer: Tokyo, Japan, 2007; pp. 221–230. [CrossRef]
62. Alotaibi, E.T.; Alqefari, S.S.; Koubaa, A. LSAR: Multi-UAV Collaboration for Search and Rescue Missions. *IEEE Access* **2019**, *7*, 55817–55832. [CrossRef]
63. Acevedo, J.J.; Arrue, B.C.; Maza, I.; Ollero, A. Cooperative Large Area Surveillance with a Team of Aerial Mobile Robots for Long Endurance Missions. *J. Intell. Robot. Syst.* **2013**, *70*, 329–345. [CrossRef]
64. Acevedo, J.J.; Arrue, B.C.; Maza, I.; Ollero, A. Distributed Approach for Coverage and Patrolling Missions with a Team of Heterogeneous Aerial Robots under Communication Constraints. *Int. J. Adv. Robot. Syst.* **2013**, *10*, 28. [CrossRef]
65. Acevedo, J.J.; Arrue, B.C.; Diaz-Bañez, J.M.; Ventura, I.; Maza, I.; Ollero, A. One-to-One Coordination Algorithm for Decentralized Area Partition in Surveillance Missions with a Team of Aerial Robots. *J. Intell. Robot. Syst.* **2014**, *74*, 269–285. [CrossRef]
66. Xu, A.; Viriyasuthee, C.; Rekleitis, I. Efficient Complete Coverage of a Known Arbitrary Environment with Applications to Aerial Operations. *Auton. Robots* **2014**, *36*, 365–381. [CrossRef]
67. Balampanis, F.; Maza, I.; Ollero, A. Area Decomposition, Partition and Coverage with Multiple Remotely Piloted Aircraft Systems Operating in Coastal Regions. In Proceedings of the 2016 International Conference on Unmanned Aircraft Systems (ICUAS), Arlington, VA, USA, 7–10 June 2016; pp. 275–283. [CrossRef]
68. Balampanis, F.; Maza, I.; Ollero, A. Coastal Areas Division and Coverage with Multiple UAVs for Remote Sensing. *Sensors* **2017**, *17*, 808. [CrossRef]
69. Kallmann, M.; Bieri, H.; Thalmann, D. Fully Dynamic Constrained Delaunay Triangulations. In *Geometric Modeling for Scientific Visualization*; Springer: Berlin/Heidelberg, Germany, 2004; pp. 241–257. [CrossRef]
70. Shewchuk, J.R. Mesh generation for domains with small angles. In Proceedings of the Sixteenth Annual Symposium on Computational Geometry, New York, NY, USA, 1 May 2000; pp. 1–10. [CrossRef]
71. Balampanis, F.; Maza, I.; Ollero, A. Spiral-like coverage path planning for multiple heterogeneous UAS operating in coastal regions. In Proceedings of the 2017 International Conference on Unmanned Aircraft Systems (ICUAS), Miami, FL, USA, 13–16 June 2017; pp. 617–624. [CrossRef]
72. Balampanis, F.; Maza, I.; Ollero, A. Area Partition for Coastal Regions with Multiple UAS. *J. Intell. Robot. Syst.* **2017**, *88*, 751–766. [CrossRef]
73. Hayat, S.; Yanmaz, E.; Brown, T.X.; Bettstetter, C. Multi-objective UAV path planning for search and rescue. In Proceedings of the 2017 IEEE International Conference on Robotics and Automation (ICRA), Singapore, 29 May–3 June 2017; pp. 5569–5574. [CrossRef]
74. Trujillo, M.M.; Darrah, M.; Speransky, K.; DeRoos, B.; Wathen, M. Optimized flight path for 3D mapping of an area with structures using a multirotor. In Proceedings of the 2016 International Conference on Unmanned Aircraft Systems (ICUAS), Arlington, VA, USA, 7–10 June 2016; pp. 905–910. [CrossRef]
75. Darrah, M.; Trujillo, M.M.; Speransky, K.; Wathen, M. Optimized 3D mapping of a large area with structures using multiple multirotors. In Proceedings of the 2017 International Conference on Unmanned Aircraft Systems (ICUAS), Miami, FL, USA, 13–16 June 2017; pp. 716–722. [CrossRef]
76. Torres, M.; Pelta, D.A.; Verdegay, J.L.; Torres, J.C. Coverage Path Planning with Unmanned Aerial Vehicles for 3D Terrain Reconstruction. *Expert Syst. Appl.* **2016**, *55*, 441–451. [CrossRef]
77. Coombes, M.; Chen, W.-H.; Liu, C. Boustrophedon Coverage Path Planning for UAV Aerial Surveys in Wind. In Proceedings of the 2017 International Conference on Unmanned Aircraft Systems (ICUAS), Miami, FL, USA, 13–16 June 2017; pp. 1563–1571. [CrossRef]
78. Coombes, M.; Fletcher, T.; Chen, W.-H.; Liu, C. Optimal Polygon Decomposition for UAV Survey Coverage Path Planning in Wind. *Sensors* **2018**, *18*, 2132. [CrossRef]
79. Cabreira, T.M.; Franco, C.D.; Ferreira, P.R.; Buttazzo, G.C. Energy-Aware Spiral Coverage Path Planning for UAV Photogrammetric Applications. *IEEE Robot. Autom. Lett.* **2018**, *3*, 3662–3668. [CrossRef]
80. Di Franco, C.; Buttazzo, G. Energy-Aware Coverage Path Planning of UAVs. In Proceedings of the 2015 IEEE International Conference on Autonomous Robot Systems and Competitions, Vila Real, Portugal, 8–10 April 2015; pp. 111–117. [CrossRef]
81. Li, D.; Wang, X.; Sun, T. Energy-Optimal Coverage Path Planning on Topographic Map for Environment Survey with Unmanned Aerial Vehicles. *Electron. Lett.* **2016**, *52*, 699–701. [CrossRef]
82. Artemenko, O.; Dominic, O.J.; Andryeyev, O.; Mitschele-Thiel, A. Energy-aware trajectory planning for the localization of mobile devices using an unmanned aerial vehicle. In Proceedings of the 2016 25th International Conference on Computer Communication and Networks (ICCCN), Waikoloa, HI, USA, 1–4 August 2016; pp. 1–9. [CrossRef]
83. Ahmadzadeh, A.; Keller, J.; Pappas, G.; Jadbabaie, A.; Kumar, V. An optimization-based approach to time-critical cooperative surveillance and coverage with UAVs. In *Experimental Robotics: The 10th International Symposium on Experimental Robotics*; Khatib, O., Kumar, V., Rus, D., Eds.; Springer: Berlin/Heidelberg, Germany, 2008; pp. 491–500. [CrossRef]
84. Araujo, J.F.; Sujit, P.B.; Sousa, J.B. Multiple UAV area decomposition and coverage. In Proceedings of the 2013 IEEE Symposium on Computational Intelligence for Security and Defense Applications (CISDA), Singapore, 16–19 April 2013; pp. 30–37. [CrossRef]
85. Majeed, A.; Lee, S. A New Coverage Flight Path Planning Algorithm Based on Footprint Sweep Fitting for Unmanned Aerial Vehicle Navigation in Urban Environments. *Appl. Sci.* **2019**, *9*, 1470. [CrossRef]

86. Majeed, A.; Hwang, S.O. A Multi-Objective Coverage Path Planning Algorithm for UAVs to Cover Spatially Distributed Regions in Urban Environments. *Aerospace* **2021**, *8*, 343. [CrossRef]
87. Cheng, C.-T.; Fallahi, K.; Leung, H.; Tse, C.K. Cooperative Path Planner for UAVs Using ACO Algorithm with Gaussian Distribution Functions. In Proceedings of the 2009 IEEE International Symposium on Circuits and Systems, Taipei, Taiwan, 24–27 May 2009; pp. 173–176. [CrossRef]
88. Kuiper, E.; Nadjm-Tehrani, S. Mobility Models for UAV Group Reconnaissance Applications. In Proceedings of the 2006 International Conference on Wireless and Mobile Communications (ICWMC'06), Bucharest, Romania, 29–31 July 2006; p. 33. [CrossRef]

Article

Landmark-Assisted Compensation of User's Body Shadowing on RSSI for Improved Indoor Localisation with Chest-Mounted Wearable Device

Md Abdulla Al Mamun, David Vera Anaya, Fan Wu and Mehmet Rasit Yuce *

Department of Electrical and Computer Systems Engineering, Clayton Campus, Monash University, Melbourne, VIC 3800, Australia; md.mamun1@monash.edu (M.A.A.M.); david.veraanaya@monash.edu (D.V.A.); fan.wu@monash.edu (F.W.)
* Correspondence: mehmet.yuce@monash.edu

Abstract: Nowadays, location awareness becomes the key to numerous Internet of Things (IoT) applications. Among the various methods for indoor localisation, received signal strength indicator (RSSI)-based fingerprinting attracts massive attention. However, the RSSI fingerprinting method is susceptible to lower accuracies because of the disturbance triggered by various factors from the indoors that influence the link quality of radio signals. Localisation using body-mounted wearable devices introduces an additional source of error when calculating the RSSI, leading to the deterioration of localisation performance. The broad aim of this study is to mitigate the user's body shadowing effect on RSSI to improve localisation accuracy. Firstly, this study examines the effect of the user's body on RSSI. Then, an angle estimation method is proposed by leveraging the concept of landmark. For precise identification of landmarks, an inertial measurement unit (IMU)-aided decision tree-based motion mode classifier is implemented. After that, a compensation model is proposed to correct the RSSI. Finally, the unknown location is estimated using the nearest neighbour method. Results demonstrated that the proposed system can significantly improve the localisation accuracy, where a median localisation accuracy of 1.46 m is achieved after compensating the body effect, which is 2.68 m before the compensation using the classical K-nearest neighbour method. Moreover, the proposed system noticeably outperformed others when comparing its performance with two other related works. The median accuracy is further improved to 0.74 m by applying a proposed weighted K-nearest neighbour algorithm.

Keywords: indoor localisation; fingerprinting; landmark; wearable device; inertial measurement device; motion mode detection; body shadowing compensation; nearest neighbour

1. Introduction

Knowledge about location information becomes the key to numerous location-based services (LBS) in various application domains including healthcare and safety, search and rescue, assisted living, robotics, shopping and museum assistance, context awareness and social networking, advertising, and marketing [1]. One of the key prerequisites to successfully empower these services is estimating the position of a subject of interest. This task can be effortlessly accomplished by employing the receivers of the Global Navigation Satellite System (GNSS) with direct line-of-sight (LOS) scenarios in the case of outdoors. The existence of the complex nature of indoors in terms of geometrical structures, presence of numerous objects made of multivariate materials, and the variations in ambient meteorological conditions lead to the reflection, refraction, or even complete blockage of the GNSS signal. Hence, the GNSS is unable to produce the desired accuracy required for the indoors [2].

Typically, an indoor localisation system utilises an infrastructure inside a building with a set of devices wirelessly connected to locate an unknown target carrying devices compatible with that network. Various technologies are used so far for indoor localisation including

Bluetooth low energy (BLE), radio frequency identification (RFID), ultra-wideband (UWB), ultrasound, wireless local area network (WLAN), and wireless sensor network (WSN) [1]. Among them, RSSI-based WSN technology has drawn massive attention of the researchers owing to the emerging usability for numerous IoT applications as well as the easiness in RSSI acquisition. RSSI is the standard to measure the received signal power, which is used by various methods, such as propagation modelling, trilateration, multidimensional scaling, DV-Hop, and fingerprinting for location estimation [1]. From them, the RSSI fingerprinting approach offers satisfactory results without the requirement of additional costs in terms of hardware and computation. The fingerprinting method comprises two major phases: the offline training phase and the online localisation phase. The training phase builds a database, named the radio map, by gathering geotagged RSSI fingerprint data from visible radio modules/anchor nodes, named reference nodes (RN), at known locations, named reference points (RP). The online phase calculates the position of an unknown target node by comparing a query fingerprint with the radio map.

Recently, WSN has become an attractive research area, especially for various monitoring applications, due to its real-time and accurate response, coverage, and simple infrastructure. With the continuous advancement and miniaturisation of sensing, as well as communication technologies, wearable devices are becoming an essential component for daily living. WSNs using wearable sensor devices are emerging for many IoT applications. Acquiring information about the location of a user is one of the key features of wearable devices, which becomes one of the major issues for WSN due to the presence of a massive number of wearable sensor nodes in modern IoT applications.

One of the major limitations of RSSI-based indoor localisation is the erroneous determination of RSSI. The main reasons for this are the abovementioned complex nature of indoor environments and the non-line-of-sight (NLOS) situations triggered by the signal blockage between a sender and a receiver. In the case of wearable devices, the user's body can introduce the NLOS scenario that leads to an additional effect on the resulted RSSI. The human body encompasses around 70% of water that can absorb part of the radio signal [3]. Moreover, the human body can scatter the longer radio signal waves while reflecting or attenuating the shorter ones due to its conductive nature [4]. Thus, the presence of the human body in between a sender and a receiver influences the propagation of radio signal that can cause an incorrect calculation of RSSI. Eventually, this circumstance leads to an erroneous position estimation in RSSI fingerprinting-based localisation when a wearable device calculates incorrect RSSI from multiple RNs. Researchers have already reported that human body shadowing could distort the RSSI by up to 5 dBm, causing a positioning performance degradation of about 67%, where there is a strong correlation between that distortion and user orientation [5]. Besides the NLOS scenarios created by the wearable user's body, there may be other sources that can introduce errors in RSSI calculation, including the presence and movement of other humans, as well as objects in between the sender and receiver. Although it is impractical to characterise all such errors precisely due to the randomness in the numbers and sizes of those humans and objects, it is realistic to deal with the systematic source of error caused by the wearable user's body [5]. In the case of RSSI fingerprint-based localisation, the user's body shadowing effect (BSE) can be mitigated explicitly by modelling and compensating this systematic error when comparing a query RSSI fingerprint with a radio map. Although there are several studies that investigate the effects of user's BSE on wireless signal transmission, there are still some challenges that require further attention, including:

- The estimation of the orientation angle between a user and an RN in real time.
- The derivation of a BSE compensation model that can mitigate the user's body effects for every single orientation angle scenario instead of some discrete orientation angles.
- The adaptation of the orientation angle estimation and BSE mitigation methods in real-life indoor localisation applications.

Knowledge of the indoor area, i.e., the spatial information, may be an assistive tool that can be leveraged to improve the indoor localisation accuracy without paying extra cost for setup. Landmark, i.e., the sensory landmark, is one such piece of spatial information that is distributed naturally to a floor plan and can be helpful to enhance localisation accuracy [2,6]. Specifically, landmarks are the markers in the indoor map that experience specific signal patterns all the time when one or more sensors meet those markers. Although some previous works utilised landmarks for robot tracking or pedestrian dead reckoning (PDR)-based positioning, this work used landmarks as a supportive tool for mitigating the human BSE on RSSI.

The aim of this study is to compensate the user's BSE on RSSI to improve the RSSI fingerprinting-based indoor localisation performance with a chest-mounted wearable device in a WSN setting. The proposed fingerprinting system composed the offline and online phases similar to the traditional fingerprint methods. However, the online phase performs several additional tasks to mitigate human body shadowing errors. To compensate the RSSIs of a query fingerprint with proper values, the angle between the wearable device and the RNs is estimated considering the user's orientation. The concept of landmark graph along with arctangent function is utilised for angle estimation. To identify a landmark, an IMU-aided decision tree-based motion mode detection classifier is implemented. Then, a human body shadowing compensation model is proposed to correct the RSSIs of the query fingerprint. Finally, both the classical k-nearest neighbour (K-NN) and weighted k-nearest neighbour (WK-NN) algorithms are employed to calculate the location of the unknown target. The main contributions of this research are as follows:

- An in-depth analysis of the behaviour of XBee RSSI is performed to investigate its effect on the user's body.
- A unique method is proposed to estimate the orientation angle between a user and the RNs.
- A new model is proposed that can compensate the user's BSE on XBee RSSI for every possible orientation angle.
- A landmark-assisted weight calculation method is used to implement the WK-NN algorithm to improve the localisation accuracy.
- Experiments were conducted in a real indoor scenario by applying the proposed method and model for real-time application.

The remainder of the paper is organised as follows: Section 2 discusses and compares the existing literature related to this study; Section 3 presents an in-depth analysis of the effect of the user's body on RSSI; Section 4 presents an overview of the proposed system; Sections 5–7 describe the details of the proposed system that includes landmark identification, user's BSE compensation, and fingerprinting localisation, respectively; Section 8 discusses the experiments and illustrates the results by comparing with other related works; finally, Section 9 concludes this study with future recommendations.

2. Related Works

Until now, there are numerous studies that investigate the effects of the human body shadowing on wireless signal transmission to characterise and model the wireless channel and antenna radiation by focusing on various aspects of body-centric radio frequency-based communication. Moreover, human BSEs have been analysed and modelled for a variety of applications, including people counting [7,8], fall detection [9,10], and activity recognition [11], as well as proximity detection for coronavirus contact tracing application [12]. Although several studies have been performed on analysing the effect of human body shadowing on radio signal transmission targeting indoor localisation applications, they have mostly neglected the derivation of a compensation model and/or the integration of a compensation model to implement a real-time localisation system. Table 1 presents an overview of the existing literature that discusses human BSEs on wireless signal transmission for indoor localisation and tracking applications.

In the literature, researchers utilised several wireless technologies; this study only focuses on the systems that exploited RSSI as their measurement approach and/or fingerprinting as their localisation method. An RFID-based system is presented in [13], where the authors demonstrated the improvement in indoor localisation accuracy by compensating the errors caused by human body shadowing. Channel models for both the LOS and NLOS cases were derived, and RFID RSSI-based Monte Carlo localisation was implemented to achieve an accuracy of 1.18 m. However, this approach has very limited applicability for real-time location tracking applications because the differentiation between the LOS and NLOS conditions were assessed manually.

Table 1. Comparison among existing studies focused on human body shadowing effects on wireless signal for indoor localisation and tracking applications.

Research Type [1]	Wireless Technology	Measurement Approach	Sensor Placement	Angle Variations (°)	Localisation Methods	Evaluation [2]	Localisation Accuracy (m)	Ref.
An, Co	2.4 GHz RF	RSSI	Handheld	90	Fingerprinting	Ex	2.94 (50th percentile)	[5]
An, Co	WiFi	RSSI	Handheld	45	Fingerprinting	Ex	1.65 (average)	[14]
Mo, Co	RFID	RSSI	Wrist	N/A	Monte Carlo	Ex	1.18	[13]
An, Mo, Co	WiFi	RSSI	Chest, Handheld	N/A	NLLS	Ex	N/A	[15]
An, Mo, Co	Zigbee	RSSI	Chest, Back	45	Fingerprinting	Ex	2.5 (median)	[16]
An, Mo, Co	Zigbee	RSSI	Chest, Back, Wrist	45	Fingerprinting	Si and Ex	2.99 (50th percentile for chest)	[17]
An, Mo	WiFi	RSSI	Handheld	N/A	Fingerprinting	Ex	N/A	[18]
An, Co	WiFi	RSSI	Handheld	90	Fingerprinting	Ex	2.00 (50th percentile)	[19]
An, Mo, Co	BLE	RSSI	Handheld	3 orientations	Ranging, Trilateration	Ex	0.77 (mean)	[20]
An, Mo, Co	Zigbee	RSSI	Chest	15	Fingerprinting	Ex	0.74 (median)	This study

[1] An = analysis; Mo = modelling; Co = compensation. [2] Ex = experiment; Si = simulation.

From Institute of Electrical and Electronic Engineers (IEEE) 802.11 family of standards, the impact of the human body on RSSI-based ranging measurements for cooperative localisation is presented in [15]. The authors investigated both the body and hand grip effects on RSSI among the neighbouring nodes. This study demonstrated that there is no significant improvement in cooperative localisation, compared to the noncooperative case, if the BSEs are not mitigated correctly. In [18], a mathematical model is proposed to mitigate the user's BSE on RSSI of WiFi signals for improving indoor positioning accuracy. Handheld mobile devices were used to collect WiFi signals in a multipath-free environment, both for the LOS and NLOS cases, to analyse the BSEs. Finally, a model was derived that can intensify the strength of the signals which are coming through the NLOS states caused by the user body. Still, the authors did not discuss the methods of orientation estimation for real-time applicability of the proposed model. Moreover, this study only considers the handheld mobile devices for the experiment, and thus the model may not be compatible with body-attached wearable devices. In [5], the authors presented the first fingerprinting-based indoor localisation system that considered the subject's BSEs on signal RSSI for position estimation. A radio map was created by using both the empirical measurements of RSS and a signal propagation model. To reduce the location estimation error that is cause by the user's body, RSSI fingerprints were collected for four orientations of the subject's body, compared to the RNs, in terms of four directions, i.e., north, south, east, or west. A K-NN search algorithm was employed, and a median accuracy of 2–3 m was achieved after compensating the human BSE. However, this solution only analyses the effect of user orientation on location estimation and falls short of proposing any compensation model with its real-time applicability to mitigate that effect. To solve the issue of estimating user orientation in real time, King et al. described an approach named COMPASS, where the authors utilised a digital compass to acquire the user's orientation during both the offline and online phases [14]. During the offline phase, radio fingerprints were collected from each RN for eight orientations in every 45° angle position. In the online phase, a subset of fingerprints from the radio map was preselected based on the user orientation, and a probabilistic algorithm was applied to the subset to calculate the user position. Results demonstrated that considering the body orientation improved the localisation accuracy, where the average accuracy was 1.65 m. Yet, the radio map becomes highly redundant as eight radio fingerprints corresponding to eight directions, i.e., in every 45°, were collected for a single RP. As a result, the search space increases by eight times, which can cause an extra burden on the system performance in terms of computation cost and memory requirement for a large environmental area. It may even become infeasible for resource-constrained wearable devices for edge computing in the case of real-world applications. Moreover, it also increases the cost of the offline phase in terms of time and labour. Additionally, using COMPASS may produce high errors in orientation estimation for indoors, especially around the objects that have electromagnetic radiations. A similar approach was applied in [19], where the authors used mobile phone integrated compass and collected radio fingerprints for four orientations in the offline phase. During the online phase, they narrowed down the search space by applying a clustering method that used both the signal domain and spatial domain. An adaptive weighted K-NN algorithm was developed, which achieved an accuracy of 2.0 m for the 50th percentile; however, this study only considers four orientations of the human body in four directions, which is not enough to explore the complete variations of RSS values around a body.

From Zigbee-based indoor localisation solutions, a body-worn device is used in [16], where the authors analysed the BSEs on RSSI of 2.4 GHz ZigBee signals. Two tags were attached on the chest and back of a wearer, and data were collected at different angular positions. The arc tangent function is used for orientation estimation, and a simple cosine model is used to compensate the user's BSE. Another improved version for BSE compensation in indoor localisation is proposed by the same group in [17], where the authors presented two solutions for improving localisation accuracy. In the first solution, a subject requires multiple wearable tags that need to be mounted in different positions to calculate RSSI. Then, their means are used as the input for fingerprint matching with reference fingerprints from a radio map generated using the WHIPP tool [21]. In the second solution, an arc tangent function is used to estimate the orientation of the target, where the target's current location is calculated by averaging four previous positions. To mitigate the BSE, two compensation models are proposed: one is a basic over/underestimation model, and the other is a simulation-based three-dimensional model. Results demonstrated that the proposed models could compensate the BSE to improve localisation accuracy from 3.48 m to 2.99 m (50th percentile for chest). However, this approach requires a subject to wear multiple tags, which may limit its scope for real-world application. Moreover, as the radio map created during the offline phase did not consider the BSE, the estimated location accuracy will be low and, eventually, the orientation estimator's performance will degrade with time. Furthermore, the system assumes a subject always walks forward, and is unable to infer rotation and moving direction, which can cause a significant difference between the estimated orientation and actual orientation.

Recently, Deng et al. reported an IMU-aided system to compensate body shadowing error for BLE-based indoor positioning [20]. The effects of the human body on BLE signal RSSI were analysed. A compensation model was proposed which considers the distance and angle between an RN and the unknown target to calculate the error. The distance is calculated from the signal propagation model, and the user's heading is approximated from the IMU. Finally, an algorithm was proposed to estimate the location of an unknown target by mitigating the body shadowing error in real time. Results demonstrated that the system could achieve an average accuracy of 0.77 m for location estimation. However, the use of IMU exclusively can produce wrong heading estimation because of error accumulation issues with IMU. Moreover, the use of a signal propagation model solely to measure distance can produce high distance error, especially indoors. Thus, the described body shadowing detection strategy can lead to erroneous output as a consequence of the errors from the heading and distance estimation.

In this study, the user's orientation is estimated by applying a unique approach using a BSE compensation model that is proposed to mitigate the user's body shadowing error in real time to improve indoor localisation accuracy.

3. Analysis of User's Body Shadowing Effect on RSSI

Several experiments were performed to investigate the effects of the user's body on radio signal RSS values. This section describes the experiments and provides observations from the experiments.

3.1. Experiments Overview

The experiments were conducted in two different indoor environments with a similar type of setup. The details of the hardware modules used for the experiments are discussed in Section 8.1. Four adults (two males and two females) with different heights, weights, and body shapes participated to allow us to collect data for the experiments. During each experiment, a participant wore a chest-mounted wearable device as shown in Figure 1a. Data were collected from six RNs installed in the ceiling, as shown in Figure 1b, and placed in the same direction 2.1 m, 5.1 m, 7.5 m, 9.9 m, 12.3 m, and 15.6 m away from the participant. There were both the LOS and NLOS communication scenarios between the participants and RNs. A participant collected the RSSI from each RN while standing and turning 360° towards the clockwise direction around the vertical axis. During data collection, a participant turned 15° in every 60 s and collected RSSI data with a frequency of 2 Hz that took 24 min for a complete rotation. The average of the collected RSSI was calculated for every angle position and stored in a database for further analysis. Each participant performed the experiment several times and at different time periods of a day for a month.

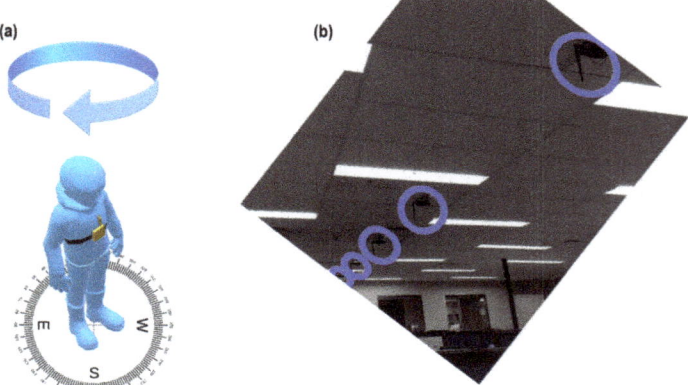

Figure 1. Experimental setup to measure user's body effect. (**a**) Wearable device attached to the chest of a user; (**b**) reference nodes attached beneath the ceiling (marked by circles).

3.2. Observations

Figure 2 shows the experimental results from two different experiments, performed by a female participant and a male participant in different indoor environments. For each case, the figure presents the average RSSI values for different angles under four specific distances between a participant and the RNs.

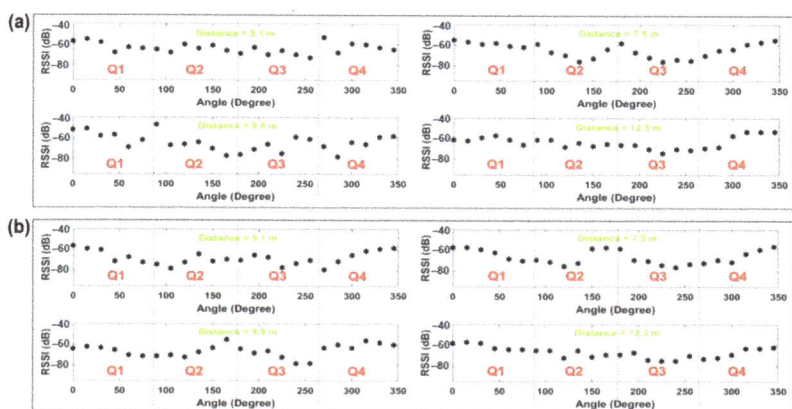

Figure 2. RSSI values versus angle for different distances. (**a**,**b**) illustrate the mean of multiple data points collected by two different participants from two separate indoor environments.

As can be observed from Figure 2, there is a similar trend in the ups and downs of the RSSI values for the different angle positions. The general trend shows the highest RSSI values when the body is placed at 0°, which is the straight LOS between the transmitter and the receiver. Then, the RSSI values start decreasing with the body's rotation throughout the first quarter (Q1) and reach the lowest at around the ending of Q1 and the starting of the second quarter (Q2). They then start rising and continue throughout Q2 and reach a small peak at around the ending of Q2 and starting of the third quarter (Q3). Then, again, the RSSI values start falling until the end of Q3 and start of the fourth quarter (Q4) where they reach the lowest once more. After that, the RSSI values continue to increase until they become straight LOS again, where the values reach the peak. Thus, for most of the cases, the lowest RSSI values are found at the angle positions just after the angle 90° and just before the angle 270°.

To further understand the reason for the results obtained above, Figure 3 illustrates the body position for each quarter, as well as a graphical representation of the electromagnetic waves when arriving at the body and the sensor.

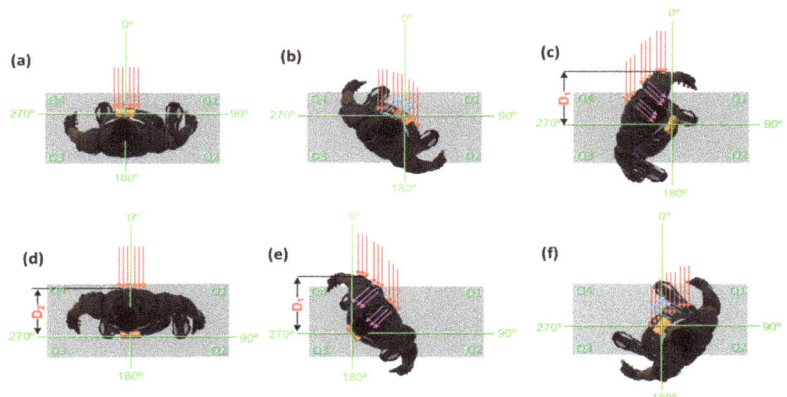

Figure 3. Body position and the status of the electromagnetic waves for different angle positions in each quarter. (**a**) at 0°, (**b**) at 45°, (**c**) just after 90°, (**d**) at 180°, (**e**) just before 270°, and (**f**) at 315°.

When the body is placed with an angle of $0°$ with respect to the LOS with the transmitter, as in Figure 3a, the electromagnetic wave arrives directly to the antenna without major interference. The antenna used for the experiments in Figure 1 was a common dipole antenna connected to the XBee receptor node, with vertical orientation. We can assume for $0°$ that the signal is mostly perpendicular to the chest surface, and therefore also the sensor surface. From $0°$ to $45°$ (clockwise orientation), the low variability among the measured RSSIs at the receptor may be explained due to their closeness to the LOS between the transmitter nodes and the body. On the other hand, after the angle values of $45°$, the variability among the consecutive RSSI values starts to increase, which can be explained using Figure 3b. The body as a transmission medium can be seen as a charged object with higher conductivity (higher loss) and low penetration depth. This low penetration depth means that the signal is highly attenuated inside the conductive body, due to muscles and tissue, and the effect of the electromagnetic wave is highly concentrated on the surface. This influences the signal in such a way that the body guides the surface wave and behaves as a reflector for space waves [22]. These surface waves are explained due to the diffraction of the electromagnetic signal. The diffracted wave's components are propagated along a curved surface, such as the body [23]. This means, as in Figure 3b, if the signal arrives first to the shoulder, the surface-propagated component can affect the direct vertical electromagnetic components that arrive at the antenna and change the RSSI value. On the other hand, some electromagnetic waves are also reflected on the surface. From all the reflected wave components, the one with the biggest amplitude has the same angle as the incident wave [24]. Such interactions between the reflected, incident, and diffracted waves influence the difference in the values obtained for specific angles and the variability of the data as well. The diffracted and reflected wave components may constructively or negatively interfere with the original signal from the transmitter. This explains the distribution of data observed in Figure 2 for some angles in Q2 and Q3.

In Figure 3c–e, it is shown, for Q1 and Q2, how the shadowing effect affects the value of the RSSI in Figure 2. Depending on the penetration depth and the shape of the human body, the signal is attenuated in a nonuniform way inside the body. This explains, then, the variability and the lower values presented for each distance in Q2 and Q3, especially at the angle positions just after $90°$ and just before $270°$. However, something important shall be mentioned. It is observed from Figure 3 that the lowest value on each of the tests is not at $180°$. As depicted in Figure 3c–e, due to the body geometry and position, the shadowing influence in $180°$ is physically lower than the shadowing in an angle position in Q2 and Q3. For $180°$ in Figure 3d, the waveform shall travel through the distance (a length from back to chest) inside the body. In the case of Figure 3c,e, the signals become more attenuated when travelling through a distance D1 (higher than D2, since it is the length from the back part of the shoulder to chest), which explains the results obtained for the values between $195°$ to $270°$. Finally, once the body is placed with a heading angle towards Q4 (Figure 3f), the reflected space components and the diffracted surface components can affect (by component cancellation) the RSSI value, with less attenuation than in Figure 3c–e. The last is proved in Figure 2, which shows the trend of increasing in Q4 and the variability of the signal being reduced. Thus, it is clear from the above observations that, on top of the other factors, the user's body has a significant impact on RF signal and RSS values, which is crucial to consider for RSSI fingerprinting-based indoor positioning applications.

4. Overview of the Proposed System

Figure 4 presents the architecture of the proposed system that follows the traditional fingerprinting scheme. It is mainly composed of two phases: offline training phase and online localisation phase. However, the core contributions lie in the profiling of the query fingerprints by correcting the RSSI and in the fusion of IMU-aided landmarks with classical K-NN method during the online localisation phase. More specifically, when matching a query from a target, the rectification of the queried fingerprint is performed to mitigate the user's BSE on signal RSSI by leveraging the geometrical features from the indoor floor plan, named landmark.

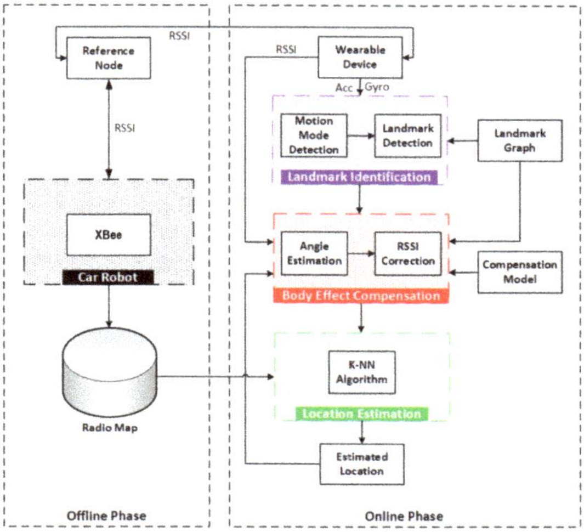

Figure 4. Proposed architecture of RSSI fingerprinting-based indoor localisation with user's body shadowing compensation.

4.1. Offline Training Phase

Although the traditional way of creating a radio map is to partition the area of interest into grids of uniform size followed by the radio fingerprint data collection along a straight path within those grids, this type of manual collection of fingerprint incurs the radio map with RSSIs which have human body shadowing error with them. To exclude this error from the collected RSSI, this study utilised a self-directed car to collect radio fingerprints for radio map. The details of the car, along with its working principle, can be found in our previous work [25]. The car can produce a radio map having nonuniform grids with curved path which have better coverage within the selected area, thus making the system more realistic for real-life localisation applications. Suppose $\widetilde{R}_{ij}$ is the set of RSS values collected from RN j in the ith entry of the radio map; therefore

$$\widetilde{R}_{ij} = \left\{ r_{ijk} : k \in N_s \right\} \quad (1)$$

where r_{ijk} is the kth sample and N_s is the total number of samples from a particular RN for a specific collection point. Together with the coordinate of the collection point, Equation (1) becomes

$$\{x_i, y_i, (\widetilde{R}_{ij} : j \in N_{RN}) : i \in N_{CP}\} \quad (2)$$

where x_i and y_i are the 2D coordinates of the ith fingerprint, N_{RN} is the total number of RN that can be accessed from i^{th} collection point, and N_{CP} is the total number of individual fingerprint tuples in the radio map. As several RSS measurements are collected from each RN for every RP, the mean value of RSS is calculated as

$$R_{ij} = \frac{1}{N_s} \sum_{k=1}^{N_s} \widetilde{R}_{ijk} \quad (3)$$

where R_{ij} is the mean RSS value recorded in the database. Therefore, the final formations of an entry in the radio map can be defined as

$$\{x_i, y_i, (R_{ij} : j \in N_{RN}) : i \in N_{CP}\} \quad (4)$$

4.2. Online Localisation Phase

As shown in Figure 4, the online phase consists of three main modules: landmark identification module, body effect compensation module, and location estimation module. The landmark identification module detects motion modes from IMU data and recognises indoor landmarks by leveraging a landmark graph. The body effect compensation module estimates the angle between a wearable device and an RN by utilising the detected landmarks, the landmark graph, and the previously estimated location. Then, this module corrects a query fingerprint by compensating the RSSI based on a user's body shadowing compensation model. Finally, the K-NN algorithm computes the current location of a target by searching the closest match from the radio map based on the corrected query fingerprint. The detailed descriptions of the three modules for the online localisation phase are presented in Sections 5–7, respectively.

5. Landmark Identification

The basic concept of landmarks and landmark graph are adopted from previous studies [2,6]. Here, landmarks are the sensory markers in the indoor floor plan that encounter specific signal patterns when one or more sensors meet those markers. A landmark can be identified by detecting a subject's motion modes and by applying a set of rules to those motion modes. Therefore, the landmark identification problem can be described as a motion mode detection problem where the key to efficiently detect a landmark depends on the accuracy of motion mode detection. If the system can detect the motion modes with high accuracy, then the detected motions can be used as inputs to a set of rule-based algorithms to identify the landmarks with high accuracy. Thus, this section mainly focuses on the motion mode detection problem. Before describing the details of the motion mode detection, this section presents an overview of the landmark types along with their detection rules, as well as the landmark graph.

5.1. Landmarks and Landmark Graph

Although there are various types of landmarks used for indoor localisation in previous works, this study only considers two types of landmarks: door landmarks and turning landmarks, which are useful for localisation in a 2D indoor environment.

Door landmarks are the sensory markers in the indoor floor plan where the state of motion modes of a subject experience a distinct change in signal patterns. In this work, accelerometer readings are utilised to perceive the door landmarks that can present a specific signal pattern. The usual change pattern in motion modes when accessing a doorway is "walking→static→walking", which is exploited to infer the door landmark. Figure 5a illustrates the concept of a door landmark that presents a typical change pattern from accelerometer measurement that can be identified by detecting the corresponding motion modes for walking and static activities, and setting a proper time threshold in between the activities. Mathematically, the rule to detect a door landmark, L_{door}, can be defined as follows [6]:

$$L_{door} = ((x_t, y_t) \mid (mm_{t-Th_{t1}:t} == walk) \land (mm_{t:t+Th_{t2}} == static) \land (mm_{t+Th_{t2}:t+Th_{t1}+Th_{t2}} == walk)) \quad (5)$$

where mm_t is the subject's motion mode at time t, and Th_{t1} and Th_{t2} are the two time thresholds that regulate the time for the corresponding motion status.

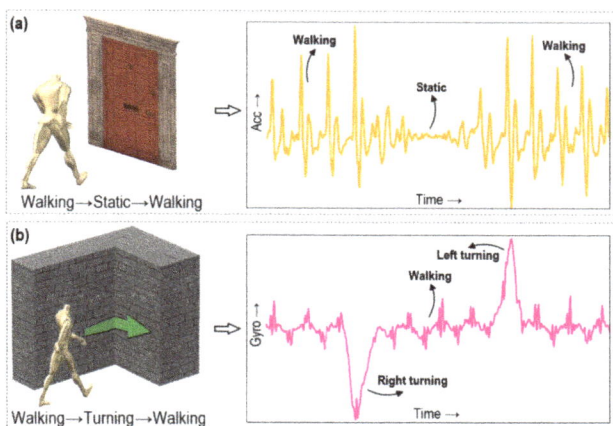

Figure 5. Illustrations and signal patterns of sensory landmarks: (**a**) door landmark; (**b**) turning landmark.

Turning landmarks are the sensory markers in the indoor map where the motion modes provide distinct signal patterns. In this study, data from both the gyroscope and accelerometer combinedly provide a specific motion pattern such as "walking→turning→walking", corresponding to the direction of turning. Figure 5b shows the idea of turning landmark when a subject takes right and left turns. The rule for detecting a turning landmark can be expressed as follows:

$$L_{turning} = ((x_t, y_t) \mid (mm_{t-Th_{t1}:t} == walk) \land (mm_{t:t+Th_{t2}} == turn) \land (mm_{t+Th_{t2}:t+Th_{t1}+Th_{t2}} == walk)) \quad (6)$$

A landmark graph is a directed graph where the landmarks act as nodes and the path segment between two adjacent landmarks acts as edges. Let $LG = (LM, TJ)$ denote a landmark graph where LM is the set of vertices $\{lm_1, lm_2, \ldots lm_N\}$, i.e., the potential set of landmarks, and TJ is the set of edges $\{tj_1, tj_2, \ldots tj_N\}$, i.e., the potential set of trajectories that connect the landmarks. Each landmark, lm_i, is represented by its location coordinates, its type, and its unique identifier using the tuple $< x_i, y_i, lt_i, id_i >$. Each trajectory tj_i is a tuple $< lm_j, lm_k, \theta_{jk}, d_i >$ that connects two adjacent landmarks, with the angle difference relative to the x-axis in anticlockwise direction and the accessible path distance between those landmarks. The locations of these landmarks are acquired from the indoor floor plan, and a landmark graph is constructed.

5.2. Motion Mode Detection

5.2.1. Motion Mode Definition

The understanding and detection of motion mode are helpful to implement IMU-assisted indoor localisation. There are different kinds of motions experienced by an IMU which mainly depend on the placement of the sensor on a user's body. Usually, motion models used by the researchers are specific to the position of IMU installation on the body, e.g., foot-mounted [26], handheld [27], head-mounted [28], etc. In this study, we use chest-mounted IMU for landmark detection; the chance of interference by motions coming from the irregular motion class is minimal because of the steadiness of the device. Moreover, the pose of the device will not vary, as in the case of handheld or head-mounted IMU. Thus, to detect a landmark precisely, the following four types of motion modes were considered in this study.

- Static motion: this type of motion mode includes all the circumstances when a subject is static. A subject will be deemed to be in static mode when his/her spatial position does not change throughout a considered time window. This mode also considers the states as static when a subject obtains slight motion that is not significant enough to infer it as typical locomotion, for example, if a subject moves slightly by stepping on the same spot while opening a door. To detect a landmark correctly, this type of movement must be identified as static.
- Striding motion: this type of motion mode involves the continuous and smooth motion states that contain periodicity and similarity characteristics for a particular time period in their feature set attributes. It includes the motions that change a subject's spatial position, e.g., plain walking, walking on stairs, or running.
- Turning motion: this refers to the motion states when a subject takes a turn while standing or walking, e.g., performing left or right turning.
- Intermittent motion: this type of motion mode refers to the cases that generate irregular motion states without having the periodicity and continuity properties. It includes all the motion states that a subject performs while remain standing and does not contribute to the change in his/her spatial position, for example, bending or shaking the subject's body while standing on the same spot.

5.2.2. Motion Mode Classification

The purpose of a typical classification system is to allocate an input pattern automatically to a known set of items based on some decision rules. As shown in Figure 6, usually, a typical classification method is performed using four steps, such as data preprocessing, data segmentation, feature extraction, and decision-making. This study adopted the motion mode classification methods as described in [29]. For data preprocessing, a band-pass filter is used that removes both the unwanted low- and high-frequency noises to focus only on the body movement-contributed signal portion. In this study, a band-pass Butterworth filter of order eight was used to remove both the low- and high-frequency noises. The power spectral analysis of the raw data collected from three-axis accelerometer and three-axis gyroscope was performed to determine the signal and noise characteristics. As analysed, most of the energy related to human motion captured by the accelerometer and gyroscope was between 0.75 Hz and 25 Hz. Thus, these two frequencies were applied as the cut-off frequencies to the filter for removing the low- and high-frequency noises, respectively. In this study, though the sensor is body fixed, it can slightly change its orientation regarding the original setup in the x-, y-, and z-axis, because of the generated motions during experiments. This change in sensor orientation affects the acceleration in various degrees towards the x, y, and z coordinate systems by decomposing the gravitational component [30]. To eliminate the dependency on device orientation, the normalised magnitude of the IMU data was considered in this study. As the human movement is a continuous process over time, an individual data point cannot reflect a complete motion mode of a user. Thus, to extract features that can characterise a motion mode, collected sensor data need to be segmented into sequences within a certain time frame, named a window. Here, a window size of 2 s with 50% overlapping was selected, which translates 200 samples for a sampling frequency of 100 Hz. The choice of this type of window is typical and often utilised for motion mode detection, which has already been validated by previous studies [29,31,32]. As this study only considers four types of motion modes, the feature set proposed by Susi et al. [29] was used, which can manage the trade-off between the classification performance and computation cost. The features were extracted from the preprocessed windowed data, which include the energy of the accelerometer and gyroscope, the variance of the accelerometer and gyroscope, and the dominant frequencies of the accelerometer and gyroscope. In addition to those features, this work considered the change of angular velocity along the vertical axis to detect the turning motion.

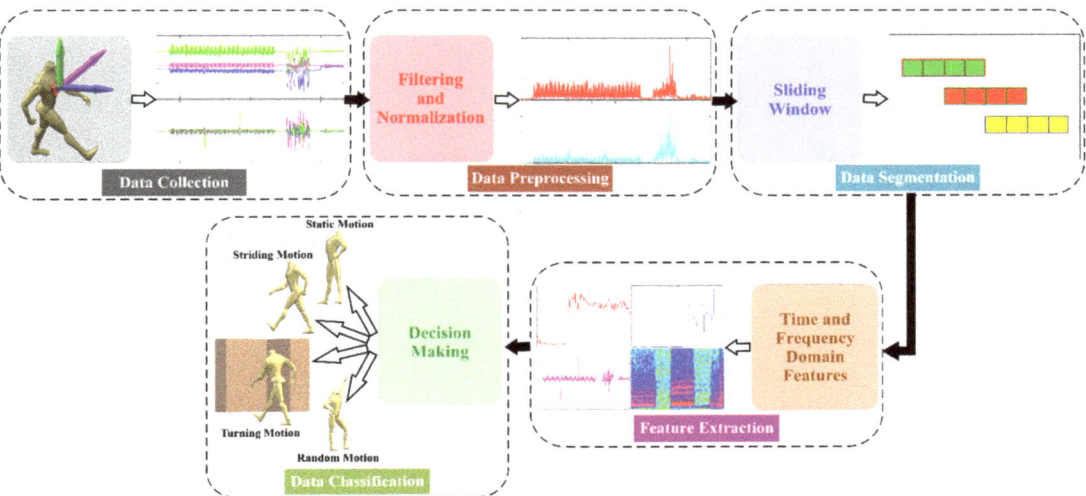

Figure 6. Motion mode detection process pipeline.

As proposed by [29], this work utilised a supervised approach of classification named decision tree for performing the classification task. A decision tree forms a tree-like mapping that comprises leaf nodes, representing the classes, and internal nodes that symbolise the tests regarding the features. The internal nodes contain one (i.e., univariate) or multiple (i.e., multivariate) conditional control statements, and traversing the tree from the root node to leaf nodes can classify a given input pattern. Figure 7 presents the decision tree that was used in this study to detect the abovementioned motion modes. The threshold values for each feature set in every internal node is set by the classifier after performing the training. Initially, the tree characterises the static and dynamic types of motion modes based on the energies and variances of the accelerometer and gyroscope, as well as the raw angular velocity along the vertical axis. As the signal variances for the random movement are significantly higher in short temporal periods than the striding motion, it is utilised to separate the striding motion from other dynamic motions. Moreover, to ascertain whether that motion is from a periodic activity, the periodicity of the dominant frequencies of the accelerometer is evaluated, which reflects the periodic motions generated from human gait. However, the striding motion can be further divided into other classes, such as plain walking, fast walking, running, walking up or down stairs, etc., for other aims that are beyond the scope of this study. Finally, the turning motions are differentiated from other random motions by evaluating the angular velocity of the raw data along the vertical axis.

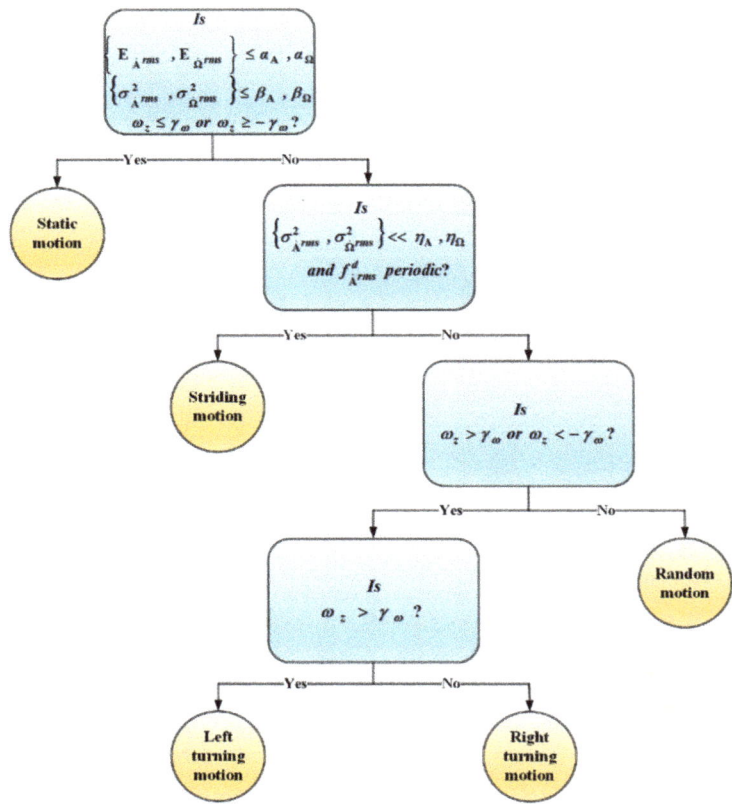

Figure 7. Structure of the decision tree to classify various motion modes.

6. User's Body Shadowing Effect Compensation

There are two main parts to compensate the effect of user's body shadowing on RSSI calculation: angle estimation and compensation model.

6.1. Angle Estimation

Because of the geometrical structures of the indoor environment, the movement of people indoors usually tends to be in the same direction at least for a few seconds, e.g., in the case of a corridor, a user can walk in two directions. As the usual direction of a chest-mounted wearable device is the same as the movement direction of a user, this behaviour, along with the concept of landmark graph, RNs location information, and indoor environmental geometrical constrain, can be exploited to calculate the angle between the chest-mounted wearable tag and the RNs. To calculate the angle between a user and an RN, the two-argument arctangent function (atan2) is used, which can estimate the angle in the Euclidean plane between the positive x-axis and a line connecting to a point, as shown in Figure 8a.

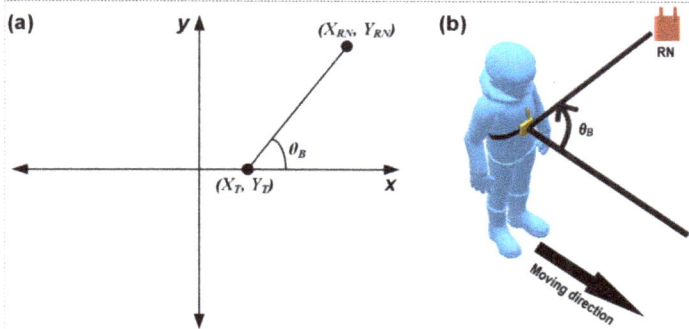

Figure 8. Angle estimation between a wearable device and an RN: (**a**) concept of atan2; (**b**) adaptation of atan2 for estimating the angle between the moving direction and the RN.

In this study, the path segment where the user is walking during the angle estimation is considered as the line corresponding to the x-axis. Thus, the angle is calculated between the moving path and the line connecting the current location of the user and the location of the RN (Figure 8b), as follows:

$$\theta_{rad} = atan2(Y_{RN} - Y_T, X_{RN} - X_T) \in (-\pi, \pi) \text{ and } (X_{RN}, Y_{RN}) \neq (0,0) \quad (7)$$

$$\theta_{deg} = \frac{180°}{\pi} \theta_{rad} \quad (8)$$

$$\theta_B = \begin{cases} \theta_{deg}, & \text{when } \theta_{deg} \geq 0 \\ 2\pi + \theta_{deg}, & \text{when } \theta_{deg} < 0 \end{cases} \quad (9)$$

where (X_T, Y_T) and (X_{RN}, Y_{RN}) are the coordinates of the current locations of the user and the RN, and θ_B is the bearing angle between the user and an RN in degrees. In this study, a coarse location of the unknown target is estimated first as the current location (X_T, Y_T) to estimate the bearing angle between the target and an RN. This coarse location is computed based on its immediate previous location, and the step length and moving direction. Here, the step length and step direction are obtained by leveraging the landmark graph. At the start, the step length for a target is initialised to a constant value. As the target progresses and passes two adjacent landmarks, the step length is updated. Let a target pass two adjacent landmarks denoted by L_1 and L_2. Then, the step length of that target can be obtained as follows:

$$l_s = \frac{\sqrt{(x_{L_1} - x_{L_2})^2 + (y_{L_1} - y_{L_2})^2}}{N_S} \quad (10)$$

where (x_{L_1}, y_{L_1}) and (x_{L_2}, y_{L_2}) are the coordinates of the landmarks L_1 and L_2, respectively, and N_S is the total number of previous location points in between the landmarks L_1 and L_2 that are estimated when the target passes the landmarks. Here, the step length will only be updated if the trajectory between the adjacent landmarks is a straight line. Otherwise, the system will retain the step length that estimated last. The estimation of step direction exploits the geometrical structures of indoor environment and infers the moving direction as the direction of the current trajectory relative to the considered x-axis. Thus, when the step length l_s and heading θ_s are known, the estimation of the next location can be obtained by using the elementary pedestrian dead reckoning (PDR) technique as follows:

$$x_{s+1} = x_s + l_s \sin(\theta_s) \quad (11)$$

$$y_{s+1} = y_s + l_s \cos(\theta_s) \tag{12}$$

where (x_s, y_s) and (x_{s+1}, y_{s+1}) are the positions of a subject at step s and $s+1$, respectively, and θ_s and l_s are the heading and displacement at step s. Therefore, (x_{s+1}, y_{s+1}) are considered as the current location (X_T, Y_T) to calculate the angle between an RN and the current coarse position of the target, considering the target is moving exactly the same direction as the trajectory's direction. However, a target can rotate his/her body while walking towards a trajectory's direction, which will eventually affect the measured θ_B in the perspective of RSSI correction. Let θ_R be the rotation angle obtained from the gyroscope. Then, the orientation angle θ_O of the target relative to an RN can be calculated as follows:

$$\theta_O = \theta_B \pm \theta_R \pm \theta_{L_1 L_2} \tag{13}$$

where $\theta_{L_1 L_2}$ is the angle difference of the path segment connecting the landmarks L_1 and L_2 relative to the x-axis in anticlockwise direction, which can be obtained from the landmark graph.

6.2. Compensation Model

To compensate the effects of the user's body on signal RSSI value, a compensation model is proposed in this study. This model can intensify the signal RSSI values that are being interrupted by the user's body. Thus, the proposed model for correcting the raw RSSI is as follows, which considers the angle between the user's body and the RN:

$$RSSI_{corr} = RSSI \left(-\frac{\sigma}{d} e^{\frac{-(\theta_O)^2}{2\theta_O^2 + 1}} + 1 \right) \tag{14}$$

where $RSSI$ and $RSSI_{corr}$ are the raw and corrected RSSI values, θ_O is the orientation angle between the user and an RN in degrees, σ is the intensification parameter, and d is the distance between the wearable tag and the edge of the far-ended shoulder. The value of σ depends both on the environment and the user orientation with respect to the RN, which needs to be chosen empirically.

However, as observed from Figure 3, the corrected RSSI will be an overestimation when applying the proposed compensation model with the straight LOS direction (e.g., around $0°$) between the wearable tag and the RN. Moreover, it will be an underestimation when the signals are being interrupted by the maximum obstacles (e.g., the angle positions just after the angles $90°$ and just before the angle $270°$). To resolve this issue, the following rules with different values of the parameter σ are chosen for different angles:

$$\sigma = \begin{cases} \sigma_1, & if \ (0° \leq \theta_B \leq 45°) \ || \ (315° \leq \theta_B \leq 360°), \\ \sigma_2, & if \ (45° < \theta_B \leq 135°) \ || \ (225° < \theta_B \leq 315°), \\ \sigma_3, & if \ (135° < \theta_B \leq 225°). \end{cases} \tag{15}$$

To estimate the values of σ, this study first calculates the amount of error (e_θ) as given in Equation (16), for a given range of angles by choosing a value of σ. Secondly, the optimum value of σ is estimated by adjusting its value untill the smallest e_θ is obtained.

$$e_\theta = \frac{\sum_{i=0}^{n} \sqrt{\left(RSSI_{\theta_s + i*m} - RSSI_{0°} \right)^2}}{\theta_e - \theta_s} \times 100\% \tag{16}$$

where $RSSI_{\theta_s + i*m}$ and $RSSI_{0°}$ are the RSSI values at angle $\theta_s + i \times m$ and $0°$, respectively, n is the total number of angle values considered to collect data within the range θ_s to θ_e, and m is the amount of angle considered to rotate in each move. As this study considers collecting RSSI data at every $15°$ rotation, the value of m is 15.

7. Location Estimation

Irrespective of the utilised features, a fingerprinting-based localisation problem is mainly a pattern matching problem. During the online phase, a target sends a query fingerprint from an unknown location that needs to be matched with the fingerprints stored in the radio map. It is very unlikely that a radio map will contain a fingerprint with an exact match. Thus, the traditional way is to find K different fingerprints closest to the queried one from the radio map, which is known as the K-nearest neighbour (K-NN) method. This study exploits the classical K-NN algorithm for location estimation. Moreover, an improved version of the classical K-NN algorithm, named weighted K-NN (WK-NN), is applied to further improve the localisation accuracy.

Let r_{uj} be the body shadowing-compensated mean RSSI collected from RN j at unknown target location u. Then, the RSSI distance between an RP i from the radio map to the target point u can be obtained as follows:

$$d_{ui} = \frac{1}{N_{RN}} \sqrt{\sum_{j=1}^{N_{RN}} (r_{uj} - r_{ij})^2} \tag{17}$$

where d_{ui} is the Euclidean distance between the RSSI of the target point and RP, r_{ij} is the mean RSSI collected from RN j at RP i, and N_{RN} is the total number of RNs considered for a fingerprint. In the case of the K-NN method, K RPs from the radio map will be selected that have the smallest distance value with the target. Therefore, the estimated location of the target can be calculated as follows:

$$LOC(x_u, y_u) = \frac{1}{K} \sum_{i=1}^{K} LOC(x_i, y_i) \tag{18}$$

where (x_i, y_i) are the locations of the RPs. Here, the spatial distances between a target location and its neighbouring RPs are usually different. Thus, the WK-NN algorithm also considers the corresponding spatial distances in terms of weight factor when selecting the K nearest RPs. As proposed in [33], the weight is inversely proportional to the spatial distance and can be calculated as follows:

$$w_{ci} = \frac{1/D_{ci}}{\sum_{i=1}^{K} 1/D_{ci}} \tag{19}$$

where $D_{ci} = \sqrt{(x_c - x_i)^2 + (y_c - y_i)^2}$ is the spatial Euclidean distance between the coarse location (x_c, y_c) of the unknown target and an RP location (x_i, y_i). Therefore, the estimated location will be

$$LOC(x_u, y_u) = \frac{\sum_{i=1}^{K} w_{ci} LOC(x_i, y_i)}{\sum_{i=1}^{K} w_{ci}} \tag{20}$$

In this study, the coarse location of an unknown target is estimated first to calculate the spatial distance between the RPs and a target location. The coarse location of the target is computed by leveraging the landmark graph and using the same technique proposed in Section 6.1. Therefore, the estimated coarse location is used to calculate the weight for the WK-NN algorithm.

8. Experimental Evaluation

This section presents the evaluation of the proposed models through quantitative experimental results.

8.1. Experimental Setup

The experiments were carried out using two customised sensor boards designed by our group. To collect the RSSI fingerprint, the XBee wireless technology-based XBee S1

802.15.4 module was used as the radio in the first sensor board, as shown in Figure 9a, which was employed both in the RNs and wearable devices. To acquire the acceleration and angular velocity data, the second sensor board was used as part of the wearable node. This sensor board is equipped with IMU and BLE, as well as some environmental sensors, as shown in Figure 9b. The RNs were attached to the ceiling, as shown in Figure 9c. The wearable sensor boards were attached to a subject's chest and fastened by an elastic strap, as presented in Figure 9d. A computer connected with an XBee module acted as a server that collected and stored both the RSSI fingerprint data through XBee and IMU data through BLE. In the case of the indoor localisation, the data collected from both the devices were synchronised by using the timestamps and stored on a database for further processing.

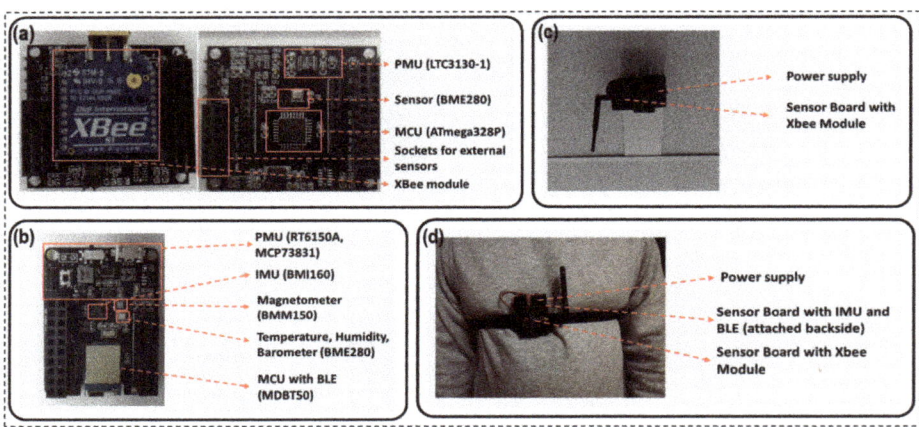

Figure 9. Experimental setup: (**a**) sensor board with XBee wireless module; (**b**) sensor board with IMU and BLE modules; (**c**) RN mounted beneath the ceiling; (**d**) wearable device mounted on chest.

8.2. Evaluation of Motion Mode Detection

8.2.1. Data Collection

Several experiments were carried out for collecting an adequate amount of data with ground truth labels to train and test the designed motion mode classifier. Data from four adults (two males and two females) of different heights and weights were collected to evaluate the performance of the proposed classifier. The data collections were conducted in an open field, and each participant was equipped with a chest-mounted IMU device. The participants were requested to walk approximately 500 m of distance while performing several activities. The experiments were performed using a predefined protocol that consisted of six activities including static, walking, turning right, turning left, opening a door, bending, and random movement. As the data were collected from an open field, the participants were asked to simulate the door-opening activity. To complete one run of the defined protocol, a participant required four minutes, and each participant repeated the protocol three times. Thus, enough data was collected for training and testing of the classifier.

To facilitate the labelling of ground truth, another IMU device, named NGIMU (https://x-io.co.uk/ngimu/ (accessed on 10 February 2021)), was utilised. Pressing the power button of the device while powered on will send a message to the receiver with timestamped information for the button-pressing event. During the experiment, the participants were asked to press the button of the NGIMU each time they began an activity. Thus, by annotating the exact number and sequence of performed activities and mapping this information with the timestamped button event's data, the ground truth labelling was performed. For precise evaluation of the classifier, the data recorded during the transition of two activities were removed manually.

8.2.2. Evaluation Metrics

To validate the skill of the proposed classification model, five-fold cross-validation was applied, which is a typical resampling technique that shuffles the dataset randomly and splits it into five equal-sized groups. From those, four groups were used for training the model, and one group was kept for testing the model. Five iterations were performed with the grouped data to cross-validate the model by using each group as a testing dataset while employing the others for training. The average of the evaluation metrics from the five iterations was taken as the final evaluation score of the model. To evaluate the motion mode classification performance, several metrics were used in this study, including accuracy, precision, sensitivity, specificity, and F-measure. The accuracy for a motion class is the ratio of correctly labelled motion modes for that class to the total number of labelled motion modes for that class. The precision for a motion class can be defined as the ratio of correctly positive-labelled motion modes for that class to the total number of positive-labelled motion modes for that class. The sensitivity (which is also known as recall) for a motion class is the ratio of correctly positive-labelled motion modes for that class to the total number of motion modes that actually belong to that class. The specificity for a motion class is the ratio of correctly negative-labelled motion modes for that class to the total number of motion modes that do not belong to that class. F-measure is the harmonic average of precision and sensitivity. These metrics are defined as follows:

$$Accuracy = \frac{TP + TN}{TP + FP + FN + TN} \tag{21}$$

$$Precision = \frac{TP}{TP + FP} \tag{22}$$

$$Sensitivity = \frac{TP}{TP + FN} \tag{23}$$

$$Specificity = \frac{TN}{TN + FP} \tag{24}$$

$$F - measure = 2 * \left(\frac{Precision * Sensitivity}{Precision + Sensitivity} \right) \tag{25}$$

8.2.3. Classification Performance

Figures 10 and 11 summarise the classification performance of the proposed motion mode classifier for detecting each motion mode. As illustrated in Figure 10, the columns of the confusion matrix refer to the ground truth motion modes performed by the participants, and the rows refer to the motion modes predicted by the classifier. The percentage of prediction accuracy, together with their actual number for each motion mode, is presented along the principal diagonal in black colour. The percentage of confused classification for the motion modes are reported along the off-diagonal sections in white colour.

As can be seen from the confusion matrix, the classifier can detect the correct motion mode in more than 95% of cases, irrespective of the type of motion class performed by a participant. The highest accuracy of 99.5% was attained by the classifier for the static motion mode. As reported, 2 and 3 segments out of 928 segments for static type motions were misclassified as striding motion and intermittent motion, respectively. The main reason behind this confusion is because of the simulated door opening activities performed by the participants, which are actually considered as static activities; however, this type of activity sometimes may generate high energy and variance in the signal which can satisfy the decision thresholds, leading to misclassification.

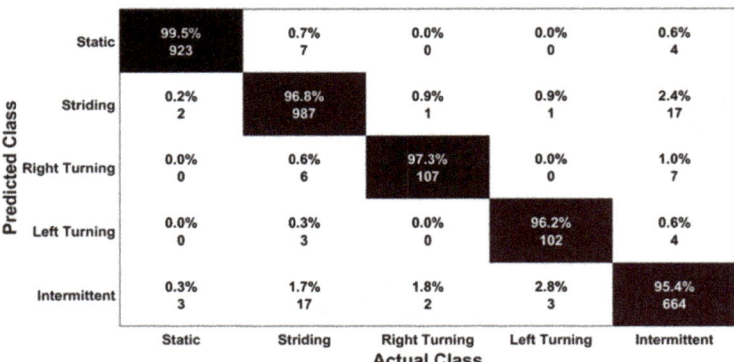

Figure 10. Confusion matrix of the proposed motion mode classifier.

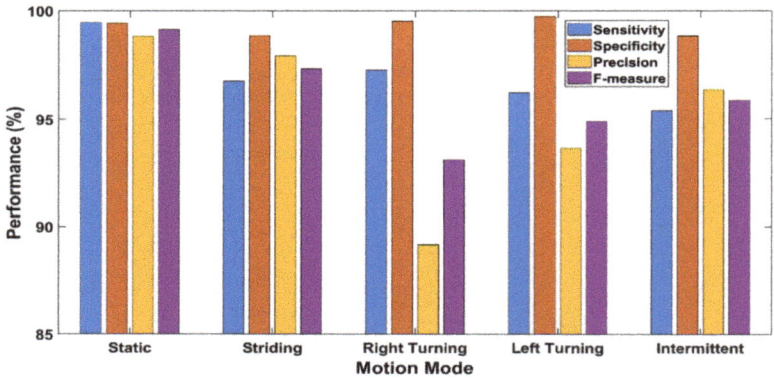

Figure 11. Classification performance of the proposed motion mode classifier for each motion mode.

The lowest accuracy of 95.4% was achieved for detecting the intermittent motion mode by the classifier. This is because intermittent motions are more likely to be confused as other types of motions. As reported in the confusion matrix, among the total 696 segments for intermittent motions, 17 were detected as striding motion. One possible reason for this confusion may be the consecutive occurrence of a similar type of movement several times (e.g., multiple bending activities), which can produce periodicity and lead to that misclassification. However, during the landmark identification phase, it is very unlikely that such intermittent movement will occur in a pattern that can satisfy the landmark rules. Thus, the defined rules for landmark identification can mitigate the impact of this type of misclassification. Moreover, 17 segments out of 1020 for the striding motion mode were misclassified as intermittent motion, causing the accuracy for that class to be 96.8%.

As presented in Figure 11, the overall sensitivity, specificity, precision, and F-measure of the proposed motion mode classifiers are 97.03%, 99.28%, 95.17%, and 96.06%, respectively, which can eventually produce high accuracy for the landmark identification task.

8.3. Evaluation of User's Body Shadowing Effect Compensation Model

The proposed BSE compensation model was evaluated firstly using the same experimental data as shown in Figure 2. The polar plot for the raw data is shown in Figure 12, where the distance between the sender and the receiver is 9.9 m. It is noticeable from the figure that the minimum and maximum attenuation of RF signals occur at the angle positions as discussed in Section 3.2, due to the effect of the user's body. The higher RSSI value is observed at approximately 0° with LOS angle positions, and the lowest values of RSSI can be observed at approximately 90° and 270° angle positions for the raw RSSI data.

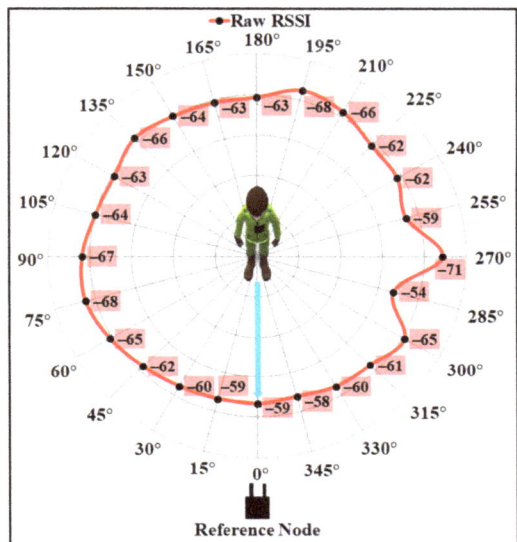

Figure 12. User's body effect on RSSI (raw RSSI vs. orientation angle).

As the human body shape is uneven, the amount of RSSI depletion, while facing the different body parts, by the signal will be different. The intensification parameter σ boosts the RSSI with an amount that can cope with the loss caused by the user's body. Using a single value of σ for every orientation angle may lead to an overestimation or an underestimation for some orientations. Thus, three different σ values were chosen based on the analysis, as described in Section 3. To evaluate the compensation model performance, firstly the values for the intensification parameters σ were empirically investigated, and optimal values were selected, as described in Section 6.2. Figures 13 and 14 present the results after applying the proposed compensation model for two different combinations of values for σ. To show the effect of σ on the proposed compensation model, first the values of σ_1, σ_2, and σ_3 were chosen as 0.1, 4.0, and 3.0, respectively. Figure 13 shows the polar plot for the compensated RSSI after applying those values to corresponding orientation angles. As can be noticed from the figure, though the model can correct some RSSI values by applying this set of σ values that are affected by the user's body shadowing, the corrected values are the underestimation of the 0° faced LOS value. Figure 14 presents the results with the values of σ_1, σ_2, and σ_3 as 0.1, 2.0, and 1.5, respectively, which were found as the optimal values for the experimented indoor environments. As can be seen from that figure, the proposed model can considerably correct the attenuated RSSI values while presenting a negligible amount of noise. Most of the corrected RSSI values are almost similar to the 0° faced LOS value with a negligible deviation. Thus, the values of σ_1, σ_2, and σ_3 were chosen as 0.1, 2.0, and 1.5 to compensate the user's body affected RSSI values for the proposed indoor localisation system.

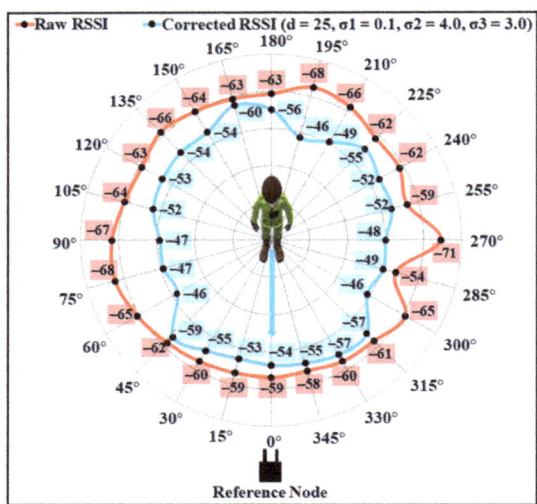

Figure 13. Results after applying the body shadowing effect compensation model with $\sigma_1 = 0.1$, $\sigma_2 = 4.0$, $\sigma_3 = 3.0$, and $d = 25$.

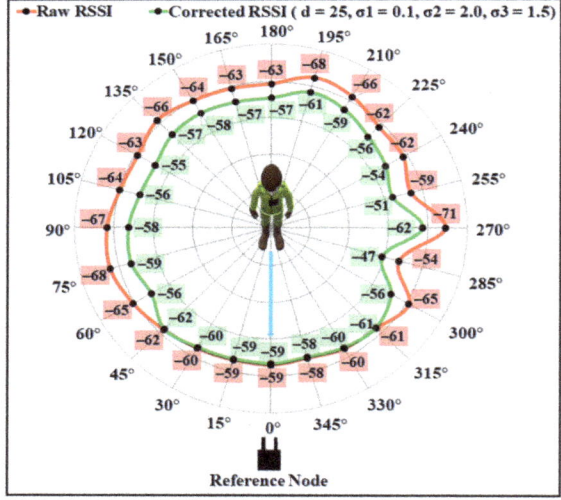

Figure 14. Results after applying the body shadowing effect compensation model with $\sigma_1 = 0.1$, $\sigma_2 = 2.0$, $\sigma_3 = 1.5$, and $d = 25$.

To validate the efficiency of the proposed model, RSSI data were also collected using the same experimental setup, except the wearable device was placed on the back of a user. Figure 15 illustrates the compensation results along with the raw RSSI for the two different sets of σ values. As can be seen from the figure, the values of σ_1, σ_2, and σ_3 as 0.1, 2.0, and 1.5 can produce the best estimates for most of the angle positions, compared to the other set.

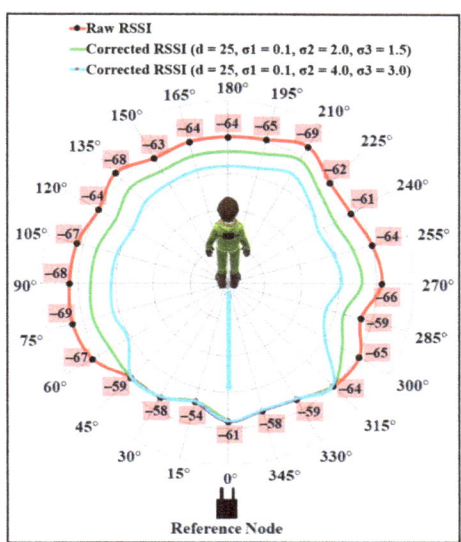

Figure 15. Raw and corrected RSSI for data collected with back-mounted wearable device.

The impacts of the parameter d, which represents the distance between the wearable tag and the edge of the far-ended shoulder, were also examined. Figure 16 presents the results of the proposed model for different d values, including 41, 25, 50, and 20. Considering the average shoulder width of 41 cm, selecting the value of d as 25 produces the best results demonstrated in the figure.

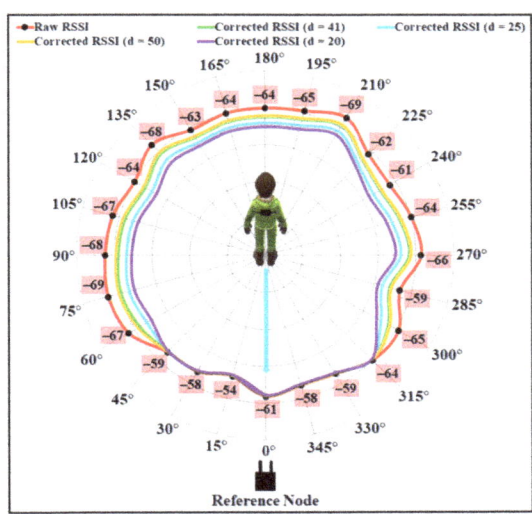

Figure 16. Results after applying different values of d with $\sigma_1 = 0.1$, $\sigma_2 = 2.0$, $\sigma_3 = 1.5$.

8.4. Evaluation of the Proposed Localisation System

To evaluate the performance of the proposed localisation system, experiments were performed in an office building. The experiment building was Building 72 at the Clayton campus of Monash University, which comprises three floors. The experiment was carried out on the second floor, with a total area of about 1475 m^2, and the length of the test path was about 150 m, represented by the green line in Figure 17. The path starts from the red circle, then follows the path, and ends at the red square.

Figure 17. Experimental area with the markings of experimented path and distribution of reference nodes.

8.4.1. Evaluation Criterion

To evaluate the system performance, 40 key points were set along the experimented path, which served as the ground truth for comparison. During the experiment, the participant was asked to stop a little at each key point for the purpose of recording the markers. Let (x_e^i, y_e^i) and (x_g^i, y_g^i) be the estimated location of an unknown target, and the ground truth location at the ith key point, respectively; then the error at ith key point is

$$\varepsilon_{ge}^i = \sqrt{\left(x_g^i - x_e^i\right)^2 - \left(y_g^i - y_e^i\right)^2} \tag{26}$$

This study reports the mean error, median error, standard deviation, 25th, 75th, and 90th percentile errors, and a cumulative distribution function (CDF) plot to analyse the localisation performance of the proposed system.

8.4.2. Localisation Performance

The localisation performance of the proposed system was compared with two other studies that proposed different models for angle estimation and BSE compensation, including the work presented in [17], referred to as MODEL I, and in [20], referred to as MODEL II. To compare the relative accuracy of the proposed system with these studies, we mainly implemented the angle estimation and compensation models as proposed by the researchers, and then applied the fingerprinting localisation method. For example, the use of multiple wearable devices and Semcad simulation parts were omitted for MODEL I, and Kalman filter and path loss model implementation were overlooked for MODEL II. Moreover, identical preprocessing of the raw data was considered, and the same weight calculation method was applied for all the cases with a value of 3 for k, both in the K-NN and WK-NN algorithms.

Figure 18 presents the routes for the experimented path estimated by the different systems using the classical K-NN algorithm. As shown in the figure, the performance of the systems with BSE compensation (WBSEC) on indoor localisation is apparent and the estimated route of the systems without BSE compensation (WOBSEC) produced the worst path. Among the other results, MODEL II generated a better path compared to MODEL I, because in MODEL I the authors only considered two groups for orientation angle (over/underestimation) when applying their compensation model. MODEL II considered three groups (front, back, side) and there was no consideration of the volume of body parts that creates the NLOS scenarios. However, this work considered three groups as well as the consideration of body volume when calculating the compensation value, and hence produced the best path.

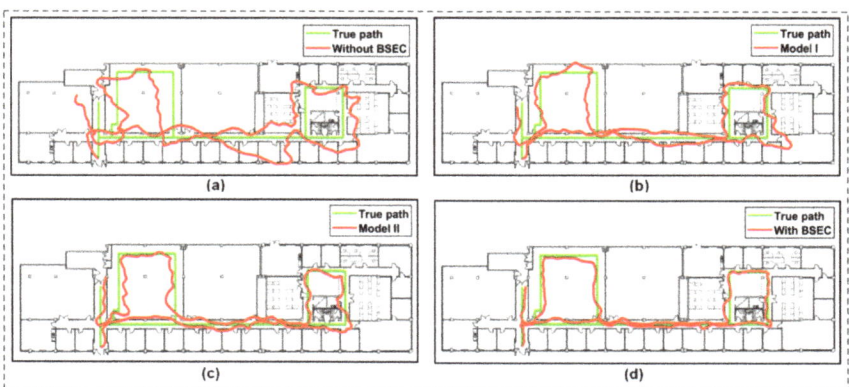

Figure 18. Routes estimated by different systems using the K-NN method: (**a**) Without BSEC; (**b**) Model I; (**c**) Model II and (**d**) With BSEC.

The details of the positioning performance are presented in Figure 19 and Table 2 for the K-NN algorithm. The boxplot indicates the summary of some error statistics including the maximum, minimum, 25th, 75th, mean, and median errors for each system. As presented in Figure 19 and Table 2, the proposed system significantly outperformed the other systems for all types of statistics. The proposed system demonstrated the best performance with a mean error of 1.62 m and median error of 1.46 m, followed by MODEL II (mean error 2.17 m and median error 2.38 m). The mean and median errors of MODEL I were 2.39 m and 1.75 m, respectively. The system without any BSE compensation achieved a mean error of 3.17 m and a median error of 2.68 m.

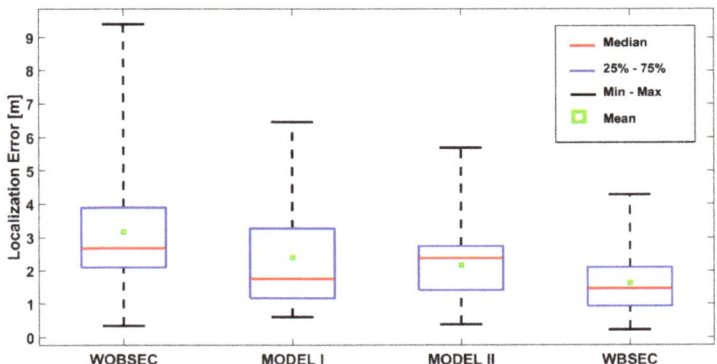

Figure 19. Localisation performance of the proposed system compared with other systems using the K-NN method.

Table 2. Localisation accuracy comparison for different systems with K-NN method.

Method	Min. (m)	Max. (m)	Mean (m)	Median (m)	STD (m)	25th (m)	75th (m)	90th (m)
WOBSEC	0.34	9.41	3.17	2.68	1.69	2.10	3.89	5.08
MODEL I	0.60	6.46	2.39	1.75	1.64	1.16	3.27	5.60
MODEL II	0.38	5.69	2.17	2.38	1.10	1.41	2.74	3.07
WBSEC	0.22	4.27	1.62	1.46	0.94	0.92	2.11	2.83

Figure 20 illustrates the accumulative distribution function of the estimated localisation errors of the different systems for the K-NN algorithm. This plot also shows the superiority of the proposed system compared to the others. More specifically, the introduced system can achieve a performance that produces localisation errors less than 2.5 m for 80% of cases, while it is around 60% for MODEL II and around 70% for MODEL I. The result that was produced without compensating the BSE outputted the worst, since the body effect errors are not considered when comparing a query fingerprint with the radio map.

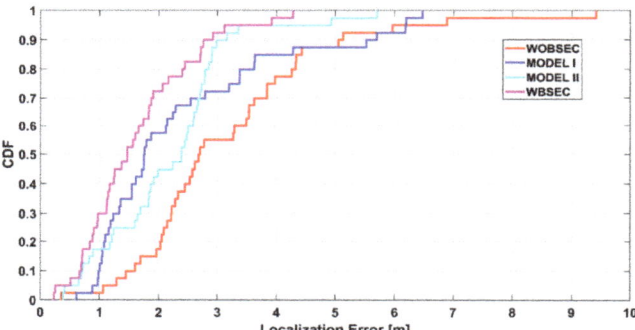

Figure 20. Accumulative distribution of positioning error of the proposed system compared with other systems using the K-NN method.

The superiority of the proposed system is further increased by applying the WK-NN algorithm when comparing a query fingerprint with the radio map. The performance is improved by including the spatial prominence of the neighbouring RPs in terms of weight factor. Figures 21 and 22, and Table 3 compare the performance of the different systems using the WK-NN algorithm. As can be seen, the WK-NN algorithm with the proposed weighting method improves the localisation accuracy for all the considered systems. The developed system yields a mean error of 1.01 m and median error of 0.74 m, while they are 1.56 m and 1.19 m for MODEL II, and 2.74 m and 2.23 m for MODEL I. The system without any BSE compensation can attain a mean error of 2.99 m and a median error of 2.98 m using the WK-NN algorithms. Moreover, as presented in Figure 22, the WK-NN algorithm produces localisation errors less than 1.5 m for 80% of cases for the proposed system, while it is around 60%, 25%, and 15% for MODEL II, MODEL I, and without applying any compensation, respectively. Overall, the proposed BSE compensation model along with the landmark-assisted WK-NN method is able to achieve sub-metre median localisation accuracy that outperforms some recent related methods.

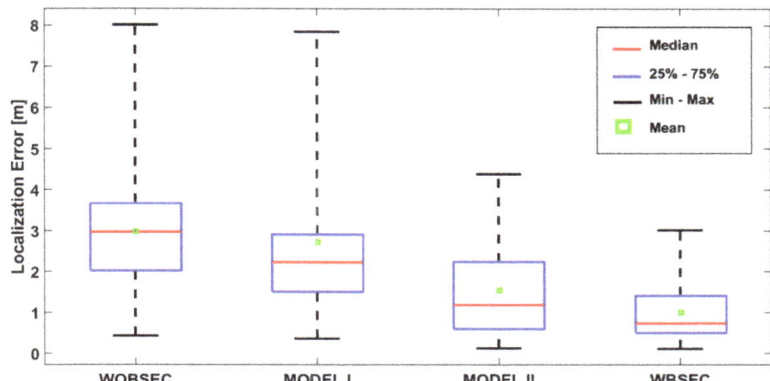

Figure 21. Localisation performance of the proposed system compared with other systems using the WK-NN method.

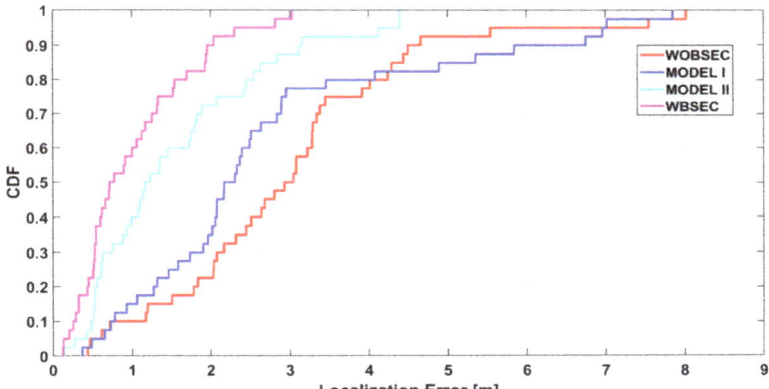

Figure 22. Accumulative distribution of positioning error of the proposed system compared with other systems using the WK-NN method.

Table 3. Localisation accuracy comparison for different systems with the proposed WK-NN method.

Method	Min. (m)	Max. (m)	Mean (m)	Median (m)	STD (m)	25th (m)	75th (m)	90th (m)
WOBSEC	0.43	8.01	2.99	2.98	1.65	2.03	3.68	4.58
MODEL I	0.36	7.85	2.74	2.23	1.92	1.51	2.92	6.29
MODEL II	0.13	4.39	1.56	1.19	1.14	0.61	2.25	3.14
WBSEC	0.12	3.02	1.01	0.74	0.73	0.51	1.42	1.99

9. Conclusions

This study describes solutions for improving the accuracy of wearable sensor-based fingerprinting indoor localisation by mitigating the user's BSE on RSSI. An in-depth analysis of RSSI for different orientations of the user's body was performed, and a body shadowing compensation model is proposed. To calculate the orientation angle between a wearable device and an RN, an IMU-aided motion mode detection technique was implemented by fusing the spatial knowledge from the indoor floor plan. The decision tree-based classifier yields an outstanding performance for motion mode detection that, in turn, accurately identifies the landmark to produce high precision for the estimation of the user's orientation angle. Results demonstrated that the implemented classifier achieves an overall accuracy of 97.31% for detecting a motion mode correctly, which eventually helps to compensate the errors caused by the user's body. To validate the proposed body shadowing compensation model, both the classical K-NN and the WK-NN methods were implemented with a unique weighting technique. For selecting the K nearest neighbours in the case of the WK-NN method, the spatial prominence of the neighbouring RPs was applied as the weights, which were calculated by using a unique landmark-assisted distance measurement method. Finally, the localisation performance of the proposed system was compared with two other recent studies that proposed different models for angle estimation and BSE compensation. The experimental results show a mean and median accuracy of 1.62 m and 1.46 m for the classical K-NN method, which is further improved to 1.01 m and 0.74 m, respectively, using the WK-NN method. Overall, the proposed BSE compensation model along with the landmark-assisted WK-NN method can realise sub-metre median localisation accuracy that noticeably outperforms the considered related studies. Although the proposed methods are intended for RSSI fingerprinting localisation, these can be adopted in other RSSI-based indoor localisation applications with body-mounted wearable devices. The main limitation of the proposed system is the dependency of the orientation angle estimation phase on the previously estimated location. Future work will include the implementation of the proposed system in multistorey buildings by addressing this issue.

Author Contributions: Conceptualisation, M.A.A.M. and D.V.A.; methodology, M.A.A.M.; software, M.A.A.M.; hardware, F.W.; investigation, M.A.A.M. and D.V.A.; resources, M.R.Y.; data curation, M.A.A.M.; writing—original draft preparation, M.A.A.M.; writing—review and editing, M.A.A.M., D.V.A., F.W., and M.R.Y.; visualisation, M.A.A.M.; supervision, M.R.Y.; project administration, M.R.Y. All authors have read and agreed to the published version of the manuscript.

Funding: This research received no external funding.

Conflicts of Interest: The authors declare no conflict of interest.

References

1. Li, Y.; Zhuang, Y.; Hu, X.; Gao, Z.; Hu, J.; Chen, L.; He, Z.; Pei, L.; Chen, K.; Wang, M.; et al. Toward Location-Enabled IoT (LE-IoT): IoT Positioning Techniques, Error Sources, and Error Mitigation. *IEEE Internet Things J.* **2021**, *8*, 4035–4062. [CrossRef]
2. Wang, H.; Sen, S.; Elgohary, A.; Farid, M.; Youssef, M.; Choudhury, R.R. No need to war-drive: Unsupervised indoor localization. In Proceedings of the 10th International Conference on Mobile Systems, Applications, and Services, Low Wood Bay, UK, 25–29 June 2012; pp. 197–210.
3. Schmitt, S.; Adler, S.; Kyas, M. The effects of human body shadowing in RF-based indoor localization. In Proceedings of the 2014 International Conference on Indoor Positioning and Indoor Navigation (IPIN), Busan, Korea, 27–30 October 2014; pp. 307–313.
4. Popleteev, A. Improving ambient FM indoor localization using multipath-induced amplitude modulation effect: A year-long experiment. *Pervasive Mob. Comput.* **2019**, *58*, 101022. [CrossRef]
5. Bahl, P.; Padmanabhan, V.N. RADAR: An in-building RF-based user location and tracking system. In Proceedings of the IEEE INFOCOM 2000, Tel Aviv, Israel, 26–30 March 2000; Volume 2, pp. 775–784.
6. Gu, F.; Valaee, S.; Khoshelham, K.; Shang, J.; Zhang, R. Landmark Graph-Based Indoor Localization. *IEEE Internet Things J.* **2020**, *7*, 8343–8355. [CrossRef]
7. Xu, C.; Firner, B.; Moore, R.S.; Zhang, Y.; Trappe, W.; Howard, R.; Zhang, F.; An, N. SCPL: Indoor device-free multi-subject counting and localization using radio signal strength. In Proceedings of the 12th International Conference on Information Processing in Sensor Networks, Philadelphia, PA, USA, 8–11 April 2013; Association for Computing Machinery: New York, NY, USA, 2013; pp. 79–90.
8. Jeong, J.; Shen, Y.; Kim, S.; Choe, D.; Lee, K.; Kim, Y. DFC: Device-free human counting through WiFi fine-grained subcarrier information. *IET Commun.* **2021**, *15*, 337–350. [CrossRef]
9. Wang, Y.; Wu, K.; Ni, L.M. WiFall: Device-Free Fall Detection by Wireless Networks. *IEEE Trans. Mob. Comput.* **2017**, *16*, 581–594. [CrossRef]
10. Kianoush, S.; Savazzi, S.; Vicentini, F.; Rampa, V.; Giussani, M. Device-Free RF Human Body Fall Detection and Localization in Industrial Workplaces. *IEEE Internet Things J.* **2017**, *4*, 351–362. [CrossRef]

11. Wang, J.; Zhang, X.; Gao, Q.; Yue, H.; Wang, H. Device-Free Wireless Localization and Activity Recognition: A Deep Learning Approach. *IEEE Trans. Veh. Technol.* **2017**, *66*, 6258–6267. [CrossRef]
12. Leith, D.J.; Farrell, S. Coronavirus contact tracing: Evaluating the potential of using bluetooth received signal strength for proximity detection. *SIGCOMM Comput. Commun. Rev.* **2020**, *50*, 66–74. [CrossRef]
13. Cully, W.P.L.; Cotton, S.L.; Scanlon, W.G.; McQuiston, J.B. Body shadowing mitigation using differentiated LOS/NLOS channel models for RSSI-based Monte Carlo personnel localization. In Proceedings of the 2012 IEEE Wireless Communications and Networking Conference (WCNC), Paris, France, 1–4 April 2012; pp. 694–698.
14. King, T.; Kopf, S.; Haenselmann, T.; Lubberger, C.; Effelsberg, W. COMPASS: A probabilistic indoor positioning system based on 802.11 and digital compasses. In Proceedings of the 1st International Workshop on Wireless Network Testbeds, Experimental Evaluation & Characterization, Los Angeles, CA, USA, 29 September 2006; Association for Computing Machinery: New York, NY, USA; pp. 34–40.
15. Della Rosa, F.; Pelosi, M.; Nurmi, J. Human-Induced Effects on RSS Ranging Measurements for Cooperative Positioning. *Int. J. Navig. Obs.* **2012**, *2012*, 959140. [CrossRef]
16. Trogh, J.; Plets, D.; Martens, L.; Joseph, W. Improved Tracking by Mitigating the Influence of the Human Body. In Proceedings of the 2015 IEEE Globecom Workshops (GC Wkshps), San Diego, CA, USA, 6–10 December 2015; pp. 1–6.
17. Trogh, J.; Plets, D.; Thielens, A.; Martens, L.; Joseph, W. Enhanced Indoor Location Tracking Through Body Shadowing Compensation. *IEEE Sens. J.* **2016**, *16*, 2105–2114. [CrossRef]
18. Rapiński, J.; Zinkiewicz, D.; Stanislawek, T. Influence of human body on Radio Signal Strength Indicator readings in indoor positioning systems. *Tech. Sci. Univ. Warm. Mazury Olszt.* **2016**, *19*, 117–127.
19. Bi, J.; Wang, Y.; Li, X.; Cao, H.; Qi, H.; Wang, Y. A novel method of adaptive weighted K-nearest neighbor fingerprint indoor positioning considering user's orientation. *Int. J. Distrib. Sens. Netw.* **2018**, *14*, 1550147718785885. [CrossRef]
20. Deng, Z.; Fu, X.; Wang, H. An IMU-Aided Body-Shadowing Error Compensation Method for Indoor Bluetooth Positioning. *Sensors* **2018**, *18*, 304. [CrossRef] [PubMed]
21. Plets, D.; Joseph, W.; Vanhecke, K.; Tanghe, E.; Martens, L. Coverage prediction and optimization algorithms for indoor environments. *EURASIP J. Wirel. Commun. Netw.* **2012**, *2012*, 123. [CrossRef]
22. Wang, J.; Wang, Q. *Body Area Communications: Channel Modeling, Communication Systems, and EMC*; John Wiley & Sons: Singapore, Singapore, 2013.
23. Hall, P.S.; Hao, Y. *Antennas and Propagation for Body-Centric Wireless Communications*; Artech House: Norwood, MA, USA, 2012.
24. Harmuth, H.F.; Hussain, M.G.; Boules, R.N. *Electromagnetic Signals: Reflection, Focusing, Distortion, and Their Practical Applications*; Springer: New York, USA, 1999.
25. Mamun, M.A.A.; Anaya, D.V.; Wu, F.; Redouté, J.M.; Yuce, M.R. Radio Map Building with IEEE 802.15.4 for Indoor Localization Applications. In Proceedings of the 2019 IEEE International Conference on Industrial Technology (ICIT), Melbourne, Australia, 13–15 February 2019; pp. 181–186.
26. Ruiz, A.R.J.; Granja, F.S.; Honorato, J.C.P.; Rosas, J.I.G. Accurate Pedestrian Indoor Navigation by Tightly Coupling Foot-Mounted IMU and RFID Measurements. *IEEE Trans. Instrum. Meas.* **2012**, *61*, 178–189. [CrossRef]
27. Antigny, N.; Servières, M.; Renaudin, V. Pedestrian track estimation with handheld monocular camera and inertial-magnetic sensor for urban augmented reality. In Proceedings of the 2017 International Conference on Indoor Positioning and Indoor Navigation (IPIN), Sapporo, Japan, 18–21 September 2017; pp. 1–8.
28. Zhang, Y.; Hu, W.; Xu, W.; Wen, H.; Chou, C.T. NaviGlass: Indoor Localisation Using Smart Glasses. In Proceedings of the 2016 International Conference on Embedded Wireless Systems and Networks, Graz, Austria, 15–17 February 2016; pp. 205–216.
29. Susi, M.; Renaudin, V.; Lachapelle, G. Motion Mode Recognition and Step Detection Algorithms for Mobile Phone Users. *Sensors* **2013**, *13*, 1539–1562. [CrossRef] [PubMed]
30. Kołodziej, M.; Majkowski, A.; Tarnowski, P.; Rak, R.J.; Gebert, D.; Sawicki, D. Registration and Analysis of Acceleration Data to Recognize Physical Activity. *J. Healthc. Eng.* **2019**, *2019*, 9497151. [CrossRef] [PubMed]
31. Reddy, S.; Mun, M.; Burke, J.; Estrin, D.; Hansen, M.; Srivastava, M. Using mobile phones to determine transportation modes. *ACM Trans. Sen. Netw.* **2010**, *6*, 1–27. [CrossRef]
32. Renaudin, V.; Susi, M.; Lachapelle, G. Step Length Estimation Using Handheld Inertial Sensors. *Sensors* **2012**, *12*, 8507–8525. [CrossRef]
33. Brunato, M.; Battiti, R. Statistical learning theory for location fingerprinting in wireless LANs. *Comput. Netw.* **2005**, *47*, 825–845. [CrossRef]

Article

Geometric Algebra-Based ESPRIT Algorithm for DOA Estimation

Rui Wang [1], Yue Wang [1], Yanping Li [1], Wenming Cao [2],* and Yi Yan [3]

[1] School of Communication and Information Engineering, Shanghai University, Shanghai 200444, China; rwang@shu.edu.cn (R.W.); wangyue8@shu.edu.cn (Y.W.); yanpingli@shu.edu.cn (Y.L.)
[2] College of Information Engineering, Shenzhen University, Shenzhen 518060, China
[3] National Space Science Center, Chinese Academy of Sciences, Beijing 100190, China; yanyi@nssc.ac.cn
* Correspondence: wmcao@szu.edu.cn

Abstract: Direction-of-arrival (DOA) estimation plays an important role in array signal processing, and the Estimating Signal Parameter via Rotational Invariance Techniques (ESPRIT) algorithm is one of the typical super resolution algorithms for direction finding in an electromagnetic vector-sensor (EMVS) array; however, existing ESPRIT algorithms treat the output of the EMVS array either as a "long vector", which will inevitably lead to loss of the orthogonality of the signal components, or a quaternion matrix, which may result in some missing information. In this paper, we propose a novel ESPRIT algorithm based on Geometric Algebra (GA-ESPRIT) to estimate 2D-DOA with double parallel uniform linear arrays. The algorithm combines GA with the principle of ESPRIT, which models the multi-dimensional signals in a holistic way, and then the direction angles can be calculated by different GA matrix operations to keep the correlations among multiple components of the EMVS. Experimental results demonstrate that the proposed GA-ESPRIT algorithm is robust to model errors and achieves less time complexity and smaller memory requirements.

Keywords: direction-of-arrival estimation; geometric algebra; ESPRIT algorithm; electromagnetic vector-sensor array

Citation: Wang, R.; Wang, Y.; Li, Y.; Cao, W.; Yan, Y. Geometric Algebra-Based ESPRIT Algorithm for DOA Estimation. *Sensors* **2021**, *21*, 5933. https://doi.org/10.3390/s21175933

Academic Editors: Zihuai Lin and Wei Xiang

Received: 28 July 2021
Accepted: 27 August 2021
Published: 3 September 2021

Publisher's Note: MDPI stays neutral with regard to jurisdictional claims in published maps and institutional affiliations.

Copyright: © 2021 by the authors. Licensee MDPI, Basel, Switzerland. This article is an open access article distributed under the terms and conditions of the Creative Commons Attribution (CC BY) license (https://creativecommons.org/licenses/by/4.0/).

1. Introduction

Direction-of-arrival (DOA) estimation of electromagnetic (EM) signals has attracted wide attention in many communication fields, such as radar [1,2], mobile networks [3] and sonar [4]. It is clear that DOA estimation is the basic and essential part in an array signal processing system. For example, a corresponding transmitting or receiving beamformer can be designed to extract signals in the direction of interest and suppress uninteresting interference signals. The electromagnetic vector sensor (EMVS) can catch polarization-related information compared to a conventional scalar sensor, which can further improve the target resolution, anti-interference ability and detection stability for DOA estimation [5–7]; therefore, the research for EMVS array direction finding has become a hotspot.

With the appearance of the Long-Vector MODEL (LV-MODEL) [5] (built for EMVS), multiple researchers have proposed various DOA estimators. The existing estimators can be summarized into three categories: (1) research on DOA estimators transplanting from scalar sensor; (2) research based on special array arrangement; (3) research based on advanced mathematical tools.

In terms of transplantation, the classic subspace-based super-resolution algorithm [8] (Multiple Signal Classification—MUSIC) was transplanted to the EMVS [9–11] array, but the algorithms often suffer high computational complexity because of the four-dimensional parameter search for two direction angles and two additional polarization angles; therefore, Weiss [12] used the polynomial root to reduce the computational complexity to a certain extent. In addition, another subspace-based super-resolution algorithm [13,14] (Estimation of Signal Parameters via Rotational Invariance Techniques—ESPRIT) was also transplanted into the EMVS array, and realized closed-form estimation of DOA. In [15,16], authors showed that the statistical performance of the maximum likelihood and subspace-fitting algorithms based on the EMVS array are better than both MUSIC and ESPRIT, but the high calculation limits its application in actual engineering.

There are few studies based on the special array arrangement because most EMVS arrays are co-centered, leading to the mutual coupling interference and spatial information loss. In [17], a double-parallel-line EMVS array whose six components are all spatially separated achieved mutual coupling reduction to refine the DOA-finding accuracy by orders of magnitude. A triangular array [18] combined with a vector cross product and interferometric angle measurement, aimed to overcome the drawback that [17] cannot achieve two-dimensional aperture expansion. In addition, a spatial expansion method of a triangle structure [19] was proposed to provide higher-precision DOA estimation.

The traditional model for EMVS is just a linear combination of each component, which somehow locally destroy the orthogonality of the signal components [20]. Meanwhile, the heavy computational efforts and memory requirements during data processing for the DOA estimation cannot be ignored [21]. Recently, the hypercomplex has been widely studied and applied in multi-dimensional parameter estimation. Miron et al. [22] first proposed a new Quaternion Model (Q-MODEL) for the two-component EMVS array. Then, many models and algorithms based on quaternion have been proposed [23–25]; however, the Q-MODEL had to discard some of the original information because the quaternion only has three imaginary parts. Further, the research has extended to bi-quaternion [26] and quad-quaternion [27,28]. These quaternion-based algorithms showed higher estimation accuracy and less complexity; however, Jiang et al. [21] found that the physical interpretations of the presented quaternion-like models have not been discussed. In order to solve the problem, they derived G-MODEL [21] by Geometric Algebra (GA) formulations of Maxwell equations. The computing technology of G-MODEL not only minimizes the memory requirements and computational complexity, but also removes the correlation of noise on different antennas.

It is easy to find that the current studies utilizing hypercomplex algebra are mainly focused on the MUSIC algorithm [22,26–28]. In fact, MUSIC greatly suffers from a heavy computational burden for its spectrum search, while the computation of ESPRIT algorithm is cheaper, and it can automatically decouple [29]; therefore, the research in this paper extends the ESPRIT algorithm using a new mathematical tool—GA. Through the new calculation rules, the physical nature of EMVS is matched with the signal processing technology, which avoids correlation loss between different components in the previous algorithms. The major contributions of this paper are as follows.

1. We incorporate the multi-dimensional consistency of GA into ESPRIT, and propose a Geometric Algebra-based ESPRIT algorithm (GA-ESPRIT) for 2D-DOA estimation.
2. We use the new calculation rules of the high-dimensional algebra system to preserve the correlation among multiple components of EMVS.
3. Experimental results demonstrate that the proposed GA-ESPRIT algorithm can achieve more accurate, stable and lighter DOA estimation.

The rest of this paper is organized as follows. Section 2 introduces the basics of GA and the EMVS model for narrow-band signals based on GA. Section 3 describes the proposed GA-ESPRIT in detail. Experimental results and analysis are provided in Section 4, followed by concluding remarks in Section 5.

2. Preliminaries

2.1. Fundamental of Geometric Algebra

The concept of GA [30] was proposed by David Hestenes in the 1960s, who combined Clifford Algebra with a physical geometric structure. After decades of research, GA has shown its absolute superiority in electromagnetism [31], cosmology [32], multi-channel image [33–35] and other physical sciences.

2.1.1. Geometric Product

The crucial product operation in GA theory is the geometric product [30]. For vectors **a** and **b**, the geometric product is denoted by

$$\mathbf{ab} = \mathbf{a} \cdot \mathbf{b} + \mathbf{a} \wedge \mathbf{b}, \tag{1}$$

where $\{\cdot\}$ and $\{\wedge\}$ denote the inner product and the outer product, respectively.

2.1.2. Multi-Vector

Let $\mathbb{G}_n = C\ell_{n,0}$, which is the real GA of the quadratic pair (V, Q) where $V = \mathbb{R}^n$ and Q is the quadratic form of signature $(n, 0)$. There is an orthogonal basis $\{\mathbf{e}_1, \mathbf{e}_2, \ldots, \mathbf{e}_n\}$ in $\mathbb{R}^n$, which generates 2^n basis elements of $\mathbb{G}_n$ via the geometric product as shown in (2):

$$\underbrace{\{1\}}_{k=0}, \underbrace{\{\mathbf{e}_i\}}_{k=1}, \underbrace{\{\mathbf{e}_{ij}, i<j\}}_{k=2}, \ldots, \underbrace{\{\mathbf{e}_1\mathbf{e}_2\cdots\mathbf{e}_n\}}_{k=n} \tag{2}$$

for $i, j = 1, 2, \ldots, n$.

The multi-vector A of $\mathbb{G}_n$ is defined as

$$A = E_0(A) + \sum_{1 \le i \le n} E_i(A)\mathbf{e}_i + \sum_{1 \le i < j \le n} E_{ij}(A)\mathbf{e}_{ij} + \ldots + E_{1\ldots n}(A)\mathbf{e}_{1\ldots n}$$
$$= \langle A \rangle_0 + \langle A \rangle_1 + \langle A \rangle_2 + \ldots + \langle A \rangle_n, \tag{3}$$

where $E_i(A), E_{ij}(A), \ldots, E_{1\ldots n}(A) \in \mathbb{R}$, and $\langle A \rangle_k$ denotes the component of A of grade k.

The reverse of multi-vector A is defined as

$$\tilde{A} = \sum_{k=0}^{n} (-1)^{k(k-1)/2} \langle A \rangle_k. \tag{4}$$

2.2. The Geometric Algebra of Euclidean 3-Space

According to the structural characteristics of EMVS, $\mathbb{G}_3$ is chosen to model and process the received signals [21]. The multiplication rule can be found in Table 1.

Table 1. The multiplication rule in $\mathbb{G}_3$.

	1	$\mathbf{e}_1$	$\mathbf{e}_2$	$\mathbf{e}_3$	$\mathbf{e}_{12}$	$\mathbf{e}_{23}$	$\mathbf{e}_{13}$	$\mathbf{e}_{123}$
1	1	$\mathbf{e}_1$	$\mathbf{e}_2$	$\mathbf{e}_3$	$\mathbf{e}_{12}$	$\mathbf{e}_{23}$	$\mathbf{e}_{13}$	$\mathbf{e}_{123}$
$\mathbf{e}_1$	$\mathbf{e}_1$	1	$\mathbf{e}_{12}$	$\mathbf{e}_{13}$	$\mathbf{e}_2$	$\mathbf{e}_{123}$	$\mathbf{e}_3$	$\mathbf{e}_{23}$
$\mathbf{e}_2$	$\mathbf{e}_2$	$-\mathbf{e}_{12}$	1	$\mathbf{e}_{23}$	$-\mathbf{e}_1$	$\mathbf{e}_3$	$-\mathbf{e}_{123}$	$-\mathbf{e}_{13}$
$\mathbf{e}_3$	$\mathbf{e}_3$	$-\mathbf{e}_{13}$	$-\mathbf{e}_{23}$	1	$\mathbf{e}_{123}$	$-\mathbf{e}_2$	$-\mathbf{e}_1$	$\mathbf{e}_{12}$
$\mathbf{e}_{12}$	$\mathbf{e}_{12}$	$-\mathbf{e}_2$	$\mathbf{e}_1$	$\mathbf{e}_{123}$	-1	$\mathbf{e}_{23}$	$-\mathbf{e}_{23}$	$-\mathbf{e}_3$
$\mathbf{e}_{23}$	$\mathbf{e}_{23}$	$-\mathbf{e}_{123}$	$-\mathbf{e}_3$	$\mathbf{e}_2$	$-\mathbf{e}_{13}$	-1	$\mathbf{e}_{12}$	$-\mathbf{e}_1$
$\mathbf{e}_{13}$	$\mathbf{e}_{13}$	$-\mathbf{e}_3$	$-\mathbf{e}_{123}$	$\mathbf{e}_1$	$\mathbf{e}_{23}$	$-\mathbf{e}_{12}$	-1	$\mathbf{e}_2$
$\mathbf{e}_{123}$	$\mathbf{e}_{123}$	$\mathbf{e}_{23}$	$-\mathbf{e}_{13}$	$\mathbf{e}_{12}$	$-\mathbf{e}_3$	$-\mathbf{e}_1$	$\mathbf{e}_2$	-1

Referring to (2) and (3), a $\mathbb{G}_3$ matrix with m-row and n-column, noted $\mathbb{G}_3^{m \times n}$, is constructed as follows [20]

$$\mathbf{A} = \mathbf{A}_0 + \mathbf{A}_1 \mathbf{e}_1 + \mathbf{A}_2 \mathbf{e}_2 + \mathbf{A}_3 \mathbf{e}_3 + \mathbf{A}_4 \mathbf{e}_{12}$$
$$+ \mathbf{A}_5 \mathbf{e}_{23} + \mathbf{A}_6 \mathbf{e}_{13} + \mathbf{A}_7 \mathbf{e}_{123}, \tag{5}$$

where $\mathbf{A}_k$ for $k = 1, 2, 3, \ldots, 7$ are all $m \times n$ real number matrices. The transpose with reversion of $\mathbf{A}$ is denoted by $\mathbf{A}^H$

$$\mathbf{A}^H = \mathbf{A}_0^T + \mathbf{A}_1^T \mathbf{e}_1 + \mathbf{A}_2^T \mathbf{e}_2 + \mathbf{A}_3^T \mathbf{e}_3 - \mathbf{A}_4^T \mathbf{e}_{12}$$
$$- \mathbf{A}_5^T \mathbf{e}_{23} - \mathbf{A}_6^T \mathbf{e}_{13} - \mathbf{A}_7^T \mathbf{e}_{123}, \tag{6}$$

where $\mathbf{A}_i^T$ for $k = 1,2,3,\ldots,7$ denotes the transpose.

2.3. G-MODEL

A compact polarized GA model for the vector-sensor array was proposed in [21], named G-MODEL, which models the six-component outputs of a vector sensor holistically using a multi-vector in $\mathbb{G}_3$. Suppose there are K narrow-band, far-field and uncorrelated sources with wavelength λ impinging on an array, which includes Q vector sensors. Define $\theta_k \in [0, 2\pi)$, $\phi_k \in [0, \pi)$, $\gamma_k \in [0, \pi/2)$ and $\eta_k \in [-\pi, \pi)$ are the azimuth angle, elevation angle, polarization amplitude angle and phase difference angle of the kth source, respectively.

Define $\boldsymbol{u}_k = \cos\theta_k \sin\phi_k \boldsymbol{e}_1 + \sin\theta_k \sin\phi_k \boldsymbol{e}_2 + \cos\phi_k \boldsymbol{e}_3$ as the unit vector (see Figure 1) of the kth source when it impinges on the sensor at the origin. $\boldsymbol{v}_{k1} = -\sin\theta_k \boldsymbol{e}_1 + \cos\theta_k \boldsymbol{e}_2$ and $\boldsymbol{v}_{k2} = \cos\theta_k \cos\phi_k \boldsymbol{e}_1 + \sin\theta_k \cos\phi_k \boldsymbol{e}_2 - \sin\phi_k \boldsymbol{e}_3$ are unit multi-vectors. The position vector of the qth sensor is $\boldsymbol{r}_q = r_{q1}\boldsymbol{e}_1 + r_{q2}\boldsymbol{e}_2 + r_{q3}\boldsymbol{e}_3$. The output of the qth vector sensor in the array is denoted by [21]

$$Y_{EH}^{(q)}(t) = \sum_{k=1}^{K} X_q(\theta_k, \phi_k) V_k P_k S_k(t) + N_{EH}^{(q)}(t), \qquad (7)$$

where $X_q(\theta_k, \phi_k) = e^{e_{123}\frac{2\pi}{\lambda}(\cos\theta_k \sin\phi_k r_{q1} + \sin\theta_k \sin\phi_k r_{q2} + \cos\phi_k r_{q3})}$ is the spatial phase factor of the kth source incident on the qth vector sensor.

$$V_k = (1 + \boldsymbol{u}_k)[\boldsymbol{v}_{k1}, -\boldsymbol{v}_{k2}],$$

$$P_k = \begin{bmatrix} \cos\gamma_k \\ \sin\gamma_k e^{e_{123}\eta_k} \end{bmatrix},$$

$$S_k(t) = |S_k(t)| \exp[e_{123}(2\pi f_k t)].$$

In next section, the GA-ESPRIT algorithm is deduced based on the G-MODEL.

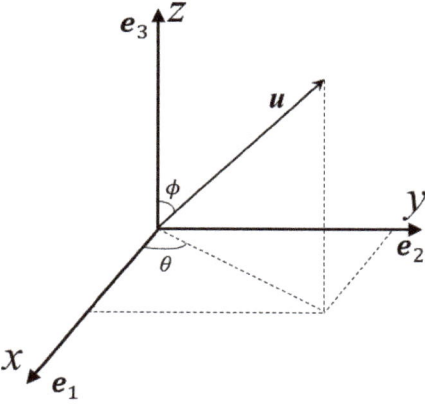

Figure 1. Direction vector of incident source.

3. Proposed Algorithm

The basic premise of the ESPRIT algorithm is that there are identical subarrays, the spacing between subarrays is known and the structure of subarrays is identical, which satisfies the rotational invariance in space [13]. Uniform linear arrays (ULAs) appear when it comes to one-dimensional DOA estimation using conventional ESPRIT [1,13]. Compared with ULAs, double parallel uniform linear arrays (DPULAs) can identify two-dimensional

DOA because of the special construction, which consists of two parallel ULAs [36–38]; therefore, the algorithm discussed in this paper is based on DPULAs.

3.1. Complex Representation Matrix and Related Calculations

In view of the paucity of research on calculations with multi-vector, the Complex Representation Matrix (CRM) [20] is introduced because of the mature matrix theories. Consider a matrix $\mathbf{A} \in \mathbb{G}_3^{m \times n}$, the CRM is defined by $\Psi(\mathbf{A})$

$$\Psi(\mathbf{A}) = \begin{pmatrix} \mathbf{A}_0 + \mathbf{A}_3 + (\mathbf{A}_7 + \mathbf{A}_4)\mathbf{e}_{123} & -\mathbf{A}_1 + \mathbf{A}_6 + (\mathbf{A}_2 - \mathbf{A}_5)\mathbf{e}_{123} \\ -\mathbf{A}_1 - \mathbf{A}_6 - (\mathbf{A}_2 + \mathbf{A}_5)\mathbf{e}_{123} & \mathbf{A}_0 - \mathbf{A}_3 + (\mathbf{A}_7 - \mathbf{A}_4)\mathbf{e}_{123} \end{pmatrix}. \tag{8}$$

Let $\nu = (-\mathbf{e}_1 + \mathbf{e}_{13})/2 \in \mathbb{G}_3$, and its reversion is $\tilde{\nu} = (-\mathbf{e}_1 - \mathbf{e}_{13})/2 \in \mathbb{G}_3$. Then,

$$\nu^2 = \tilde{\nu}^2 = 0 \quad \text{and} \quad \nu\tilde{\nu} + \tilde{\nu}\nu = 1, \tag{9}$$

which imply $\nu\tilde{\nu}\nu = \nu$, $\tilde{\nu}\nu\tilde{\nu} = \tilde{\nu}$, $(\nu\tilde{\nu})^2 = \nu\tilde{\nu}$, $(\tilde{\nu}\nu)^2 = \tilde{\nu}\nu$.

It immediately follows that, for every $\mathbf{A} \in \mathbb{G}_3$, we have

$$\mathbf{A} = \mathbf{E}_{2m}\Psi(\mathbf{A})\mathbf{E}_{2n}^H, \tag{10}$$

$$\Psi(\mathbf{A}) = \mathbf{Q}_{2m}\begin{bmatrix} \mathbf{A} & 0 \\ 0 & \mathbf{A} \end{bmatrix}\mathbf{Q}_{2n}, \tag{11}$$

where in (10) and (11) we have

$$\mathbf{E}_{2k} = [\nu\tilde{\nu}\mathbf{I}_k \quad \tilde{\nu}\mathbf{I}_k] \in \mathbb{G}_3^{k \times 2k}, \tag{12}$$

$$\mathbf{Q}_{2k} = \begin{bmatrix} \nu\tilde{\nu}\mathbf{I}_k & \tilde{\nu}\mathbf{I}_k \\ \nu\mathbf{I}_k & \tilde{\nu}\nu\mathbf{I}_k \end{bmatrix} \in \mathbb{G}_3^{2k \times 2k}. \tag{13}$$

$\mathbf{I}_k$ denotes the $k \times k$ identity matrix. It is not difficult to prove that

$$\mathbf{Q}_{2k} = \mathbf{Q}_{2k}^H = \mathbf{Q}_{2k}^{-1}, \tag{14a}$$

$$\Psi(\mathbf{A}^H) = (\Psi(\mathbf{A}))^H, \tag{14b}$$

$$\Psi(\mathbf{A}^+) = (\Psi(\mathbf{A}))^+, \tag{14c}$$

where $\{+\}$ denotes the pseudo-inverse. Referring to (10) and (14c), the pseudo-inverse of any $\mathbf{A} \in \mathbb{G}_3$ is

$$\mathbf{A}^+ = \mathbf{E}_{2n}(\Psi(\mathbf{A}))^+ \mathbf{E}_{2m}^H. \tag{15}$$

Since $\mathbf{e}_{123}^2 = -1$ and $\mathbf{e}_{123}$ commutes with all elements in $\mathbb{G}_3$, one can identify it with the complex imaginary unit j [20], and so we can view $\Psi(\mathbf{A})$ given in (8) as a complex matrix.

3.2. Model for DPULAs

Consider a DPULA with $2M + 2$ sensors, as shown in Figure 2, in which d and M refer to the spacing between two adjacent sensors and the number of sensors in per subarray, respectively. The array is divided into three subarrays. The 1st to Mth sensors on the x-axis compose the first subarray, the 2nd to $(M + 1)$th sensors form the second subarray and the $(M + 2)$th to $(2M + 1)$th that located on a straight line parallel to the x-axis make up the third subarray. The reason for the division can be found in Figure 3, that is, there are two unknown DOA parameters in the model, which need two rotational invariance relations.

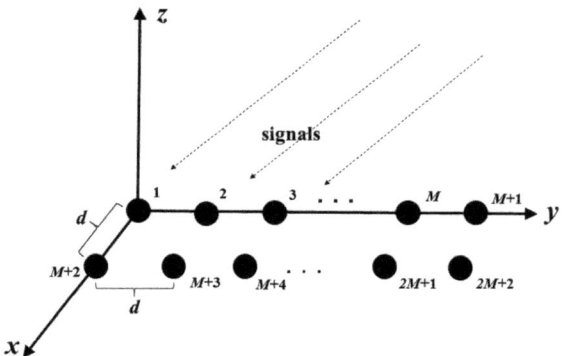

Figure 2. Double parallel uniform linear array.

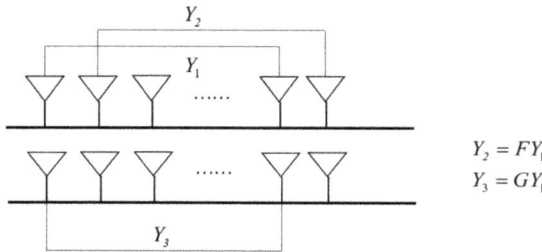

Figure 3. Schematic diagram of GA-ESPRIT.

Since the three subarrays have the same structure and the same number of sensor, each output of them has only one phase difference for the same signal. Signals received by subarray one, two and three are defined as $\mathbf{Y}_{EH}^1$, $\mathbf{Y}_{EH}^2$ and $\mathbf{Y}_{EH}^3$, respectively. According to the above array model, the outputs of the three subarrays at time t are as follows

$$\begin{aligned}\mathbf{Y}_{EH}^1(t) &= \mathbf{A}\mathbf{S}(t) + \mathbf{N}^1(t),\\ \mathbf{Y}_{EH}^2(t) &= \mathbf{A}\mathbf{F}\mathbf{S}(t) + \mathbf{N}^2(t),\\ \mathbf{Y}_{EH}^3(t) &= \mathbf{A}\mathbf{G}\mathbf{S}(t) + \mathbf{N}^3(t),\end{aligned} \qquad (16)$$

where

$$\begin{aligned}\mathbf{Y}_{EH}^1(t) &= \left[Y_{EH}^{(1)}(t),\ldots,Y_{EH}^{(M)}(t)\right]^T,\\ \mathbf{Y}_{EH}^2(t) &= \left[Y_{EH}^{(2)}(t),\ldots,Y_{EH}^{(M+1)}(t)\right]^T,\\ \mathbf{Y}_{EH}^3(t) &= \left[Y_{EH}^{(M+2)}(t),\ldots,Y_{EH}^{(2M+1)}(t)\right]^T,\end{aligned} \qquad (17)$$

and

$$\begin{aligned}\mathbf{A} &= [a(\Gamma_1),\ldots,a(\Gamma_K)],\\ a(\Gamma_k) &= \left[1, x(\theta_k,\phi_k),\ldots, x^{M-1}(\theta_k,\phi_k)\right]^T V_k P_k,\\ x(\theta_k,\phi_k) &= e^{e_{123}\frac{2\pi}{\lambda}d\cos\theta_k\sin\phi_k},\\ \mathbf{F} &= \mathrm{diag}(f_1,\ldots,f_K), \quad \mathbf{G} = \mathrm{diag}(g_1,\ldots,g_K).\end{aligned} \qquad (18)$$

According to (18), we find that the DOA information is contained in matrix $\mathbf{A}$, $\mathbf{F}$ and $\mathbf{G}$. Because $\mathbf{F}$ and $\mathbf{G}$ are diagonal matrices that only contain direction information of incident signals, the focus is the two matrices, i.e.,

$$\begin{aligned} f_k &= e^{e_{123}\frac{2\pi}{\lambda}d\cos\theta_k\sin\phi_k}, \\ g_k &= e^{e_{123}\frac{2\pi}{\lambda}d\sin\theta_k\sin\phi_k}. \end{aligned} \tag{19}$$

Clearly, it is easy to figure out the DOA in the light of (19) if we obtain the two ideal matrices $\mathbf{F}$ and $\mathbf{G}$. From the rules of subarray division, we can see that the latter $(M-1)$ sensors of subarray one and the former $(M+1)$ sensors of subarray two are overlapped. Thus, in order to reduce the computational complexity, subarray one and subarray two can be merged to form a new matrix $\mathbf{Y}_{EH}$, that is,

$$\mathbf{Y}_{EH}(t) = [y_1(t), y_2(t), \ldots, y_{M+1}(t)]^T. \tag{20}$$

After merging, the $(2M+2)th$ redundant sensor is added to subarray three to form a new subarray $\mathbf{P}_{EH}$, so that the third subarray has the same dimension as $\mathbf{Y}_{EH}$

$$\mathbf{P}_{EH}(t) = [y_{M+2}(t), y_{M+3}(t), \ldots, y_{2M+2}(t)]^T. \tag{21}$$

Let $\overline{\mathbf{A}}$ be the array flow pattern of $\mathbf{Y}_{EH}$, then

$$\begin{aligned} \overline{\mathbf{A}} &= [\bar{a}(\Gamma_1), \bar{a}(\Gamma_2), \ldots, \bar{a}(\Gamma_K)], \\ \bar{a}(\Gamma_k) &= [1, x(\theta_k, \phi_k), \ldots, x^M(\theta_k, \phi_k)]^T V_k P_k. \end{aligned} \tag{22}$$

$\mathbf{Y}_{EH}$ and $\mathbf{P}_{EH}$ can be written as

$$\begin{aligned} \mathbf{Y}_{EH}(t) &= \overline{\mathbf{A}}\mathbf{S}(t) + \mathbf{N}_a(t), \\ \mathbf{P}_{EH}(t) &= \overline{\mathbf{A}}\mathbf{G}\mathbf{S}(t) + \mathbf{N}_b(t), \end{aligned} \tag{23}$$

where

$$\mathbf{N}_a(t) = \begin{bmatrix} \mathbf{N}^1(t) \\ n_{M+1}(t) \end{bmatrix}, \quad \mathbf{N}_b(t) = \begin{bmatrix} \mathbf{N}^3(t) \\ n_{2M+2}(t) \end{bmatrix}.$$

Then, $\mathbf{B}(t)$ is defined as

$$\mathbf{B}(t) = \begin{bmatrix} \mathbf{Y}_{EH}(t) \\ \mathbf{P}_{EH}(t) \end{bmatrix} = \mathbf{C}\mathbf{S}(t) + \mathbf{N}(t), \tag{24}$$

where

$$\mathbf{C} = \begin{bmatrix} \overline{\mathbf{A}} \\ \overline{\mathbf{A}}\mathbf{G} \end{bmatrix}, \quad \mathbf{N}(t) = \begin{bmatrix} \mathbf{N}_a(t) \\ \mathbf{N}_b(t) \end{bmatrix}.$$

Finally, the output of the whole array is denoted by

$$\mathbf{B}(t) = \mathbf{C}\mathbf{S}(t) + \mathbf{N}(t). \tag{25}$$

3.3. Algorithm Details

It is assumed that the sources received by the vector-sensor array are random signals which are independent and uncorrelated. In the same way, the measuring noise on six antennas of each sensor is white noise with the same power.

3.3.1. Subspace Separation

Under the above assumption, theoretically, the covariance matrix of the array output is given by

$$\mathbf{R} = \mathbb{E}\left\{\mathbf{B}\mathbf{B}^H\right\} = \mathbf{C}\mathbf{R}_s\mathbf{C}^H + 6\sigma^2\mathbf{I}_{2M+2}, \tag{26}$$

where $\mathbb{E}\{\cdot\}$ stands for the mathematical expectation operator, σ^2 is the noise power on each vector antenna, $\mathbf{R}_S = \mathbb{E}\{\mathbf{S}(t)\mathbf{S}^H(t)\}$.

Since the geometric product is non-commutativity, the Eigenvalue Decomposition (ED) is different from the conventional real methods but similar to the quaternion case. In other words, there are two possible eigenvalues, namely the left and the right eigenvalue for $\mathbb{G}_3$ matrix. In the proposed algorithm, the right eigenvalue is selected because the right ED of $\mathbb{G}_3$ matrix can be converted to the right ED of its CRM [20].

The ED of $\mathbf{R}$ is denoted by

$$\mathbf{R} = \mathbf{U}_s \mathbf{\Sigma}_s \mathbf{U}_s^H + \mathbf{U}_n \mathbf{\Sigma}_n \mathbf{U}_n^H. \tag{27}$$

According to the principle of subspace separation, $\mathbf{U}_s$ is the signal subspace corresponding to K larger eigenvalues, and $\mathbf{\Sigma}_s$ is a diagonal matrix composed of K larger eigenvalues. In addition, $\mathbf{U}_n$ is orthogonal to $\mathbf{U}_s$ and it is the noise subspace corresponding to the remaining $4(M+1) - K$ small eigenvalues. Similarly, $\mathbf{\Sigma}_n$ is a diagonal matrix composed of the remaining small eigenvalues.

In the actual processing, the received signal is usually sampled. So, for a certain number of snapshots N, (26) and (27) can be rewritten as

$$\hat{\mathbf{R}} = \frac{1}{N} \sum_{i=1}^{N} \mathbf{B}(t_i)\mathbf{B}^H(t_i),$$
$$\hat{\mathbf{R}} = \hat{\mathbf{U}}_s \hat{\mathbf{\Sigma}}_s \hat{\mathbf{U}}_s^H + \hat{\mathbf{U}}_n \hat{\mathbf{\Sigma}}_n \hat{\mathbf{U}}_n^H. \tag{28}$$

Because the space formed by the eigenvectors corresponding to the larger eigenvalues is the same as the space formed by the steering multi-vectors of the incident signals, that is, span$\{\mathbf{U}_s\}$ = span$\{\mathbf{C}\}$, there exists a unique non-singular matrix $\mathbf{T}$, which satisfies

$$\mathbf{U}_s = \mathbf{CT}. \tag{29}$$

The rotational invariance relations exist among three subarrays, but $\mathbf{U}_s$ is the signal subspace of the whole array; therefore, after obtaining $\mathbf{U}_s$, the signal subspace of three subarrays must be separated. By the arrangement of sensor array, we find that the signal subspace of three subarrays can be calculated by

$$\begin{aligned}\mathbf{U}_{s1} &= \mathbf{K}_1 \mathbf{U}_s = \mathbf{CT}, \\ \mathbf{U}_{s2} &= \mathbf{K}_2 \mathbf{U}_s = \mathbf{CFT}, \\ \mathbf{U}_{s3} &= \mathbf{K}_3 \mathbf{U}_s = \mathbf{CGT},\end{aligned} \tag{30}$$

where $\mathbf{U}_{s1}$, $\mathbf{U}_{s2}$ and $\mathbf{U}_{s3}$ are signal subspaces of subarray one, subarray two and subarray three, respectively.

$$\begin{aligned}\mathbf{K}_1 &= \begin{bmatrix} \mathbf{I}_M & \mathbf{0}_{M\times(M+2)} \end{bmatrix}_{M\times(2M+2)}, \\ \mathbf{K}_2 &= \begin{bmatrix} \mathbf{0}_{M\times 1} & \mathbf{I}_M & \mathbf{0}_{M\times(M+1)} \end{bmatrix}_{M\times(2M+2)}, \\ \mathbf{K}_3 &= \begin{bmatrix} \mathbf{0}_{M\times(M+1)} & \mathbf{I}_M & \mathbf{0}_{M\times 1} \end{bmatrix}_{M\times(2M+2)}.\end{aligned} \tag{31}$$

3.3.2. Rotation Invariance

From (30), the pivotal matrices $\mathbf{F}$ and $\mathbf{G}$ can be found. So, let

$$\mathbf{U}_{s2} = \mathbf{U}_{s1}\mathbf{\Psi}_x \tag{32}$$

in the same way,

$$\mathbf{U}_{s3} = \mathbf{U}_{s1}\mathbf{\Psi}_y \tag{33}$$

It is discovered that the eigenvalues of $\boldsymbol{\Psi}_x$ and $\boldsymbol{\Psi}_y$ are diagonal elements of $\mathbf{F}$ and $\mathbf{G}$, respectively.

Equations (32) and (33) are equations themselves, and are usually solved by the Least Squares (LS) method [29,36,38,39]; however, LS only takes the error on the left side of the equation into account, it ignores that the coefficient matrix also has an error; therefore, in order to reduce the error caused by solving the equation as much as possible, this paper considers a more accurate method—TLS [13]. Next, the solution of the equation is obtained by taking (32) as an example.

Combining the idea of TLS with the orthogonal property of subspace, we define a new matrix $\mathbf{U}_{s12} = [\mathbf{U}_{s1} \quad \mathbf{U}_{s2}]$. In fact, the main aim is to seek a unitary matrix $\mathbf{D} \in \mathbb{G}_3^{M \times 2K}$, which is orthogonal to $\mathbf{U}_{s12}$. In other words, the space formed by $\mathbf{D}$ is orthogonal to the space formed by the column vectors of $\mathbf{U}_{s1}$ or $\mathbf{U}_{s2}$. So the $\mathbf{D}$ can be obtained from the ED of $\mathbf{U}_{s12}^H \mathbf{U}_{s12}$ [40]

$$\mathbf{U}_{s12}^H \mathbf{U}_{s12} = \mathbf{E}\mathbf{\Lambda}\mathbf{E}^H, \tag{34}$$

where $\mathbf{\Lambda}$ is the diagonal matrix whose diagonal elements are composed by K multi-vectors that only have 0-grade-vector (can regard as non-zero real number) and $3K$ multi-vectors that equal to 0. $\mathbf{E}$ can be written as

$$\mathbf{E} = \begin{bmatrix} \mathbf{E}_{11} & \mathbf{E}_{12} \\ \mathbf{E}_{21} & \mathbf{E}_{22} \end{bmatrix}. \tag{35}$$

Let $\mathbf{E}_N = \begin{bmatrix} \mathbf{E}_{12} \\ \mathbf{E}_{22} \end{bmatrix}$, which is composed by eigenvectors whose eigenvalues are 0 and form the noise subspace. Since $\mathbf{U}_{s12}$ is signal subspace, we find that $\mathbf{D} = \mathbf{E}_N$, i.e.,

$$\mathbf{U}_{s12}\mathbf{D} = [\mathbf{U}_{s1} \quad \mathbf{U}_{s2}] \begin{bmatrix} \mathbf{E}_{12} \\ \mathbf{E}_{22} \end{bmatrix} = 0. \tag{36}$$

Then,

$$\boldsymbol{\Psi}_x = -\mathbf{E}_{12}\mathbf{E}_{22}^+. \tag{37}$$

The pseudo-inverse of $\mathbb{G}_3$ matrix $\mathbf{E}_{22}$ can be found in (15).

3.3.3. Angle Estimation

The azimuth and elevation angle of K signals are included in $\mathbf{F}$ and $\mathbf{G}$. In theory, the eigenvectors obtained by ED of these two matrices are both $\mathbf{T}$; however, in the actual calculation process, the two eigenvalue decomposition operations are carried out independently, which can not ensure that the arrangement of eigenvectors in them is reflected well; therefore, the diagonal elements of $\mathbf{F}$ and $\mathbf{G}$ should be matched.

Suppose that $\mathbf{T1}$ and $\mathbf{T2}$ are eigenvector matrices derived from GA-ED of $\boldsymbol{\Psi}_x$ and $\boldsymbol{\Psi}_y$, respectively. Then

$$\mathbf{O} = |\mathbf{T2}^H\mathbf{T1}| \tag{38}$$

where $\{|\cdot|\}$ is the operator that gets magnitude of every multi-vector in a matrix. For the same signal, the eigenvectors in $\mathbf{T1}$ and $\mathbf{T2}$ corresponding to matched f_k and g_k are related; therefore, the order of diagonal elements in $\mathbf{F}$ and $\mathbf{G}$ can be adjusted by the coordinate of the largest element in each row (or column) of $\mathbf{O}$ to complete matching.

After observing (19), f and g are multi-vectors that only have scalar and 3-grade-vector, if we replace e_{123} with the imaginary unit j of complex number, f and g can be regarded as complex numbers. Finally, we calculate θ_k and ϕ_k with f_k and g_k, that is,

$$\begin{aligned} \theta_k &= \tan^{-1}\left[\tfrac{\text{angle}(g_k)}{\text{angle}(f_k)}\right], \\ \phi_k &= \sin^{-1}\left\{\tfrac{\lambda}{2\pi}\text{sqrt}\left[\text{angle}(g_k)^2 + \text{angle}(f_k)^2\right]\right\}, \end{aligned} \tag{39}$$

where angle(·) is the operator for getting phase angle. In conclusion, the steps of the GA-ESPRIT algorithm are:

1. The original data received from three subarrays are integrated into the measurement model of the whole array according to (25);
2. Calculate the covariance matrix $\hat{\mathbf{R}}$, and then the ED in GA of $\hat{\mathbf{R}}$ is performed and the signal subspace $\mathbf{U}_s$ can be obtained by the larger eigenvalues;
3. According to (30), the signal subspace $\mathbf{U}_s$ of the whole array is divided into three subspaces $\mathbf{U}_{s1}$, $\mathbf{U}_{s2}$ and $\mathbf{U}_{s3}$;
4. $\mathbf{\Psi}_x$ and $\mathbf{\Psi}_y$ can be obtained using TLS in GA, and details can be found in (34)–(37);
5. The ED of $\mathbf{\Psi}_x$ and $\mathbf{\Psi}_y$ is performed to obtain matrices $\mathbf{F}$ and $\mathbf{G}$;
6. The eigenvalues are matched in line with (38) and then taken them into Equation (39) to calculate K pairs direction angles.

Further, the corresponding relationship between the logic flow and steps of GA-ESPRIT is shown in Figure 4.

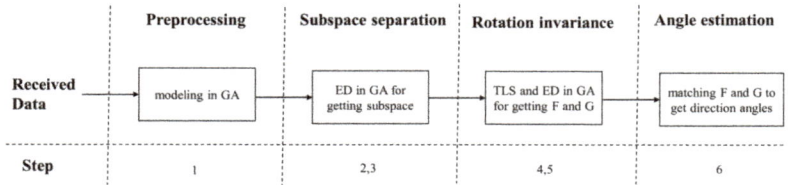

Figure 4. Logic flow diagram of GA-ESPRIT.

3.4. Complexity Analysis

As discussed in [21,22,27], the estimation of the data covariance matrix is an important factor to illustrate the complexity of ESPRIT algorithm and another one is ED, because they imply many repetitive operations and results, which mean heavy computational burden and memory requirements. Thus, we evaluate the time complexity of the two processes and space complexity in terms of real value memory requirements.

Suppose that an array composed of M vector sensors, and N snapshots are taken. LV-ESPRIT [13] and GA-ESPRIT consider six-component measurements of each vector sensor, whereas Q-ESPRIT [25] only records two-component measurements (electric field on x-axis and y-axis.); therefore, we compare the complexity between LV-ESPRIT and GA-ESPRIT. The output of each vector sensor for each signal consists of six complex numbers in LV-ESPRIT, while GA-ESPRIT only has one multi-vector with vector and bivector parts.

The geometric product of two multi-vectors received by the array output implies 36 real multiplications [21], which is nine times as many real multiplications as two complex numbers. As mentioned in Section 2, the ED of a $\mathbb{G}_3$ matrix is calculated by its CRM; therefore, the time complexity of the two algorithms is shown in Table 1. As for space complexity, the memory requirements of a real number is used to measure [21]. In the following two tables, CM is the Covariance Matrix, R represents real number.

The complexity comparison of these two algorithms can be found in Table 2, where CM and R represent covariance matrix and real number, respectively. Observing the time complexity in Table 2, it is not difficult to find that the computational burdens of CM and ED in GA-ESPRIT are a quarter and 1/27 of these in LV-ESPRIT, respectively. As for space complexity, GA-ESPRIT achieves such a significant reduction, more than 1.5 times compared to LV-ESPRIT, which means that the memory pressure is alleviated, especially for the large data size. The reason for the above comparison results is the natural advantage of GA matrix operations. In detail, because the six-dimensional measurement data in LV-MODEL (stored as 12 real numbers) are mapped into a multi-vector in the G-MODEL (stored as six real numbers), the amount of calculation will be reduced to varying degrees with different matrix operations, which will also bring fewer data storage requirements.

The superior description and calculation ability of GA for multi-dimensional signals make GA-ESPRIT a very notable method for direction finding.

Table 2. Complexity of GA-ESPRIT and LV-ESPRIT.

Method	Time Complexity		Space Complexity		
	CM	ED	CM (R)	Eigenvalue (R)	Eigenvectors (R)
LV-ESPRIT	$O\{N \cdot (6M)^2\}$	$O\{(6M)^3\}$	$72M^2$	$6M$	$36M^2$
GA-ESPRIT	$O\{9N \cdot M^2\}$	$O\{(2M)^3\}$	$8M^2$	$2M$	$64M^2$

4. Simulation Results and Analysis

In this section, we simulate and analyze the proposed GA-ESPRIT based on DPULAs with $d = \lambda/2$, discuss its feasibility and performance compared with LV-ESPRIT [14] (in complex number field) and Q-ESPRIT [24] (in quaternion field). The estimation accuracy is evaluated by Root Mean Square Error (RMSE), which is calculated by the average of 200 Monte Carlo simulation experiments.

The RMSE of DOA estimation is defined as

$$\text{RMSE} = \frac{1}{K}\sum_{k=1}^{K}\sqrt{\frac{1}{200}\sum_{k=1}^{200}[(\Delta\theta_k)^2 + (\Delta\phi_k)^2]}, \quad (40)$$

where K, $\Delta\theta_k$ and $\Delta\phi_k$ denote the number of incident signals and errors between the result calculated by DOA algorithm and direction angle initially defined in the experiment, respectively.

In actual applications, the sensor model errors [27,41–43] cannot be ignored, which main include sensor-position error, gain error and phase error. The sensor-position error, as defined in [27], is the error between the actual position and the ideal position of each vector sensor. In the simulation experiment, the sensor-position error is modeled as additive noise with uniform distribution in a certain range, that is,

$$\overline{\mathbf{k}}_m = \mathbf{k}_m + d\sqrt{P_{pe}}\left[\varepsilon_{mx}, \varepsilon_{my}, \varepsilon_{mz}\right]^T \quad (41)$$

where $\overline{\mathbf{k}}_m$ and $\mathbf{k}_m$ are the actual position and ideal position of the mth sensor in vector-sensor array, respectively. ε_{mx}, ε_{my} and ε_{mz} are uniformly distributed noise terms. P_{pe} represents the perturbation power of sensor-position error and the larger P_{pe} means the greater deviation of the sensor from its ideal position. Further, referring to [43], the array output with the gain and phase error is denoted by

$$\mathbf{B}(t) = (\mathbf{I} + \mathbf{\Pi}\mathbf{\Xi})\mathbf{CS}(t) + \mathbf{N}(t), \quad (42)$$

where

$\mathbf{\Pi} = \text{diag}(\eta_1, \eta_2, \ldots, \eta_{2M+2})$,
$\mathbf{\Xi} = \text{diag}(\exp(e_{123}\xi_1), \ldots, \exp(e_{123}\xi_{2M+2}))$,

in which η_i and ξ_i ($i = 1, 2, 3, \ldots, 2M + 2$) are gain error and phase disturbance, respectively. In this paper, we also model them as additive noise. In addition, the six components of all EMVSs are added with noise according to the Signal-to-Noise ratio (SNR) in the following experiments. The SNR is defined as $SNR = 10lg(P_s/P_n)$, in which P_s and P_n are the power of signal and noise on each component, respectively.

In the first experiment, we consider three far-field, narrow-band and uncorrelated signals with parameters $\Gamma = \{160°, 80°, 35°, -60°\}$, $\{60°, 50°, 35°, 60°\}$ and $\{20°, 110°, 45°, 80°\}$ with respect to Signal-to-Noise ratio (SNR) vary -10 dB to 20 dB in two different cases. In addition, we set $M = 7$ and the snapshot number is 200. The aim of the first experiment was to examine the performance of GA-ESPRIT, LV-ESPRIT and Q-ESPRIT under different noise statistical characteristics. Figure 5a shows the estimation results of three algorithms when ideal Gaussian white noise is added, whereas, the noise in Figure 5b is related. It can be concluded that the three algorithms have very close accuracy of calculating DOA at high levels of SNR from Figure 5a,b, while with the lower SNR, GA-ESPRIT has higher accuracy over the other two and can achieve remove the correlation of noise partially.

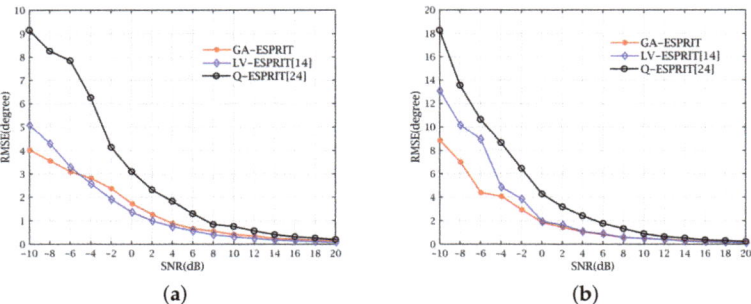

Figure 5. RMSE versus SNR with different noise. (**a**) RMSE versus SNR with uncorrelated noise. (**b**) RMSE versus SNR with correlated noise.

In the second experiment, we compare the performance of GA-ESPRIT, LV-ESPRIT and Q-ESPRIT when the sensor-position error exists. Assume that two signals with $\Gamma = \{58°, 77°, 35°, -60°\}$ and $\{136°, 50°, 35°, 60°\}$ impinge on a DPULA with $M = 9$. Figure 6a shows the performance of the three algorithms when sensor-position error exists with different intensities. Meanwhile, we set SNR to 10 dB and the snapshot number is 200. The sensor-position error of the array sensor is changed by the value of P_{pe}, whose range is 0–0.07. It can be seen in Figure 6b that we fix $P_{pe} = 0.02$ to observe the estimation of the three algorithms by altering SNR from -10 dB to 20 dB. Figure 6a,b both imply that accuracy of GA-ESPRIT is highest in the presence of the sensor-position error, so the conclusion is that GA-ESPRIT has the strongest robustness against sensor-position errors among the three algorithms.

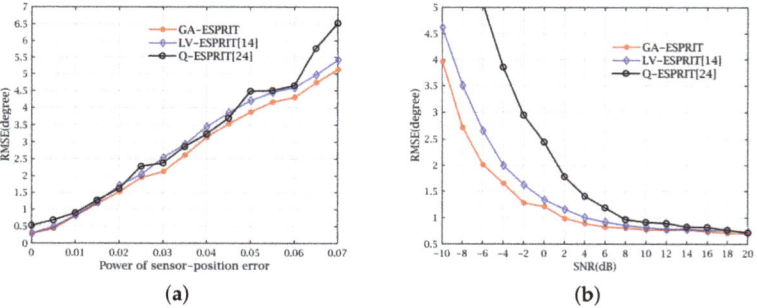

Figure 6. RMSE with sensor-position error. (**a**) RMSE versus the power of sensor-position error. (**b**) RMSE versus SNR in the presence of sensor-position error.

The third experiment is also designed for two cases. Case one is that only gain error exists (see Figure 7a), while for case two, only phase error exists (see Figure 7b). Other conditions are the same as experiment two except that there is no position error. The gain error is constructed by the random numbers, whose mean value is 1 and variance is 0.2, and the phase error is constructed by the random numbers with zero-mean and 0.005 variance. We can learn from Figure 7a,b that, whether there is gain error or phase error, GA-ESPRIT can maintain the estimation accuracy very well, especially in low SNR.

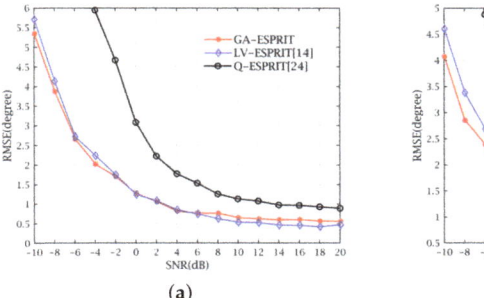 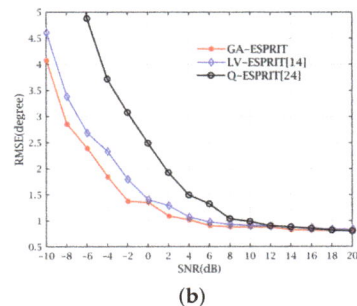

(a) (b)

Figure 7. RMSE with gain or phase error. (a) RMSE versus SNR in the presence of gain error. (b) RMSE versus SNR in the presence of phase error.

In general, it is because Q-ESPRIT only takes part of the array output information into consideration that makes large RMSE. The reason for LV-ESPRIT's poor accuracy in the face of the sensor-model error would be that its "long vector" destroys the orthogonality of the signal components. The improvement of detection robustness of GA-ESPRIT largely results from the fact that it can effectively preserve the orthogonality of the signal components and guarantee the completeness of the information.

5. Conclusions

In this paper, considering that the GA representation contains physical interpretations and complete information of incident signals, we use the idea of the traditional ESPRIT algorithm to find multiple EM signals in the direction finding method in GA. In particular, the model for DPULAs was built in GA and GA-ESPRIT was successfully derived using new calculation rules to achieve the two-dimensional DOA estimation. Compared with the previous ESPRIT algorithms, due to the robustness to sensor-model error and correlated noise, our proposed approach has potential in many practical situations, such as military radar in difficult environments. According to the experimental results, we have confirmed that the GA-ESPRIT has improved accuracy in two-dimension DOA estimation and can resist environmental interference to some extent. More importantly, the proposed algorithm achieves a reduction of more than 1/3 of the memory requirements while the time complexity is also greatly decreased.

Future works on the GA-ESPRIT will include polarization parameter estimation by optimizing matrix operations in GA and the ability of DOA recognition when facing coherent EM signals. It is expected that the proposed GA-ESPRIT will be an efficient DOA estimator.

Author Contributions: Conceptualization, R.W. and W.C.; methodology, R.W. and Y.W.; software, Y.W.; data curation, Y.W. and Y.Y.; writing—original draft preparation, Y.W. and Y.L.; project administration and funding acquisition, R.W. All authors have read and agreed to the published version of the manuscript.

Funding: This research was funded by National Natural Science Foundation of China (NSFC) under Grant No. 61771299, 61771322 and Shenzhen foundation for basic research JCYJ 20190808160815125.

Institutional Review Board Statement: Not applicable.

Informed Consent Statement: Not applicable.

Data Availability Statement: Not applicable.

Conflicts of Interest: The authors declare no conflict of interest.

References

1. Chen, J.L.; Gu, H.; Su, W.M. Angle estimation using ESPRIT without pairing in MIMO radar. *Electron. Lett.* **2008**, *44*, 1422–1423.
2. Xu, B.Q.; Zhao, Y.B.; Cheng, Z.F.; Li, H. A novel unitary PARAFAC method for DOD and DOA estimation in bistatic MIMO radar. *Signal Process.* **2017**, *138*, 273–279. [CrossRef]
3. Rzymowski, M.; Trzebiatowski, K.; Nyka, K.; Kulas, L. Doa estimation using reconfigurable antennas in millimiter-wave frequency 5G systems. In Proceedings of the 2019 17th IEEE International New Circuits and Systems Conference (NEWCAS), Munich, Germany, 23–26 June 2019; pp. 1–4.
4. Saucan, A.; Chonavel, T.; Sintes, C.; Le Caillec, J. CPHD-DOA tracking of multiple extended sonar targets in impulsive environments. *IEEE Trans. Signal Process.* **2016**, *64*, 1147–1160. [CrossRef]
5. Nehorai, A.; Paldi, E. Vector-sensor array processing for electromagnetic source localization. *IEEE Trans. Signal Process.* **1994**, *42*, 376–398. [CrossRef]
6. Li, J. Direction and polarization estimation using arrays with small loops and short dipoles. *IEEE Trans. Antennas Propag.* **1993**, *41*, 379–387. [CrossRef]
7. Guo, X.; Wan, Q.; Chang, C.; Lam, E.Y. Source localization using a sparse representation framework to achieve superresolution. *Multidimens. Syst. Signal Process.* **2010**, *21*, 391–402. [CrossRef]
8. Schmidt, R.O. Multiple emitter location and signal parameter estimation. *IEEE Trans. Antennas Propag.* **1986**, *34*, 276–280. [CrossRef]
9. Miron, S.; Bihan, N.L.; Mars, J.I. Vector-sensor MUSIC for polarized seismic sources localization. *Eurasip J. Adv. Signal Process.* **2005**, *2005*, 74–84. [CrossRef]
10. Guo, H.S.; Yan, B.; Wu, Z.D.; Li, X. Two-dimensional DOA estimation by seismic sensor in shallow water multi-path environment. *J. Electron. Inf. Technol.* **2014**, *36*, 988–992.
11. Yuan, Q.W.; Chen, Q.; Sawaya, K. MUSIC based DOA finding and polarization estimation using USV with polarization sensitive array antenna. In Proceedings of the IEEE Radio and WirelessSymposium, San Diego, CA, USA, 17–19 October 2006; pp. 339–342.
12. Weiss, A.J.; Friedlander, B. Direction finding for diversely polarized signals using polynomial rooting. *IEEE Trans. Signal Process.* **1993**, *41*, 1893–1905. [CrossRef]
13. Kailath, T.; Paulraj, A.; Roy, R. ESPRIT-Estimation of signal parameters via rotational invariance techniques. *IEEE Trans. Acoust. Speech Signal Process.* **1989**, *37*, 984–995. [CrossRef]
14. Gao, F.; Gershman, A.B. A generalized ESPRIT approach to direction-of-arrival estimation. *IEEE Signal Process. Lett.* **2005**, *12*, 254–257. [CrossRef]
15. Li, J.; Stoica, P.; Zheng, D.M. Efficient direction and polarization estimation with a COLD array. *IEEE Trans. Antennas Propag.* **1996**, *44*, 539–547.
16. Li, J.; Stoica, P. Efficient parameter estimation of partially polarized electromagnetic waves. *IEEE Trans. Signal Process.* **1994**, *42*, 3114–3125.
17. Wong, K.T.; Yuan, X. "Vector cross-product direction-finding" with an electromagnetic vector-sensor of six orthogonally oriented but spatially noncollocating dipoles/loops. *IEEE Trans. Signal Process.* **2011**, *59*, 160–171. [CrossRef]
18. Luo, F.; Yuan, X. Enhanced "vector-cross-product" direction-finding using a constrained sparse triangular-array. *Eurasip J. Adv. Signal Process.* **2012**, *2012*, 115. [CrossRef]
19. Zheng, G.M. A novel spatially spread electromagnetic vector sensor for high-accuracy 2-D DOA estimation. *Multidimens. Syst. Signal Process.* **2015**, *28*, 23–48. [CrossRef]
20. Meng, T.Z.; Wu, M.J.; Yuan, N.C. DOA estimation for conformal vector-sensor array using geometric algebra. *Eurasip J. Adv. Signal Process.* **2017**, *2017*, 64. [CrossRef]
21. Jiang, J.F.; Zhang, J.Q. Geometric algebra of euclidean 3-Space for electromagnetic vector-sensor array processing, part I: Modeling. *IEEE Trans. Antennas Propag.* **2011**, *58*, 3961–3973. [CrossRef]
22. Miron, S.; Bihan, N.L.; Mars, J.I. Quaternion-MUSIC for vector-sensor array processing. *IEEE Trans. Signal Process.* **2006**, *54*, 1218–1229. [CrossRef]
23. Zhao, J.C.; Tao, H.H. Quaternion based joint DOA and polarization parameters estimation with stretched three-component electromagnetic vector sensor array. *J. Syst. Eng. Electron.* **2017**, *28*, 1–9. [CrossRef]
24. Chen, H.; Wang, W.; Liu, W. Augmented Quaternion ESPRIT-Type DOA Estimation With a Crossed-Dipole Array. *IEEE Commun. Lett.* **2020**, *24*, 548–552. [CrossRef]
25. Li, Y.; Zhang, J.Q.; Hu, B.; Zhou, H.; Zeng, X.Y. A novel 2-D quaternion ESPRIT for joint DOA and polarization estimation with crossed-dipole arrays. In Proceedings of the 2013 IEEE International Conference on Industrial Technology (ICIT), Cape Town, South Africa, 25–28 February 2013; pp. 1038–1043.
26. Gou, X.M.; Liu, Z.W.; Xu, Y.G. Biquaternion cumulant-MUSIC for DOA estimation of noncircular signals. *Signal Process.* **2013**, *93*, 874–881. [CrossRef]
27. Gong, X.; Liu, Z.; Xu, Y. Quad-Quaternion MUSIC for DOA estimation using electromagnetic vector sensors. *Eurasip J. Adv. Signal Process.* **2008**, *2008*, 1–14. [CrossRef]
28. Xiao, H.K.; Zou, L.; Xu, B.G.; Tang, S.L.; Wan, Y.H.; Liu, Y.L. Direction and polarization estimation with modified quadquaternion music for vector sensor arrays. In Proceedings of the 2014 12th International Conference on Signal Processing (ICSP), Hangzhou, China, 19–23 October 2014; pp. 352–357.

29. Ko, C.; Lee, J. Performance of ESPRIT and Root-MUSIC for angle-of-arrival(AOA) Estimation. In Proceedings of the 2018 IEEE World Symposium on Communication Engineering (WSCE), Singapore, 28–30 December 2018; pp. 49–53.
30. David, H. *New Foundations for Classical Mechanics*; D. Reidel Publishing Company: Boston, MA, USA; Kluwer: Alfen am Rhein, The Netherlands, 1986; pp. 10–34.
31. Arthur, J.W. Understanding geometric algebra for electromagnetic theory. *IEEE Antennas Propag. Mag.* **2011**, *56*, 292. [CrossRef]
32. Lasenby, A.N. Grassmann, geometric algebra and cosmology. *Ann. Phys.* **2010**, *19*, 161–176. [CrossRef]
33. Jorge, R.R.; Eduardo, B.C. Medical image segmentation, volume representation and registration using spheres in the geometric algebra framework. *Pattern Recognit.* **2007**, *40*, 171–188. [CrossRef]
34. Shen, M.; Wang, R.; Cao, W. Joint sparse representation model for multi-channel image based on reduced geometric algebra. *IEEE Access* **2018**, *6*, 24213–24223. [CrossRef]
35. Cao, W.M.; Lyu, F.F.; He, Z.H.; Cao, G.T.; He, Z.Q. Multi-modal medical image registration based on feature spheres in geometric algebra. *IEEE Access* **2018**, *6*, 21164–21172. [CrossRef]
36. Xia, T.; Zheng, Y.; Wan, Q.; Wang, X.; Roy, R. Decoupled estimation of 2-D angles of arrival using two parallel uniform linear arrays. *IEEE Trans. Antennas Propag.* **2007**, *55*, 2627–2632. [CrossRef]
37. Zheng, Z.; Li, G.; Teng, Y. 2D DOA estimator for multiple coherently distributed sources using modified propagator. *Circuits Syst. Signal Process.* **2012**, *31*, 255–270. [CrossRef]
38. Li, J.; Zhang, X.; Chen, W.; Tong, H. Reduced-dimensional ESPRIT for direction finding in monostatic MIMO radar with double parallel uniform linear arrays. *Wirel. Pers. Commun.* **2014**, *77*, 1–19. [CrossRef]
39. Roy, R.; Paulraj, A.; Kailath, T. ESPRIT—A subspace rotation approach to estimation of parameters of cisoids in noise. *IEEE Trans. Acoust. Speech Signal Process.* **1986**, *34*, 1340–1342. [CrossRef]
40. Wang, Y.L. *Theory and Algorithm of Spatial Spectrum Estimation*; Tsinghua University Press: Beijing, China, 2004; pp. 186–191.
41. Kintz, A.L.; Gupta, I.J. A modified MUSIC algorithm for direction of arrival estimation in the presence of antenna array manifold mismatch. *IEEE Trans. Antennas Propag.* **2016**, *64*, 4836–4847. [CrossRef]
42. He, X.; Zhang, Z.; Wang, W. DOA estimation with uniform rectangular array in the presence of mutual coupling. In Proceedings of the 2016 2nd IEEE International Conference on Computer and Communications (ICCC), Chengdu, China, 14–17 October 2016; pp. 1854–1859.
43. Lu, R.; Zhang, M.; Liu, X.; Chen, X.; Zhang, A. Direction-of-arrival estimation via coarray with model errors. *IEEE Access* **2018**, *6*, 56514–56525. [CrossRef]

Article

Unpacking the '15-Minute City' via 6G, IoT, and Digital Twins: Towards a New Narrative for Increasing Urban Efficiency, Resilience, and Sustainability

Zaheer Allam [1,2,*], Simon Elias Bibri [3,4], David S. Jones [5,6], Didier Chabaud [1] and Carlos Moreno [1]

[1] Chaire Entrepreneuriat Territoire Innovation (ETI), IAE Paris—Sorbonne Business School, Université Paris 1 Panthéon-Sorbonne, 75013 Paris, France; chabaud.iae@univ-paris1.fr (D.C.); carlos.moreno@univ-paris1.fr (C.M.)

[2] Live+Smart Research Laboratory, School of Architecture and Built Environment, Deakin University, Geelong, VIC 3220, Australia

[3] Department of Computer Science, Norwegian University of Science and Technology, Sem Saelands veie 9, NO-7491 Trondheim, Norway; simoe@ntnu.no

[4] Department of Architecture and Planning, Norwegian University of Science and Technology, Alfred Getz vei 3, Sentralbygg 1, 5th Floor, NO-7491 Trondheim, Norway

[5] Wadawurrung Traditional Owners Aboriginal Corporation, 86 Mercer Street, Geelong, VIC 3220, Australia; davidsjones2020@gmail.com

[6] Cities Research Institute, Griffith University, 170 Kessels Road, Nathan, QLD 4111, Australia

* Correspondence: zaheerallam@gmail.com

Abstract: The '15-minute city' concept is emerging as a potent urban regeneration model in post-pandemic cities, offering new vantage points on liveability and urban health. While the concept is primarily geared towards rethinking urban morphologies, it can be furthered via the adoption of Smart Cities network technologies to provide tailored pathways to respond to contextualised challenges through the advent of data mining and processing to better inform urban decision-making processes. We argue that the '15-minute city' concept can value-add from Smart City network technologies in particular through Digital Twins, Internet of Things (IoT), and 6G. The data gathered by these technologies, and processed via Machine Learning techniques, can unveil new patterns to understand the characteristics of urban fabrics. Collectively, those dimensions, unpacked to support the '15-minute city' concept, can provide new opportunities to redefine agendas to better respond to economic and societal needs as well as align more closely with environmental commitments, including the United Nations' Sustainable Development Goal 11 and the New Urban Agenda. This perspective paper presents new sets of opportunities for cities arguing that these new connectivities should be explored now so that appropriate protocols can be devised and so that urban agendas can be recalibrated to prepare for upcoming technology advances, opening new pathways for urban regeneration and resilience crafting.

Keywords: smart cities; Internet of Things (IoT); sensors; 6G; wireless communications; resilience; sustainability; climate change; connectivity; data

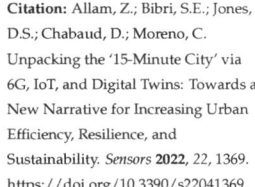

Citation: Allam, Z.; Bibri, S.E.; Jones, D.S.; Chabaud, D.; Moreno, C. Unpacking the '15-Minute City' via 6G, IoT, and Digital Twins: Towards a New Narrative for Increasing Urban Efficiency, Resilience, and Sustainability. *Sensors* **2022**, *22*, 1369. https://doi.org/10.3390/s22041369

Academic Editors: Zihuai Lin and Wei Xiang

Received: 21 December 2021
Accepted: 9 February 2022
Published: 10 February 2022

Publisher's Note: MDPI stays neutral with regard to jurisdictional claims in published maps and institutional affiliations.

Copyright: © 2022 by the authors. Licensee MDPI, Basel, Switzerland. This article is an open access article distributed under the terms and conditions of the Creative Commons Attribution (CC BY) license (https://creativecommons.org/licenses/by/4.0/).

1. Introduction

Cities across the globe have undergone notable transformations, especially following the different waves of the industrial revolutions witnessed since the 18th century. Underpinned by these waves, contemporary cities are now experiencing a transformation hinged on widespread technological integration, with diverse and city-specific outcomes being expected. Of this, the most notable objective and outcome being pursued by these cities is the increasing efficiency and performance in different urban frontiers [1]. Technological integration in different elements of cities is enveloped within the Smart City concept. Proponents of the concept envision an urban environment characterised by reduced human interventions as a result of automation of different urban elements in diverse geographical locations globally. However, automation, being one aspect of 'smartness', is dependent on the amount, quality, and type of data that different urban elements generate [2]. Thus, it is credible to argue that data are becoming a cornerstone of urban planning practice. Therefore, data collection, storage, analysis, and interpretation is critical, especially in helping to understand different urban dynamics and in scaffolding more informed decision making [3].

In addition to data, the smart city concept is further grounded upon the availability of other diverse and advanced technologies that not only allow for data collection, storage, and exploitation but also permit for technologies that help in implementing decisions and insights after data are synthesised, analysed, and interpreted. Such technologies include Artificial Intelligence (AI), Machine Learning (ML), Crowd Computing (CC), connectivity technologies such as 5G (including anticipated 6G), Robotics, and many others [4]. When compounded, the bulk of these technologies, plus different urban elements are seen to make smart cities and their market attractiveness very lucrative.

Currently, the smart cities 'industry' is valued at approximately USD 741.6 billion. With the attention on its sustained implementation amid the prevailing global pandemic challenges, this industry is expected to grow substantially to over USD 2.5 trillion by 2026 [5]. While there are other sources that estimate the market value of this industry differently (for example, Marekts and Markets [6] estimates it to be currently worth USD 457 billion and to grow to USD 873.7 billion by 2026), it is evident that regardless of different market valuations, the concept is promising and is expected to contribute to strategic urban planning trends globally. In particular, beyond the economic frontier, the smart city concept is expected to continue providing opportunities for increased liveability standards, increased sustainability prospects, and improved social dimensions amongst other unparralleled benefits [7–9].

While the smart cities concept is still gaining traction, the COVID-19 pandemic has prompted the emergence of a new urban planning concept—the '15-minute city'. This city type focuses more on promoting social dimensions, urban proximity, and diversity via increasing use of technologies [10]. While the '15-minute city' concept will be described comprehensively in the next section, it is worth noting that the concept has only been described in the literature since 2016, with authors noting commonalities including social distancing, work-from-home concepts, and reduced travel movements, and the concept is increasingly becoming associated in tandem with the smart city concept in crafting and structuring more liveable and human scale cities [11]. However, this urban planning model promises to introduce new characteristics, such as proximity-based approaches to the planning of urban amenities, and urban restructuring, especially in relation to existing urban infrastructures and other elements, so that ultimately, there is a need to mediate these new aspects to human dimensions and values.

The 15-minute city concept, when unpackaged, both in the global north and south, is expected to transform urban areas and allow them to become more human-friendly, especially in the shadow of the post-pandemic 'new normal'. Furthermore, this concept will prompt urban areas to better align for prospective future post-pandemic urban morphologies, especially with prospects that new concepts and technologies (e.g., metaverse) may have in prompting urban restructuring, greenfield policy reinvention, and regeneration generally. The contributions of this paper to the knowledge on 15-minute city concept include the following:

- Comprehensively showcases that the adoption of smart technologies can render more inclusive urban fabrics.
- Explores the adoption of the concept within the new realities (new normal) prompted by COVID-19.
- Explains in detail how emerging technologies such as the Digital Twins and the anticipated 6G technology will further accelerate the adoption and success of this new planning model.

- Highlights some possible obstacles that will need to be overcome to ensure that the anticipated benefits, especially on the social front, are realised.

Within this context, this paper seeks to explore the technological dimensions that are influencing the adoption and implementation of the 15-minute city concept in different global cities. This appraisal includes introducing the '15-minute city' concept in Section 2; Section 3 situates its relationship within Smart Cities practices and relevant techniques and technologies; Section 4 unpacks the concept's technological dimensions; and Section 5 presents a discussion and conclusions.

2. The '15-Minute City'

The '15-minute city' concept is a new urban planning model conceived in 2016 by Franco-Colombian scientist Carlos Moreno, a specialist in intelligent control of complex systems, who envisioned the need for urban environments to be people-centred [12,13]. Moreno acknowledges that he drew inspiration from Jane Jacobs' writings [14]. His model gained prominence with its electoral advancement by Paris Mayor Anne Hidalgo within her "living smart city" initiative called the "Ville du quart d'heure"—the 15-minute city [15].

Moreno's concept is that within an urban area, where human aspects such as socialisation, self-actualisation, cultural demand, and health, among others, the time required for people to access different nodes within the space is given precedence and priority during city planning. This policy empowers that the placement of essential urban amenities, infrastructures, and opportunities is deliberately actioned to facilitate enhanced accessibility. With policy implementation, it becomes possible for residents within given urban areas to comfortably walk or cycle to any given node within a city in a timeframe not exceeding 15 minute [16]. Thus, the demand for the use of automobiles to travel within the city is reduced, providing room for opportunities to create walkways and bicycle lanes that would have been otherwise suppressed in conventional urban planning models that prioritise vehicular flows' efficiencies. Therefore, the 15-minute city concept seeks to bring a paradigm shift in the way urban planning has been previously practiced, shifting it from one focused upon vehicular flows, resulting in gridlocked cities, being a deterrent to the human societal endeavours and city liveability.

Moreno's '15-minute city' concept is inspired from 'Chrono-urbanism', where the aspect of time is believed to be a key factor to consider relative to space [10,13]. That is, the act of placing of different urban amenities and different elements needs to be guided by how much time it would take a walking or a cycling resident to move from one node to the next. In essence, even in urban areas endowed with maximum space, the proximity of different urban elements needs to be a critical consideration.

Within this concept, it is possible to structure a number of nodes within a city, as long as all these observe the four key characteristics (as shown in Figure 1 below) that Moreno, Allam, Chabaud, Gall and Pratlong [11] argue are key in driving urban liveability. These include proximity, diversity, ubiquitousness, and density. In regard to diversity, the vision is to render urban areas accommodating of people from given backgrounds, thus promoting cultural vibrancy, while ensuring that there is diversity in terms of urban structures. That is, planners need to ensure that each of the urban structures, infrastructure, and elements could be used utilised for multiple purposes, hence allowing for their maximum utilisation [12]. For instance, in the case of neighbourhoods, there are opportunities for building urban structures such as car parks that would have capacities for multiple use. Such a move would ensure that there is maximum utility derived from buildings and urban public spaces. Therefore, there is the capacity to craft sufficient urban spaces for the creation of other critical amenities within the same neighbourhoods. Another example is the utilisation of school playing grounds for other purposes including parking, and recreation centres, especially external to school time whether on the surface, above, or below.

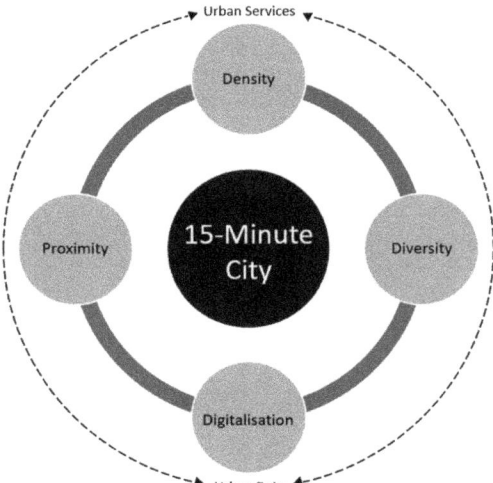

Figure 1. The 15-minute city framework as introduced by Moreno, Allam, Chabaud, Gall, and Pratlong [11].

In terms of density, the 15-minute city concept espouses that cities should have an optimal number of residents. This assumption thereby ensures that it is possible to facilitate quality resource and service provision, without over-consumption or under-utilisation. In respect to the theme of this paper, the density dimension is critical in terms of data generation, which in turn helps not only with influencing how resources are utilised but also in feeding the virtual model of the city, thereby helping in rendering improved urban dimensions. The ubiquitousness principle advances the need for 15-minute cities' requirements to be in large supply in all geographies, thereupon making them available for everyone and at an affordable cost. This aspect will greatly benefit from the deployment of the three technologies (IoT, Digital Twins, and 6G) being advanced in this paper. We argue that with these technologies, it will be possible for urban planners to contextualise and implement 15-minute city models. Additionally, through using the aforementioned technologies, it will be possible to customise each model to varying geographies to fully address place-relevant human dimensions. Thus, such technologies will help in fast tracking those customisations and implementation as is already being done in cities such as Paris, where the agenda is to reduce private cars with a target of 50%, while ensuring cycling-friendly environments through the creation of more bicycle lanes in the city. Another city implementing this concept is Bellevue through its *Environmental Stewardship Plan 2021–2025*.

3. The '15-Minute City' via the Smart City Network

One of the key dimensions of the 15-minute city is digitalisation. This entails the use of digital technologies to influence how the city functions and thus deliver services as well as provide value-producing opportunities in relation to various urban systems and domains. The latter relate to the overall landscape of the 15-minute city in terms of its underlying components, including ICT infrastructure, built infrastructure, green/blue infrastructure, transport infrastructure, energy infrastructure, economic infrastructure, and social infrastructure. Against the backdrop of this perspective paper, the focus is mainly on the ICT infrastructure and the built infrastructure given their particular relevance to the dimensions of the 15-minute city. The built infrastructure denotes the following:

> "... the patterns of the physical objects in the city pertaining to the built-up areas as well as those areas planned for new development and redevelopment ... The compact and ecological dimensions of urban design characterize most of the built

infrastructure as regards its buildings, blocks, streets, open space, public space, green space, and essential infrastructure" [17].

The core design strategies shared by both the compact city and the eco-city are density, diversity, and proximity enabled by mixed land use—which are strategies that also characterise the 15-minute city. ICT infrastructure enables a 15-minute city to move to a data-driven form of urbanism by leveraging advanced data and information technologies to entirely transform its processes and practices—evaluating, analysing, re-engineering, and envisioning the way urban infrastructures and services can be designed, developed, managed, and planned in line with the vision of sustainability. ICT infrastructure digitally consists of those components that power the technology that pervades the fabric of the city (e.g., sensors, smart devices, systems, software programs, networks, data storage facilities, data processing platforms, cloud and fog computing, policies, and standards) and thus permeates urban life, providing support for the management of the city. To coordinate the many different components that comprise the digitalisation dimension of the 15-minute city requires a much stronger function of intelligence. This brings together what government and business have to offer in terms of engaging users of services and communities and providing hardware, software, and solutions enabling smartness, respectively. In this respect, among the issues deemed important are the ways in which ICT infrastructure can be integrated and coordinated, how the data can be analysed and harnessed, and how services can be delivered in a more efficient way. With respect to the latter, digital infrastructure is critical in remote areas in improving not only the efficiency of infrastructure networks but also their sustainability and services (e.g., energy, mobility, transport).

ICT infrastructure can be deployed within the 15-minute city's own facilities (or cloud computing) in order to deliver solutions to different stakeholders with respect to services and applications. In this regard, ICT infrastructure should initiate innovative approaches to the use and integration of IoT, AI, AIoT, big data analytics, simulation models, and intelligent decision-support systems as part of urban computing to enable urban intelligence (e.g., enhancing mobility, reducing congestion, lowering energy use, reducing air pollution, improving planning, optimising governance, etc.). The purpose of this approach will be towards solving problems and issues related to the 15-minute city's operational management and development planning. As related to urban computing, the efforts " . . . dedicated to connecting unobtrusive and ubiquitous sensing technologies, advanced data management and analytics models, and novel visualisation methods to structure intelligent urban computing systems for smart cities . . . " [18] can be utilised to develop innovative solutions in the form of applied urban intelligence for the management, planning, and governance of the 15-minute city.

Unsurprisingly, urban computing and intelligence is increasingly gaining momentum in academic circles and policy debates as a policy agenda for integrated advanced technologies and their novel smart applications for tackling many of the contemporary complex problems and challenges associated with urbanisation and sustainability.

[As] " . . . a process of acquisition, integration, and analysis of big and heterogeneous data generated by a diversity of sources in urban spaces, such as sensors, devices, vehicles, buildings and humans, to tackle the major issues that cities face . . . " [urban computing] " . . . create win–win–win solutions that improve urban environment, human life quality, and city operation systems . . . [and] also helps us understand the nature of urban phenomena [and urban dynamics] and even predict the future of cities . . . " [19].

As an integrated and holistic approach, urban computing and intelligence makes it possible to generate well-informed decisions concerning a wide range of city services and operations, and it can also enable feedback loops between urban environments, human activities, and physical movements [20]. The analytical process in this approach enables the creation of knowledge services required for enhancing decision making based on the design of the components and their relationships, as illustrated in Figure 2.

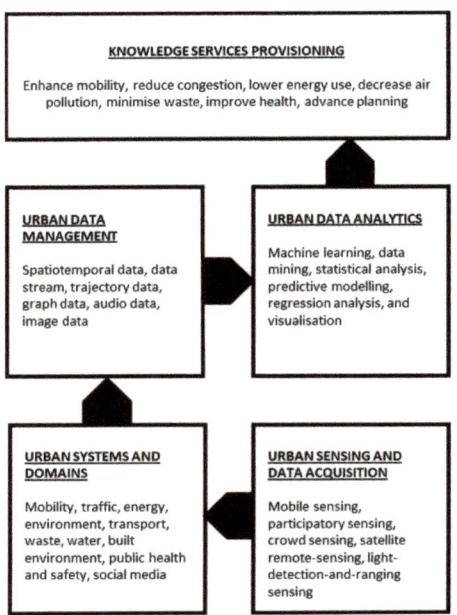

Figure 2. General architecture for urban computing and intelligence based on big data analytics. Illustration by authors.

With the escalating rate of urbanisation and mounting challenges of sustainability, it has become of crucial importance to develop a new urban fabric that can deal effectively with urban development in regard to its dimensions—namely land use change, population increase, cultural change, and economic growth, through such design strategies as compactness, density, diversity, and mixed land use. In this context, an urban fabric refers to

> " ... the physical characteristics of urban areas in terms of components, buildings, spatial patterns, scales, streetscapes, infrastructure, networks, and functions, as well as socio-cultural, ecological, economic, and organizational structures ... " [21].

This also involves making the best use of the digital and informational assets to ensure that the city is sustainable in its approaches to integrating new technologies and their novel applications with compact design strategies. This requires implementing an advanced form of urban computing for monitoring, measuring, analysing, evaluating, designing, and planning urban systems, thereby enabling many functions of urban intelligence for the purpose of improving the sustainability, efficiency, resilience, and life quality in the 15-minute city.

Within this context, IoT has recently become the predominant paradigm of urban computing and intelligence, shifting from a vision of ICT of ubiquitous computing towards one of a deployable paradigm. This shift heralds a new wave of city analytics whose basic ingredient is big data analytics [22–24], which is fostered by the proliferation and widespread diffusion of wireless communication technologies on a hard-to-imagine scale. This is manifested mainly in the quantity and scale of Wi-Fi hotspots covering many urban areas to form a dense multi-faceted IoT network necessitating a large number of sensors exhaustively deployed across the city in order to enhance their communication capabilities and data transfer processes.

Given the wide array of its network in urban areas, via smart city networks, IoT has been extensively installed and used in cities without many engineering obstacles as regards resources, buildings, and infrastructures. IoT infrastructure, involving a myriad of devices seamlessly connected for information exchange, is used to collect vast troves of data to aid in enhancing and optimising urban operations, functions, designs, and policies in relation to various urban domains. IoT when coupled with the data deluge flowing through its multiple networks of sensors plays a key role in the development and implementation of the 15-minute city as a new concept, serving as a technological backbone to the city's attempts to address its goals of sustainability with respect to its underlying dimensions. As an unprecedented planning effort, the 15-minute city initiative is a response to the deconcentration of land use, and as such, it emphasises density, diversity, and mixed land use as key strategies for ensuring liveability, vitality, affordability, energy conservation, and environmental quality. This emerging approach to urban development seeks to deliver more efficient land use, build a resilient and adaptable urban community, lower per capita rates of energy usage and per capita infrastructure provision, and thereby reduce pollution thanks to density and proximity.

Further, IoT infrastructure is necessary to fulfil the needs and visions of the 15-minute city as a smart sustainable approach to urban development. IoT is seen as key to enabling both the smart city [1] and sustainable city infrastructure [25], as it provides a flexible infrastructure that is of crucial importance to deal with the myriad of interconnected devices. It is important for the 15-minute city to have IoT infrastructure in place, where end-device connectivity is monitored, communication reliability is assured, and its sub-systems are intelligent enough to communicate and exchange information with one another while forming a large-scale digital system with widely deployed devices to enable services [26–28] associated with sustainable urban living. A successful implementation of IoT in the 15-minute city means

> "... supporting the complexity of different sensors and their networks set up in urban environments as well as simplifying the composition of interoperable services and applications. Sensor–enabled smart objects are regarded as the essential feature of the interconnected infrastructures of the future" [29].

IoT is an advanced form of ICT of ubiquitous computing. It includes an array of ICT architectures that are fundamentally aimed at describing and providing the relevant infrastructure that underlie the functioning of the digital ecosystem of the city—urban computing and intelligence—within both smart cities [30] and sustainable cities [29]. Thus, ICT architecture denotes a framework for the design of the components and their relationships, functioning as a kind of a roadmap to a city's ICT aspects: for example, what needs to be done to respond to the city's digital needs. ICT infrastructure, in contrast, includes the assets themselves that are used in the city, such as hardware, software, networks, computers, towers, servers, and so forth. Accordingly, the architectural design of ICT determines the variety and number of technologies that can be included in the ICT infrastructure. In essence, a digital ecosystem is built on an infrastructure that has a particular architecture. Therefore, it is impossible to use a particular architecture or infrastructure as a blueprint for all possible implementations in real-world settings. In other words, there is no single consensus on architecture for ICT or infrastructure for ICT that can be agreed upon universally. Different cities have different architectures and different infrastructures, such as planning-based architecture, governance-based architecture, operations-based architecture, healthcare-based architecture, and smart home-based architecture and others.

Bibri and Krogstie [29] offer a detailed review of the key technological and computational components of IoT, including its relationship with big data technology and analytics, sensors and things, big data analytics as a holistic digital system, the core enabling technologies of big data ecosystem, big data analytics solutions, ICT architecture, and IoT infrastructure. Nevertheless, as an advanced approach to ICT design, IoT architecture tends to converge on the number of layers with regard to the design of the components that make up a technological system and their relationships. This still depends on the application domain [24,31–34]. Sometimes, the architectural layers are combined depending on the complexity of the application domain while using different, and sometimes overlapping, labels, such as the physical layer, perception layer, information source layer, middleware layer, network layer, technology layer, application layer, service layer, and domain layer. In the context of the 15-minute city, the four layers of IoT architecture include the following:

(1) Physical/Perception Layer: Urban sensing and data acquisition;
(2) Network/Transmission Layer: Data transfer and communication;
(3) Middleware/Technology Layer: Data hosting, management, processing, and analysis;
(4) Application/Service Layer: Service provisioning specific to urban domains.

3.1. Physical/Perception Layer

This layer works with various types of sensors to generate and collect the data from different sources from across urban systems and domains based on sensor-centric and human-centric sensing mode. A sensor is a device that converts signals from one energy domain to an electrical domain.

Urban sensing includes mobile sensing, participatory sensing, crowd sensing, satellite remote-sensing, and light-detection-and-ranging (LIDAR) sensing. For example, these pertain to mobility, traffic, energy, environment, road networks, transport systems, and built objects. Accordingly, the billions of connected devices forming the IoT infrastructure across a city are equipped with sensors to collect data about the way they are used as well as about the environment surrounding them, with built-in wireless connectivity and communication capabilities enabling them to exchange the generated data. One of the key features of sensors, once deployed, is their ability to interpret the data received from the surrounding environment and generate an output. Sensors measure physical input and send signals to the processor, converting it into data that can be interpreted by a machine or a human. With respect to the former, sensors send their readings to a backend system with humans being left out of the loop. In essence, sensors have the ability to convert data obtained from the outside world into a format that can be preprocessed and further processed and analysed. Sensors are the core enabling technology of IoT, providing an automated approach to urban data generation and thus serve as the main source for big data management and analytics as a form of large-scale computation through middleware. However, among the challenges of urban sensing are resource deployment, implicit and noisy data, skewed sample data, and data sparsity and missing data [35–37].

Raw data are generated about a city in terms of its urban environments, human activities, and physical movements by means of a broad network of sensors spread across the city, including radio frequency identification (RFID) tags, near-field communication (NFC), accelerometers, surveillance cameras, LIDAR, transponders, smart metres, global positioning system (GPS), transduction loops, smartphones, and a number of other digital platforms generating ranges of real-time data. In terms of sensors associated with human mobility and activity, sensors leverage humans as data agents to investigate urban phenomena and dynamics during their movements in urban areas for the purposes of solving and servicing urban problems collectively.

3.2. Network/Transmission Layer

This layer acts as a bridge between the physical layer and the application layer through a middleware layer. It includes a dispersed network as a set of technologies and solutions that allows the transmission of the acquired data for further processing and analysis. As such, it carries and transfers the data collected from the physical objects through sensors by means of multiple wireless networking technologies that provide continuous data regarding the physical and social forms of the city, including Wi-Fi, Bluetooth, satellite, cellular (4G/5G/6G), Local Area Network (LAN), Low Power Wide Area Networks (LPWANs), Zigbee (a low-cost, low-power, wireless mesh network standard targeted at battery-powered devices in wireless control and monitoring applications), and other mesh protocols. While IT devices traditionally are connected to a central access point, satellite, or cell tower somewhere, relying upon expensive hardware infrastructure, mesh networking devices connect directly to each other. Thus, IoT is predicated upon making it possible for about anything to be wirelessly connected and to communicate data over a multiplicity of networks.

3.3. The Middleware/Technology Layer

This layer provides a connectivity layer for the physical layer and for the application layer. As such, it serves as an interface between the varied components of the IoT architecture, making communication and possible among different, often complex, and already existing elements and programs. It contains the main framework for organising and centralising the data collected from the sensor network. One of the key functions of middleware, which operates on cloud computing, is handling the distribution, heterogeneity, interoperability, dynamicity, and scalability of computing resources and systems related to the logic of IoT applications. Middleware is at the core of IoT as a pervasive computing environment, distributing applications across urban domains. It empowers distributed processing for information fusion from multiple components of the physical layer [38]. It is the logic glue with respect to the functionality of distributed applications by connecting and coordinating many components of the IoT as a complex distributed computing system. This involves a variety of the heterogeneous hardware and software elements that are highly interoperable and dynamic, involving a myriad of embedded devices and information processing units required for scaffolding the IoT environment and its proper functioning.

Middleware is necessary for bridging the gap between the massively embedded and networked devices and systems elevating IoT as a form of urban computing and intelligence as it allows multiple processes to run on various sensors, computers, and networks to link up, interact, and communicate to support and maintain the operation of IoT applications. The scope is the ability of an ensemble of devices, smartphones, computers, databases, data warehouses, application integration methods, application servers, application networks, web servers, content management systems, messaging systems, routing, and message transformation to cooperate, interconnect, and communicate seamlessly across disparate networks that create the IoT environment rather than their pervasiveness and extensiveness. Furthermore, middleware supports and deploys numerous applications networks across large geographical areas that are created by sensor networks, network–monitoring systems, and dynamic Web, and that collaborate with, or leverage services from, other disparate systematically tied applications using integration approaches [24].

Middleware is a multi-layered architecture in itself, which comprises four distinct sub-layers, namely [39]:

1. Host–infrastructure middleware or infrastructure and communication;
2. Semantic services and agents or distribution middleware;
3. Common middleware services or services for software environment;
4. Intelligence or domain-specific services related to application action coordination.

Considering the last two sub-layers, the middleware layer also provides processing and analytics procedures to obtain the meaningful information or extract the useful knowledge for numerous applications [23]. This pertains to both urban data management using cloud computing platforms, indexing structures, and retrieval algorithms in regard to spatio-temporal data [19], as well as urban data analytics in terms of adopting machine-learning and data-mining algorithms and models to extract useful knowledge from data across different urban domains using such supervised and unsupervised techniques such as classification, clustering, regression, causal modelling, predictive modelling, and profiling. The process of urban data analytics also fuses the knowledge from multiple disparate datasets across domains [40], using such methods as deep learning-based [35], multi-view-based, transfer learning-based data fusion, similarity-based, and probabilistic dependency-based [19].

3.4. The Application/Service Layer

This layer involves a varied set of applications that use the meaningful information obtained from the physical and middleware layers. Accordingly, this layer provides a wide range of knowledge services for a city in the form of applied intelligence functions. The categories of services provided in this regard are based on the common types of big data analytics, namely diagnostic, descriptive, predictive, and prescriptive, and the domain for which these services are created. Furthermore, this layer offers interfaces that allow urban domain systems to call the knowledge from an IoT application through a city's own facilities (or cloud computing platforms), where the knowledge extracted from data must be integrated into decision-support processes in existing urban domain systems to inform their decision making. This includes visual analytics for model exploration, simulation and prediction methods, and distributed data mining or knowledge discovery strategies. These processes in turn entail distributed data mining and network analytics, extracted models underpinning management and evaluation, and model construction for making assumptions and powerful predictions enabled by mining through AI to improve decision-making processes. In particular, it is of vital importance to enable interactive visual analytics [41], which " ... combine human wisdom with machine intelligence by keeping domain experts in a learning loop" [19]. This layer also relates to urban dashboards and smart boards with visualisation in relation to management system control, automated response systems, and other types of applications. It offers a connectivity of an extensive and transversal resources to multiple users and consumers enabling the adoption of data-driven smart solutions.

Accordingly, 15-minute city architecture is organised into four layers: (1) data generation, (2) data transmission, (3) data management and analytics, and (4) smart services for applications. The services supporting the 15-minute city architecture benefit from the analytical outcome of urban data. Thus, they are provided based upon sensor-based data that is abundant, holistic, dynamic, fine-grained, relational, resolute, and actionable thanks to the intensification of datafication of contemporary cities, allowing for real-time analysis and innovative and adaptive forms of city management and planning. Concerning the latter, IoT architecture for the 15-minute city involves a whole collection of data-driven smart solutions for various urban systems and domains. Such solutions can be adopted by city management agencies and city planning centres to serve various stakeholders by improving sustainability, optimising efficiency, strengthening resilience, and enhancing life quality.

4. Unpacking the '15-Minute City' via Tech-Centric Approaches

4.1. The '15-Minute City' and Digital Twins

Digital twins is an emerging technology that allows for the virtual representation of the physical objects, processes, and services in the virtual environment [4]. With this technology, it has become possible to create replicas that different stakeholders are able to interact within the virtual world, just as in the real world. This then makes it possible to virtualise, analyse, simulate, test, and map different aspects and scenarios in the virtual world to help in making informed decisions, including predictions and modelling that ultimately influence situations in the physical world [42]. In this context, the backbone of the Digital Twins (DT) technology, as argued by Dontha [43], is the availability of real-time, massive, and quality data of a given object, or process that is collected from multiple sources and later fed into the virtual model; hence, it allows us to understand the real situation in the physical world. With data, Qi and Tao [44] note that it is becoming possible for different players in the global spheres to predict future scenarios as well optimise performances.

For cities, especially those that have already embraced some aspects of the smart city concept, they now have the capacity and means to not only generate enough data but also have systems to collect and store the data. Accordingly, DT technology is expected to play a critical role in urban regeneration and its implementation [4]. Indeed, DT could help in the introduction and fashioning of improved regeneration models, as DT technology would allow for the simulation and testing of different scenarios. DT technology would in particular benefit from dynamic models of decision making that allow for participatory planning models that in turn make it possible for data to be collected even from different communities, especially in regard to their expectations and aspirations [45]. With such data, it would then be possible to simulate the impacts of those expectations on the urban areas in case they are factored in, in the regeneration models, including in the adoption of the '15-minute city' concept. Ultimately, it becomes possible for the urban planners and designers to incorporate urban dwellers' expectations, which guarantees optimal impacts and also allows for maximum acceptance and increases 'ownership' of the regeneration model. For instance, in the pursuit of the '15-minute city' concept, the integration of DT technology in the urban planning approaches would allow residents to visualise, and where possible interact virtually, with the different anticipated scenarios and outcomes. Such opportunities would allow them to participate and propose the elements and components that they would wish to be included in their proposed urban regeneration model. However, on a negative tone, the integration of DT technology in the 15-minute city concept may prompt a number of disadvantages. For instance, this will render higher initial costs of implementation of the planning model, noting that this technology is also equally expensive. Furthermore, due to the financial needs to establish relevant infrastructures for DT technology, some cities may take longer to achieve as already, most cities do not have the capacities to finance such projects from their public budgets and have to rely on PPP programs and external loans. The DT technology is also in its infancy stage and will take time before it matures complete to be successfully implemented in the 15-minute cities.

4.2. The '15-Minute City' and the Internet of Things (IoT) Network

The '15-minute city' concept, as advanced by its proponents, may overly concentrate on the human-centric agendas and therefore, conceivably, could work sufficiently without the need for digital solutions. However, there is evidence that cities that have deployed elements of technology are able to improve aspects including quality of life for its residents by 10% to 30% [46], and they are also far better placed to achieve global accords and policy expectations including the United Nations' (UN) Sustainable Development Goals (especially Goal 11 that seeks to "Make cities and human settlements inclusive, safe, resilient and sustainable") [47] and the Glasgow Climate Pact emanating from the 26th UN Climate Change Conference (COP26) [48].

For the '15-minute city' concept, while the vision is to render neighbourhoods that are compact and diverse, if not all services and amenities available are within reach, the integration of IoT networks and devices would further underpin the benefits being sought. On this, Turner and Townsend [49] note that '15-minute cities' would benefit from the use of smart technologies in areas such as pollution warning, controlling, managing, and implementing local traffic and parking policies, and in providing real-time information on issues such as weather and emissions.

The massive data generated from different smart IoT devices in different nodes within '15-minute cities' would further help in the adoption of modern solutions including smart lighting, especially for street lighting projects, smart parking, traffic light coordination, bike sharing, and others. With the data, for instance, it would be possible to effectively estimate local energy consumption rates enabling the adoption of economic and energy-efficient lighting programs that meet place-specific needs without unnecessary generic consumption [50]. Similar benefits would be derived in respect to the demand and consumption of other resources and the utilisation of available public spaces and infrastructures, thereby allowing for informed decision making in the provision of services. On this, one of the positive aspects of the '15-minute city' concept is the prospect of addressing car dependency [51], by reducing the number of vehicles in cities through the provision of bicycle and walk-friendly environments for urban dwellers. Such can equally be rolled out in monitoring and measuring the health and oxygen generating of street trees and park trees, and their corresponding capacity in hosting animal and avian species offering new avenues to monitor and measure the biodiversity (and biophilic [52]) ecological health of a city [53]. With the adoption of smart technologies, Woods [50] showcases that such objectives can be achieved even quicker, as the availability of data could help identify priority areas as well as be used to help convince local residents of the benefits they would derive from adopting the new planning methods. However, to derive those benefits, a number of limitations associated with IoT and its applications will need to be addressed. First, the issue of data quality, data quantity, and privacy is critical. For instance, the case of privacy has been reported to prompt apprehension on the part of residents to embrace and participate in projects, and such predicaments could befall the 15-minute city concept if there are pitfalls regarding the guarantee for privacy and security. IoT technologies are also expensive; hence, they will require urban managers to commit extra financial resources, which might, in some scenarios, lead to plunging cities into increasing debt margins where the projects are already credit financed.

4.3. The '15-Minute City', 6G, and the Data-Driven Cities

As the number of IoT networks and devices continue to increase and become more advanced, we argue that there will be an increased demand for wireless connectivity and speed to facilitate real-time and fast connections [54]. However, although 5G networks have the capacity to help meet our current demands for connectivity, they are still not widespread, and most cities are only covered by the 4G networks, which are limited in their capacity to handle the demand for huge data transfers, real-time recordings, and other connectivity needs [55]. For 5G, Barakat et al. [56] argue that substantial resources including financing will need to be availed to ensure sufficient infrastructures are installed to facilitate the increasing demand emanating from both subscribers and the new devices and innovations that are powered by this technology. For subscribers, it is noted that by the first quarter of 2021, the numbers of those already subscribed had increased by 41%, and more will join following the realities of COVID-19 [57]. This means that in the near future, 5G technology will not be sufficient to support all the connection needs, especially in cities where the number of IoT devices are projected to increase to over 25.4 billion devices by 2030 from the current 8.74 billion devices available globally [58].

Then, the 6th generation (6G) of wireless communication will be inevitable, as the large number of smart IoT devices and networks and the subsequent increase in data will require a wireless technology that can facilitate quicker and real-time transfers from the point of generation to the relevant networks [4]. According to Nguyen et al. [59], 6G technology will bring to life the prospects of full intelligence and automations of systems. 6G technology will help actualise concepts such as the smart city, and in this case, it will be instrumental in helping realise the '15-minute city' concept, which will also ride on the power of technology. In particular, 6G will allow for more user participation and efficiency in the correction and transfer of real-time data to the central networks for analysis and decision making. 6G will further allow for the emergence of new technologies such as the anticipated metaverse (a combination of multiple elements of technology, including virtual reality, augmented reality, and video where users "live" within a digital universe) that have prospects to help actualise the '15-minute city', especially due to reduction of the need to travel, through its emphasis on remote working, and increase social aspects.

However, the 6G technology might take time before the prerequisite infrastructures are put in place before the technology undergoes the relevant piloting and testing, prior to eventually being launched. Whereas the 15-minute city concept is also still very new, it might not immediately benefit from the prospects of 6G as expressed here; thus, it will have to content with available technologies and find alternative ways of overcoming shortcomings that would otherwise be eased by the 6G technology.

5. Discussions and Conclusions

The smart city concept emerged and gained popularity in parallel to the equally exponential growth and increase in the use of advanced technologies riding on internet connections that came about with the fourth industrial revolution wave [60]. In particular, the technologies that emerged and continue to advance to date have initiated a new wave of data generation and also correction, and as such, they have been viewed to increase the prospect of efficiency in many frontiers such as decision making, improving performance, and allowing for real-time predictions, among others [1]. In urban areas as explained above, data mining prospects are seen to be propitiously play critical roles in shaping discourses and trends in topics including climate change, biodiversity health, cultural heritage conservation, traffic management, First Nations' Country Plans [61], adoption of alternative energy such as rooftop solar energy, etc. However, while those benefits amass and will continue to emerge, it is worthwhile to note that the smart city concept attracts notable and genuine criticism from different quarters, especially in regard to the collection, storage, analysis, management, and control of data [62] enveloping also human privacy concerns echoing Orwell's dystopian social science fiction novel and cautionary tale [63]. Such criticisms, especially associated with handling the urban data by the private entities, are well placed, particularly in view of privacy and security of personal data, and also the fear of monetisation thereof that is synonymous with most profit-oriented private entities [64].

Empowering the local government is particularly important as the '15-minute city' concept promises to enhance people-centric dimensions and thereby is expected to generate large private datasets that would need high levels of privacy and security. With the assurance of data safety, it would be possible to convince urban residents to embrace the concept, enabling wider opportunities for urban managers to pursue the project, including adopting the smart urban technologies such as the DT, IoT, and the anticipated 6G technology.

From the literature above, it has been expressed how the DT technology holds unmatched potentials in helping actualise the '15-minute city' concept. In particular, the ability to create a virtual replica of the city model will be critical, noting that the '15-minute city' concept is an emerging technology that may face objections from people who do not understand how it may pan out. In a similar way, the DT technology has had successes in other sectors including manufacturing, the transport industry, and others, especially in providing opportunities for simulations and predictions [65]; it would be expected that DT could help actualise the '15-minute city' concept and its implementation. As noted in the literature, this will be influenced by the availability of real-time data that is also quality and in emanating from different urban fabrics to ensure it allows for a real replica of the '15-minute city' concept. Areas that would gain from this would include sustainability, resilience and quality of life, and the ability to predict and monitor urban health dimensions, such as air quality. These topics are increasingly the subject of new research activities leading to the emergence of varying air monitoring tools [66]. Such will be facilitated by the availability of IoT networks and devices that, as Liu [67] notes, will continue to increase, and new ones will emerge as more demand increases. Therefore, it will be safe to argue here that the prospect of the '15-minute city' concept will not only benefit from existing IoT networks devices but will also play a critical role in influencing the emergence of new ones that help in its actualisation. It is possible that the increase in IoT devices, especially in advanced forms, will pose some connection challenges, and this then justifies the need for the 6G technology explained above. However, it is worth noting that while 6G technology is still in the pipeline, the early piloting of the '15-minute city' concept can still benefit from the 5G technology, but the launch of 6G will revolutionise this concept, especially in view of the speed of transfer of data and increased capacities for communication between different smart devices installed.

The increased connectivity of devices, and the data gathered and processed, can further aid in automating urban dimensions [2,68,69], which can lead from 'automation' to 'autonomy' [70,71]. Of interest to this concept would be the need to replicate those automated features digitally for testing in varying scenarios [72], supporting the need for Digital Twins and the concept of 'City Brains'. This convergence of AI-driven data processing and simulation can further aid in the better planning and implementation of the 15-minute city concept, leading to the contextual solutions, supporting the identity and character of neighbourhoods.

The need to align the '15-minute city' concept with smart city networks would not need to be overemphasised. This is because it is already evident from the realities prompted by the global COVID-19 pandemic that data-driven cities hold the future for the urban areas. However, data alone are not enough. It will need to be exploited to help align urban areas with people-centric dimensions, which have not been sufficiently explored in existing smart cities [63]. With this element catered for, the '15-minute city' concept is poised to be a successful urban planning model, with human scale elements sufficiently provided. However, this success will depend on how well this concept will be aligned with existing and emerging 'smart' technologies. Therefore, it could be safe to argue that the '15-minute city' concept is emerging as a by-product, or an evolution, of the smart city narrative, and going into the future, it might become even more prominent and widely embraced within smart city models.

While this paper has anticipated some of the benefits that cities would accrue from a widespread acceptance and implementation of the 15-minute city concept, a follow-up study will be necessary to evaluate how the concept is taking shape. In particular, this will be relevant as some of the technologies appraised in this article are still very new, or some such as the 6G are still in the pipeline, and it will be prudent to report how those will eventually influence the 15-minute city model once they are deemed mature enough.

Author Contributions: Conceptualisation, D.S.J., Z.A., D.C. and C.M.; methodology, Z.A. and S.E.B.; writing—original draft preparation, Z.A. and S.E.B.; writing—review and editing, Z.A., S.E.B. and D.S.J. All authors have read and agreed to the published version of the manuscript.

Funding: This research received no external funding.

Institutional Review Board Statement: Not applicable.

Informed Consent Statement: Not applicable.

Conflicts of Interest: The authors declare no conflict of interest.

References

1. Allam, Z. *Cities and the Digital Revolution: Aligning Technology and Humanity*; Springer International Publishing: Berlin/Heidelberg, Germany, 2020.
2. Allam, Z. *The Rise of Autonomous Smart Cities: Technology, Economic Performance and Climate Resilience*; Springer International Publishing: Berlin/Heidelberg, Germany, 2020.
3. Ersoy, A.; Alberto, K.C. Understanding urban infrastructure via big data: The case of Belo Horizonte. *Reg. Stud. Reg. Sci.* **2019**, *6*, 374–379. [CrossRef]
4. Allam, Z.; Jones, D.S. Future (post-COVID) digital, smart and sustainable cities in the wake of 6G: Digital twins, immersive realities and new urban economies. *Land Use Policy* **2021**, *101*, 105201. [CrossRef]
5. Report Linker. Global Smart Cities Market to Reach $2.5 Trillion by 2026. Available online: https://www.globenewswire.com/news-release/2021/07/12/2260896/0/en/Global-Smart-Cities-Market-to-Reach-2-5-Trillion-by-2026.html (accessed on 10 December 2021).
6. Marekts and Markets. Smart Cities Market by Focus Area, Smart Transportation, Smart Buildings, Smart Utilities, Smart Citizen Services (Public Safety, Smart Healthcare, Smart Education, Smart Street Lighting, and E-Governance), and Region—Global Forecast to 2026. Available online: https://www.marketsandmarkets.com/Market-Reports/smart-cities-market-542.html (accessed on 10 December 2021).
7. Zhao, Z.; Zhang, Y. Impact of Smart City Planning and Construction on Economic and Social Benefits Based on Big Data Analysis. *Complexity* **2020**, *2020*, 8879132. [CrossRef]
8. Richter, C.; Kraus, S.; Syrjä, P. The Smart City as an opportunity for entrepreneurship. *Int. J. Enterp. Ventur.* **2015**, *7*, 211–226. [CrossRef]
9. Anthopoulos, L.G. *Understanding Smart Cities: A Tool for Smart Government or an Industrial Trick?* Springer International Publishing: Berlin/Heidelberg, Germany, 2017; Volume 22, p. 293.
10. Allam, Z.; Moreno, C.; Chabaud, D.; Pratlong, F. Proximity-Based Planning and the "15-Minute City": A Sustainable Model for the City of the Future. In *The Palgrave Handbook of Global Sustainability*; Springer International Publishing: Cham, Switzerland, 2020; pp. 1–20. [CrossRef]

11. Moreno, C.; Allam, Z.; Chabaud, D.; Gall, C.; Pratlong, F. Introducing the "15-Minute City": Sustainability, Resilience and Place Identity in Future Post-Pandemic Cities. *Smart Cities* **2021**, *4*, 6. [CrossRef]
12. Moreno, C. *Droit de Cité*; Humensis: Paris, France, 2020.
13. Moreno, C. La Ville du Quart D'heure: Pour un Nouveau Chrono-Urbanisme. Available online: https://www.latribune.fr/regions/smart-cities/la-tribune-de-carlos-moreno/la-ville-du-quart-d-heure-pour-un-nouveau-chrono-urbanisme-604358.html (accessed on 3 December 2020).
14. Jacobs, J. *The Death and Life of Great American Cities*; Random House: New York, NY, USA, 1961.
15. WIllsher, K. Paris Mayor Unveils '15-Minute City' Plan in Re-Election Campaign. Available online: https://www.theguardian.com/world/2020/feb/07/paris-mayor-unveils-15-minute-city-plan-in-re-election-campaign (accessed on 1 February 2022).
16. White, N. Welcome to the '15-Minute City'. Available online: https://www.ft.com/content/c1a53744-90d5-4560-9e3f-17ce06aba69a (accessed on 7 November 2020).
17. Bibri, S.; Krogstie, J. A Novel Model for Data-Driven Smart Sustainable Cities of the Future: A Strategic Roadmap to Transformational Change in the Era of Big Data. *Future Cities Environ.* **2021**, *7*, 3. [CrossRef]
18. Liu, D.; Weng, D.; Li, Y.; Bao, J.; Zheng, Y.; Qu, H.; Wu, Y. SmartAdP: Visual Analytics of Large-scale Taxi Trajectories for Selecting Billboard Locations. *IEEE Trans. Vis. Comput. Graph.* **2017**, *23*, 1–10. [CrossRef]
19. Zheng, Y. Urban computing: Enabling urban intelligence with big data. *Front. Comput. Sci.* **2017**, *11*, 1–3. [CrossRef]
20. Bibri, S.E. Data-driven smart sustainable cities of the future: Urban computing and intelligence for strategic, short-term, and joined-up planning. *Comput. Urban Sci.* **2021**, *1*, 8. [CrossRef]
21. Bibri, S.E. Eco-districts and data-driven smart eco-cities: Emerging approaches to strategic planning by design and spatial scaling and evaluation by technology. *Land Use Policy* **2021**, *113*, 105830. [CrossRef]
22. Allam, Z.; Dhunny, Z.A. On big data, artificial intelligence and smart cities. *Cities* **2019**, *89*, 80–91. [CrossRef]
23. Bibri, S.E. Big Data Analytics and Context-Aware Computing: Core Enabling Technologies, Techniques, Processes, and Systems. In *Smart Sustainable Cities of the Future: The Untapped Potential of Big Data Analytics and Context–Aware Computing for Advancing Sustainability*; Bibri, S.E., Ed.; Springer International Publishing: Cham, Switzerland, 2018; pp. 133–188. [CrossRef]
24. Rathore, M.M.; Ahmad, A.; Paul, A.; Rho, S. Urban planning and building smart cities based on the Internet of Things using Big Data analytics. *Comput. Netw.* **2016**, *101*, 63–80. [CrossRef]
25. Bibri, S.E. Data Science for Urban Sustainability: Data Mining and Data-Analytic Thinking in the Next Wave of City Analytics. In *Smart Sustainable Cities of the Future: The Untapped Potential of Big Data Analytics and Context–Aware Computing for Advancing Sustainability*; Bibri, S.E., Ed.; Springer International Publishing: Cham, Switzerland, 2018; pp. 189–246. [CrossRef]
26. Corici, A.; Steinke, R.; Magedanz, T.; Coetzee, L.; Oosthuizen, D.; Mkhize, B.; Catalan, M.; Fontelles, J.C.; Paradells, J.; Shrestha, R.; et al. Towards programmable and scalable IoT infrastructures for smart cities. In Proceedings of the 2016 IEEE International Conference on Pervasive Computing and Communication Workshops (PerCom Workshops), Sydney, Australia, 14–18 March 2016; pp. 1–6.
27. Joseph, T.; Jenu, R.; Assis, A.K.; Kumar, V.A.S.; Sasi, P.M.; Alexander, G. IoT middleware for smart city: (An integrated and centrally managed IoT middleware for smart city). In Proceedings of the 2017 IEEE Region 10 Symposium (TENSYMP), Cochin, India, 14–16 July 2017; pp. 1–5.
28. Cheng, B.; Solmaz, G.; Cirillo, F.; Kovacs, E.; Terasawa, K.; Kitazawa, A. FogFlow: Easy Programming of IoT Services Over Cloud and Edges for Smart Cities. *IEEE Internet Things J.* **2018**, *5*, 696–707. [CrossRef]
29. Bibri, S.E.; Krogstie, J. Environmentally data-driven smart sustainable cities: Applied innovative solutions for energy efficiency, pollution reduction, and urban metabolism. *Energy Inform.* **2020**, *3*, 29. [CrossRef]
30. Allam, Z.; Newman, P. Redefining the Smart City: Culture, Metabolism and Governance. *Smart Cities* **2018**, *1*, 2. [CrossRef]
31. Berkel, A.; Singh, P. *An Information Security Architecture for Smart Cities*; Springer: Cham, Switzerland, 2018; pp. 167–184. [CrossRef]
32. Mocnej, J.; Pekár, A.; Seah, W.K.G.; Papcun, P.; Kajati, E.; Upková, D.; Koziorek, J.; Zolotová, I. Quality-enabled decentralized IoT architecture with efficient resources utilization. *Robot. Comput. Integr. Manuf.* **2021**, *67*, 102001. [CrossRef]
33. Park, E.; Del Pobil, A.P.; Kwon, S.J. The Role of Internet of Things (IoT) in Smart Cities: Technology Roadmap-oriented Approaches. *Sustainability* **2018**, *10*, 1388. [CrossRef]
34. Allam, Z. Achieving Neuroplasticity in Artificial Neural Networks through Smart Cities. *Smart Cities* **2019**, *2*, 9. [CrossRef]
35. Zheng, Y.; Capra, L.; Wolfson, O.; Yang, H. Urban Computing: Concepts, Methodologies, and Applications. *ACM Trans. Intell. Syst. Technol.* **2014**, *5*, 38. [CrossRef]
36. Ji, S.; Zheng, Y.; Li, T. Urban sensing based on human mobility. In Proceedings of the 2016 ACM International Joint Conference on Pervasive and Ubiquitous Computing, Heidelberg, Germany, 12–16 September 2016; pp. 1040–1051.
37. Zheng, Y. Trajectory Data Mining: An Overview. *ACM Trans. Intell. Syst. Technol.* **2015**, *6*, 29. [CrossRef]
38. Azodolmolky, S.; Dimakis, N.; Mylonakis, V.; Souretis, G.; Soldatos, J.; Pnevmatikakis, A.; Polymenakos, L. *Middleware for In-Door Ambient Intelligence: The PolyOmaton System*; Waterloo Press: East Sussex, UK, 2021; pp. 2–6.
39. Schmidt, D.C. Middleware for real-time and embedded systems. *Commun. ACM* **2002**, *45*, 43–48. [CrossRef]
40. Zheng, Y. Methodologies for Cross-Domain Data Fusion: An Overview. *IEEE Trans. Big Data* **2015**, *1*, 16–34. [CrossRef]
41. Liu, W.; Cui, P.; Nurminen, J.K.; Wang, J. Special issue on intelligent urban computing with big data. *Mach. Vis. Appl.* **2017**, *28*, 675–677. [CrossRef]

42. Dembski, F.; Wössner, U.; Letzgus, M.; Ruddat, M.; Yamu, C. Urban Digital Twins for Smart Cities and Citizens: The Case Study of Herrenberg, Germany. *Sustainability* **2020**, *12*, 2307. [CrossRef]
43. Dontha, R. Data and Trending Technologies: Role of Data in Digital Twin Technology. Available online: https://tdan.com/data-and-trending-technologies-role-of-data-in-digital-twin-technology/23630 (accessed on 10 December 2021).
44. Qi, Q.; Tao, F. Digital Twin and Big Data Towards Smart Manufacturing and Industry 4.0: 360 Degree Comparison. *IEEE Access* **2018**, *6*, 3585–3593. [CrossRef]
45. Clément, F.; Kabamdana, D.G.; Food and Agriculture Organization of the United Nations. *Decentralized and Participatory Planning*; Food and Agriculture Organization of the United Nations: Rome, Italy, 1995.
46. Woetzel, J.; Remes, J.; Boland, B.; Katrina, L.; Sinha, S.; Strube, G.; Means, J.; Law, J.; Cadena, A.; Tann, V.V.D. Smart Cities: Digital Solutions for a More Livable Future. 2018. Available online: https://www.mckinsey.com/business-functions/operations/our-insights/smart-cities-digital-solutions-for-a-more-livable-future (accessed on 1 December 2021).
47. UN. U.N. Sustainable Development Goals. Available online: https://www.un.org/sustainabledevelopment/sustainable-development-goals (accessed on 16 October 2020).
48. UNFCCC. *Glasgow Climate Pact*; UNFCCC: Glasgow, UK, 2021; pp. 1–10.
49. Turner & Townsend. Why Birmingham's 15-Minute City Must Be Smart Too. Available online: https://www.greaterbirminghamchambers.com/latest-news/blogs/2021/6/why-birmingham-s-15-minute-city-must-be-smart-too (accessed on 15 December 2021).
50. Woods, E. The 15-Minute City Can Also Be a Smart City. Available online: https://guidehouseinsights.com/news-and-views/the-15-minute-city-can-also-be-a-smart-city (accessed on 15 December 2021).
51. C40 Cities Climate Leadership Group. How to Build Back Better with A 15-Minute City. Available online: https://www.c40knowledgehub.org/s/article/How-to-build-back-better-with-a-15-minute-city?language=en_US#:~{}:text=In%20a%20\T1\textquoteright15%2Dminute%20city\T1\textquoteright%2C%20all%20citizens%20are,and%20sustainable%20way%20of%20life (accessed on 10 November 2020).
52. Beatley, T.; Newman, P. Biophilic Cities Are Sustainable, Resilient Cities. *Sustainability* **2013**, *5*, 3328–3345. [CrossRef]
53. DELWP. Central Geelong: Draft Framework Plan. Available online: https://www.revitalisingcentralgeelong.vic.gov.au/projects/underway-projects/central-geelong-framework-plan (accessed on 8 February 2022).
54. Saad, W.; Bennis, M.; Chen, M. A Vision of 6G Wireless Systems: Applications, Trends, Technologies, and Open Research Problems. *IEEE Netw.* **2020**, *34*, 134–142. [CrossRef]
55. Peters, M.A.; Besley, T. 5G transformational advanced wireless futures. *Educ. Philos. Theory* **2019**, *53*, 847–851. [CrossRef]
56. Barakat, B.; Taha, A.; Samson, R.; Steponenaite, A.; Ansari, S.; Langdon, P.M.; Wassell, I.J.; Abbasi, Q.H.; Imran, M.A.; Keates, S. 6G Opportunities Arising from Internet of Things Use Cases: A Review Paper. *Future Internet* **2021**, *13*, 159. [CrossRef]
57. Steers, S. 5G Americas Notices "Rapid Growth" in 5G Adoption. Available online: https://mobile-magazine.com/5g-and-iot/5g-americas-notices-rapid-growth-5g-adoption (accessed on 15 December 2021).
58. Holst, A. Number of IoT Connected Devices Worldwide 2019–2030. Available online: https://www.statista.com/statistics/1183457/iot-connected-devices-worldwide/#:~{}:text=Number%20of%20IoT%20connected%20devices%20worldwide%202019%2D2030&text=The%20number%20of%20Internet%20of,billion%20IoT%20devices%20in%202030 (accessed on 15 December 2021).
59. Nguyen, D.C.; Ding, M.; Pathirana, P.N.; Seneviratne, A.; Li, J.; Niyato, D.; Dobre, O.; Poor, H.V. 6G Internet of Things: A Comprehensive Survey. *IEEE Internet Things J.* **2021**, *9*, 359–383. [CrossRef]
60. Bibri, S.E.; Krogstie, J. Smart sustainable cities of the future: An extensive interdisciplinary literature review. *Sustain. Cities Soc.* **2017**, *31*, 183–212. [CrossRef]
61. WTOAC. *Paleert Tjaara Dja—Let's Make Country Good Together 2020–2030*; Wadawurrung Country Plan: Geelong, VIC, Australia, 2020. Available online: https://s3.ap-southeast-2.amazonaws.com/hdp.au.prod.app.vic-engage.files/5516/1543/1669/3.8_Wadawurrung_Healthy_Country_Plan_Wadawurrung_Traditional_Owners_Aboriginal_Corporation_2020.pdf (accessed on 1 December 2021).
62. Allam, Z. The Emergence of Anti-Privacy and Control at the Nexus between the Concepts of Safe City and Smart City. *Smart Cities* **2019**, *2*, 7. [CrossRef]
63. Orwell, G. *Nineteen Eighty-Four: A Novel*; Harvill Secker: London, UK, 1949.
64. Li, W.C.; Nirei, M.; Yamana, K. Value of Data: There's No Such Thing as a Free Lunch in the Digital Economy. 2019. Available online: https://ideas.repec.org/p/eti/dpaper/19022.html (accessed on 1 December 2021).
65. Theo. Digital Twins, IoT and the Metaverse. Available online: https://medium.com/@theo/digital-twins-iot-and-the-metaverse-b4efbfc01112 (accessed on 3 December 2021).
66. Ojagh, S.; Cauteruccio, F.; Terracina, G.; Liang, S.H.L. Enhanced air quality prediction by edge-based spatiotemporal data preprocessing. *Comput. Electr. Eng.* **2021**, *96*, 107572. [CrossRef]
67. Liu, S. Global IoT Market Size 2017–2025. Available online: https://www.statista.com/statistics/976313/global-iot-market-size (accessed on 13 December 2019).
68. Macrorie, R.; Marvin, S.; While, A. Robotics and automation in the city: A research agenda. *Urban Geogr.* **2021**, *42*, 197–217. [CrossRef]
69. While, A.H.; Marvin, S.; Kovacic, M. Urban robotic experimentation: San Francisco, Tokyo and Dubai. *Urban Stud.* **2021**, *58*, 769–786. [CrossRef]

70. Cugurullo, F. *Frankenstein Urbanism: Eco, Smart and Autonomous Cities, Artificial Intelligence and the End of the City*; Routledge: London, UK, 2021. [CrossRef]
71. Cugurullo, F. Urban Artificial Intelligence: From Automation to Autonomy in the Smart City. *Front. Sustain. Cities* **2020**, 2. [CrossRef]
72. Barns, S. Out of the loop? On the radical and the routine in urban big data. *Urban Stud.* **2021**, *58*, 3203–3210. [CrossRef]

Article

Designing a Reliable and Low-Latency LoRaWAN Solution for Environmental Monitoring in Factories at Major Accident Risk

Dinesh Tamang [1], Alessandro Pozzebon [2], Lorenzo Parri [1], Ada Fort [1] and Andrea Abrardo [1,*]

[1] Department of Information Engineering and Mathematical Science, University of Siena, 53100 Siena, Italy; dinesh.tamang@student.unisi.it (D.T.); parri@diism.unisi.it (L.P.); ada@diism.unisi.it (A.F.)

[2] Department of Information Engineering, University of Padova, 35131 Padova, Italy; alessandro.pozzebon@unipd.it

* Correspondence: abrardo@diism.unisi.it

Abstract: In this article, we propose a reliable and low-latency Long Range Wide Area Network (LoRaWAN) solution for environmental monitoring in factories at major accident risk (FMAR). In particular, a low power wearable device for sensing the toxic inflammable gases inside an industrial plant is designed with the purpose of avoiding peculiar risks and unwanted accidents to occur. Moreover, the detected data have to be urgently and reliably delivered to remote server to trigger preventive immediate actions so as to improve the machine operation. In these settings, LoRaWAN has been identified as the most proper communications technology to the needs owing to the availability of off the shelf devices and software. Hence, we assess the technological limits of LoRaWAN in terms of latency and reliability and we propose a fully LoRaWAN compliant solution to overcome these limits. The proposed solution envisages coordinated end device (ED) transmissions through the use of Downlink Control Packets (DCPs). Experimental results validate the proposed method in terms of service requirements for the considered FMAR scenario.

Keywords: LoRaWAN; reliability; downlink; safety; IoT

1. Introduction

Safety can be defined as the state of being free from unacceptable risks which can potentially cause damages to humans, environment or properties [1]. Public safety is very important in industrial environments especially in high-risk fields such as oil and gas, chemical industry or nuclear reactor plants where any event or accident can lead to catastrophic consequences both for humans, i.e., the workers and/or those who are living in the surrounding areas, and for the environment. These kinds of scenarios are typically categorized as Factories at Major Accident Risk (FMAR). A mapping of the most frequent major accidents in this context has been produced in Italy in 2013 by the ISPRA (Istituto superiore per la protezione e la ricerca ambientale—High Institute for Environmental Protection and Research) [2] and they are mainly due to release of dangerous substances (liquid or gas), fires, and explosions. The possible causes are several, such as machinery or piping break up, watertight loss, tanks overfill, mistakes during the handling procedures, submission of material on already filled cans, blending of incompatible materials, valves break up, accidental falls or vehicle collisions, and, more generally, spillover of cisterns.

Activities of ensuring compliance with procedures to prevent major accidents are traditionally in charge of humans control, and, as such, they are naturally error prone. On the other hand, with the rapid development of the numerous innovations in manufacturing and in information and wireless communications technologies, we have assisted in the last years to the birth of the new Industry 4.0 era [3]. In this scenario, the Internet of Things (IoT) is a hot topic nowadays either from a research point of view and from an application perspective as well [4]. Needless to say, IoT may provide great support to safety in the considered FMAR scenario in which workers and machines operate in a shared environment and a great amount of heterogeneous information can be collected by sensors and automatically delivered to a Central Controller (CC) allowing a real time control of the whole activity chain (and, as a consequence, of the associated risks). More specifically, the most important sensing technologies in the considered application scenario are represented by sensors for detecting gases with the purpose of avoiding fires and explosions involving flammable gas leakages as well as for controlling the level of oxygen in the atmosphere. In particular, electrochemical and catalytic gas sensors were chosen in this work, whose main focus is the implementation of a reliable limited time delay transmission system using a LoRa radio channel to transmit dangerous gas concentration detected by the sensors. For this purpose, we refer to the wearable device embedding sensing and communications capabilities described in Section 3, which is under development within a collaboration between the University of Siena and INAIL (Istituto nazionale per l'assicurazione contro gli infortuni sul lavoro—Italian National Institute for Insurance against Accidents at Work). Regarding in particular the communication aspects, the considered IoT scenario is characterized by low power devices transmitting infrequent short bursts of data over a low power wide area network. Indeed, the devices cannot have any external power supply (i.e., they run on batteries and they are installed in areas where frequent batteries substitution cannot be always guaranteed). Accordingly, they are expected to be very power efficient. Moreover, most of the detected data are not critical and, as such, can be referred to as Regular Packets (RPs). However, in some cases, i.e., when the concentration of gases crosses a pre-defined threshold, the detected data have to be urgently and reliably delivered and, accordingly, they are referred to as Urgent Packets (UPs). In this setting, one of the most promising communications technologies is represented by the Long Range (LoRa) one, together with the associated LoRa Wide Area Network (LoRaWAN) protocol.

When designing a LoRa-based network infrastructure, the adoption of LoRaWAN provides several advantages with respect to other customized Media Access Control (MAC) protocols. First of all, there is a large availability of open source LoRaWAN servers which can be easily installed and employed to rapidly set-up LoRaWAN networks. As a matter of fact, most of the Gateways (GWs) currently available on the market support the LoRaWAN protocol. In terms of range and coverage, this technology provides a far longer range than Wireless Fidelity (WiFi) or Bluetooth connections [5], applicable for indoor as well as outdoor scenarios, especially in remote areas where the cellular networks have poor connections. Moreover, when setting up dense network infrastructures, upper layer protocols like MAC ones are developed in LoRaWAN with the aim of managing the whole network infrastructure as well as the coexistence of large quantity of end devices (EDs). Moreover, LoRaWAN provides several built-in features that can be very important for the scenario at hand such as: (i) security in the transmission by means of Advanced Encryption Standard (AES-128) end-to-end data encryption; (ii) Possibility of boosting the capacity through the use of multiple channels; (iii) Adaptive Data Rate (ADR) and power consumption by controlling the Spreading Factor (SF), the Bandwidth (BW) and the Coding Rate (CR).

Finally, LoRaWAN provides a number of different message types which allow to set up unconfirmed (UNCONF) or confirmed (CONF) data transmissions (by means of Acknowledgement (ACK)) and a downlink (DL) channel to send information back to ED as well as the Over-The-Air Authentication (OTAA), a procedure that simplifies the association of an ED to the network by means of Join request messages.

The rest of the paper is structured as follows. In Section 2, the state of the art of LoRa solutions for reliable communication and related works are presented. In Section 3, we introduce the sensing and communication parts describing the main sensor characteristics, power consumption analysis and server architecture. Then, the proposed approach is described in Section 4. The main experimental results and discussions are presented in Section 5. Finally, concluding remarks are given in Section 6.

2. State of the Art

Various works have investigated the reliability of LoRa networks. In [6], the authors investigate the problem of collision and propose two distinct mechanisms for collision free transmission, namely TDMA (Time-Division Multiple Access)-based and FDMA (Frequency Division Multiple Access)-based with an ultimate aim of increasing the reliability of the service. The first mechanism allows all clusters to transmit in sequence where up to six EDs belonging to the same cluster can transmit using different SFs in parallel whereas the latter allows all clusters to transmit in parallel, each cluster on its own frequency. However, within each cluster, all EDs transmit in sequence. The simulation results provide better performances than standard LoRaWAN in terms of Packet Delivery Rate (PDR) even if the number of EDs is high. Similarly in [7], a two-step lightweight scheduling is proposed to divide nodes into groups where similar transmission powers are used in each group to reduce the capture effect. The nodes are guided by the GW coarse-grained scheduling to use different SFs to enable simultaneous transmissions through the use of beacon signals at every pre-defined interval, thus reducing packet collisions. The validity of the proposed scheme is assessed using NS-2 simulations showing better performance than legacy LoRaWAN in terms of packet error ratio, throughput, and energy efficiency. However, inter-SF transmission is still a problem due to the loss of the orthogonality between the two signals [8,9]. Since the ALOHA mechanism of the LoRaWAN drastically decreases the performance because of the non-negligible on-air collision probabilities, some authors in literature have proposed the synchronization of LoRa networks by assigning slots to each node using fine-grained scheduling [10].

In order to overcome the problems of classical ALOHA in LoRaWAN, Zorbas et al. [11] propose a time-slotted mechanism where data are buffered locally and transmitted whenever a GW is available by avoiding bursts of collisions. Similarly, in [12] a Time Slotted (TS)-LoRa that allows the nodes to organize autonomously and determine their slot positions in a frame is proposed. This is achieved by sharing an easy-to-compute hash algorithm between the network server and the nodes able to map the nodes' addresses that are assigned during the join phase into unique slot numbers. Moreover, this mechanism ensures backward compatibility with legacy LoRaWAN nodes and liberates TS-LoRa from the huge schedule dissemination overhead. The last slot in each frame is used for sending synchronization ACK responsible for handling time synchronization and ACKs. The considered TS-LoRa achieves a very high packet delivery ratio for all the tested SFs.

The availability of CONF messages is one of the important features in the LoRaWAN networks that is not available in some of its competing low power technologies like Sigfox, Bluetooth Low Energy (BLE) etc. This feature can be used in those scenarios where data reliability is concerned. However, very few works in literature have investigated the CONF traffic, its effects and, more in general, the DL viability. Marais et al. [13] provides an analysis of use cases requiring CONF traffic and concludes that CONF traffic is viable in small networks, especially when data transfer is infrequent. Additionally, aspects likes duty-cycle regulations, SF12 for RX window 2, maximum re-transmission numbers and ACK_TIMEOUT transmission back-off interval negatively impact the viability of the CONF traffic. Similarly, Capuzzo et al. [14] conclude that the performance of a single LoRaWAN cell can significantly degrade when the fraction of nodes that require CONF traffic grows excessively. Moreover, they also suggest that it is necessary to carefully choose the maximum number of transmission attempts for CONF packets, based on the node density and traffic load to get the best performances. In addition, various works in the literature [15–18] investigate the applicability as well as criticism of DL in LoRaWAN networks and its possible negative impact on performances when not well implemented. To sum up, LoRaWAN technology has limitations that need to be carefully considered for its use in the considered FMAR scenario. In particular, one of the most critical issues is related to the use of the CONF mode to provide link reliability. Indeed, as discussed in [19], the use of ACK in DL can significantly drain the network capacity since GWs must be compliant with duty-cycle regulations. This problem is also studied in detail in [20] where the authors propose a solution called *sub-band swapping* where a first receiving window is opened on the dedicated downlink (DL) channel and a second one on the uplink (UL) channel to alleviate the duty cycle's bottleneck. Another line of investigation deals with the use of ad-hoc control schemes which deviate from the standard LoRaWAN ACK mechanism [21,22].

One of the most closely paper to ours is [6], where collision free mechanism based on clustering of EDs is provided with the aim of increasing the reliability. The authors provide various optimized solutions by maximizing the number of EDs in a service area via maximum possible channel utilization. However, the considered test scenario is built with the aim of enabling massive IoT (mIoT) and it differs then from ours (FMAR) in the following aspects:

- The considered solution is more generally focused on a air pollution monitoring system where there is the requirement of sending RPs containing the measurement report at every pre-defined interval within a given time window (400 s–1600 s).
- The EDs should be perfectly synchronized with the GW in time so that the transmission times for the users are accurate. In this case, they have introduced some guard times which require major infrastructural modifications both at ED and GW level.
- It does not consider the FMAR scenario; especially the situations of UP delivery is not considered in their work: this aspect deals not only with the reliability but also with the stringent latency constrains.
- The EDs are assumed fixed.

In this paper, we propose a reliable and low-latency communications solution which is suitable for the considered FMAR scenario and that is fully LoRaWAN compliant. On the other hand, to the best of our knowledge all the previous works in the literature addressing the problem of reliability in LoRa deviate from LoRaWAN standard and, as such, would require a brand-new firmware update at both nodes' and GW level. Conversely, the aim of the solution proposed in this work is to integrate the standard LoRaWAN configurations: this would make the implementation of this system almost straightforward since only minor modifications on the server side of the network infrastructure are required. As a matter of fact, the proposed solution may exploit already existing LoRaWAN networks and thus may be installed without any need for major infrastructural integration. Accordingly, this work proposes a fully operating experimental setup and the viability of the proposed scheme is demonstrated by means of a fully operating LoRaWAN network infrastructure whereas in all the referenced works the results could be achieved only by means of simulations. To sum up, this work does not require any modification in the current LoRaWAN protocol and is fully compliant with the current standard.

3. The Integrated Sensing and Communication Platform

3.1. Sensor Node Architecture

The sensor node was fully custom designed to fulfill the requirements of the application scenario and to comply with the constrains in terms of physical dimensions, measurement accuracy and energy consumption. Its architecture is reported in Figure 1 and it encompasses 3 gas sensors, two electrochemical amperometric gas sensors [23] and a catalytic gas sensor. The sensors used in this application, all manufactured by Alphasense, Braintree, UK, are the CO-A4 for the measurement of carbon monoxide concentration (CO) and the O2-A1 for the measurement of oxygen concentration (O_2), while the catalytic sensor, for the detection of potentially explosive atmospheres, is the CH-A3.

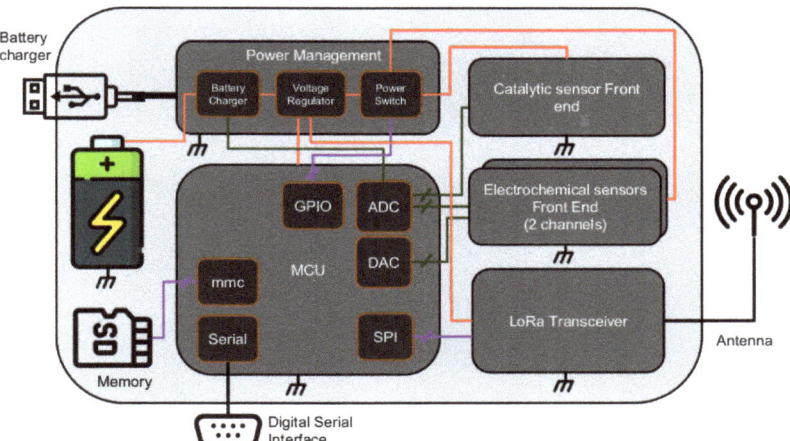

Figure 1. Sensor node architecture.

Electrochemical and catalytic gas sensors, as well as many other chemical sensors, behave linearly in the working range. The average response time (t90) for all the used sensors is less than than 15 s from producer specifications. From tests reported in Figures 2–4 we verified the sensor responses. The response times, for the CO sensor tested from 0 ppm to 28 ppm of CO in air, for the O_2 sensor tested from 20.9% to 10% of O_2 in nitrogen and for the explosive gas sensor tested with steps of 0.2% of methane in air, do not exceed what expected and for all the tested sensors is in general around 10 s. This induces a time delay which is relatively small and aligned with commercial gas detecting systems performance. Considering the application scenario, for example a small room or an operator entering into a tank, the sensor response time is smaller than the time required to a gas to fill the room volume or the time an operator takes to enter into a tank. Hence, this aspect does not represent a critical issue for the developed architecture as it is actually a delayed or unreliable data transmission.

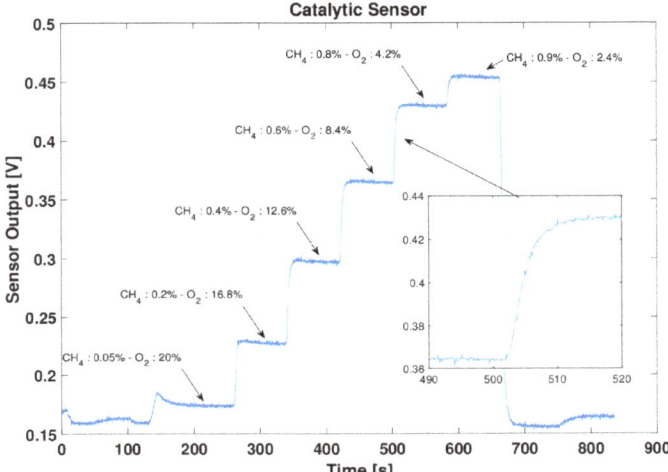

Figure 2. Catalytic Sensor test.

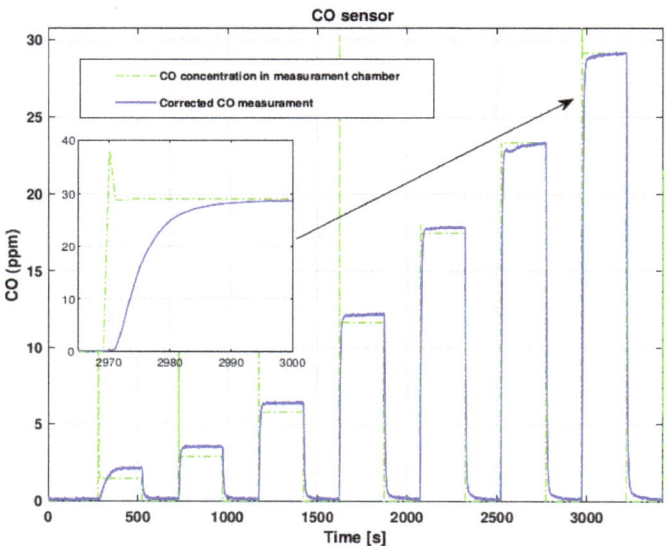

Figure 3. CO Sensor test.

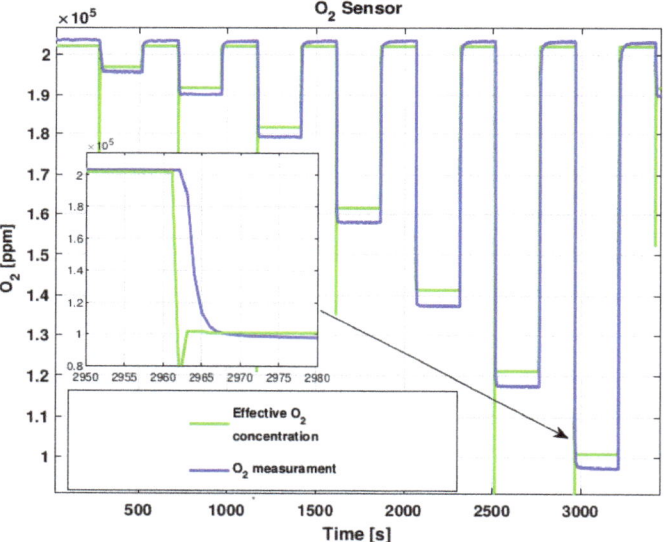

Figure 4. O_2 Sensor test.

The digital part of the node is based on a low power ARM microcontroller from STMicroelectronics (STM32LQT5), that embeds 12 bits Analog to Digital Converters (ADCs) and Digital to Analog Converters (DACs) used to acquire signals from the sensors front end and to provide the correct biasing of electrochemical sensors. The microcontroller can save and read data from an Secure Digital (SD) card memory storage for logging purposes and to load sensors calibration parameters. The node sends data through the LoRa radio channel exploiting an RFM95 transceiver by HopeRF, interfaced through an SPI bus.

In Figure 5, the current absorption of the sensor node microcontroller and LoRa transceiver parts is shown. The measurement was taken during a sensors data acquisition phase followed by a 20 bytes LoRaWAN packet transmission using SF = 7. From this plot, it is possible to measure the maximum time delay from the sensor data acquisition phase end and the radio transmission start. The measurement was obtained by clocking the microcontroller at 32 MHz, acquiring data from the ADC channels and processing them. The acquisition and processing time (few milliseconds) is negligible with respect to the LMIC stack packet elaboration, however, this time delay is below 50 ms.

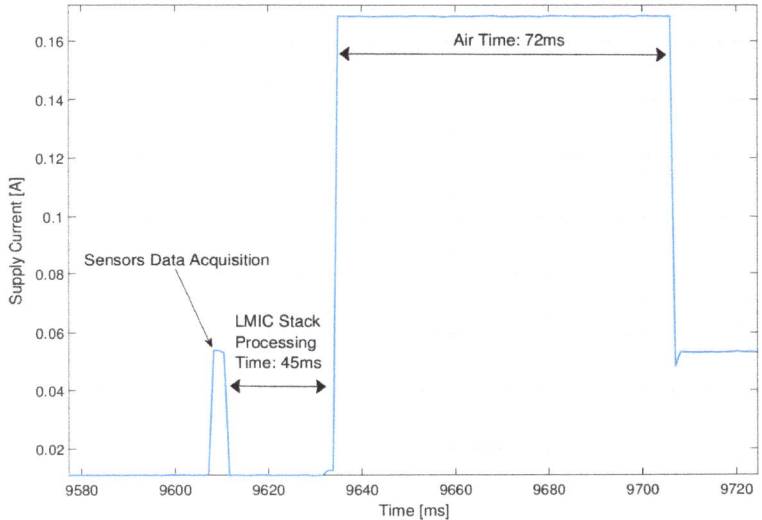

Figure 5. Sensor Node supply current at 4.2V showing sensors data acquisition followed by LoRaWAN message transmission.

The final prototype of the node with gas sensors is reported in Figure 6.

Figure 6. Sensor node device final prototype.

3.2. LoRaWAN Communication Platform: LoRa Node, GW and Server

In the following we provide a brief outline on the overall technology and the network architecture considered in this work.

LoRaWAN protocol is based on the LoRa transmission technology, a proprietary modulation patented by Semtech. LoRa operates in the unlicensed Sub-GHz (below 1 GHz) Industrial, Scientific and Medical (ISM) bands, with three operating frequencies: 433 MHz, 868 MHz and 915 MHz. It exploits the Chirp Spread Spectrum (CSS) modulation technique which allows to achieve extremely high receiver sensitivity values (up to −146 dBm): in these conditions, very long transmission ranges are obtained (up to some kms in urban areas and some tens of kms in rural areas) with limited power consumption. LoRa is then the ideal candidate for a plethora of wide area IoT applications, with a large number of connected devices. LoRaWAN networks adopt a star-of-stars topology, which enables multiple GWs to receive packets from a large quantity of EDs: GWs are in charge to transfer the packets to a central network server which manages the network aspects related to security, scalability and reliability. EDs are divided in 3 Classes (Class A, Class B and Class C) which differ for their ability to receive DL packets from the GW. Class A devices are the simplest and less power hungry ones and, as such, they are by far the most common ones. However, all the three ED typologies are bi-directional in operation. Every ED must be registered with a network before performing communication. These activation processes are of two types: (a) OTAA; the most secure and recommended for EDs and (b) Activation by Personalization (ABP); less secure and requires hardcoding the device address as well as the security keys in the device. Moreover, one of the important aspects of LoRaWAN is the use of frequency plan and its duty cycle regulations. More specifically, duty cycle indicates the fraction of time a resource is busy. As an example, when a ED transmits on a channel for 3 time units every 10 time units, the device has a duty cycle of 30%. As for the transmission frequencies, they are specified in the LoRaWAN regional parameters document [24]. Note that the words frequency and channel are used interchangeably throughout the paper. The duty cycle policy is often regulated by the government and it applies to an entire sub-band. This means that if a user transmits in one of the channels of a given sub-band, it cannot use any of the frequencies of the same sub-band for a time interval regulated by the duty cycle policy. Specifically, the duty cycle values for different sub-bands are regulated by the ETSI EN300.220 standard and are reported below [25].

- g (863.0—868.0 MHz): 1%
- g1 (868.0—868.6 MHz): 1%
- g2 (868.7—869.2 MHz): 0.1%
- g3 (869.4—869.65 MHz): 10%
- g4 (869.7—870.0 MHz): 1%

The above regulations apply to both EDs and GWs. In the followings, we briefly describe the main components of overall system.

3.2.1. RFM95 Radio Transceiver

RFM95 LoRa module is a radio transceiver manufactured by HopeRF [26]. It has a receiver sensitivity of −148 dBm and power amplifier of +20 dBm. Consequently, it has a maximum link budget of 168 dBm. It requires 3.3 V of voltage supply and draws a minimum RX current of 10.3 mA. The choice of this transceiver is due to its low cost and its low power consumption with a very good receiver sensitivity, suitable for the proposed application scnario.

3.2.2. LPS8 GW

LPS8 is an open source LoRaWAN GW [27] which acts as a bridge between the ED and the network infrastructure. It has a backhaul Internet connectivity that connects it to the remote network server. The LPS8 uses a Semtech packet forwarder, a software responsible for forwarding packets to the server and includes a SX1308 LoRa concentrator. It allows users to send data and reach extremely long ranges at low data-rates providing 10 parallel demodulation paths. The receiver has a sensitivity of up to −140 dBm with SX1257 Tx/Rx front-end.

3.2.3. Chirpstack LoRaWAN Server

We consider ChirpStack open-source LoRaWAN Network Server stack [28] for the server side. Chirpstack provides open-source components to form the network infrastructure. Any instance of each component can be installed locally or in a cloud platform to construct the overall network infrastructure. Moreover, this infrastructure provides a user-friendly web-interface for device management and Application Programming Interfaces (APIs) for integration.

It is also important to highlight that by default Chirpstack uses Message Queue Telemetry Transport (MQTT) protocol for publishing and receiving application payloads. MQTT is used by ChirpStack GW Bridge, ChirpStack Network Server, and ChirpStack Application Server. Figure 7 provides the overall network architecture including the sensor node/ED described in the previous subsection.

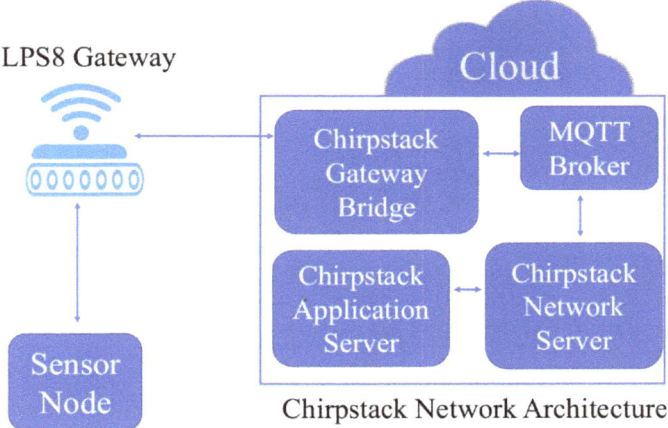

Figure 7. LoRaWAN Infrastructure together with Sensor Node shown in Figure 1.

3.3. Communication Service Requirements

As discussed in detail in Section 1, in order to avoid fires and explosions in the presence of faults and gas leakages, it is necessary to promptly communicate to the CC any anomalous concentration of gases. In the considered system, this kind of data is referred to as UPs. Accordingly, UPs are characterized by stringent requirements in terms of reliability and latency. More specifically, in the scenario at hand we have identified as minimum requirements a packet loss rate (PLR) and end-to-end latency (l) of 0.1% and 500 ms, respectively. Moreover, basing on the required resolution of the data collected from the sensors, the UPs payload is set to 20 bytes. Basing on these requirements, we have limited our LoRaWAN system to use only SF7-10, since with SF11 and SF12 the time on air exceeds 500 ms [29]. Finally, packets retransmissions are not allowed for UP packets which, as such, can be transmitted in UNCONF mode only.

4. The Proposed Communication Strategy

In the following, we will focus on the approach proposed in this paper to provide the required reliability. To this aim, we exploit the DL communication scheme provided by standard LoRaWAN: DL packets are sent by a Network Server to only one ED through one or more GWs. To elaborate, in the proposed system a remote central LoRaWAN server shown in Figure 7 is capable of performing various tasks such as reception of data from the EDs forwarded by GWs, exploitation of the collected data with further processing, and more importantly scheduling of DL messages to the EDs for enabling coordinated transmissions. We describe each aspect in detail in the following subsections.

4.1. Clustering of EDs

In our system, the central server not only collects the sensor data from the EDs, but it is also responsible of forming clusters of users in close proximity. This task is achieved assuming that the system is equipped with a localization system where the server is aware of users' locations. More specifically, the creation of clusters of users is performed with the aim of coordinating the transmissions of close users to avoid possible packets collisions. Indeed, it is highly probable that a critical event such as the presence of gas is jointly detected by all the EDs of the cluster.

The clustering algorithm and the specific localization technology to be adopted in the system are still under investigation and are beyond the scope of this paper. Figure 8 depicts the overall vision of the system where several GWs are installed inside the service area and the clusters of users are associated to the closest GW (represented by different colors). In the figure DCP stands for downlink control packets, which are regularly transmitted by GWs to the associated EDs as detailed in the next section. Note that ED-GWs association is fully in charge of the server and is only to provide a separated control mechanisms, i.e., it is completely transparent to the EDs which are actually connected to each GW as in standard LoRaWAN.

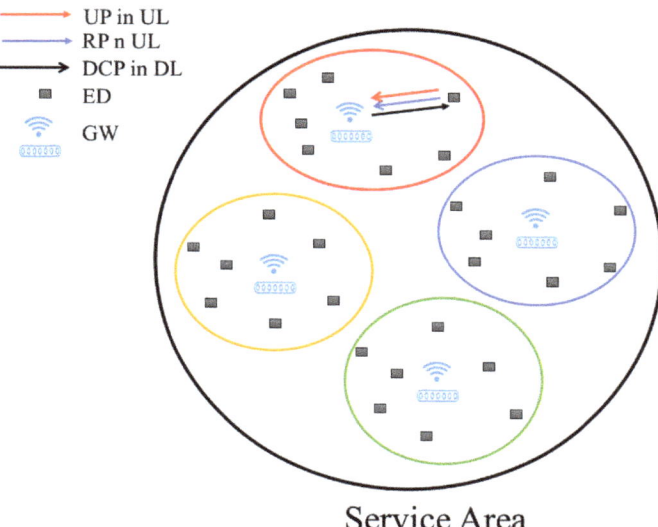

Figure 8. A service area with different clusters represented by different colors.

4.2. Coordination of ED Transmissions through Downlink Control Packets (DCPs)

We refer to the Class A DL operation in which the Network Server transmits a DL packet to an ED after the reception of an UP packet precisely at the beginning of one of two possible receiving windows. More precisely, the ED opens Class A RX1 and RX2 receiving windows after RECEIVE_DELAY1 and RECEIVE_DELAY2 secs respectively. The DL data rate for RX1 depends on the corresponding UL whereas RX2 uses a fixed data rate depending on the region.

In the considered scenario, each node transmits RPs periodically every predetermined (long) time intervals (e.g., several minutes) in UNCONF mode. We program the server in such a way that for every received RP a corresponding DCP is scheduled. Specifically, the DCP message is intended to control the eventual transmission of UPs. The adopted control mechanism, which is discussed in detail in the next section, acts independently on each cluster since it is highly unlikely that in the considered scenario EDs of different clusters have to transmit an UP at the same time (the potentially dangerous event is local and infrequent).

Hence, upon the necessity of delivering an UP to the system, the ED transmits according to the control information specified in the last received DCP. More specifically, we opted to choose the UNCONF mode also for UPs. The rationale for this choice will be given in the next section.

One of the important aspects that has to be taken into consideration while designing any LoRaWAN system is to comply with the duty cycle regulations as discussed in Section 3. This poses some stringent constrains in the process of allocating the resources to the EDs. Owing to the per sub-band duty cycle regulations, we have various possibilities to assign the resources to the EDs for the next UPs. In particular, one of the feasible choice is to differentiate the sub-bands for the two types of packets, i.e., allocating fixed non-overlapping sub-bands for the RPs and UPs. Indeed, this not only allow the isolation in terms of frequencies but also address the issue of duty cycling in the case when it is necessary to transmit an UP when the time elapsed form the last RP is lower than the minimum time established by duty cycling restrictions. In particular, the server assign different sub-bands for RPs and UPs so that duty cycling restrictions are independently established for the two kinds of transmissions. An illustrative example is given in Figure 9, where 5 EDs in close proximity, i.e., belonging to the same cluster, are allocated sub-band g (Channel (Ch0-4) with frequencies 867.1, 867.3, 867.5, 867.7, 867.9 MHz) and g1 (Channel (Ch5-7) with frequencies 868.1, 868.3, 868.5 MHz) for UPs and RPs respectively. To elaborate, each EDs transmit RPs by randomly selecting one of the available frequencies from g1 whereas the UPs are transmitted using different frequencies in sub-band g to avoid collisions. Such frequencies can be selected by each ED according to the last DCP received from the server which is in charge of isolating the UP transmissions of the same cluster. Considering 5 channels in sub-band g, it is worth noting that we can allocate a maximum of 5 different channels to 5 EDs in each cluster. However, we also have the possibility to accommodate more users in the cluster by assigning different SFs as shown in the next section.

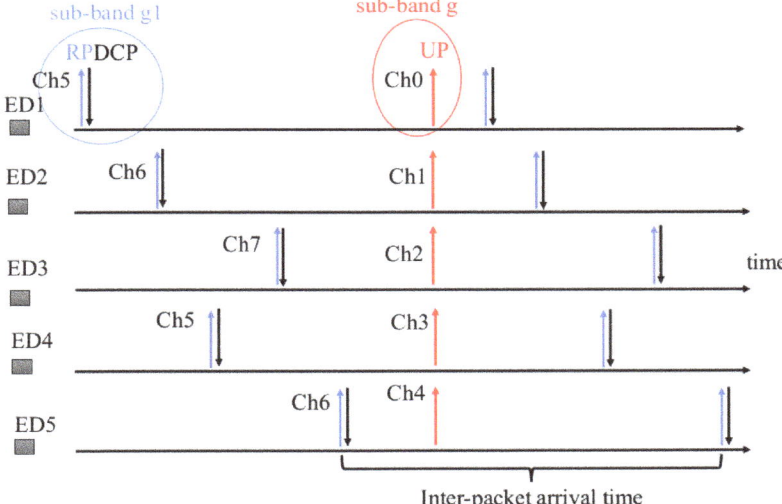

Figure 9. A schematic diagram of transmission of RP, UPs and DCP for 5 EDs belonging to the same cluster where each ED transmits UPs at the same time using different channels from sub-band g.

4.3. The Problem of DL Priority

In standard LoRaWAN, the GWs work in half duplex mode only, i.e., they cannot receive and transmit simultaneously. Moreover, in commercial GWs, if there is the need to send a DL message, the reception of any incoming signal is interrupted, i.e., the concurrent UL packet is lost. Accordingly, the mechanism proposed in this paper for coordinating simultaneous UL UPs, which is based on periodic delivery of DCPs, could dramatically affect the PLR of UPs.

It is then of paramount importance to evaluate the PLR due to GW transmissions. To this aim, it is worth noting that the UL packet is lost by the concurrent DL transmission either when the DL packet is ongoing at the UL packet arrival time, or if it is started during the reception of the UL packet, since the GW gives priority to transmission anyway. Accordingly, denoting by τ and D the durations of a DCP and of an UP, respectively, the PLR of UPs is equal to the probability that at least one DCP is generated in the interval $\Delta = \tau + D$. To elaborate, in the considered setting DCPs are created as a response of RPs transmitted in the UL by each node. Accordingly, the DCPs arrival process statistics is equivalent to that of RPs generation process. Let then denote by T the rescheduling period set by each ED. Owing to inevitable clock drifts, the actual rescheduling time can be modeled as a random variable (rv) $r = T + \delta$, where δ is the clock error.

In the considered scenario we deal with internal clocks which are natively embedded inside the ED microcontroller. This choice allows to save cost, energy, and complexity with respect to external clocks. In this case, it is shown in [30] that the clock errors are unbiased and that they can be reasonably modeled as independent and identically distributed (IID) Gaussian rvs, i.e., $\delta \sim \mathcal{N}(0, \sigma)$. Accordingly, also the interarrival times r are IID rvs, i.e., the arrival process of DCPs belong to the class of renewal processes [31]. More specifically, we have $r \sim \mathcal{N}(T, \sigma)$.

From the theory of renewal processes, it is possible to evaluate the time asymptotic density $w(x)$ for the the time elapsed from a generic time till the next arrival, i.e.,

$$w(x) = \tfrac{1}{T}(1 - F_r(x)) \tag{1}$$

where $F_r(x)$ is the cdf of r. In the following, we are interested in evaluating the probability that an UP does not experience any collision with DCPs. To this aim, it is reasonable to assume that different nodes are characterized by independent clocks and independent time delays, and, hence, the probability P_C that the an UP does not experience any collision can be evaluated as the product of individual probabilities of all independent events. To elaborate, let us denote by N the number of EDs and by $\boldsymbol{\tau} = \{\tau_1, \tau_2, \ldots, \tau_N\}$ the DCP time duration of each node. Such terms depends on the SF used by the correspondent nodes to transmit DCPs, i.e., the higher SF the longer τ. Since the adopted SFs in the UL depend on the channel conditions of each ED, e.g., the distance from the GW, it is reasonable to consider $\boldsymbol{\tau}$ as a set of i.i.d. rvs with individual pdf $f_\tau(t)$. Similarly, also D depends on the SF adopted by the ED to transmit an UP and, hence, it can be characterized by a given pdf $f_D(y)$.

Accordingly, the probability P_C that the an UP does not experience any collision with DCPs for given $\boldsymbol{\tau} = \{\tau_1, \tau_2, \ldots, \tau_N\}$ and D is:

$$P_C(\boldsymbol{\tau}, D) = \prod_{n=1}^{N} \left(1 - \frac{1}{T} \int_0^{\tau_n + D} (1 - F_r(x)) dx\right) \tag{2}$$

and the marginal probability is:

$$P_C = \left[\iint_{t\ y} \left(1 - \frac{1}{T} \int_0^{t+y} (1 - F_r(x)) dx\right) f_\tau(t) f_D(y) dt dy\right]^N \tag{3}$$

with $PLR = 1 - P_C$.

In the interesting case where $PLR \ll 1$, the expression in (3) can be manipulated to get an easy to understand approximation of the PLR. To elaborate, when $T \gg \tau + D$ (i.e., small PLR), we have $1 - F_r(x) \approx 1$ thus yielding:

$$P_C \approx \left(1 - \tfrac{\mathbb{E}(\tau) + \mathbb{E}(D)}{T}\right)^N \tag{4}$$

$$PLR \approx N\frac{\mathbb{E}(\tau)+\mathbb{E}(D)}{T} \quad (5)$$

5. Experimental Results and Discussion

In the following we describe the experimental testbed to assess the possibility of achieving the required service requirements using the proposed LoRaWAN solution based on DL control and clustering.

5.1. Test Scenario

We consider an indoor testbed at the premises of the Department of Information Engineering and Mathematical Science (DIISM) of the University of Siena. The environment is made of several rooms at the same level and includes the presence of machinery and obstacles, movements of objects and people. In this setting, we deployed several EDs and GWs inside the building for different test scenarios. In particular, we have conducted a preliminary set of tests to evaluate the PLR in the presence of a single ED transmitting in UNCONF mode. In this case, we have verified that the PLR is always well below the required limit of 0.1% even in the SF7 case. These results are in line with the LoRa coverage expectations and are not reported here for the sake of brevity. Hence, we focus on the results obtained in three different experimental setup characterized by the presence of possible collisions and of concurrent DL transmissions. In all the reported results, we consider only the PLR of UP packets, since RPs are not critical in the considered scenario. We report the parameter settings for the three different scenarios in Table 1 where N_{ED}, N_{GW}, N_{UP}, Δt_{RP}, and Δt_{UP} denote the number of EDs, GW, and UPs, and the RPs and UPs inter-packet arrival times, respectively. In order to achieve consistent statistics and reduce the experimental time, we have kept the RPs inter-packet arrival times to relatively low values. Moreover, the UPs are generated by forcing the triggering of gas sensors at random intervals. In the following, we report the description of each scenario, rationale for performing the particular test, the results obtained and the corresponding discussion.

Table 1. Parameter settings for experimental tests.

Test	N_{ED}	N_{GW}	N_{UP}	Δt_{RP}	Δt_{UP}
1	2	1	20,000	70 s	Random (120–130 s)
2	8	1	20,000	70 s	Random (120–130 s)
3	8	2	20,000	70 s	Random (120–130 s)

5.2. Test 1: Analysis of PLR in the Presence of Collisions

In the first set of experiments, as reported in Table 1, we deployed two EDs, namely ED1 and ED2 nearly close to each other at distance of approximately 20 m from the GW. Both nodes asynchronously transmit RPs every 70 s using Ch5. In addition, both nodes are triggered to transmit synchronously the UPs at random intervals using Ch0. The SFs adopted for transmitting the UPs are set according to the DCP commands. In this case, we force the users to transmit at the same time using the same channel to assess the possibility of isolating the two transmissions in the SF domain.

We summarize the results in Tables 2–4 where we report the PLRs for different cases. We neglect the PLR due to DL transmissions discussed in Section 4.3 and which will be separately assessed in Test 2. As expected, the results reveal significant packet losses when transmitting with the same SF. On the other hand, in the case of different SFs, the node transmitting with higher SF has a much lower PLR. i.e., it is able to often capture the packet even in the presence of interference owing to quasi inter-SF orthogonality. Nevertheless, in the SF7-8 case there is a residual probability (slightly higher than the constraint of 0.1%) that the packet is lost by both EDs, while this probability goes to zero for the SF8-9 and SF9-10 cases. Since it is highly probable that concurrent UPs will report the same information to the server, e.g., a gas concentration is higher than the threshold, in many cases it could be sufficient to receive only one UP to prevent the accident. Under this hypothesis, the 0.1% constraint can be satisfied when a maximum of 3 nodes in a cluster are assigned the same frequency and different SFs, namely, SFs 8–10. Considering the 5 channels in sub-band g, a cluster can be composed of 15 EDs.

Table 2. Analysis of PLR between SF7 and SF8.

Node	SF	PLR (%)
ED1	7	94.3
ED2	7	35
Both	7	29.66
Node	**SF**	**PLR (%)**
ED1	7	4.7
ED2	8	3.23
Both	7 and 8	0.12

Table 3. Analysis of PLR between SF8 and SF9.

Node	SF	PLR (%)
ED1	8	17.96
ED2	8	84.88
Both	8	3.46
Node	**SF**	**PLR (%)**
ED1	8	32
ED2	9	0

Table 4. Analysis of PLR between SF9 and SF10.

Node	SF	PLR (%)
ED1	9	5.18
ED2	10	0

5.3. Test 2: Analysis of the PLR in the UL Due to DL

We have deployed 8 EDs where ED1-7 transmit asynchronous RPs every T = 70 s (with SF7) whereas ED8 is triggered to transmit also the UP at a random interval of 120–130 s with SF9. According to the notation used in (5) the considered scenario corresponds to N = 8, $\mathbb{E}(\tau) = 72$ ms, $\mathbb{E}(D) = 247$ ms, yielding a predicted PLR of 3.6%.

In the considered setting we force ED1-7 to use a random channel from sub-band g1 whereas ED8 is allowed to transmit using Ch0 from sub-band g. Table 5 reports a PLR of 3.66% which is indeed a very high value and is certainly unacceptable in our case. It is worth noting that this value almost perfectly matches the PLR predicted by (5).

A possible way for overcoming this problem is to duplicate each GW. More specifically, only one of the 2 GWs is configured to transmit the DCPs, so that the other is always free to receive UPs.

As a matter of fact, there are several papers that propose full-duplex or multi-cast GWs to overcome this problem. However, as discussed in Section 2, such feature is not present in off-the-shelf GW solutions.

Table 5. Analysis of PLR in UL due to DL transmission.

Node	SF	PLR (%)
ED8 (UPs)	9	3.66

5.4. Test 3: Analysis of Residual Loss with Two GWs

The final test is performed with considering the same scenario of Test 2 in the presence of an additional GW configured to receive only UP packets. Table 6 reports the total number of lost packets for ED8 for two different SFs. It is worth noting that in this case the residual PLR is by far less than 0.1% even for small SFs (7–8), which confirms that the double GW solution provides the required service levels in the considered scenario. It is also important to stress that the use of double GW is effective if and only if the proposed transmission scheme is adopted: this highlights the viability and importance of such scheme. In particular, the required service levels cannot be achieved just using two GWs.

Table 6. Analysis of residual loss using double GW for ED8.

Test Set	SF	Number of Loss Packets
1	7	10
2	8	8

5.5. Discussions

In Table 7 we have summarized some of the performances of various schemes from the literature related to reliability of LoRaWAN networks. Even though there is a huge gap between these papers and ours, we did our best to compare the performance levels with the ones achieved in this work.

In particular, the achieved reliability values are reported with the corresponding number of nodes used in the simulations. As mentioned in Section 2, the closely paper to ours is [6] which provides significant results in terms of reliability with the accommodation of nearly 2000 users. However, this scheme doesn't provide any specific solution to address the problem of urgent traffic delivery, i.e., the latency problem is not specifically addressed. As a matter of fact, the authors just speculate that the proposed solution can handle urgent traffic if the number of EDs is small without actually simulating this scenario. Moreover, the nodes are considered to be fixed so that it is possible to design a fixed frequency/time scheduling procedure in which all the nodes are perfectly synchronized with the GW, a situation which is not verified in the FMAR time varying scenario considered in our paper. Finally, the solution provided in [6,7] deviates from standard LoRaWAN and requires modification of the MAC layer. To sum up, the results obtained through our scheme not only validate the proposed algorithm but also provide important intuitions in enabling a real FMAR scenario where both reliability and latency requirements are very critical.

Table 7. Performance comparison with other works from the literature.

Paper/Scenario	Proposed Scheme (LoRaWAN Compliance)	Reliability (PDR)	Latency Constraint	Number of Nodes Simulated
[15]/General	Standard LoRaWAN MAC (✓)	0.6 (CONF) 0.8 (UNCONF)	Not discussed	100
[7]/General	New MAC protocol RS-LoRa (✗)	0.84	Not discussed	100
[6]/Air pollution monitoring (mIoT)	TDMA-based FDMA-based (✗)	≈1	Not available	<2000
Our/FMAR	Coordinated transmission through DCP (✓)	>0.999	<500 ms	15/cluster can be accommodated

6. Conclusions

In this work, we have proposed a LoRaWAN compliant reliable and low-latency solution to fulfil the requirements of a FMAR scenario. To this aim, a low-cost and low-power wearable device was developed to detect the leakage of hazardous and flammable gases. The proposed approach allows to reliably transmit urgent data to the central server. This goal is achieved by leveraging the transmission of DL control messages aimed at avoiding collisions among concurrent transmissions. Finally, we validated the proposed approach with extensive experimental tests in an industrial-like scenario. Numerical results suggest that LoRaWAN can be exploited to obtain the required level of reliability in the considered scenario.

Author Contributions: Conceptualization, A.A.; methodology, D.T., A.P., L.P., A.F. and A.A.; software, D.T. and L.P.; validation, D.T., A.P., L.P., A.F. and A.A.; formal analysis, A.A.; investigation, D.T.; resources, D.T.; data curation, A.A. and A.F.; writing—original draft preparation, D.T.; writing—review and editing, D.T., A.P., L.P. and A.A.; visualization, D.T. and A.A.; supervision, A.A.; project administration, A.F. and A.A.; funding acquisition, A.A. All authors have read and agreed to the published version of the manuscript.

Funding: This work has been supported by INAIL, the Italian National Institute for Insurance against Accidents at Work, within the framework of the CP-SEC project: Cyber-Physical system (CPS) for the safety of factories at major accident risk.

Institutional Review Board Statement: Not applicable.

Informed Consent Statement: Not applicable.

Data Availability Statement: Not applicable.

Conflicts of Interest: The authors declare no conflict of interest.

Abbreviations

The following abbreviations are used in this manuscript:

ABP	Activation by Personalization
ACK	Acknowledgement
ADR	Adaptive Data Rate
ADC	Analog to Digital Converter
AES	Advanced Encryption Standard
API	Application Programming Interface
BLE	Bluetooth Low Energy
BW	Bandwidth
CC	Central Controller
Ch	Channel
CO	Carbon Monoxide
CONF	CONFirmed
CR	Coding Rate
CSS	Chirp Spread Spectrum
DAC	Digital to Analog Converter
DCP	Downlink Control Packet
DL	Downlink
ED	End Device
FDMA	Frequency-Division Multiple Access
FMAR	Factories at Major Accident Risk
GW	Gateway
IoT	Internet of Things
ISM	Industrial, Scientific and Medical
ISPRA	Istituto superiore per la protezione e la ricerca ambientale
KPI	Key Performance Indicator
LoRa	Long Range
LoRaWAN	Long Range Wide Area Network
LEL	Lower Explosive Level
MAC	Medium Access Control
mIoT	massive IoT
MQTT	Message Queue Telemetry Transport
OTAA	Over-The-Air-Activation

O_2	Oxygen
PDR	Packet Delivery Rate
PLR	Packet Loss Rate
QoS	Quality of Service
RF	Radio Frequency
RP	Regular Packet
SD	Secure Digital
SF	Spreading Factor
TDMA	Time-Division Multiple Access
TS	Time Slotted
UL	Uplink
UNCONF	UNCONFirmed
UP	Urgent Packet
WiFi	Wireless Fidelity

References

1. International Electrotechnical Commission. *IEC/TR 61508-0. Functional Safety Electrical/Electronic/Programmable Electronic Safety-Related Systems—Part 0: Functional Safety and IEC 61508 (see Functional Safety and IEC 61508)*; International Electrotechnical Commission: Geneva, Switzerland, 2005.
2. ISPRA. Mappatura dei Pericoli Di Incidente Rilevante in Italia. 2013. Available online: https://www.isprambiente.gov.it/files/pubblicazioni/rapporti/rapporto_181_2013.pdf (accessed on 30 January 2022).
3. Chen, B.; Wan, J.; Shu, L.; Li, P.; Mukherjee, M.; Yin., B. Smart Factory of Industry 4.0: Key Technologies, Application Case, and Challenges. *IEEE Access* **2017**, *6*, 2169–3536. [CrossRef]
4. Palattella, M.R.; Dohler, M.; Grieco, A.; Rizzo, G.; Torsner, J.; Engel, T.; Ladid, L. Internet of things in the 5G era: Enablers, architecture, and business models. *IEEE J. Sel. Areas Commun.* **2016**, *34*, 510–527. [CrossRef]
5. Mekki, K.; Bajic, E.; Chaxel, F.; Meyer, F. A comparative study of LPWAN technologies for large-scale IoT deployment. *ICT Express* **2019**, *5*, 1–7. [CrossRef]
6. Haiahem, R.; Minet, P.; Boumerdassi, S.; Azouz Saidane, L. Collision-Free Transmissions in an IoT Monitoring Application Based on LoRaWAN. *Sensors* **2020**, *20*, 4053. [CrossRef] [PubMed]
7. Reynders, B.; Wang, Q.; Tuset-Peiro, P.; Vilajosana, X.; Pollin, S. Improving reliability and scalability of lorawans through lightweight scheduling. *IEEE Internet Things J.* **2018**, *5*, 1830–1842. [CrossRef]
8. Mikhaylov, K.; Petajajarvi, J.; Janhunen, J. On LoRaWAN scalability: Empirical evaluation of susceptibility to inter-network interference. In Proceedings of the 2017 European Conference on Networks and Communications (EuCNC), Oulu, Finland, 12–15 June 2017; pp. 1-6.
9. Croce, D.; Gucciardo, M.; Mangione, S.; Santaromita, G.; Tinnirello, I. Impact of LoRa imperfect orthogonality: Analysis of link-level performance. *IEEE Commun. Lett.* **2018**, *22*, 796–799. [CrossRef]
10. Abdelfadeel, K.Q.; Zorbas, D.; Cionca, V.; O'Flynn, B.; Pesch, D. *FREE*—Fine-Grained Scheduling for Reliable and Energy-Efficient Data Collection in LoRaWAN. *IEEE Internet Things J.* **2019**, *7*, 669–683. [CrossRef]
11. Zorbas, D.; Caillouet, C.; Abdelfadeel Hassan, K.; Pesch, D. Optimal Data Collection Time in LoRa Networks—A Time-Slotted Approach. *Sensors* **2021**, *21*, 1193. [CrossRef] [PubMed]
12. Zorbas, D.; Abdelfadeel, K.; Kotzanikolaou, P.; Pesch, D. TS-LoRa: Time-slotted LoRaWAN for the industrial Internet of Things. *Comput. Commun.* **2020**, *153*, 1–10. [CrossRef]
13. Marais, J.M.; Abu-Mahfouz, A.M.; Hancke, G.P. A Survey on the Viability of Confirmed Traffic in a LoRaWAN. *IEEE Access* **2020**, *8*, 9296–9311. [CrossRef]
14. Capuzzo, M.; Magrin, D.; Zanella, A. Confirmed traffic in LoRaWAN: Pitfalls and countermeasures. In Proceedings of the 2018 17th Annual Mediterranean Ad Hoc Networking Workshop (Med-Hoc-Net), Capri, Italy, 20–22 June 2018; pp. 1–7.
15. Farhad, A.; Kim, D.H.; Pyun, J.Y. Scalability of LoRaWAN in an urban environment: A simulation study. In Proceedings of the 2019 Eleventh International Conference on Ubiquitous and Future Networks (ICUFN), Zagreb, Croatia, 2–5 July 2019; pp. 677–681.
16. Varsier, N.; Schwoerer, J. Capacity limits of LoRaWAN technology for smart metering applications. In Proceedings of the 2017 IEEE international conference on communications (ICC), Paris, France, 21–25 May 2017; pp. 1–6.
17. Markkula, J.; Mikhaylov, K.; Haapola, J. Simulating LoRaWAN: On importance of inter spreading factor interference and collision effect. In Proceedings of the ICC 2019 IEEE International Conference on Communications (ICC), Shanghai, China, 20–24 May 2019; pp. 1–7.
18. Pop, A.I.; Raza, U.; Kulkarni, P.; Sooriyabandara, M. Does bidirectional traffic do more harm than good in LoRaWAN based LPWA networks? In Proceedings of the GLOBECOM 2017 IEEE Global Communications Conference, Singapore, 4–8 December 2017; pp. 1–6.

19. Adelantado, F.; Vilajosana, X.; Tuset-Peiro, P.; Martinez, B.; Melia-Segui, J.; Watteyne, T. Understanding the limits of LoRaWAN. *IEEE Commun. Mag.* **2017**, *55*, 34–40. [CrossRef]
20. Magrin, D.; Capuzzo, M.; Zanella, A. A thorough study of LoRaWAN performance under different parameter settings. *IEEE Internet Things J.* **2019**, *7*, 116–127. [CrossRef]
21. Hasegawa, Y.; Suzuki, K. A multi-user ack-aggregation method for large-scale reliable lorawan service. In Proceedings of the ICC 2019 IEEE International Conference on Communications (ICC), Shanghai, China, 20–24 May 2019; pp. 1–7.
22. Centenaro, M.; Vangelista, L. Time-power multiplexing for LoRa-based IoT networks: An effective way to boost LoRaWAN network capacity. *Int. J. Wirel. Inf. Netw.* **2019**, *26*, 308–318. [CrossRef]
23. Fort, A.; Landi, E.; Mugnaini, M.; Parri, L.; Pozzebon, A.; Vignoli, V. A LoRaWAN Carbon Monoxide Measurement System With Low-Power Sensor Triggering for the Monitoring of Domestic and Industrial Boilers. *IEEE Trans. Instrum. Meas.* **2020**, *70*, 5500609. [CrossRef]
24. RP2-1.0.2 LoRaWAN Regional Parameters. Available online: https://lora-alliance.org/resource_hub/rp2-102-lorawan-regional-parameters/ (accessed on 30 January 2022).
25. ETSI, ETSI EN 300 220-2 V3.2.1 (2018-04) Short Range Devices (SRD) Operating in the Frequency Range 25 MHz to 1000 MHz; Part 2: Harmonised Standard for Access to Radio Spectrum for Non Specific Radio Equipment. Available online: https://www.etsi.org/deliver/etsi_en/300200_300299/30022002/03.02.01_60/en_30022002v030201p.pdf (accessed on 30 January 2022).
26. Hope RF. Hope RF RFM95/96/97/98(W)—Low Power Long Range Transceiver Module. Available online: https://cdn.sparkfun.com/assets/learn_tutorials/8/0/4/RFM95_96_97_98W.pdf (accessed on 30 January 2022).
27. Dragino. LPS8 LoRaWAN Indoor GW. Available online: https://www.dragino.com/downloads/downloads/LoRa_GW/LPS8/Datasheet_LPS8_LoRaWAN%20Pico%20Station.pdf (accessed on 30 January 2022).
28. Chirpstack. ChirpStack open-source LoRaWAN Network Server. Available online: https://www.chirpstack.io/ (accessed on 30 January 2022).
29. LoRaWAN Air Time Calculator. Available online: https://avbentem.github.io/airtime-calculator/ttn/eu868 (accessed on 30 January 2022).
30. Abrardo, A.; Pozzebon, A. A Multi-Hop LoRa Linear Sensor Network for the Monitoring of Underground Environments: The Case of the Medieval Aqueducts in Siena, Italy. *Sensors* **2019**, *19*, 402. [CrossRef] [PubMed]
31. Parzen, E. *Stochastic Processes*; Dover Publications: San Francisco, CA, USA, 1965.

Article

Monitoring Soil and Ambient Parameters in the IoT Precision Agriculture Scenario: An Original Modeling Approach Dedicated to Low-Cost Soil Water Content Sensors

Pisana Placidi [1,*], Renato Morbidelli [2], Diego Fortunati [1], Nicola Papini [1], Francesco Gobbi [1] and Andrea Scorzoni [1]

[1] Dipartimento di Ingegneria, University of Perugia, 06125 Perugia, Italy; diego.fortunati@studenti.unipg.it (D.F.); nicola.papini@studenti.unipg.it (N.P.); francesco.gobbi@studenti.unipg.it (F.G.); andrea.scorzoni@unipg.it (A.S.)
[2] Dipartimento di Ingegneria Civile e Ambientale, University of Perugia, 06125 Perugia, Italy; renato.morbidelli@unipg.it
* Correspondence: pisana.placidi@unipg.it

Abstract: A low power wireless sensor network based on LoRaWAN protocol was designed with a focus on the IoT low-cost Precision Agriculture applications, such as greenhouse sensing and actuation. All subsystems used in this research are designed by using commercial components and free or open-source software libraries. The whole system was implemented to demonstrate the feasibility of a modular system built with cheap off-the-shelf components, including sensors. The experimental outputs were collected and stored in a database managed by a virtual machine running in a cloud service. The collected data can be visualized in real time by the user with a graphical interface. The reliability of the whole system was proven during a continued experiment with two natural soils, Loamy Sand and Silty Loam. Regarding soil parameters, the system performance has been compared with that of a reference sensor from Sentek. Measurements highlighted a good agreement for the temperature within the supposed accuracy of the adopted sensors and a non-constant sensitivity for the low-cost volumetric water contents (VWC) sensor. Finally, for the low-cost VWC sensor we implemented a novel procedure to optimize the parameters of the non-linear fitting equation correlating its analog voltage output with the reference VWC.

Keywords: soil water content; sensor networks; distributed sensing; IoT measurements; Precision Agriculture; moisture sensor; wireless communication; LoRa; LoRaWAN™

1. Introduction

In recent years, the rapid development and broad application of the IoT (Internet of Things) concept pushed towards the improvement of best practices in Wireless Sensor Networks (WSNs) [1] in Precision Agriculture (PA) applications, also relevant to Greenhouses [2,3]. Smart, cheap, and powerful connected sensor nodes (things) are transforming from stand-alone devices to parts of collaborative systems [4,5]. Data are stored, aggregated, and analyzed to improve the precision of temporal-spatial parameters on croplands [6,7]. WSN could be made of simple and cheap components: the results provided by complex technology systems are not necessarily significantly better than the results derived from a combination of descriptive statistics and simple sensors: intrinsic limitations of the sensing element could be overcome [8] also providing the measurement readout in a digital format [9].

Currently, the sensor networks that characterize the IoT technology have the main purpose of collecting data from the surrounding world on intelligent systems for environmental applications [10,11]. Additionally, in cloud computing approaches, the collected data are analyzed, processed, and used to undertake the correct decisions to optimize natural resources: it follows that the set of sensors, devices, and storage systems, by which the IoT is composed, is very similar to a huge, distributed measurement system, as clearly outlined in [12]. The management of such complex systems is part of the present Big Data paradigm. Details on sampling techniques, distributed smart monitoring, and mathematical theories of distributed sensor networks can be found in [1,13].

In [14] the authors made a very good literature review on the use of machine learning (ML), a subset of artificial intelligence having a considerable potential to handle numerous challenges in the establishment of knowledge-based farming systems. In the paper, the authors considered four main generic categories of applications: crop, water, soil, and livestock management. In the paper the authors underlined also that (i) the majority of the journal papers focused on crop management [15,16]; (ii) several ML algorithms have been developed to handle the heterogeneous data coming from agricultural fields [17]; (iii) multispectral or RGB images constituted the most common input for ML algorithms, thus justifying the broad usage of Convolutional Neural Networks due to their ability to handle this type of data more efficiently [18,19]. Moreover, a wide range of parameters regarding the weather as well as the soil, water, and crop quality was used. The most common means of acquiring measurements for ML applications was remote sensing, including imaging from satellites, unmanned vehicles (both ground (UGV) and aerial (UAV)), while in situ and laboratory measurements were also used [20].

Very good reviews of the most common sensors used in agriculture applications are reported in [21,22]. In [23], agricultural sensors have been divided into three main classes: physical property type sensors, biosensors, and micro-electro-mechanical system (MEMS) sensors. Near and remote sensing techniques use IoT sensors for monitoring multiple parameters, such as soil water content, temperature, and pH level, air humidity, temperature, light, and pressure [23–26]. The determination of soil water content is a subject of great value in different scientific fields, such as agronomy, soil physics, geology, soil mechanics, and hydraulics. Physical, mineralogical, chemical, and biological properties are also involved. Moreover, soil water content measurements could be affected by soil temperature [27]. Ambient Relative Humidity (RH) affects leaf growth, photosynthesis, pollination rate, and finally crop yield. A prolonged dry environment or high temperature can make the delicate sepals dry quickly and cause the death of flowers before maturity. Hence it is very crucial to control air humidity and temperature. Recent technological advances have enabled real-time sensors to be used directly in the soil, wirelessly transmitting data without the need for human intervention. It is now possible to set up a large number of low-cost devices not only capable of transducing a physical quantity of interest but also of performing some post-processing on raw data to extract useful information, fully complying with current regulations [27–29]. Due to the rapid advancement of technologies, the size and the cost of sensors have been reduced, making WSN the foremost driver of PA [30].

While most previously cited parameters (including soil temperature) can reliably be monitored through low-cost sensors available in the market, the experimental and accurate determination of soil water content with low-cost sensors is still an issue. A summary of state of the art on soil water content measurement techniques has been reported in [31]. The prices of the most reliable soil water content sensors range between USD 150 and USD 5000, thus positioning these sensors far from the IoT world. Instead, the reliability of very low-cost soil water content sensors easily purchasable in the worldwide internet market is still a matter of scientific debate [8,32–37] as further highlighted in the next sections.

In this scenario, the objectives of the present work can be summarized as follows:

- Acquisition of basic physical parameters of plants and ambient with low-cost sensors: soil water content and temperature, greenhouse ambient RH, temperature, and light. Even if the present paper will mostly be focused on soil water content and most parameters will not be discussed, the availability of multiple parameters could be exploited in the future to build a more intelligent system by using machine learning algorithms.
- Availability of a modular system built with cheap off-the-shelf components also providing capabilities for automation and management of plant irrigation.

- Comparison of the performance of a very low-cost soil moisture sensor with a commercially available expensive system using two different types of soil with an original modeling approach which helps us to compare measurement results taken at different soil depths.

2. IoT Architecture in Precision Agriculture Scenario

2.1. Water Waste and Agriculture

The integration of information and control technologies in agriculture processes is known as Precision Agriculture. To obtain the greatest optimization and profitability PA adapts common farming techniques to the specific conditions of each point of the crop, by applying different technologies: micro-electro-mechanical Systems, Wireless Sensor Networks (WSN), computer systems, and enhanced machinery. PA optimizes production efficiency, increases quality, minimizes environmental impact, and reduces the use of resources (energy, water) [38].

The application of IoT allows farmers to boost the production process through plantation monitoring, soil and water management, irrigation scheduling, fertilizer optimization, pest control through chemicals as herbicides, delivery tracking. These tasks can be accomplished by using data from sensors, images, agricultural information management systems, global positioning systems (GPS), and communication networks. This integration results in the optimization of scarce resources [39].

Atmospheric changes and, in particular, the sudden rise in temperatures worsen the problem of searching for fresh water and water storage resources [40]. These problems are exacerbated in countries characterized by drought and rare rainfall, where the difficulty in finding the raw material prevents the development of crops (e.g., the California drought [41]). The scarcity of water in some regions of the world has led farmers to re-evaluate conventional agricultural methods to reduce waste. To this purpose, innovative systems and methods aimed at PA are needed, where sensor technology, electronic and communication engineering, and farming machinery are blended with cloud storage and computing. If on one hand, there is a tendency to optimize traditional irrigation systems using intelligent drip systems [42–44], on the other hand, systems and sensors [8] are sought to measure the soil water content in real-time [45]. In this way, it is possible to know the exact time and the specific position of soil that requires irrigation. However, regardless of all the advances in the IoT domain, the adoption of PA has been limited to some developed countries. Because of the lack of resources, remote sensing-based techniques to monitor crop health are uncommon in developing countries, thus resulting in a loss of yield. [25]

The development of WSN applications in PA makes it possible to increase efficiency, productivity, and profitability in many agricultural production systems while minimizing unintended impacts on wildlife and the environment. The real-time information obtained from the fields can provide a solid base for farmers to adjust strategies at any time. Instead of making decisions based on some hypothetical average conditions, which may not exist anywhere, a precision farming approach recognizes differences and adjusts management actions accordingly [46].

The combination of WSN, which are cheaper to implement than wired networks [29], with intelligent embedded systems and applying on this combination the technology of ubiquitous systems [40], leads to the development of the design and implementation of low-cost systems for monitoring agricultural environments, suitable for developing countries and difficult access areas.

2.2. IoT Architectures

Wireless Sensor Networks (WSNs) have extensively been adopted in agriculture [47] as well as in livestock farming [48] due to installation flexibility especially when wireless transmission introduces a significant reduction and simplification in wiring and harness [30,49]. In addition, greenhouse technology profits from this technology through automation and informatization. In [46] an intelligent system, controlling and monitoring greenhouse temperature has been described, aiming at reducing consumed energy while maintaining good conditions that improve productivity. A review of the common wireless nodes and sensors capturing environmental parameters related to crops in the agriculture domain is reported in [29]. Agriculture in the Internet era is quickly becoming a data-intensive industry. Farmers need to gather and evaluate a massive amount of information from meteorological and physical sensors to increase production efficiency [50]. In [51] a description of a modular IoT architecture for several applications including but not limited to healthcare, health monitoring, and PA is reported. All the proposed subsystem choices used in that research are cheap off-the-shelf components with open-source software libraries.

2.3. Radio and Wireless Protocols in PA

The goal of optimizing water use for crops leads also to the development of automated irrigation systems. In [52] wireless sensors are linked by ZigBee radio transceivers, implementing a WSN where soil water content and temperature data are transferred. The wireless information unit also features a GPRS module that connects to a web server via the public mobile network. An online graphical application through Internet access devices allows operators to remotely monitor the information data. The feasibility of the implemented automated irrigation system was demonstrated. However, the total cost was high for some applications. The cost of each wireless sensor unit was ∼USD 100, whereas the wireless information unit cost was ∼USD 1800.

In addition to ZigBee, the IoT world is pushing new technologies. The Long-Range (LoRa) technology, originally developed for IoT, is investigated in [10] to demonstrate its use for implementing Distributed Measurement Systems. The LoRa wireless technology is designed for sending small packets at a low data rate (0.3–5.5 kbps) at relatively long distances. The protocol can be used in IoT nodes where energy efficiency is considered the most critical parameter.

The LoRaWAN™ protocol exploits the unlicensed radio spectrum in the Industrial, Scientific, and Medical band. Operating frequencies (433 MHz, 868 MHz, or 915 MHz) depend on the particular geographical region. Formally, LoRaWAN™ is a member of the low power LPWAN family, i.e., WAN wireless communications that are designed to minimize the power consumption while covering large areas but offering a relatively small bit rate. The specification defines the device-to-infrastructure of LoRa physical layer parameters and the LoRaWAN™ protocol. The LoRa physical layer or PHY exploits a Chirp Spread Spectrum (CSS) modulation. Fundamental keywords are low power transmission, low throughput, and optimum coverage. The LoRaWAN™ network architecture is deployed in a star-of-stars topology in which gateways relay messages between end-devices and a central network server. The gateways are connected to the network server via standard IP connections and act as a transparent bridge, simply converting RF packets to IP packets and vice versa [53].

LoRa is having success, as confirmed by the high number of papers adopting it (see e.g., [54–58] and references therein). The LoRa alliance sponsors the integration of LoRaWAN™ into the IoT, and some open implementations of network servers are available helping the constant growth of the LoRaWAN™ ecosystem. A fairly complete analysis of the scalability of networks based on LoRaWAN™ is reported in [58]. Employing analytic and simulation-based approaches, the authors explore the dimensions of the LoRa network configuration. The chosen spreading factor, a parameter directly related to the bitrate of the LoRa message, significantly depends on the number of sensors deployed in the field and on the transmission rate, given in packets/day.

3. System Architecture

The modular design of the proposed approach splits the architecture into different layers (Figure 1): (*i*) wireless nodes (encompassing sensors, actuators, low-power embedded processor, battery), (*ii*) internet gateway/concentrator, and The Things Network (TTN) [59], a worldwide open-access LoRaWAN™ network, (*iii*) uplink and downlink connection, database applications, and user interface placed in a virtual machine in the cloud. Our layer structure is a simplified version for what is reported in [23] where our layer "*i*" corresponds to the perception layer, layer "*ii*" merges the *network* and the middleware layers, while our layer "*iii*" combines the common platform and the application layers. Details on the different blocks of Figure 1 will be given in the following sections.

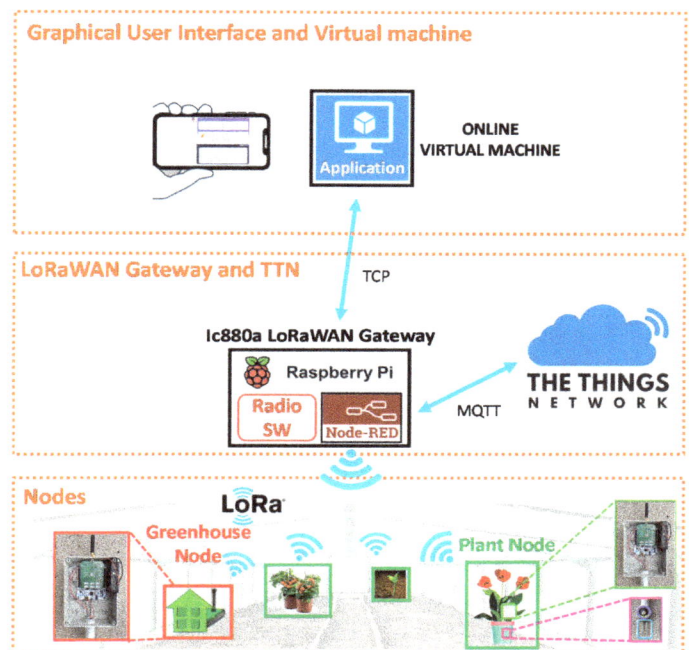

Figure 1. System architecture.

3.1. Nodes

Two basic wireless nodes have been envisaged, each one equipped with Semtech SX1272 LoRa Radio: "Greenhouse Node" and "Plant Node". A single "Greenhouse Node" is needed for a greenhouse. Instead, every plant to be monitored will feature a "Plant Node".

Both node types share the same structure (Figure 2a), i.e., (1) low power, ARM-based STM32L152RE microcontroller hosted in a NUCLEO_L152RE board, (2) chirp spread spectrum SX1272 LoRa Radio, and (3) sensor shield.

The sensor shield of the Greenhouse Node (Figure 2b) provides the connections to (1) a Si7021, a common and widely used RH and temperature sensor, (2) a Photoresistor for light detection, and (3) a 4.8 V battery.

The Plant Node is dedicated to measuring the fundamental parameters of the soil, i.e., soil water content and soil temperature, and to soil watering. Its dedicated shield (Figure 2c) hosts a BD6212HFP H-bridge used for driving a bistable solenoid valve and a power feed interconnection to turn the sensors on and off. Moreover, it is connected to (1) a TMP36 or an LM35 temperature sensor, (2) a "Capacitive Soil Moisture Sensor v1.2" for measuring water content, (3) a bistable solenoid valve, and (4) a 4.8 V battery. Finally, it includes a 1 MΩ shunt resistor useful to correct a fabrication defect of the batch of sensor we received.

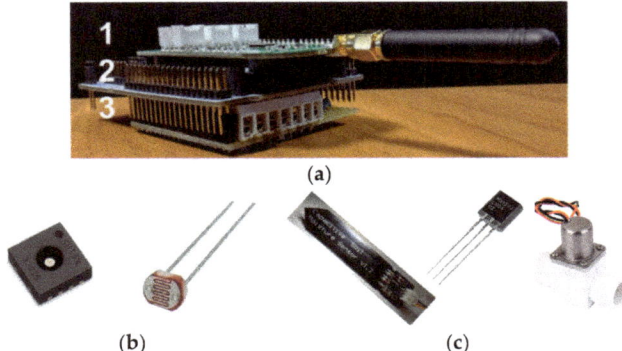

Figure 2. (a) Plant and Greenhouse node electronics and their actuator and sensors. (1) Semtech SX1272 LoRa Radio, (2) L152RE low power microcontroller, (3) Sensor/Actuator Interface Shield; (b) from left to right for the Greenhouse Node: Si7021 and Photoresistor. (c) from left to right for the Plant Node: "Capacitive Soil Moisture Sensor v1.2", TMP36 or LM35 temperature sensor, and bistable solenoid valve.

It is worth giving some details here on the design choices of this Plant Node. The use of an H-bridge and a bistable valve greatly helps in minimizing power consumption, as the valve will drain current only during switching. Power supply for temperature and water content sensors was delivered through the GPIOs of the microcontroller to feed the sensors only when a measurement must be accomplished. In fact, the maximum allowed current delivered by the GPIOs of the STM32L152RE microcontroller is more than enough for the low power sensors we adopted. However, it must be pointed out the present version of the system is designed to provide maximum flexibility and is not conceived for power optimization. For example, the STM NUCLEO boards still include the ST-LINK/V2-1 programming and debugging tool, whose power consumption is way larger than that of the main STM32L152RE microcontroller.

Several plant nodes have been manufactured (Figure 3). After some design steps, the present version of the Plant Node is composed of (*i*) a waterproof junction box (including all electrical and electronic components) connected with (*ii*) a 3D printed PET-G shell which protects the soil sensor and the temperature sensor.

Figure 3. The final version of the Plant Node.

Regarding the soil water content sensor, most papers presented in the literature either measure the capacitance of the soil, which, of course, depends on the water content, or adopt high frequency (around 100 MHz) AC measurements to characterize the dielectric constant of the soil [58]. Other papers describing low-cost IoT nodes adopt fork-like metallic sensors that, when used with a DC bias, mainly characterize the ionic content of the soil by measuring the electrical resistance between the two arms of the fork. AC could also be used with fork-like sensors to limit electrolysis and consequent metal electrode etching/degradation and DC ion currents in the soil. Examples can be found in GardenBot literature [60] or [27] and references therein.

For this prototype of the system, we use the commercial, blade-shaped, "Capacitive Soil Moisture Sensor v1.2" (also dubbed SKU: SEN0193 in its version 1.0 by DFROBOT [61]) for sensing water content in the soil. This sensor is undoubtedly the least expensive water content sensor in the market and was also exploited in other scientific papers (see for example [32,33]) and well-documented internet projects [62]. In [32] the authors found that this sensor did not perform acceptably in predicting soil moisture content in a laboratory soil mixture prepared by mixing organic-rich soil and vermiculite, while it can estimate soil water in gardening soil in the so-called "field capacity" range. In [33] the author linearly correlates the voltage provided by the sensor reading to the gravimetric moisture approximations, providing an effective relationship between the reading from the capacitive sensor and the water content in the soil. This calibration procedure demonstrated that low-cost capacitive-type soil moisture sensors are capable of predicting the water content in soils to a high degree of accuracy, with little required outside of the device itself, which is in direct contrast to the time it takes to traditionally measure the water content in soils.

Being that the water content sensor is the hearth of our plant nodes and since a detailed data sheet is not available for the sensors, an accurate study of the sensor electronics was initially accomplished to get acquainted with the operation of the sensors [8]. A low dropout 3.3 V voltage regulator (omitted in a very recent version–v1.2 and v2.0–of this sensor) feeds a TL555I CMOS timer (Figure 4) which generates a trapezoidal waveform in astable mode running at about 1.5 MHz. The trapezoidal shape is because the operating frequency of the timer is pushed beyond the physical limit for the TL555I device, specified in the datasheet as guaranteed for 1.2 MHz in astable mode. On the other hand, the non-steep rising and falling edges of the waveform help in minimizing the electromagnetic interference possibly generated by the sensor and would be beneficial in the case of "CE" or "FCC" compliance certification.

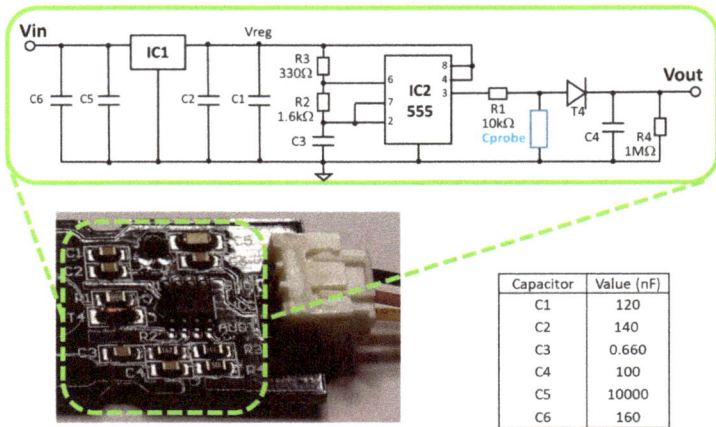

Figure 4. A detailed view of the printed circuit board component section with the electrical schematic of the "Capacitive Soil Moisture Sensor v1.2". The reported resistance values are taken from the component labels while the capacitance values were measured using an HP4275A LCR meter. Cprobe is the variable capacitance of the coplanar capacitor printed on the circuit board. Due to a missing grounding line of the printed circuit board [8], in our measurements, a 1 MΩ shunt resistor has been directly connected to the Sensor/Actuator Interface Shield.

The trapezoidal waveforms of nine sensors (S1, S2, S5, S6, S7, S9, S10, S13, and S14) were initially characterized to assess their uniformity. We discarded sensor S1 since its measured frequency f and duty cycle DC (1.22 MHz and 37.12% respectively) were very far from the average operating frequency and duty cycle of the other eight sensors (1.53 MHz and 34.48%, with sample standard deviations of 1% and 2.2%, respectively) [8], as reported in Table 1.

Table 1. Selected sensors characteristics.

Sample ID	DC/%	f/MHz
S1	37.12	1.221
S2	35.58	1.533
S5	34.36	1.533
S6	32.93	1.524
S7	34.78	1.552
S9	34.36	1.533
S10	35.00	1.510
S13	35.12	1.535
S14	34.02	1.527

After this initial screening of the available samples of Capacitive Soil Moisture Sensor v1.2, we are confident that the measurement results of a single sensor chosen among other homogeneous samples represent the expected behavior of the whole family "S2, S5, S6, S7, S9, S10, S13, and S14".

The TL555I timer supplies a passive circuit shown in Figure 4, composed of a first stage where the coplanar capacitor of the sensor Cprobe is low-pass connected with a 10 kΩ resistor. Then a peak detector provides the analog output signal that we acquire through the ADC of the microcontroller. Regarding sensor settling time, in [33] it was asserted this sensor should settle in 1–5 min, depending on the saturation level of the soil and how well the wet soil was mixed. We accomplished measurements with the Capacitive Soil Moisture Sensor v1.2 immersed in tap water and found that the output voltage could take up to one hour to reach the regime value. This could be due to a non-complete waterproofing of the sensor materials that likely incorporate water molecules. Therefore, the behavior of the Capacitive Soil Moisture Sensor v1.2 after initial watering could not be completely reproducible.

We underline that other more documented and reliable but also more expensive blade-shaped moisture sensors have been commercialized. Examples are the dielectric capacitance sensors ECH2O probe (Decagon Devices, Inc. Pullman, WA USA, now discontinued [63,64]) and the PROBE sensor [65], then modified to SMT100 ring-oscillator sensor (Truebner GmbH, Neustadt, Germany [66]) operating at approximately 150 MHz in water and 340 MHz in air.

A worst-case estimation of the overall cost of our plant node is roughly USD 60, where the most impacting figures are the microcontroller board and the LoRa shield.

3.1.1. Soil Volumetric Water Content Fitting Equations

Water content measurements were previously accomplished in silica sandy soil with the Capacitive Soil Moisture Sensor v1.2 in conditions such that the dry unit weight $\gamma_{dry} = W_s/V$ (W_s = dry soil weight, V = total volume of the soil) could be assumed as a constant [8]. It was demonstrated that this condition guarantees a monotonically decreasing V_s output voltage as a function of gravimetric water content (GWC), which was approximated using a 2nd order polynomial or an exponential function. In this paper, we will deal with volumetric water content (VWC) instead of GWC. However, the two parameters are proportional to each other for a given soil where the dry unit weight is constant. In the remainder of this paper, we will use the following exponential fitting equation between the output voltage V_s of the Capacitive Soil Moisture Sensor v1.2 and the VWC:

$$V_s = A \exp\left(-\frac{VWC}{B}\right) + C, \qquad (1)$$

$$\text{VWC} = B \, \ln\left(\frac{A}{V_s - C}\right) \tag{2}$$

being A, B, and C suitable constants. Other fitting equations were also adopted in the literature for the same sensor. In [32] a 3rd order polynomial function $\text{VWC} = f(V_s)$ was implemented. In [33] the following equation was used:

$$\text{VWC} = \frac{P}{V_s} - Q \tag{3}$$

being $P = 2.48$ V and $Q = 0.72$ for a soil composed of dried coconut coir. In the remainder of this paper, we will mainly deal with fitting Equations (2) and (3).

3.1.2. Embedded Software Implementation of Nodes

The C++ code exploits ARM Mbed OS libraries. Mbed OS is an open-source Real-Time Operating System (RTOS) for the creation and deployment of IoT devices based on ARM processors. The code structure is outlined in Figure 5 for the case of the Plant Node.

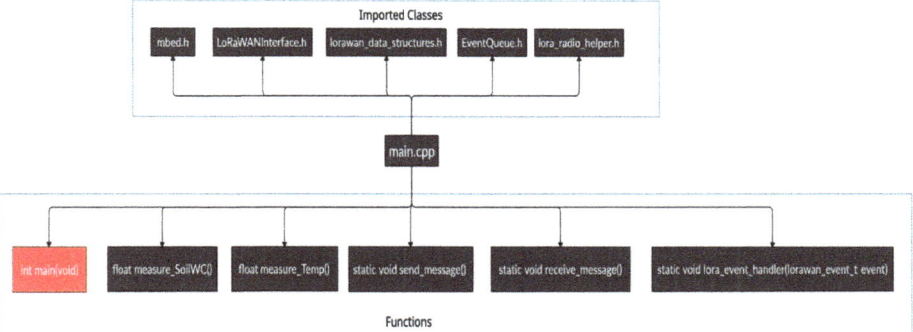

Figure 5. Embedded software implementation for the Plant Node.

The heart of the firmware is the *main.cpp* file. The header of main.cpp includes Mbed libraries (e.g., **EventQueue.h**) and LoRaWAN™ libraries. In particular, **LoRaWANInterface.h** encompasses the prototypes of the member functions managing the upper level of the LoRaWAN™ protocol stack, **lorawan_data_structures.h** includes LoRaWAN parameters, e.g., network and application key, datarate, duty cycle, antenna gain, buffer size, and SNR while **lora_radio_helper.h** regards the physical layer and selects the type of shield adopted in our system. The functions of **main.cpp** dedicated to the Plant Node are listed in the lower part of Figure 5. Among them, we cite the **lora_event_handler()** which manages the state machine of the LoRa events, the measuring functions for water content (**measure_SoilWC()**) and temperature (**measure_Temp()**), and the LoRa send and **send_message()** and **receive_message()** functions. The receive function also handles the bistable irrigation solenoid valve. In the case of the Greenhouse Node, the actual measuring functions regard ambient RH, temperature, and the ambient luminous flux.

Every node transmits a packet conforming to the structure defined in Figure 6. Depending on the node type (plant or greenhouse node), it will include different values. For example, the Plant Node features node type = 1 and transmits soil water content and temperature, while the greenhouse node is characterized by node type = 0 and transmits ambient RH, temperature, and light intensity.

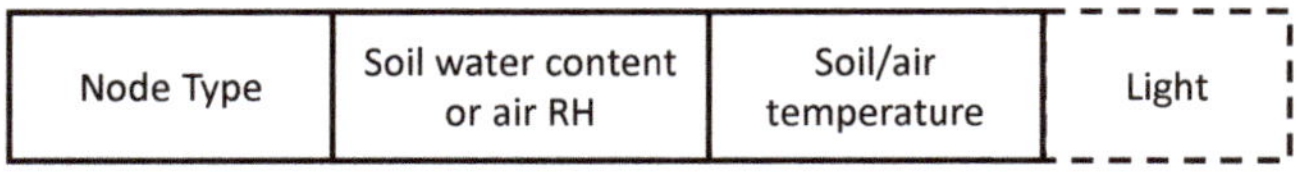

Node Type	Soil water content or air RH	Soil/air temperature	Light

Figure 6. Packet structure.

3.2. The Things Network and Connection to the LoRaWAN™ Gateway

Routing and processing procedures of the LoRaWAN™ network are managed by The Things Network (TTN), acting as an active crossroad between the gateway and the application. For our application, we extensively use the TTN, a network server whose aim is building a global, worldwide open LoRaWAN™ network. They provide a set of open tools and a global, open network to build an IoT application at low cost, featuring maximum security and ready to scale. A secure and collaborative Internet of Things network is built through robust end-to-end encryption, spanning many countries around the globe. A network server does the complicated part in creating a LoRaWAN™ network (handling duplicate packets from multiple gateways, shunting data to servers, handling joins, etc.).

As shown in Figure 1, in the network architecture The Things Network is located between the LoRa concentrator/gateway and the applications. TTN is composed of three main structures: Router, Brokers, and Handler. The Router is in charge of managing the gateway's status and of planning transmissions. Each Router is associated with one or more Brokers. The assignment of Brokers is to map a device to an application, to forward uplink messages to the proper application, and to forward downlink messages to the correct Router-Gateway path. A Handler is responsible for treating the data of different Applications. To do so, it deals with a Broker where it registers devices and applications. The Handler is also in charge of encrypting and decrypting data.

In our system, the Uplink connection to TTN is carried out by the Radio SW of the gateway (Figure 1) that publishes the node sensor data on a specific uplink topic of the TTN MQTT broker using an internet connection.

Then, through the well-known flow-based programming tool Node-RED [67] running on the Gateway, a specific device is allowed to communicate with the database installed in the virtual machine, as sketched in Figure 1.

The Node-RED flow is composed of two sub-flows, an uplink, and a downlink flow, respectively (Figure 7a).

The uplink sub-flow, after subscribing to the same uplink topic of the TTN broker, is in charge of:

- Retrieving through the internet the data received and published by the TTN broker exploiting the light blue TTN Uplink Node producing an output Node.js buffer;
- Converting this Node.js buffer to a string;
- Parsing this string by exploiting two function nodes featuring JavaScript codes, dedicated to Water Content and Temperature, respectively, which also compose the query for the database;
- Sending the query to the MySQL database running on the Virtual Machine through a dedicated TCP port (internet connection through MySQL 3306 port) employing the orange node.

Moreover, in the second Node-RED sub-flow the application is allowed to transmit downlinks to TTN (i.e., to the device) when the bistable solenoid valve must be actuated. This is accomplished by publishing on a specific downlink topic of the TTN broker using the internet again. In the NodeRED flow, the first "TCP in" node is ready to receive messages on a given unassigned TCP port, then a "Reply" JavaScript function returns an object which contains the ID of the target node and the payload, i.e., the message sent by the server. The last light blue node is a TTN Downlink Node which publishes these data on the TTN broker.

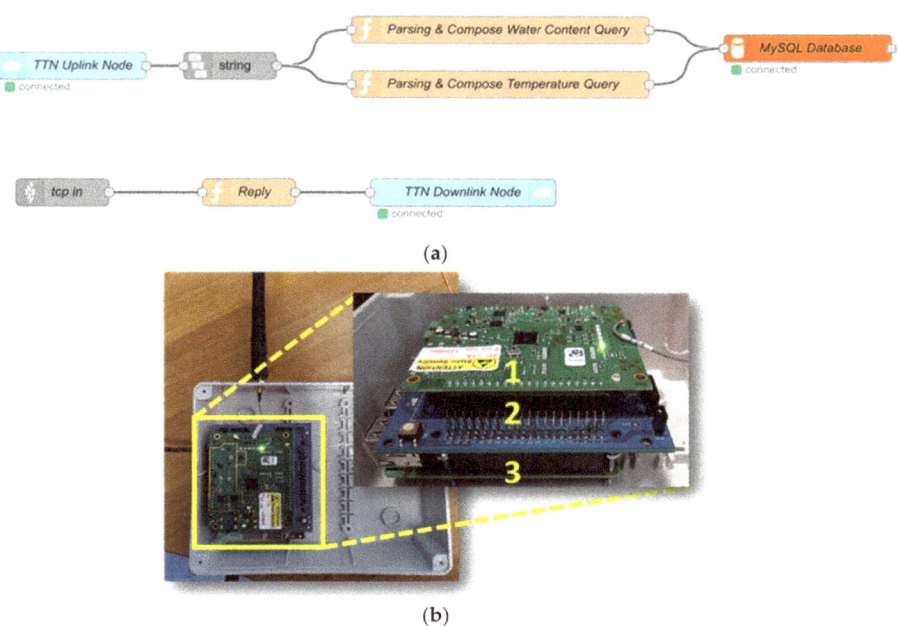

Figure 7. (**a**) The Node-RED flow running in the Raspberry of the Gateway. (**b**) The 8-channel iC880a /Concentrator, with interconnection backplane and Raspberry PI3, is mounted in a plastic box.

Our LoRaWAN™ Gateway is composed of a Raspberry Pi, an iC880a concentrator able to receive packets of different end devices simultaneously sent with different spreading factors on up to 8 different channels in parallel, and an interconnecting backplane (Figure 7b). The embedded software of the Gateway is proprietary and supplied by TTN. The gateway receives LoRa packets from nodes and forwards them to The Things Network [59] through the MQTT protocol thanks to a wideband network, typically WiFi or Ethernet build. On the other hand, it is well known that for data transmission, MQTT could rely on the TCP protocol but a variant, MQTT-SN, is used over other transports such as UDP (or even Bluetooth). However, TTN does not specify which transport protocol is exploited in its Raspberry Pi firmware.

3.3. The Virtual Machine in the Cloud, Database Application, and Graphical User Interface

At the application level, we installed a Linux virtual machine (Figure 8) that includes:

- Web site (HTML, PHP, CSS, and JavaScript) within a web server;
- MySQL Database Management System (DBMS) server.

The database is divided into two units: (i) node data section and (ii) web application user data section (e.g., username and password). The node data section is further composed of two tables: the first one identifies the node and the second one the sensor with its data. Finally, the webserver fetches data in the database using PHP and shows them on a web page. CanvasJS is used for the Graphical User Interface (GUI). CanvasJS is described as a JavaScript Charting Library for High Performance and ease of use. It is built using the Canvas element and it can render thousands of data points in a matter of milliseconds. CanvasJS is also interactive and can be updated dynamically. Examples of the GUI, operating both from a PC and a smartphone, can be found in the following sections.

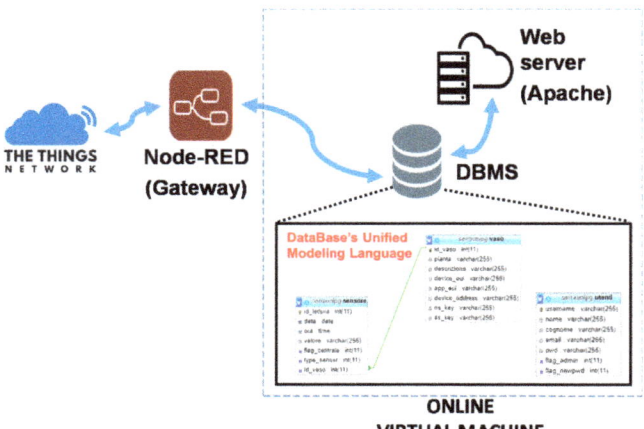

Figure 8. Block diagram of the virtual machine.

The information contained in the database of our virtual machine could represent a starting point for decision-making processes supporting smart monitoring in the frame of PA. A possible implementation could be to develop and enhance the PHP code, used until now to retrieve information from the database, adding a new section where data are analyzed by a dedicated algorithm. The decisions made by the algorithm could be directly sent to the nodes through TTN and the gateways. As an alternative, the watering decision could be directly issued by an application running on the mobile device of the greenhouse manager.

A plant node was placed in a pot hosting a daisy plant, while a greenhouse node was acquiring data in the ambient. Figure 9 includes three plots of our JavaScript GUI showing the soil water content recorded by the Plant Node together with the ambient relative humidity and temperature recorded by the greenhouse node in the same room where the plant pot was located.

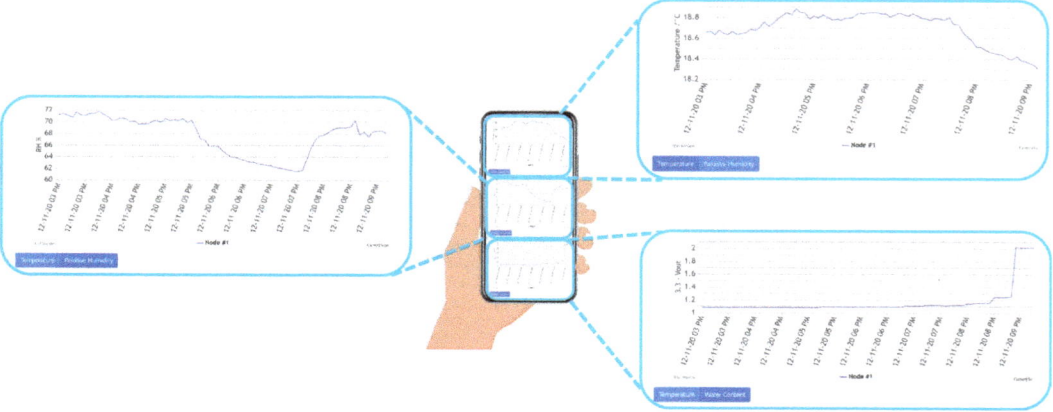

Figure 9. Simultaneous acquisition of ambient temperature and ambient relative humidity (RH) of the greenhouse node and soil water content of a single plant node, shown on the display of a portable device.

4. Materials

The functionality and reliability of the whole system were proven during two continued experiments with two different natural soils, characterized by very different soil hydraulic properties (see Table 2).

Table 2. Main hydraulic properties of study soils. K_s = saturated hydraulic conductivity; θ_s and θ_r = saturated and residual water content, respectively; b_d = bulk density.

	Fine-Textured Soil (Silty Loam)	Coarse-Textured Soil (Loamy Sand)
K_s (mmh^{-1})	10.0	30.0
θ_s	0.420	0.295
θ_r	0.057	0.035
b_d (gcm^{-3})	2.628	2.669

The fine-textured soil (a Silty Loam, according to the United States Department of Agriculture, USDA, classification [68,69]) was composed of 1% gravel, 22% sand, 54% silt, and 23% clay (Figure 10a), while the coarse-textured soil (a Loamy Sand, according to USDA) was composed of 4% gravel, 79% sand, 11% silt, and 6% clay (Figure 10b).

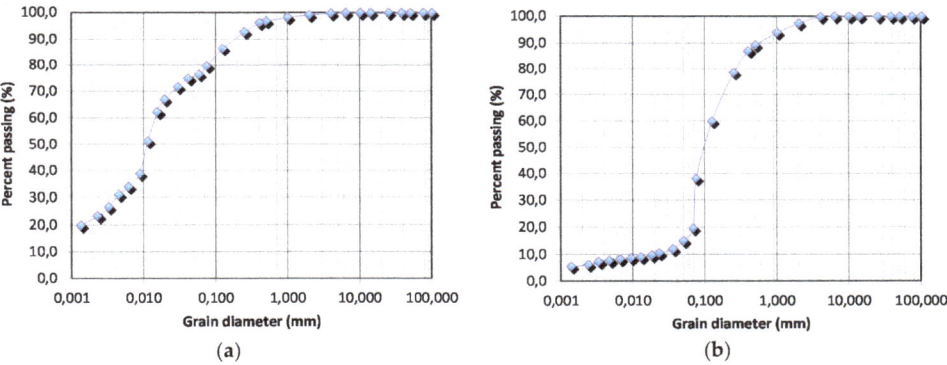

Figure 10. Grain size distribution of the soils used for the experiments: (**a**) fine-textured soil (Silty Loam, according to USDA); (**b**) coarse-textured soil (Loamy Sand, according to USDA).

5. Methods, Tests, and Results

Focusing on the Plant Nodes, the system has been tested during a continued experiment where the two different greenhouse soils were watered several times, to verify if the sensor was able to reliably acquire, transmit, and store the ambient temperature and the soil water content parameters in real time and to show them on the custom GUI.

The measurements were made in a plastic box initially filled with expanded clay aggregate which allowed percolated water to outflow and where a Sentek Drill & Drop Probe (hereafter named "reference sensor") was driven (Figure 11a). Then the remaining top 30 cm of the box was filled with the chosen soil, either Loamy Sand or Silty Loam. Both soils were packed in 0.05 m lifts and gently tapped into place. This accurate packing mechanism was adopted to achieve homogeneity vertically, to keep perfect contact at the interface, and to minimize preferential flow along the sides of the box. The reference sensor was placed at the center of the box. It features an array of water content and temperature sensors placed at 5 cm, 15 cm, 25 cm, 35 cm, and 45 cm from the top surface.

(a) (b)

Figure 11. The plastic box used for our experiments. (**a**) Expanded clay aggregate bottom filler, together with a Sentek Drill & Drop Probe. (**b**) Plant Node #1 and Sentek sensor during acquisition.

Then our Plant Node #1 was inserted in the soil at a distance of 10 cm from the Sentek sensor (Figure 11b, where Node #1 is shown without the lid and connected to a 230 ACV-5 DCV adapter during a test measurement). Since the reference sensor and the Plant Node #1 are installed in different positions/depths, this has an impact on the measurements, as explained in the following sections. Data of Plant Node #1 were collected every 5 min for several days while the automatic acquisition system of the reference sensor stored the measurement results every minute. In carrying out the measurements, the soil was watered in consecutive steps.

Before and after each measurement a calibration was performed on the Capacitive Soil Moisture Sensor v1.2 measuring water content, exposing it for 15 min. to air, then dipping it for 15 min in tap water. The reproducibility of these measurements certifies that the low-cost water content is in working order.

5.1. Measurements in Silty Loam

In Figure 12 we compare the water content measured by the reference sensor at a depth of 5 and 15 cm in Silty Loam. After installing the sensors in a uniformly and slightly moistured Silty Loam (initial volumetric water content of 10%), then four synchronous waterings, clearly visible at a depth of 5 cm, were performed during the last two days of this measurement. The two plots witness the strong dependence of the water content on the soil depth in Silty Loam.

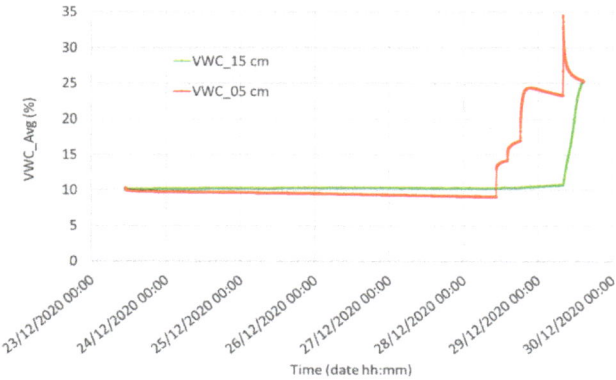

Figure 12. Water content measured by the reference system at two different depths in Silty Loam: red, 5 cm underground, and green, 15 cm underground.

Figure 13 shows the water content measured by the reference system (5 cm underground) and the output voltage of the Capacitive Soil Moisture Sensor v1.2 in Silty Loam (Node #1). The qualitative correlation between the two plots is evident: each watering causes an increase in the measured water content of the reference sensor and a decrease in the output voltage of Node #1. Moreover, the long time elapsed in soil with VWC of about 10% before the first watering guarantees the Capacitive Soil Moisture Sensor v1.2 had plenty of time to reach its settling time. However, we note the lack of linearity of the Capacitive Soil Moisture Sensor v1.2: its sensitivity is too high for small values of water content and it is substantially reduced for volumetric water contents greater than about 15%. The low draining capability of this soil which maintains its water content during the time causes high values of water content and for this reason, the Capacitive Soil Moisture Sensor v1.2 works most of the time almost in saturation. Future improvements for the sensor should be directed towards the linearization of the input/output curve to obtain a constant sensitivity. A detailed discussion of the correlation between the results of the two sensors is reported below.

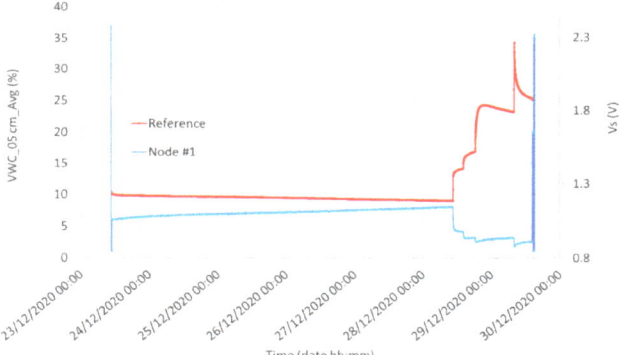

Figure 13. Water content measured by the reference system (red, 5 cm underground) and the Capacitive Soil Moisture Sensor v1.2 (blue) in Silty Loam. Initial and final peaks of the Node #1 plot represent a calibration of the Capacitive Soil Moisture Sensor v1.2 obtained by placing the sensor for 15 min in the air (maximum peak) and 15 min in tap water (minimum peak).

Figure 14 shows the reference temperature compared to the temperature of Node #1. In addition, in this case a slight difference is detected, most likely due to the distance of the two sensors and the intrinsic measurement error. Indeed, the LM35 declares a 0.5 °C ensured accuracy (at 25 °C) while the reference Sentek system has a temperature error of 0.1 °C.

5.2. Measurements in Loamy Sand

In Figure 15 we compare the water content measured by the reference sensor at a depth of 5 and 15 cm in Loamy Sand. The water content curve at 5 cm clearly shows five consecutive waterings performed during the 2 days of this measurement. On the other hand, the water content curve at 15 cm shows an increase only after the 3rd watering, clearly witnessing the dependence of water content on the soil depth. Furthermore, due to the alternation of rainfall and water redistribution periods, the evolution in time of the wetting front is very complex, as [70,71] showed in their schemes with compound profiles.

Figure 16 shows the water content measured by the reference system (5 cm underground) and the output voltage of the Capacitive Soil Moisture Sensor v1.2 in Loamy Sand. In addition, for this soil, we obtain a "first sight" reasonable qualitative agreement between the results of the two sensors. Again, we note that the sensitivity of the Capacitive Soil Moisture Sensor v1.2 is too high for small values of water content and it is substantially reduced for volumetric water contents greater than about 10% for this soil material. A detailed discussion of the correlation between the results of the two sensors is reported below.

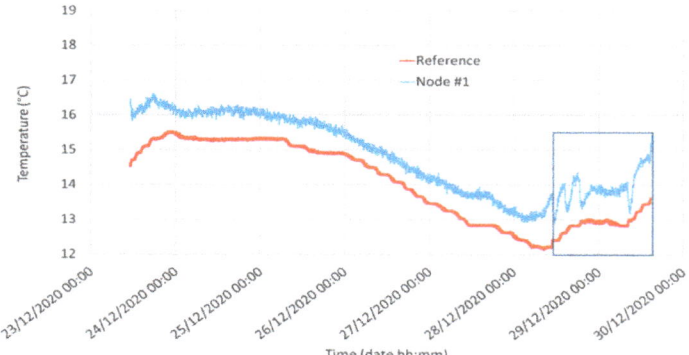

Figure 14. Temperature measured by the reference system (red, 5 cm underground) and the LM35 mounted in Node #1 (blue) in Silty Loam. The rectangular area shows the region where we realized the calibrations of Section 6.2.

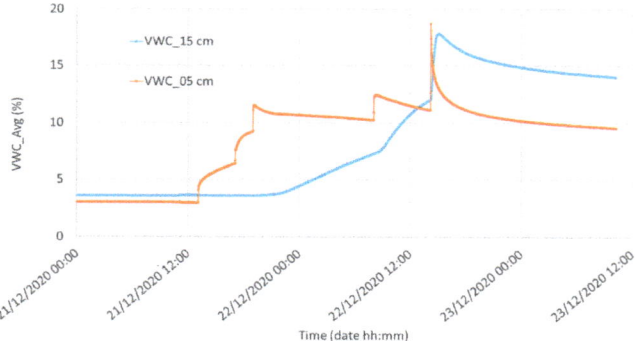

Figure 15. Water content measured by the reference system at two different depths in Loamy Sand: orange, 5 cm underground, and blue, 15 cm underground.

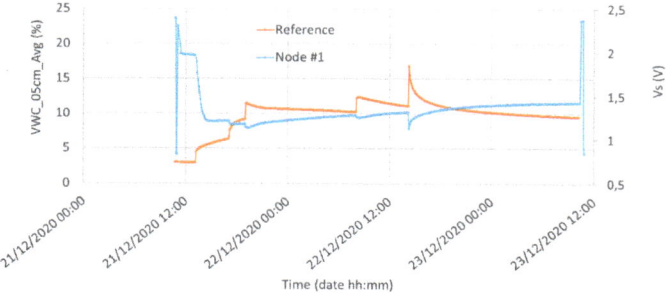

Figure 16. Water content measured by the reference system (orange, 5 cm underground) and the Capacitive Soil Moisture Sensor v1.2 (blue) in Loamy Sand. Initial and final peaks of the Node #1 plot represent a calibration of the Capacitive Soil Moisture Sensor v1.2 obtained by placing the sensor for 15 min in the air (maximum peak) and 15 min in water (minimum peak).

Figure 17 shows the reference temperature compared to the temperature of Node #1. A slight difference is detected, most likely due to the distance of the two sensors and the intrinsic measurement error.

Figure 17. Temperature measured by the reference system (orange, 5 cm underground) and the LM35 mounted in Node #1 (blue) in Loamy Sand.

6. Discussion

Measurement outcomes have been discussed in the relevant sections. In this section, we add some comments and considerations that will help to clarify the experimental observations and will allow us to extract the VWC from the output voltage of the Capacitive Soil Moisture Sensor v1.2.

In principle, the results of the reference sensor and the Capacitive Soil Moisture Sensor v1.2 could not exactly be correlated since:

- Node #1 is 10 cm far from the reference sensor and soil compaction and watering could not be perfectly uniform in that area;
- Measurement results from Node #1 could be influenced by temperature variations;
- The Capacitive Soil Moisture Sensor v1.2 measures an average water content of approximately the first 5 cm of the soil where it is inserted, while the reference sensor is placed at 5 cm from the soil surface with a wider thickness of influence (spanning a depth between 0 and 10 cm).

Regarding the first observation, sample preparation described in Section 4 included an accurate packing mechanism to achieve vertical homogeneity. However, this sample preparation does not guarantee uniform compaction of the soil. In [8] we demonstrated that compaction has an obvious strong influence on the results of the Capacitive Soil Moisture Sensor v1.2. The 10 cm distance between the reference sensor and the Capacitive Soil Moisture Sensor v1.2 could affect water content measurement accuracy and cause a discrepancy between the reference and the low-cost sensor. Non-uniform watering is a second source of non-uniformity, even if watering was manually performed trying to evenly distribute water. Therefore, non-uniform soil compaction and watering represent a random added error to our measurements, which should be kept at a minimum using experience and best practices.

Regarding the temperature variations during the measurements, temperature compensation of VWC could be feasible. This task has been demonstrated to be necessary in the case of a temperature spanning about 20 °C [27]. In that case, backpropagation neural networks have been successfully adopted for correcting the soil moisture information from a low-cost sensor using soil temperature data. However, in our experiment, the temperature variations are significantly smaller, about 2 °C, and a correction was not implemented.

Differently from the previous sources of error, the possible error due to different depths of the reference and the low-cost sensor could be taken into account by properly modeling infiltration and redistribution of water during and after rainfall [69–71] as explained in the next subsections. In detail, in Section 6.1 we introduce a consolidated infiltration model available in the literature to obtain the soil water content at any depth, whereas in Section 6.2 we correlate the described Capacitive Soil Moisture Sensor v1.2 output voltage with the prediction of the infiltration model for the two different soils.

6.1. The Modeling Infiltration and Redistribution of Water

To obtain the soil water content at any depth, z, the Corradini et al. [72] infiltration model was used (hereafter named "C et al. (97)"). As shown by Melone et al. [70,71], this model can accurately represent the infiltration process during complex rainfall patterns involving rainfall hiatus periods.

The model was derived considering a constant value of the initial soil water content, θ_i, and combining the depth-integrated forms of the Darcy law and continuity equation [72]. In addition, as the event progresses in time, t, a dynamic wetting profile, of the lowest depth Z and represented by a distorted rectangle through a shape factor $\beta(\theta_0) \leq 1$, was assumed. The resulting ordinary differential equation is

$$\frac{d\theta_0}{dt} = \frac{(\theta_0 - \theta_i)\beta(\theta_0)}{F'\left[(\theta_0 - \theta_i)\frac{d\beta(\theta_0)}{d\theta_0} + \beta(\theta_0)\right]} \left[q_0 - K_0 - \frac{(\theta_0 - \theta_i)G(\theta_i,\theta_0)\beta(\theta_0)pK_0}{F'}\right] \quad (4)$$

where p is a parameter linked with the profile shape of the soil water content, θ, θ_0 is the soil water content at the surface, K_0 is the hydraulic conductivity at the soil surface, F' is the cumulative dynamic infiltration amount, and $G(\theta_i,\theta_0)$ is expressed by the following equation:

$$G = \frac{1}{K_s} \int_{\theta_i}^{\theta_s} D(\theta) d\theta \quad (5)$$

where θ_0 and K_0 were replaced by θ_s and K_s, with K_s the saturated hydraulic conductivity. $D(\theta)$ is the soil water diffusivity, defined by $D(\theta) = K(\theta)\frac{\partial \psi}{\partial \theta}$, where K is the hydraulic conductivity and ψ the soil water matric potential. Equation (4) can be applied until a second rainfall pulse happens, with the profile shape of $\theta(z)$ approximated [72] by

$$\frac{\theta(z) - \theta_i}{\theta_0 - \theta_i} = 1 - exp\left[\frac{\beta z(\theta_0 - \theta_i) - F'}{(\beta - \beta^2) - F'}\right] \quad (6)$$

Functional forms for β and p were obtained by calibration using results provided by the Richards equation applied to a generic silty loam soil, specifically:

$$\beta(\theta_0) = 0.6\frac{\theta_s - \theta_i}{\theta_s - \theta_r} + 0.4 \quad (7)$$

$$\beta \cdot p = 0.98 - 0.87 \, exp\left(-\frac{r}{K_s}\right) \quad \frac{d\theta_0}{dt} \geq 0 \quad (8)$$

$$\beta \cdot p = 1.7 \quad \frac{d\theta_0}{dt} < 0 \quad (9)$$

Equation (4) can be solved numerically. For $q_0 = r$, with $F' = (r - K_i)t$, it gives $\theta_0(t)$ until time to ponding, t_p, corresponding to $\theta_0 = \theta_s$ and $d\theta_0/dt = 0$, then after t_p, with $\theta_0 = \theta_s$ and $d\theta_0/dt = 0$, it provides the infiltration capacity ($q_0 = f_c$) and for the period with $r = 0$, with $q_0 = 0$, it gives $d\theta_0/dt < 0$ thus describing the redistribution process.

The involved parameters were estimated through the volume balance criterion along with a best-fit procedure for the water content measured at 5 cm depth by the reference sensor. The initial water contents were set equal to those observed before each experiment, which was found to be almost invariant with depth. Figure 18 shows the results of the model calibration for both the study soils at different depths.

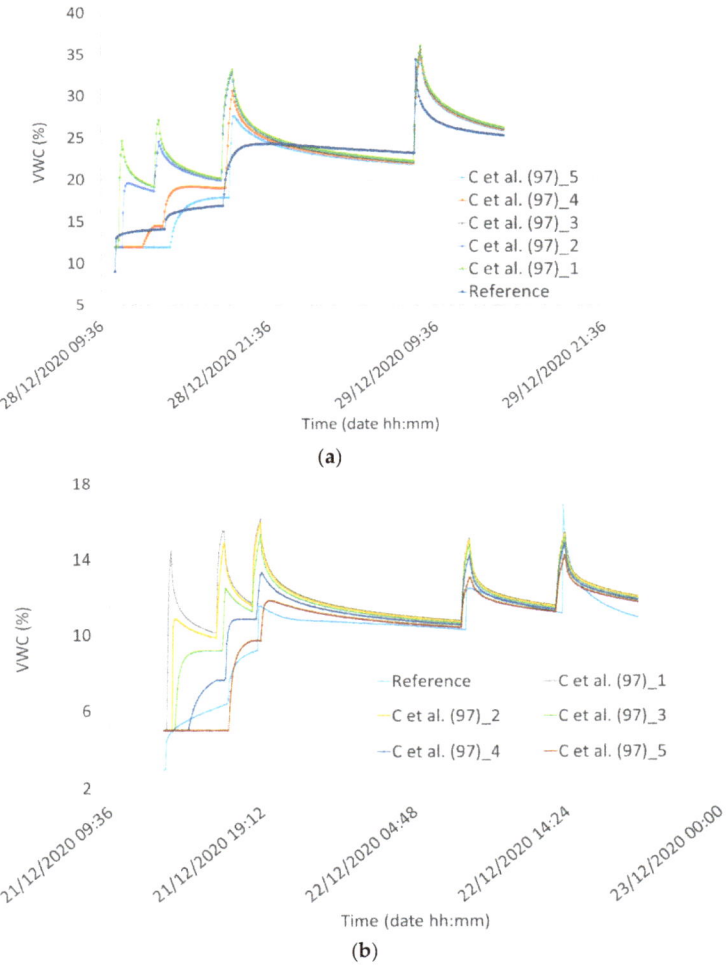

Figure 18. Modeling results for infiltration and redistribution of water ("C et al. (97)" model) during and after rainfall at different depths (1, 2, 3, 4, and 5 cm) for (**a**) Silty Loam and (**b**) Loamy Sand. The model was calibrated with respect to the measured values at a 5 cm depth (Reference sensor). For example, in the figure "C et al. (97)_1" stands for the modeling result at a 1 cm depth.

6.2. Correlation of the Capacitive Soil Moisture Sensor v1.2 Output Voltage with the Prediction of the Hydraulic Model

In Section 3.1.1 we listed two equations used to correlate and calibrate the output voltage of the Capacitive Soil Moisture Sensor v1.2 with certified water content for two different types of soil: Silty Loam and Loamy Sand. However, the experiments described in the present paper provide a reference water content at an average depth of 5 cm, which is for sure greater than the average detection depth of 2 or 3 cm of the Capacitive Soil Moisture sensor v1.2, which spans a depth from 0 to 5 cm. A possible solution to this problem is to correlate the VWC from Equations (2) and (3) with the extrapolations of the infiltration and distribution Corradini et al. model at a depth of 2 or 3 cm.

In the remainder of the paper, we show the results obtained for the two different soils.

6.2.1. Water Content in Silty Loam

A three-parameter least-square best fit was calculated between the VWC function obtained using the Corradini model (hereafter indicated as "C et al. (97)" [72] at different depths of 2 and 3 cm, with Equation (1), obtaining two triplets of A, B, and C values shown in Table 3 where the Placidi model "P et al. (20)" [8] is referred to different depths. Similarly, a two-parameter least-square best fit was calculated with the Hrisko model "H (20)" at 2 and 3 cm depths and the model from "C et al. (97)", obtaining two couples of P and Q values shown in Table 3.

Table 3. Least-square best-fit parameters of Equations (2) and (3) in Silty Loam.

	P et al. (20)_3	P et al. (20)_2
A	0.711	0.731
B	9.72	10.2
C	0.864	0.859
	H (20)_3	H (20)_2
P	73.4	75.8
Q	55.1	57.2

The plots reporting the water content obtained by using the three models for the two different depths are reported in Figure 19.

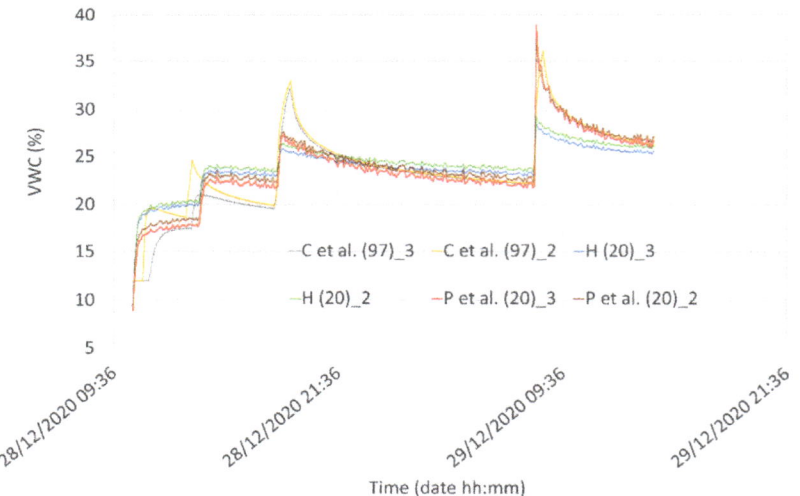

Figure 19. Comparison among the water infiltration and redistribution "C" model for Silty Loam at a depth of 2 and 3 cm with the VWC obtained from the voltage measured by Node #1 using Equation (2) ("P" curves) and Equation (3) ("H" curves).

In Figure 20 a statistical analysis between all the possible couples of the "C et al. (97)" model and voltage measured by Node #1 using Equation (2) ("P" curves) and Equation (3) ("H" curves) at different depths has been reported. The analysis has been performed by using scattering plots, cross-correlation values, and kernel density estimation accomplished by using the Seaborn Python3 tool [73]. In the figure, the eight plots in the main diagonal are the calculated histograms of the corresponding eight quantities, together with the estimated Gaussian mixture probability density function. The plots in the lower triangular part represent the scattering plots of each couple of quantities, together with the locally weighted regression curve whereas the values in the upper triangular part, instead, represent the correlation coefficients between each couple of quantities.

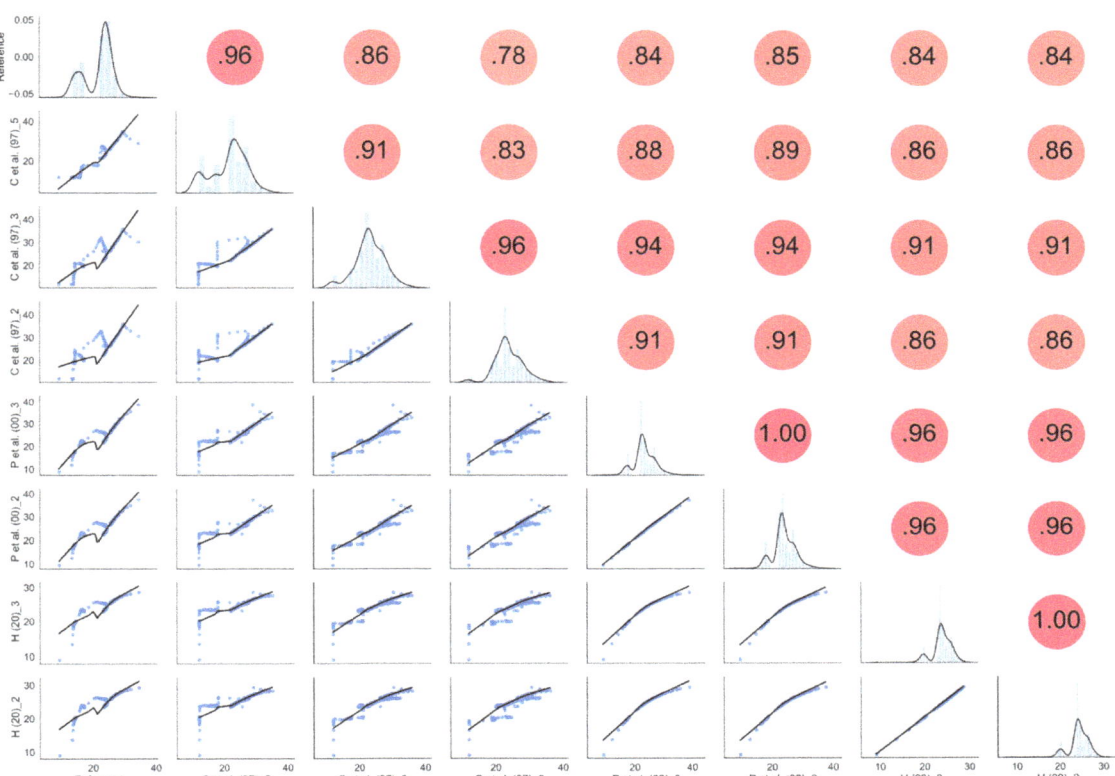

Figure 20. Statistical analysis with scattering plots, cross-correlation values, and kernel density estimation (KDE) obtained by using the Seaborn Python3 tool [73–75] in Silty Loam.

Looking at homogeneous values (i.e., correlation data obtained at the same depth), The best correlation values we obtained (0.94) are between "C et al. (97)" and "P et al. (20)" at a depth of 3 cm. The 0.94 correlation coefficient is slightly greater than the value of 0.91 obtained between "C et al. (97)" and "P et al. (20)" at a depth of 2 cm. In Figure 21 the comparison among the best results obtained from the correlation are reported for the three models. Even if peaks and valleys of the hydraulic model are not always perfectly reproduced by the Capacitive Soil Moisture Sensor v1.2 fitting equations, the overall behavior of the "P et al. (20)" model can capture the main features of the VWC at a shallow depth. We note that, due to the peculiarities of experimental systems involving natural soils, it is impossible to obtain results that are completely reproducible from mathematical schemes. For example, inserting different sensors into the soil produces different preferential waterways that can turn out in minimally different results, especially when the experimental behavior is compared with mathematical model performances.

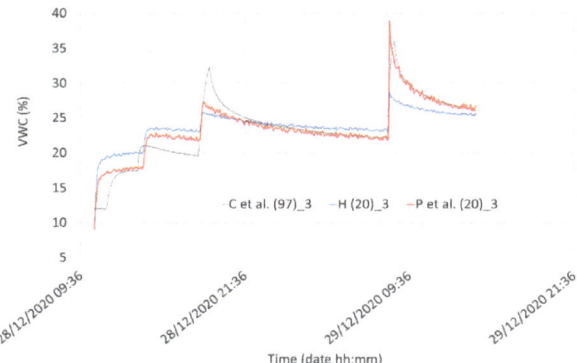

Figure 21. Comparison among the best results obtained from the correlation procedure for the three considered models.

6.2.2. Water Content in Loamy Sand

A three-parameter least-square best fit was also calculated in Loamy Sand between Equation (1) and the VWC function obtained using the "C et al. (97)" model at different depths of 2 and 3 cm. The two triplets of A, B, and C values are reported in Table 4 where the model "P et al. (20)" is referred to different depths. Similarly, a two-parameter least-square best fit was calculated with the Hrisko model "H (20)" at 2 and 3 cm depths and the model from "C et al. (97)", obtaining two couples of P and Q values shown in Table 4.

Table 4. Least-square best-fit parameters of Equation (2) and Equation (3) in Loamy Sand.

	P et al. (20)_3	P et al. (20)_2
A	1.64	1.65
B	8.16	8.41
C	0.85	0.85
	H (20)_3	H (20)_2
P	17.44	17.54
Q	2.37	2.1

The plot with the water content for the three models for the two different depths is reported in Figure 22.

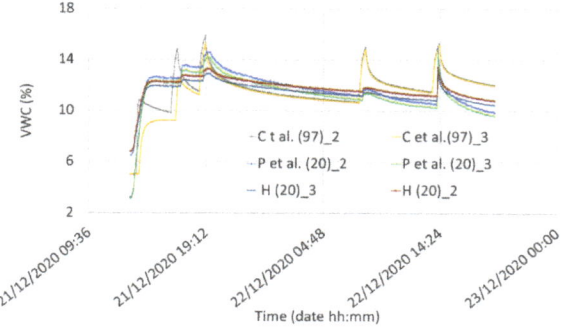

Figure 22. Comparison among the water infiltration and redistribution "C" model for Loamy Sand at a depth of 2 and 3 cm with the VWC obtained from the voltage measured by Node #1 using Equation (2) ("P" curves) and Equation (3) ("H" curves).

Then a statistical analysis with scattering plots, cross-correlation values, and kernel density estimation was accomplished by using the Seaborn Python3 tool (Figure 23) between all the possible couples of models at different depths. Looking at homogeneous values (i.e., correlation data obtained at the same depth), the best correlation values we obtained (0.58) are between "C et al. (97)" and the "Hrisko model". A much worse correlation was obtained for Loamy Sand compared to Silty Loam. However, as highlighted in [54], the behavior of coarse-textured soil (as the Loamy Sand) can be mathematically modeled with greater difficulty than that of fine-textured soil (as the Silty Loam).

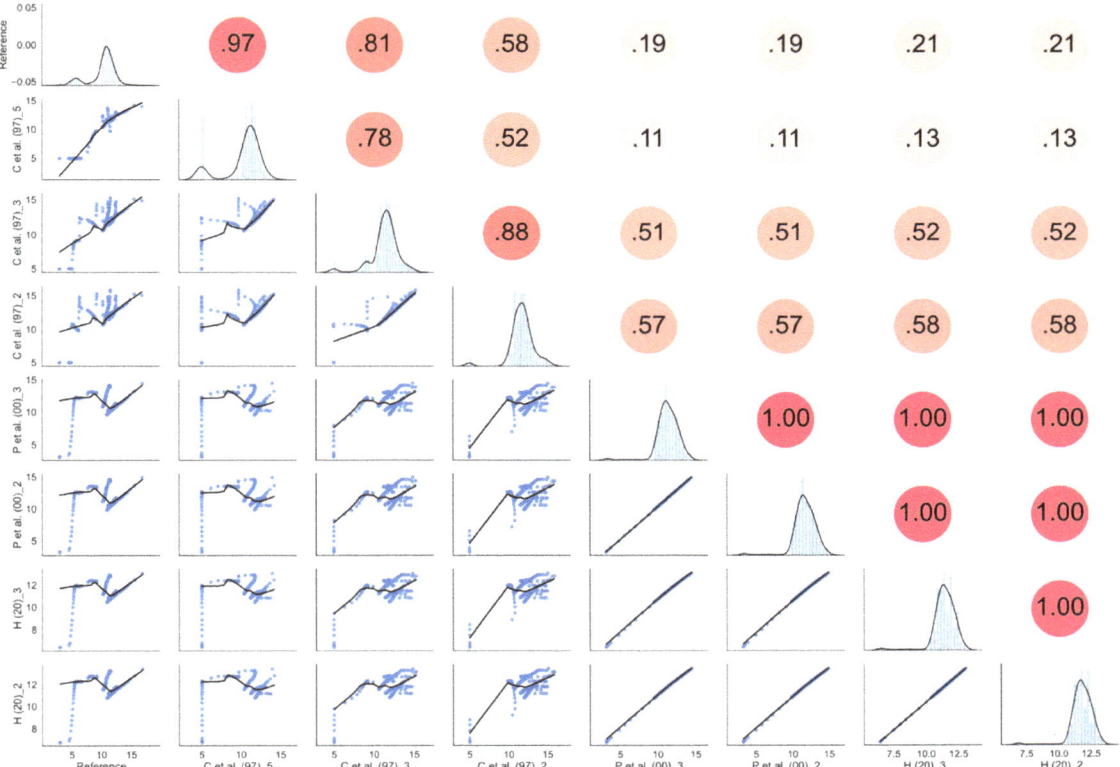

Figure 23. Statistical analysis with scattering plots, cross-correlation values, and kernel density estimation (KDE) in Loamy Sand.

Figure 24 shows the comparison among the best results obtained from the correlation procedure. Even if a first sight comparison of the three curves shows significant differences, it should be noted that for practical applications of sensors for measuring the soil water content, differences of a few percent are often irrelevant.

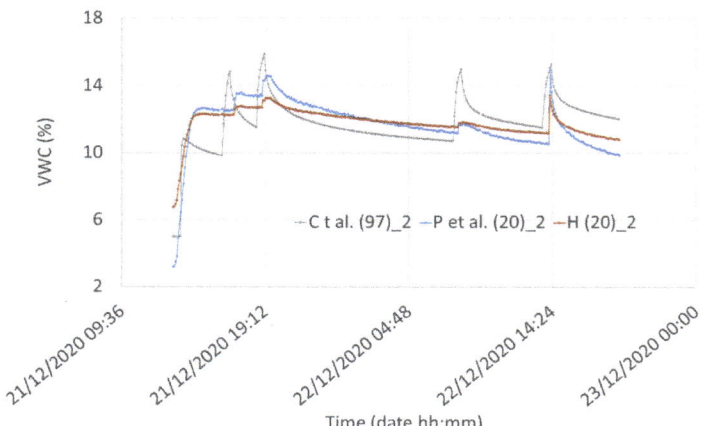

Figure 24. Comparison among the best results obtained from the correlation procedure for the three considered models.

7. Conclusions

A low-power WSN based on LoRaWAN™ was designed with a focus on low-cost PA applications, such as greenhouse sensing and actuation. Two types of wireless nodes were envisaged, greenhouse node and plant node, and the whole LPWAN was designed and implemented, including an 8-channel gateway/concentrator. The first experimental results were collected and stored in a database managed by a virtual machine running in a cloud service. Since all subsystems adopted in this research are off-the-shelf elements with available open-source software libraries, only a minimal effort is needed when the system is implemented for a different application.

Measurement results were focused on measurements of water content and were collected using plant nodes in Loamy Sand and Silty Loam, proving the functionality and reliability of the whole system (sensor nodes, gateway, GUI, Node-RED, and Cloud) and comparing the system behavior with a reference sensor from Sentek. Temperature measurements of our plant nodes compare as expected with the reference sensor within the supposed accuracy of the adopted sensors. Regarding water content measurements, a correlation was attempted between the results of the cheap Capacitive Soil Moisture Sensor v1.2 and those of the Sentek reference sensor. We realized the low-cost water content sensor suffers from a non-constant sensitivity; therefore, non-linear fitting equations are necessary for correlating its voltage output with the VWC. We adopted for the Capacitive Soil Moisture Sensor v1.2 two VWC fitting equations taken from the literature. Since the reference sensor and the cheap water content sensor span different soil depths (5 cm and 2-3 cm, respectively), we first modeled the theoretical VWC profiles at different depths using a proven water infiltration and redistribution model, calibrating the model on the reference sensor results at a depth of 5 cm. Then we used the two fitting equations for the Capacitive Soil Moisture Sensor v1.2 and calculated multi-parameter least squares fit to the hydraulic model at 2 and 3 cm depths. A very satisfactory correlation coefficient of 0.94 was obtained for Silty Loam using the exponential/logarithmic "P" model at a depth of 3 cm. Instead, the best correlation value we obtained using the same fitting procedure applied to the results in Loamy Sand was 0.58 at a depth of 2 cm using the hyperbolic "H" model. Despite the low correlation coefficient, the VWC values we obtained with the hyperbolic "H" model can be considered as representative of the real VWC at a depth of 2 cm, since differences of a few percent are often irrelevant for practical applications of sensors for measuring the soil water content.

In this work, we demonstrated the lack of linearity of the adopted soil water content sensor. Notwithstanding this lack of linearity, the introduction of the infiltration model and of a dedicated statistical analysis allowed us to extract reliable values of the volumetric water content for both Silty Loam and Loamy Sand. This procedure represents the novelty and the potential of the proposed approach. Therefore, future work will address the optimization of the sensor performance. To this purpose, it will be useful to better understand the behavior of the sensor from simulations and to optimize the layout of the sensor without impacting significantly on the cost, also considering the mechanical integration constraints needed for industrialization.

Author Contributions: Conceptualization, P.P., R.M., and A.S.; data curation, P.P., R.M., D.F., N.P., F.G., and A.S.; formal analysis, P.P., R.M. and A.S.; funding acquisition, P.P.; investigation, R.M., D.F., N.P. and F.G.; methodology, P.P., and A.S.; Project administration, P.P. and A.S.; Software, D.F., and F.G.; supervision, P.P., R.M., and A.S.; validation, P.P., D.F., N.P. and A.S.; visualization, D.F., and F.G.; writing—original draft, P.P., and A.S.; writing—review and editing, P.P., R.M., and A.S. All authors have read and agreed to the published version of the manuscript.

Funding: This research was partly funded by the DEPARTMENT OF ENGINEERING OF THE UNIVERSITY OF PERUGIA, Italy, grant "Ricerca di base 2018-RICBA18-PP", "Ricerca di base 2019-RICBA19-PP", "Ricerca di base 2020-RICBA20-PP".

Acknowledgments: Thanks are due to colleagues D. Grohmann, G. Marconi, and M. Cecconi for helpful discussions and technical support.

Conflicts of Interest: The authors declare no conflict of interest. The funders had no role in the design of the study; in the collection, analyses, or interpretation of data; in the writing of the manuscript, or in the decision to publish the results.

References

1. Banđur, Đ.; Jakšić, B.; Banđur, M.; Jović, S. An analysis of energy efficiency in Wireless Sensor Networks (WSNs) applied in smart agriculture. *Comput. Electron. Agric.* **2019**, *156*, 500–507. [CrossRef]
2. Kochhar, A.; Kumar, N. Wireless sensor networks for greenhouses: An end-to-end review. *Comput. Electron. Agric.* **2019**, *163*, 104877. [CrossRef]
3. Castañeda-Miranda, A.; Castaño, V.M. Internet of things for smart farming and frost intelligent control in greenhouses. *Comput. Electron. Agric.* **2020**, *176*, 105614. [CrossRef]
4. Pradeepkumar, D.; Ravi, V. Soft computing hybrids for FOREX rate prediction: A comprehensive review. *Comput. Oper. Res.* **2018**, *99*, 262–284. [CrossRef]
5. Panigrahi, S.; Behera, H.S. A hybrid ETS–ANN model for time series forecasting. *Eng. Appl. Artif. Intell.* **2017**, *66*, 49–59. [CrossRef]
6. Fan, Y.; Zhang, Y.; Chen, Z.; Wang, X.; Huang, B. Comprehensive assessments of soil fertility and environmental quality in plastic greenhouse production systems. *Geoderma* **2021**, *385*, 114899. [CrossRef]
7. Guo, Y.; Zhao, H.; Zhang, S.; Wang, Y.; Chow, D. Modeling and optimization of environment in agricultural greenhouses for improving cleaner and sustainable crop production. *J. Clean. Prod.* **2021**, *285*, 124843. [CrossRef]
8. Placidi, P.; Gasperini, L.; Grassi, A.; Cecconi, M.; Scorzoni, A. Characterization of Low-Cost Capacitive Soil Moisture Sensors for IoT Networks. *Sensors* **2020**, *20*, 3585. [CrossRef]
9. Syafrudin, M.; Alfian, G.; Fitriyani, N.L.; Rhee, J. Performance Analysis of IoT-Based Sensor, Big Data Processing, and Machine Learning Model for Real-Time Monitoring System in Automotive Manufacturing. *Sensors* **2018**, *18*, 2946. [CrossRef] [PubMed]
10. Castañeda, A.; Castaño, V.M. Smart frost measurement for anti-disaster intelligent control in greenhouses via embedding IoT and hybrid AI methods. *Measurement* **2020**, *164*, 108043. [CrossRef]
11. Rizzi, M.; Ferrari, P.; Flammini, A.; Sisinni, E. Evaluation of the IoT LoRaWAN Solution for Distributed Measurement Applications. *IEEE Trans. Instrum. Meas.* **2017**, *66*, 12. [CrossRef]
12. Zamora-Izquierdo, M.A.; Santa, J.; Martínez, J.A.; Martínez, V.; Skarmeta, A.F. Smart farming IoT platform based on edge and cloud computing. *Biosyst. Eng.* **2019**, *177*, 4–17. [CrossRef]
13. Iyengar, S.S.; Boroojeni, K.G.; Balakrishnan, N. *Mathematical Theories of Distributed Sensor Networks*; Springer: Berlin/Heidelberg, Germany, 2014.
14. Benos, L.; Tagarakis, A.; Dolias, G.; Berruto, R.; Kateris, D.; Bochtis, D. Machine Learning in Agriculture: A Comprehensive Updated Review. *Sensors* **2021**, *21*, 3758. [CrossRef]
15. Patrício, D.I.; Rieder, R. Computer vision and artificial intelligence in precision agriculture for grain crops: A systematic re-view. *Comput. Electron. Agric.* **2018**, *153*, 69–81. [CrossRef]
16. van Klompenburg, T.; Kassahun, A.; Catal, C. Crop yield prediction using machine learning: A systematic literature review. *Comput. Electron. Agric.* **2020**, *177*, 105709. [CrossRef]

17. Liakos, K.G.; Busato, P.; Moshou, D.; Pearson, S.; Bochtis, D. Machine learning in agriculture: A review. *Sensors* **2018**, *18*, 2674. [CrossRef] [PubMed]
18. Yuan, Y.; Chen, L.; Wu, H.; Li, L. Advanced agricultural disease image recognition technologies: A review. *Inf. Process. Agric.* **2021**, *8*, 1–12. [CrossRef]
19. Akbar, A.; Kuanar, A.; Patnaik, J.; Mishra, A.; Nayak, S. Application of Artificial Neural Network modeling for optimization and prediction of essential oil yield in turmeric (*Curcuma longa* L.). *Comput. Electron. Agric.* **2018**, *148*, 160–178. [CrossRef]
20. Vij, A.; Vijendra, S.; Jain, A.; Bajaj, S.; Bassi, A.; Sharma, A. IoT and Machine Learning Approaches for Automation of Farm Irrigation System. *Procedia Comput. Sci.* **2020**, *167*, 1250–1257. [CrossRef]
21. Rehman, A.U.; Abbasi, A.Z.; Islam, N.; Shaikh, Z.A. A review of wireless sensors and networks' applications in agriculture. *Comput. Stand. Interfaces* **2014**, *36*, 263–270. [CrossRef]
22. Farooq, M.S.; Riaz, S.; Abid, A.; Umer, T.; Bin Zikria, Y. Role of IoT Technology in Agriculture: A Systematic Literature Review. *Electronics* **2020**, *9*, 319. [CrossRef]
23. Shi, X.; An, X.; Zhao, Q.; Liu, H.; Xia, L.; Sun, X.; Guo, Y. State-of-the-Art Internet of Things in Protected Agriculture. *Sensors* **2019**, *19*, 1833. [CrossRef] [PubMed]
24. Sagheer, A.; Mohammed, M.; Riad, K.; Alhajhoj, M. A Cloud-Based IoT Platform for Precision Control of Soilless Greenhouse Cultivation. *Sensors* **2020**, *21*, 223. [CrossRef] [PubMed]
25. Shafi, U.; Mumtaz, R.; García-Nieto, J.; Hassan, S.A.; Zaidi, S.A.R.; Iqbal, N. Precision Agriculture Techniques and Practices: From Considerations to Applications. *Sensors* **2019**, *19*, 3796. [CrossRef]
26. Messina, G.; Modica, G. Applications of UAV Thermal Imagery in Precision Agriculture: State of the Art and Future Research Outlook. *Remote. Sens.* **2020**, *12*, 1491. [CrossRef]
27. Gnecchi, J.A.G.; Tirado, L.F.; Campos, G.M.C.; Ramirez, R.D.; Gordillo, C.F.E. Design of a Soil Moisture Sensor with Temperature Compensation Using a Backpropagation Neural Network. In Proceedings of the 2008 Electronics, Robotics and Automotive Mechanics Conference, Cuernavaca, Mexico, 30 September–3 October 2008; pp. 553–558.
28. Danita, M.; Mathew, B.; Shereen, N.; Sharon, N.; Paul, J.J. IoT Based Automated Greenhouse Monitoring System. In Proceedings of the 2018 Second International Conference on Intelligent Computing and Control Systems (ICICCS), Madurai, India, 14–15 June 2018; pp. 1933–1937.
29. Ruiz-Garcia, L.; Lunadei, L.; Barreiro, P.; Robla, J.I. A Review of Wireless Sensor Technologies and Applications in Agriculture and Food Industry: State of the Art and Current Trends. *Sensors* **2009**, *9*, 4728–4750. [CrossRef]
30. BeechamRes. Towards Smart Farming Agriculture Embracing the IoT Vision. Available online: http://www.beechamresearch.com/files/BRL%20Smart%20Farming%20Executive%20Summary.pdf (accessed on 12 May 2020).
31. SU, S.L.; Singh, D.N.; Baghini, M.S. A critical review of soil moisture measurement. *Measurement* **2014**, *54*, 92–105. [CrossRef]
32. Nagahage, E.A.A.D.; Nagahage, I.S.P.; Fujino, T. Calibration and Validation of a Low-Cost Capacitive Moisture Sensor to Integrate the Automated Soil Moisture Monitoring System. *Agriculture* **2019**, *9*, 141. [CrossRef]
33. Hrisko, J. Capacitive Soil Moisture Sensor Theory, Calibration, and Testing. *Tech. Rep.* **2020**, *2*, 1–12. [CrossRef]
34. Morais, R.; Mendes, J.; Silva, R.; Silva, N.; Sousa, J.; Peres, E. A Versatile, Low-Power and Low-Cost IoT Device for Field Data Gathering in Precision Agriculture Practices. *Agriculture* **2021**, *11*, 619. [CrossRef]
35. Figorilli, S.; Pallottino, F.; Colle, G.; Spada, D.; Beni, C.; Tocci, F.; Vasta, S.; Antonucci, F.; Pagano, M.; Fedrizzi, M.; et al. An Open Source Low-Cost Device Coupled with an Adaptative Time-Lag Time-Series Linear Forecasting Modeling for Apple Trentino (Italy) Precision Irrigation. *Sensors* **2021**, *21*, 2656. [CrossRef]
36. Escriba, C.; Bravo, E.G.A.; Roux, J.; Fourniols, J.-Y.; Contardo, M.; Acco, P.; Soto-Romero, G. Toward Smart Soil Sensing in v4.0 Agriculture: A New Single-Shape Sensor for Capacitive Moisture and Salinity Measurements. *Sensors* **2020**, *20*, 6867. [CrossRef] [PubMed]
37. Kojima, Y.; Shigeta, R.; Miyamoto, N.; Shirahama, Y.; Nishioka, K.; Mizoguchi, M.; Kawahara, Y. Low-Cost Soil Moisture Profile Probe Using Thin-Film Capacitors and a Capacitive Touch Sensor. *Sensors* **2016**, *16*, 1292. [CrossRef]
38. Ferrández-Pastor, F.J.; García-Chamizo, J.M.; Nieto-Hidalgo, M.; Pascual, J.M.M.; Mora-Martínez, J. Developing Ubiquitous Sensor Network Platform Using Internet of Things: Application in Precision Agriculture. *Sensors* **2016**, *16*, 1141. [CrossRef]
39. Ferrández-Pastor, F.J.; García-Chamizo, J.M.; Nieto-Hidalgo, M.; Mora-Martínez, J. Precision Agriculture Design Method Using a Distributed Computing Architecture on Internet of Things Context. *Sensors* **2018**, *18*, 1731. [CrossRef] [PubMed]
40. Coping with Water Scarcity. UN—Water Thematic Initiatives. A Strategic Issue and Priority for System-Wide Action. 2006. Available online: http://www.unwater.org/publications/coping-water-scarcity/ (accessed on 3 January 2021).
41. Hanak, E.; Mount, J.; Chappelle, C.; Lund, J.; Medellín-Azuara, J.; Moyle, P. Public Policy Institute of California, Water Policy Center. What If California's Drought Continues? 2015. Available online: http://www.ppic.org/publication/what-if-californias-drought-continues/ (accessed on 3 January 2021).
42. Yuanyuan, C.; Zuozhuang, Z. Research and Design of Intelligent Water-saving Irrigation Control System Based on WSN. In Proceedings of the 2020 IEEE International Conference on Artificial Intelligence and Computer Applications (ICAICA), Dalian, China, 27–29 June 2020.
43. Mohanraj, I.; Gokul, V.; Ezhilarasie, R.; Umamakeswari, A. Intelligent drip irrigation and fertigation using wireless sensor networks. In Proceedings of the 2017 IEEE Technological Innovations in ICT for Agriculture and Rural Development (TIAR), Chennai, India, 7–8 April 2017.

44. Walker, J.P.; Willgoose, G.R.; Kalma, J.D. In Situ measurement of soil moisture: A comparison of techniques. *J. Hydrol.* **2004**, *293*, 85–99. [CrossRef]
45. Tremsin, V.A. Real-Time Three-Dimensional Imaging of Soil Resistivity for Assessment of Moisture Distribution for Intelligent Irrigation. *Hydrology* **2017**, *4*, 54. [CrossRef]
46. Subahi, F.; Bouazza, K.E. An Intelligent IoT-Based System Design for Controlling and Monitoring Greenhouse Temperature. *IEEE Access* **2020**, *8*, 125488–125500. [CrossRef]
47. Pawlowski, A.; Guzman, J.L.; Rodríguez, F.; Berenguel, M.; Sánchez, J.; Dormido, S. Simulation of Greenhouse Climate Monitoring and Control with Wireless Sensor Network and Event-Based Control. *Sensors* **2009**, *9*, 232–252. [CrossRef]
48. Germani, L.; Mecarelli, V.; Baruffa, G.; Rugini, L.; Frescura, F. An IoT Architecture for Continuous Livestock Monitoring Using LoRa LPWAN. *Electronics* **2019**, *8*, 1435. [CrossRef]
49. Wang, N.; Zhang, N.; Wang, M. Wireless sensors in agriculture and food industry—Recent development and future perspective. *Comput. Electron. Agric.* **2006**, *50*, 1–14. [CrossRef]
50. Haseeb, K.; Ud Din, I.; Almogren, A.; Islam, N. An Energy Efficient and Secure IoT-Based WSN Framework: An Application to Smart Agriculture. *Sensors* **2020**, *20*, 2081. [CrossRef] [PubMed]
51. Yelamarthi, K.; Aman, M.S.; Abdelgawad, A. An Application-Driven Modular IoT Architecture. *Wirel. Commun. Mob. Comput.* **2017**, *2017*, 1350929. [CrossRef] [PubMed]
52. Gutiérrez, J.; Villa-Medina, J.F.; Nieto-Garibay, A.; Porta-Gándara, M.Á. Automated Irrigation System Using a Wireless Sensor Network and GPRS Module. *IEEE Trans. Instrum. Meas.* **2014**, *63*, 1. [CrossRef]
53. LoRa Alliance. What is LoRaWAN® Specification. Available online: https://lora-alliance.org/about-lorawan/ (accessed on 1 February 2021).
54. Zourmand, A.; Hing, A.L.K.; Hung, C.W.; Abdulrehman, M. Internet of Things (IoT) using LoRa technology. In Proceedings of the Conference I2CACIS 2019, Selangor, Malaysia, 29 June 2019.
55. Gu, C.; Tan, R.; Huang, J. SoftLoRa—A LoRa-Based Platform for Accurate and Secure Timing. In Proceedings of the IPSN 19, Montreal, QC, Canada, 16–18 April 2019; pp. 309–310.
56. Van Torre, P.; Ameloot, T.; Rogier, H. Long-range body-to-body LoRa link at 868 MHz. In Proceedings of the 13th European Conference on Antennas and Propagation (EuCAP 2019), Krakow, Polland, 31 March–15 April 2019.
57. Lavric, A.; Popa, V. Internet of Things and LoRa™ Low-Power Wide-Area Networks: A survey. In Proceedings of the International Symposium on Signals, Circuits and Systems (ISSCS), Iasi, Romania, 3–14 July 2017.
58. Mikhaylov, K.; Petaejaejaervi, J.; Haenninen, T. Analysis of capacity and scalability of the LoRa low power wide area network technology. In Proceedings of the 22nd European Wireless Conference, Oulu, Finland, 1–6 May 2016.
59. Thethingsnetwork.org. Available online: https://www.thethingsnetwork.org/docs/network/architecture.html (accessed on 2 May 2019).
60. FarmBot. Available online: https://farm.bot/ (accessed on 2 May 2019).
61. SKU:SEN0193. Available online: https://wiki.dfrobot.com/Capacitive_Soil_Moisture_Sensor_SKU_SEN0193 (accessed on 10 December 2020).
62. A WiFi Enabled Soil Moisture Sensor. Available online: https://www.hackster.io/rbaron/a-wifi-enabled-soil-moisture-sensor-49912b (accessed on 2 May 2021).
63. Munyaradzi, M.; Rupere, T.; Nyambo, B.; Mukute, S.; Chinyerutse, M.; Hapanga, T.; Mashonjowa, E. A Low Cost Automatic Irrigation Controller Driven by Soil Moisture Sensors. *Int. J. Agric. Innov. Res.* **2013**, *2*, 1–7.
64. Nemali, K.S.; Montesano, F.; Dove, S.K.; van Iersel, M.W. Calibration and performance of moisture sensors in soilless substrates:ECH2O and Theta probes, Scientia. *Horticulturae* **2007**, *112*, 227–234. [CrossRef]
65. Qu, W.; Bogena, H.R.; Huisman, J.A.; Vereecken, H. Calibration of a Novel Low-Cost Soil Water Content Sensor Based on a Ring Oscillator. *Vadose Zone J.* **2013**, *12*, vzj2012.0139. [CrossRef]
66. Bogena, H.R.; Huisman, J.A.; Schilling, B.; Weuthen, A.; Vereecken, H. Effective Calibration of Low-Cost Soil Water Content Sensors. *Sensors* **2017**, *17*, 208. [CrossRef] [PubMed]
67. Node-RED. Available online: https://nodered.org/ (accessed on 3 January 2021).
68. Linsley, R.K.; Franzini, J.B.; Freyberg, D.L.; Tchobanoglous, G. *Water–Resources Engineering*; McGraw-Hill Singapore: Singapore, 1992.
69. Morbidelli, R.; Saltalippi, C.; Flammini, A.; Cifrodelli, M.; Picciafuoco, T.; Corradini, C.; Govindaraju, R.S. In-situ measurements of soil saturated hydraulic conductivity: Assessment of reliability trough rainfall-runoff experiments. *Hydrol. Process.* **2017**, *31*, 3084–3094. [CrossRef]
70. Melone, F.; Corradini, C.; Morbidelli, R.; Saltalippi, C. Laboratory experimental check of a conceptual model for infiltration under complex rainfall patterns. *Hydrol. Proc.* **2006**, *20*, 439–452. [CrossRef]
71. Melone, F.; Corradini, C.; Morbidelli, R.; Saltalippi, C.; Flammini, A. Comparison of theoretical and experimental soil moisture profiles under complex rainfall patterns. *J. Hydrol. Eng.* **2008**, *13*, 1170–1176. [CrossRef]
72. Corradini, C.; Melone, F.; Smith, E.R. A unified model for infiltration and redistribution during complex rainfall patterns. *J. Hydrol.* **1997**, *192*, 104–124. [CrossRef]

73. Waskom, M.; Botvinnik, O.; O'Kane, D.; Hobson, P.; Ostblom, J.; Lukauskas, S.; Gemperline, D.C.; Augspurger, T.; Halchenko, Y.; Cole, J.B.; et al. mwaskom/seaborn: v0.9.0 (July 2018). Available online: http://seaborn.pydata.org/ (accessed on 25 November 2019).
74. Silverman, B.W. *Density Estimation for Statistics and Data Analysis*; Routledge: New York, NY, USA, 1998.
75. Fan, J. *Local Polynomial Modelling and Its Applications: Monographs on Statistics and Applied Probability 66*; Routledge: New York, NY, USA, 1996.

Article

An Application of a LPWAN for Upgrading Proximal Soil Sensing Systems

Yonghui Tu [1], Haoye Tang [1] and Wenyou Hu [2,*]

- [1] State Key Laboratory of Soil and Sustainable Agriculture, Institute of Soil Science, Chinese Academy of Sciences, Nanjing 210008, China; yhtu@issas.ac.cn (Y.T.); hytang@issas.ac.cn (H.T.)
- [2] Key Laboratory of Soil Environment and Pollution Remediation, Institute of Soil Science, Chinese Academy of Sciences, Nanjing 210008, China
- * Correspondence: wyhu@issas.ac.cn

Abstract: In recent years, the Internet of Things (IoT), based on low-power wide-area network (LPWAN) wireless communication technology, has developed rapidly. On the one hand, the IoT makes it possible to conduct low-cost, low-power, wide-coverage, and real-time soil monitoring in fields. On the other hand, many proximal soil sensor devices designed based on conventional communication methods that are stored in an inventory face elimination. Considering the idea of saving resources and costs, this paper applied LPWAN technology to an inventoried proximal soil sensor device, by designing an attachment hardware system (AHS) and realizing technical upgrades. The results of the experimental tests proved that the sensor device, after upgrading, could work for several years with only a battery power supply, and the effective wireless communication coverage was nearly 1 km in a typical suburban farming environment. Therefore, the new device not only retained the original mature sensing technology of the sensor device, but also exhibited ultralow power consumption and long-distance transmission, which are advantages of the LPWAN; gave full play to the application value and economic value of the devices stored in inventory; and saved resources and costs. The proposed approach also provides a reference for applying LPWAN technology to a wider range of inventoried sensor devices for technical upgrading.

Keywords: IoT; LPWAN; proximal soil sensor device; conventional communication methods; ultralow power consumption; long-distance transmission; economic value; inventoried sensor devices

1. Introduction

Soil is an important natural resource and the most critical material basis for agricultural production. The acquisition and analysis of data related to soil moisture, salt, pH, and other physicochemical properties is an important basis for land resource utilization and agricultural production activities. Conventional soil sampling and laboratory analyses have long sampling and analysis cycles and high labor costs; therefore, various proximal soil sensor devices and data acquisition systems have been widely used in fields [1,2]. Proximal soil sensing mainly refers to the use of field sensors to acquire information proximal to the ground or in the soil. This concept was first proposed by Viscarra Rossel and McBratney in 1998 [3] and further developed in 2010 [4]. At present, various proximal ground sensor devices based on different working principles have been developed. For example, the EM38 conductivity meter developed by Geonics Inc. (Mississauga, ON, Canada) is an instrument used to obtain soil comprehensive apparent electrical conductivity (ECa) based on the principle of electromagnetic induction. Myers et al. used this instrument to combine conductivity data from the soil surface and soil profiles for high-resolution ECa soil digital mapping [5]. Besson et al. used MUCEP (multi-continuous electrical profiling) to measure soil resistance coefficient and monitor the temporal and spatial changes in soil moisture at the field scale [6]. Electrochemical sensors based on ion selective electrodes (ISEs) and ion sensitive field effect transistors (ISFETs) are mainly used for the determination of soil pH and nitrate and potassium ion concentrations. Adsett and Thottan designed a real-time automatic nitrate content measurement system using ISFETs and a nitrate detector [7]. Similar instruments include the pH meters produced by Veris Technologies, Inc. (Salina, KS, USA) and Spectrum Technologies, Inc. (Aurora, IL, USA). These proximal soil sensor devices or systems are usually deployed in field environments and connected to data acquisition devices with RS-485, RS-232, or SDI-12 cables, and manually obtain data on-site. In practical applications, such systems face the problems of limited bearing capacity, cumbersome wiring, high operation costs, and inconvenient installation and maintenance. With the rapid development of information technology, the IoT has been widely used in various industries, promoting their rapid development and extension. Low-power wide-area networks (LPWANs) form one of the main hotspots of IoT access technology [8]. Compared with conventional wired communication technologies (such as RS-485 and SDI-12), mobile cellular technologies (such as 2G, 3G, 4G, etc.), and short-range wireless communication technologies (such as Bluetooth, ZigBee, etc.), an LPWAN has the advantages of low cost, low power consumption, wide coverage, and strong connection [9], and can effectively achieve the application of a proximal soil sensing system. An LPWAN is an important technical support tool for promoting the transformation from conventional laboratory-based physicochemical soil analyses to field-based measurements. Vu et al. designed an automatic irrigation system for greenhouses based on LoRa technology [10]. Rachmani et al. designed an IoT monitoring system based on LoRa technology for a starfruit plantation [11]. Co et al. designed and developed the hardware and software components of a wireless sensor network (WSN) for soil monitoring [12]. In these applications, the LoRa communication module is usually independent of the sensor device, but the sensor device is still based on the conventional application design, and its power requirements cannot be met by a long-term battery power supply. Thus, the deployment of the system is troublesome, and a lot of maintenance work is required in the later stages of use.

To meet the needs of long-term field work, and considering the limitations of sensor battery power supply, LPWAN technology needs to be integrated and applied in proximal soil sensor devices. However, such devices generally need to be redesigned at a high cost, and this leads to the elimination of the previous generation of inventoried devices, due to their outdated technology. At the same time, the redesigned sensor not only needs to have its communication function tested, but also requires more practice to verify its sensing technology [13]. Therefore, this paper selected an inventoried soil moisture sensor based on an RS-485 interface as the research object; designed and adapted the attachment hardware system (AHS), according to its electrical specifications and communication protocol, integrated LPWAN technology; and realized the technological upgrade of the sensor, so that it not only retained the function and performance of the original sensor, but also had the attributes of ultralow power consumption and long-distance transmission, while supporting long-term battery power supply, easy deployment, and simple management. At the same time, the elimination of an inventoried device due to the application of new technology was avoided, and resources and costs were saved, because the design was based on the inventoried device.

The main contributions of this paper are as follows:

1. This paper put forward a new idea for applying the emerging LPWAN technology in proximal soil sensing systems and carried out engineering practice.
2. Instead of directly eliminating the inventoried proximal soil sensor device with outdated technology, this paper upgraded it by designing an AHS; the new device not only retained the original mature sensing technology of the sensor device, but also exhibited ultralow power consumption and long-distance transmission. In addition, this paper gave full play to the application value and economic value of the devices stored in the inventory.
3. The proposed approach also provides a reference for applying LPWAN technology to a wider range of inventoried sensor devices for technical upgrading.

The rest of this paper is organized as follows: Section 2 presents the overall architecture design of the system. Section 3 describes the hardware design of the AHS. Section 4 introduces the software design of the system. Section 5 tests and analyzes the sensor device, after loading the AHS, and also discusses the relevant factors affecting the communication quality. The last section summarizes this paper and discusses its significance.

2. Design of the System Architecture

The AHS, which was designed to adapt inventoried sensor devices, mainly included an ultralow-power MCU system, a communication module, and a power module. The overall architecture is shown in Figure 1. The AHS took the ultralow-power MCU system as its core, and enabled and controlled the boost chip to turn the working power supply of the sensor device on or off. It obtained the data acquired by the sensor device or configured its relevant parameters by adapting the 485 interface communication protocol; connected and controlled its communication module through UART; and exchanged data with the server through ultralow power wireless transmission, which included uploading the data acquired by the sensor device and receiving the control command parameters sent by the server to control the workflow of the system.

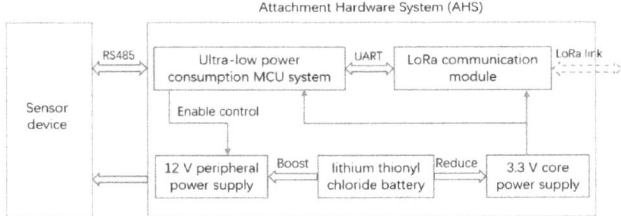

Figure 1. Overall structure of the system.

3. Hardware Design of the AHS

3.1. Proximal Soil Sensor Device and Power Supply

Soil moisture is not only an important part of soil fertility and an important factor affecting plant growth and development, but it is also an important parameter for studying agricultural drought and crop drought. Therefore, data acquisition devices and systems based on various soil moisture sensors have been widely used [14]. In this paper, a commercial soil moisture sensor was used as the research object and was taken as the sensing module of the AHS. Its accuracy and reliability have been tested in practice and in the market for a long time. The volumetric moisture content of the soil was measured with an RS-485 standard communication interface, with a working voltage of 12 V and a response time of less than 1 s. Under the condition of no external load, the maximum working current was less than 25 mA, and the average was no more than 10 mA. More parameters are shown in Figure 2.

Figure 2. Soil moisture sensor.

The system power supply adopted a lithium thionyl chloride battery with an output voltage of 3.6 V and a battery capacity of 3500 mAh. It has the characteristics of high energy density, long service life, and excellent low-temperature performance. It is especially suitable for all-weather battery-based power supply devices in the field [15]. To simplify the hardware structure and facilitate application deployment, the system adopted a single-battery global power supply and an efficient power management scheme. The sensor device adopted a 12 V DC power supply, and the MCU, flash chip, RS-485 transceiver, and other chips adopted 3.3 V power supplies. Therefore, the battery voltage was boosted to 12 V in the circuit hardware, to supply power to the sensor. A stable 3.3 V was output by the multichannel linear voltage regulator to supply power to the chips, in which the main controller (3.3 V) and the peripheral circuit (3.3 V) were independently supplied to eliminate the interaction between loads.

3.2. Ultralow-Power MCU System

MCUs typically use CMOS technology, and their power consumption mainly includes static power consumption and dynamic power consumption. Static power consumption mainly consists of the energy consumed by transistors, which is almost constant, most of the time. Dynamic power consumption includes switching power consumption, short-circuit power consumption, and burr power consumption. In general, especially when working at a high frequency, dynamic power consumption plays a major role, which can be approximately expressed as the following Equation (1) [16]:

$$P = C_L \times V_{DD}^2 \times f \qquad (1)$$

where C_L is the load capacitance, V_{DD} is the supply voltage, and f is the clock frequency. The total power consumption is the sum of the static power consumption and dynamic power consumption. Therefore, to reduce the total power consumption, we can reduce the size of the MCU chip or the number of transistors; reducing the MCU supply voltage can reduce power consumption at the square level and reduce the clock frequency to just meet the application needs. In addition, a reasonable choice of working mode, such as entering sleep mode after working at full speed for a very short time, can also greatly save energy [17–19].

In this paper, the ultralow-power MCU adopted the MSP430 series, which was specially designed for battery-powered devices in field environments [20]. It adopted a low-power supply voltage of 1.8–3.6 V. When operating under the clock condition of 1 MHz, the power consumption in active mode was only approximately 280 µA, in standby mode it was approximately 1.6 µA, and the minimum power consumption in RAM hold mode was only 0.1 µA. In addition, the MSP430 integrated rich on-chip resources and had multiple interrupt sources, which could be arbitrarily nested and used in a flexible and convenient manner. When the system was in a low-power state, the wake-up interrupt took only 5 µs. The minimum ultralow-power MCU system of the AHS is shown in Figure 3.

3.3. Communication Module Based on LoRa

LPWANs have attracted extensive attention, mainly because they can provide affordable connections for low-power devices distributed in very large geographical areas. When realizing the vision of the IoT, LPWAN technologies complement and sometimes even replace conventional wired communication and cellular and short-range wireless technologies, in terms of their performance for various emerging smart city and machine-to-machine applications [21]. Sigfox, LoRa, and NB-IoT are the three leading LPWAN technologies that compete for large-scale IoT deployment, and they have different characteristics that affect the performance of IoT solutions; device connectivity, information delay, and even device battery life [22]. Some of their key characteristics are shown in Table 1.

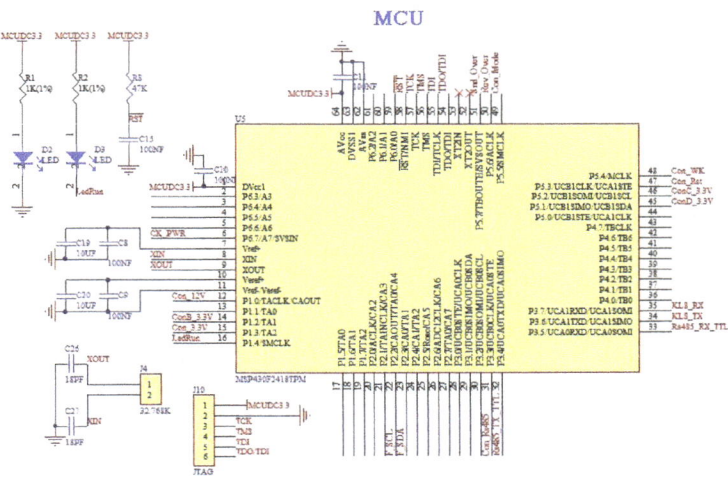

Figure 3. Minimum ultralow-power MCU system.

Table 1. The key characteristics of LPWAN technologies: Sigfox, LoRa, and NB-IoT.

	Sigfox	LoRa	NB-IoT
Frequency	Unlicensed sub-1 GHz ISM bands	Unlicensed sub-1 GHz ISM bands	Licensed LTE frequency bands
Range	10 km (urban), 40 km (rural)	5 km (urban), 20 km (rural)	1 km (urban), 10 km (rural)
Bandwidth	100 Hz	250 kHz and 125 kHz	200 kHz
Maximum data rate	100 bps	50 kbps	200 kbps
Interference immunity	Very high	Very high	Low
Adaptive data rate	No	Yes	No
Allow private network	No	Yes	No

LoRa has the characteristics of long-distance and low power consumption, which can prolong the battery life. It uses the unlicensed Sub-1GHz ISM bands and does not need to pay additional licensing fees. In addition, LoRa can adapt the data rate and allow private networks, while Sigfox and NB-IoT cannot [23]. LoRa, as a representative LPWAN technology, has emerged as an attractive communication platform for the IoT [24,25]. Therefore, in this paper, the mature commercial LoRa module, which was designed based on SemTech sx1278 (Camarillo, CA, USA), was used as the communication module of the AHS, with an adjustable transmission power and a maximum transmission power of 20 dBm; it supported remote wake-up in sleep mode and adopted advanced channel coding technology. Its receiving sensitivity could reach −142 dBm, enabling it to realize long-distance communication under ultralow power consumption. The LoRa gateway was designed based on a sx1301 transceiver controller. The gateway has a higher receiving sensitivity than other technologies, its sight distance coverage radius can reach 5 km, it includes eight receiver channels and one transmission channel (among which 8 receiver channels can receive data simultaneously), and it supports up to 10,000 LoRa terminals, which are convenient for building a massive connection network. It can also support LTE (4G/3G/2G), connect to servers without wiring, and adapt to the multiple access modes of PAAS platforms, such as MQTT, TCP, and Modbus [26].

4. Software Design

4.1. Software Design of the MCU

To reduce power consumption, in addition to selecting low-power devices in the hardware design, the key modules also adopted an efficient power management algorithm in the software design of the MCU. The working voltage of the soil moisture sensor device, which was one of the main energy consumption components of the system, was high. However, in practical applications, the sensor device does not need to work continuously for a long time. In a data acquisition and transmission cycle, it would be idle most of the time. Therefore, a 12 V power supply that enabled control was designed for the hardware. When the sensor device worked and effectively outputted data, the MCU controlled the MOS tube to be in the cut-off state and turned off the power supply of the sensor device to avoid continuous power consumption after data acquisition. The LoRa RF communication module was also a main energy-consuming unit, and the working currents corresponding to 5 dB and 20 dB transmission powers were 75 ma and 130 mA, respectively. When the module did not need to work, the MCU would put it to sleep.

The working flow of the MCU software is shown in Figure 4. After the initialization of the MCU and each module, the MCU controlled the MOS tube to be in the cut-off state, turned off the sensor power supply to reduce energy consumption, enabled the MCU to interrupt, and then entered the low power consumption mode. The timer interrupt function was used to realize the acquisition of sensor data, and the MCU timer could automatically overload the time constant, thereby accurately controlling each module to complete different tasks during different timer cycles (250 ms).

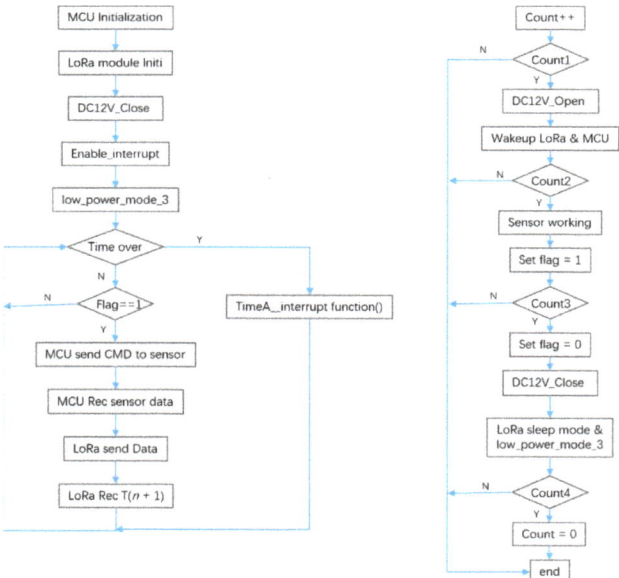

Figure 4. Software flow diagram and timer interrupt function.

4.2. Server Design

The data acquired by the sensor device were finally transmitted to the server for storage and user access. The LPWAN system was composed of a sensor device loaded with the AHS as a node. Its network structure diagram is shown in Figure 5.

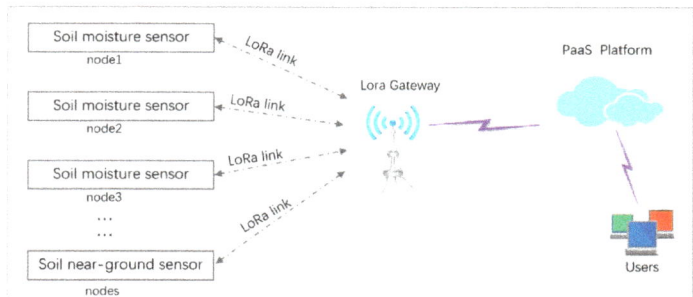

Figure 5. Overall system network structure.

Although a server built based on a private cloud can control all resources, such as computing and storage resources, and enjoy exclusive use rights, it also faces high design, installation, deployment, and upgrading costs, and cannot meet the connection requirements of an increasing number of sensor devices and the management requirements of data for multiple future applications [27]. Therefore, this paper used the operator's IoT platform (OneNet) based on a PAAS as the service end, which was efficient, stable, and safe; could adapt to a variety of network environments and common transmission protocols; provided a fast access scheme, a management service, and data storage capacity for terminal devices; facilitated data storage and querying; and had flexible on-demand payments and controllable costs [28]. Its architecture is shown in Figure 6.

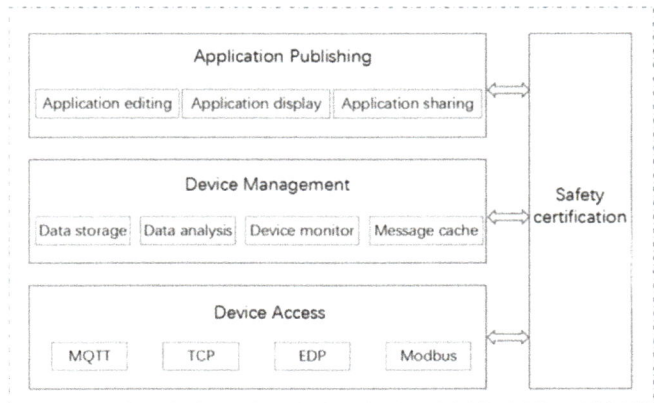

Figure 6. IoT platform architecture based on a PaaS.

In this paper, the sensor node used TCP-based transparent communication to access the IoT platform of the server. We customized the protocol content, wrote the protocol analysis script in the Lua scripting language, and uploaded the analysis script to complete the protocol analysis. The application interface is shown in Figure 7.

5. System Test and Analysis

5.1. Actual Energy Consumption Test and Analysis

The physical object of the AHS and the encapsulated soil moisture sensor loaded with the AHS are shown in Figure 8.

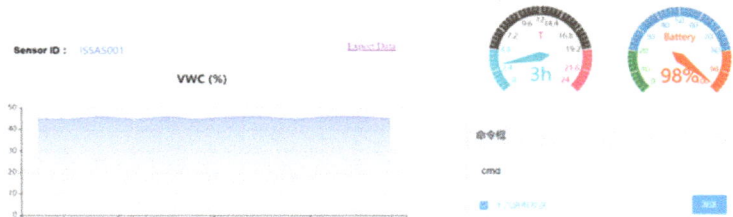

Figure 7. Server application interface.

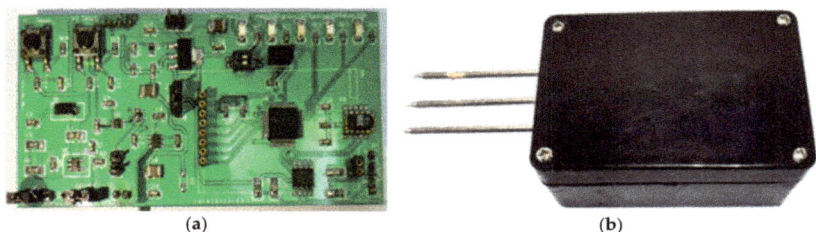

Figure 8. Hardware of the system. (**a**) Physical object of the AHS; (**b**) Soil moisture sensor after loading the AHS.

Energy consumption is a major problem for battery-powered devices. Once the power is exhausted, the device will "strike". Although the system minimized the energy consumption of device selection and algorithm design at the beginning of the design process, there needed to be a gap between the actual energy consumption and the theoretical value [29]. To analyze the actual energy consumption performance of the system, the energy consumption of the sensor device after loading the AHS was tested by connecting a high-precision multimeter in series in the system; the real current of the system in each state was measured, and its single service life was estimated according to the battery capacity. When designing the hardware circuit of the AHS system, a special current test interface was reserved so that the jumper would be used for the short circuit during operation, and the multimeter could be directly connected in series during measurement. In this experiment, the DC micro-ampere mode of a Fluke (18B+) multimeter was used. The interrupt timing cycle was set to 250 ms, the system was initialized within the initial 2 s, and the MCU entered the low power consumption mode after configuring the LoRa module. At the 4th second, the MCU exited the low power consumption mode, the ADC started sampling the battery voltage, and the MCU entered low power consumption mode again after sampling. At the 10th second (COUNT1), the MCU turned on the 12 V power supply of the sensor, the MCU exited low power consumption mode and woke up the LoRa. At the 13th second (COUNT2), the sensor started working, the MCU acquired the sensor data, and LoRa started sending and receiving data. At 15 s (COUNT3), the sensor power supply was forcibly turned off, the MCU entered low power consumption mode, and the LoRa entered sleep mode.

The energy consumption of each main state of the system is shown in Table 2. If the sampling period T was 2 h, i.e., 2 × 3600 s, the energy consumption in one cycle can be expressed as E0 = 2.99 J. If the battery capacity P1 = 3500 mAh, then the single battery energy was E = P1 × 3.6 V = 45,360 J, and the battery life was 2 × E/E0 = 30,340 h; approximately 3.46 years. When working with ultralow power consumption, if effective data were acquired every two hours, a single-battery power supply could work for more than 3 years without considering natural attenuation, thus meeting the requirements of general applications. Additionally, flame-retardant epoxy resin could be used for integral molding and pouring; this would make the system more compact as a whole, with high mechanical strength, strong heat resistance, and easy deployment, as well as being maintenance-free, waterproof, and anti-corrosion.

Table 2. Energy consumption of each main state of the system.

	ADC Sampling Battery Voltage	Sensor Work	Data Sending and Receiving	Low Power Mode
Voltage(V)	3.3	12	3.3	3.3
Current (mA)	15.4	40.9	168.3	4.3×10^{-3}
Duration (s)	6	3	2	T-11

5.2. Channel Characteristics and Gateway Capacity Analysis

The key parameter settings of the node are shown in Table 3. When setting the parameters of radio device, on the basis of meeting the radio management specifications, we optimized the LoRa modulation and demodulation technology through designing the key parameters, such as modulation spread factor, modulation bandwidth, and error correction coding rate, to make the system reach an optimal state, as far as possible [30,31]. The spread spectrum LoRa modulation is performed by representing each bit of the payload information using multiple chips of information. The rate at which the spread information is sent is referred to as the symbol rate; while, the ratio between the nominal symbol rate and chip rate is the spreading factor and represents the number of symbols sent per bit of information. Spread spectrum transmission can reduce the bit error rate; that is, the SNR, as shown in Table 4. Under the condition of a negative signal-to-noise ratio, the signal can be received normally, which improves the sensitivity, link budget, and coverage of the LoRa receiver, but reduces the actual data that can be transmitted under the condition of the same amount of data [32]. Therefore, the larger the spread spectrum factor, the smaller the number rate (bit rate) of the transmitted data. In this paper, we set the spreading factor (SF) as 12 to maximize the signal coverage, under the condition of meeting the transmission rate. The LoRa modem employs cyclic error coding to perform forward error detection and correction. Such error coding incurs a transmission overhead, but it can further improve the robustness of the link. Therefore, we set the coding rate (CR) as 4/5. An increase in signal bandwidth (BW) permits the use of a higher effective data rate; thus, reducing transmission time at the expense of a reduced sensitivity improvement [33]. Apparently, there are regulatory constraints in most countries on the permissible occupied bandwidth. As it is stipulated in China that the power in 470~−510 mHz frequency band shall not exceed 50 mW (17 dBm (ERP)) and the occupied bandwidth shall not exceed 200k [34], we set the bandwidth to 125 k; considering the cable loss and air path loss, we set the transmission power of the node to 20 dBm. In short, these parameters were closely related to the range and robustness of radio communication links. Changing the BW, SF, and CR would change the link budget and transmission time. It was necessary to have a trade-off between battery life and distance.

For large-scale LoRa connection applications, gateway capacity is an important characteristic [35,36], especially in a typical suburban farming environment; and whether the gateway is sufficient for the determined number of nodes is an important concern. In the same application scenario, for a certain gateway, the maximum number of packets that can be received per day is also determined. However, different packet forms and sending rates will change the total number of packets. The LoRa standard data frame format is shown in Figure 9.

Table 3. The key parameter settings of the node.

TFREQ	RFREQ	POW	BW	TSF	RSF	CR
475.5 MHz	506.5 MHz	20 dBm	125 kHz	12	12	4/5

Table 4. Range of spreading factors.

Spreading Factor	7	8	9	10	11	12
Demodulator SNR	−7.5 dB	−10 dB	−12.5 dB	−15 dB	−17.5 dB	−20 dB

Note that the spreading factor must be known in advance on both transmit and receive sides of the link, as different spreading factors are orthogonal to each other.

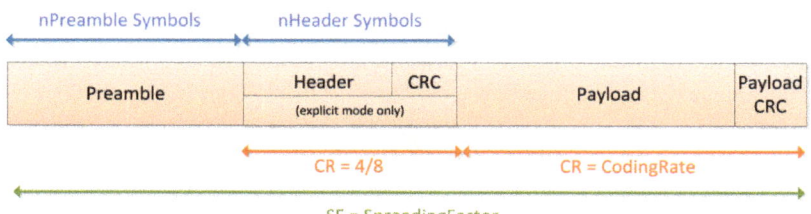

Figure 9. LoRa packet structure.

The data frame includes a preamble byte, a header byte, a payload, and an optional CRC byte for synchronization. Although the number of preamble bytes can be programmable, the number of remaining bytes depends on the coding rate and spreading factor used in other parameters. The number of preamble symbols is generally set to $M_{preamble} = 4.25 + N_{prog}$, where N_{prog} is the programmed preamble length. The total number of bytes of the physical layer data frame is calculated using Equation (2) [37].

$$M = \left[M_{preamble} + 8 + M_{SF} * (CR + 4) \right] \quad (2)$$

$$M_{SF} = max\left(\left[\frac{8PL - 4SF + 28 + 16CRC - 20IH}{4(SF - 3DE)} \right], 0 \right) \quad (3)$$

Equation (3) gives M_{SF}, which mainly gives the number of payload symbols, where $CR \in \{1, 2, 3, 4\}$ represents the coding rate of $4/(CR + 4)$; PL is the MAC layer, including MAC header and application data payload (in bytes); SF is the spread spectrum factor. If the optional function CRC is enabled, $CRC = 1$; $IH = 1$ indicates that the implicit header function is enabled (i.e., the physical layer header is not transmitted); and $DE = 1$ indicates that the data optimization function is activated. For a given combination of spreading factor (SF), coding rate (CR), and signal bandwidth (BW) the total on-the-air (ToA) transmission time of a LoRa packet can be calculated using Equation (4), where Ts is the transmission time of one symbol, which is calculated using Equation (5).

$$ToA = Ts * M \quad (4)$$

$$Ts = 2^{SF}/BW \quad (5)$$

For a LoRa gateway with eight channels, Equation (6) calculates the channel capacity (i.e., number of nodes) without LBT (listen before talk) [38].

$$S = 8T/(2e * ToA) \quad (6)$$

where 8 represents eight channels, T represents the transmission interval, which is related to the packet length and rate. While, $1/2e$ is the maximum throughput of the basic Aloha algorithm and e is a constant, equal to 2.718. Under the premise of 10-byte preload, the relationship between different *SF* and *BW* and their theoretical gateway capacity are shown in Figures 10 and 11.

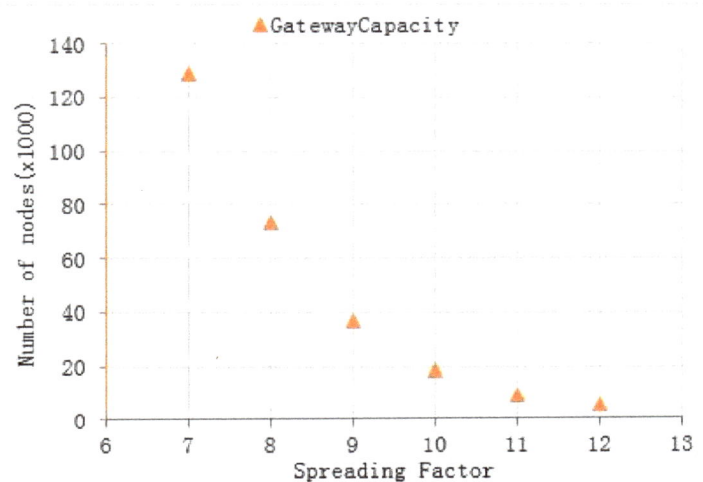

Figure 10. When BW = 125 kHz, T = 3600 s, the gateway capacity at different *SF*.

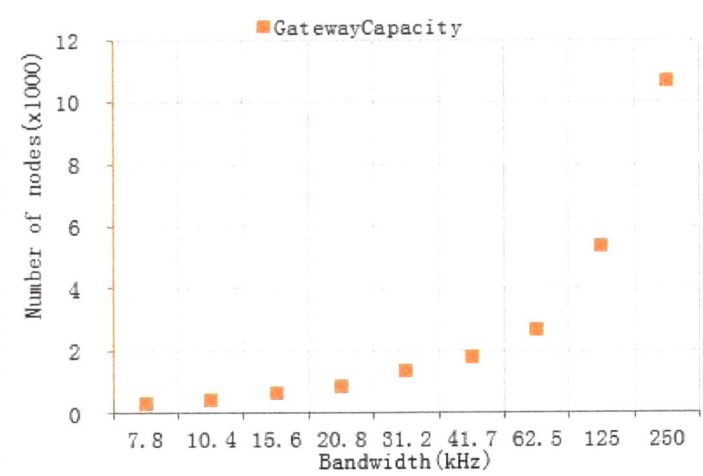

Figure 11. When *SF* = 12, T = 3600 s, the gateway capacity at different bandwidths.

If different algorithms are adopted, this will also lead to a change of maximum throughput, resulting in a change of theoretical capacity. For example, if the precondition is modified so that each node has a LBT function and the slot Aloha algorithm is used instead of the previous basic Aloha algorithm, the maximum throughput is different, due to different algorithms. At this time, the maximum throughput is $1/e$, so the theoretical capacity of the channel will be doubled. It can be seen that under the condition of setting parameters as shown in Table 3, a single LoRa gateway can theoretically connect 5345 nodes. In practical applications, the gateway can receive SF7–SF12 signal data at the same time.

Due to the limited demodulation and coverage capacity of a single gateway, in reality, it is actually difficult to meet the requirements of the theoretical capacity, but it can be deployed with multiple gateways to maximize the network capacity.

5.3. Communication Test and Discussion

In principle, a wireless communication gateway should be deployed at the highest possible position, such as a communication operator's iron tower or the roof of a high-rise building, to improve the communication distance and signal quality. In practical applications, the site environment, operating conditions, economic cost, and other factors need to be fully considered [39]. This test took the farms around the Red Azalea Agricultural Ecological Park (RAAEP) in the Baguazhou area as the test site, to evaluate the communication distance and signal coverage between the gateway and the nodes in a typical suburban natural farmland environment. No tower or high-rise building was available for the operators in the area, no advantageous terrain was available, and certain obstacles were contained in the communication space. The test took the RAAEP as the starting point, and considering the implementation difficulty and cost control, the LoRa gateway device was deployed on a billboard approximately 2.5 m above the ground (Figure 12c), while the mobile power supply was used to power the LoRa gateway (Figure 12b). A communication test route diagram is shown in Figure 13. The AHS was specially programmed for the data transmission test as a terminal node (Figure 12a). We drove along the lane with the terminal node for the communication test, and several test points were placed in the southwest direction. Tall and dense trees were located on both sides of the road, but there were relatively open road areas at 450–500 m and 750–800 m in front of the starting point. The system started to enter a village at 1000 m, passed through the village at 1100 m, and entered woods on the two sides of the road at the same time. A highway bridge was located at 2200 m, and the test route crossed under the highway bridge.

(a) (b) (c)

Figure 12. Communication test: the LoRa gateway was placed on a billboard 2.5 m above the ground. (**a**) LoRa terminal node; (**b**) LoRa gateway with the DTU; (**c**) Gateway placement location.

At each test location, the terminal node sent a group of sequentially numbered data every 3 s, for a total of no less than 20 groups. The gateway received the data, printed the received signal strength indication (RSSI) and signal-to-noise ratio (SNR) information of the data, and uploaded this information to the cloud through a data transfer unit (DTU). We calculated the average RSSI and SNR of the test points at the same distance, which are shown in Figure 14.

Figure 13. Communication test route for each designated test location, and the node used to send data in the simulations. The red star symbol indicates the starting point of the test.

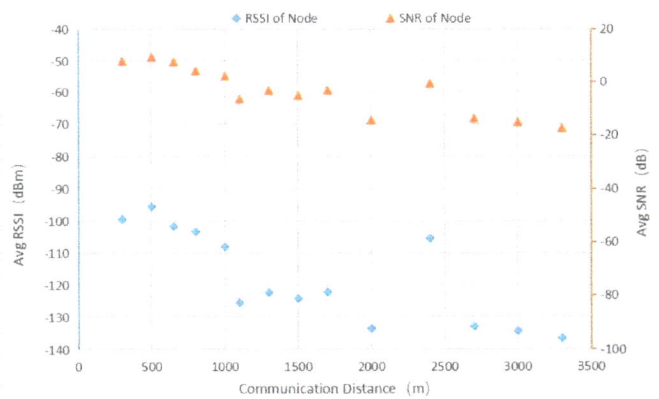

Figure 14. RSSI and SNR values.

Figure 15 shows the data packet loss rate. As seen from the figure, with the increase in the communication distance, the RSSI and SNR gradually decreased, and the packet loss rate gradually increased. At 500 m and 800 m from the test point, the area was relatively open, the influence of tree shielding was small, and the received signal improved. Although signals were still received at 1100 m, the packet loss rate was too high, and the communication accuracy was lost. Therefore, in practical applications, the communication quality evaluation should focus on more than the RSSI, and the packet loss rate was the prerequisite for the evaluation of communication quality.

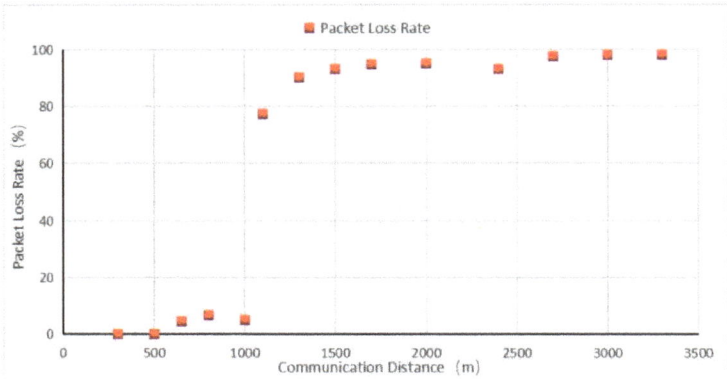

Figure 15. Packet loss rate.

In a communication system, if the signal power value in the communication link is equal to, or greater than, the sensitivity of the receiver, the receiver can normally obtain the information contained in the transmitted signal; the communication is successful. On the contrary, if the signal power is lower than the sensitivity, the quality of information obtained will be far lower than the specified requirements [40].

Figure 16 shows the obtained signal strength distribution. We selected a set of test data under relatively poor test conditions (such as antennas without enhanced gain), where the analyzed system performance would have more redundancy space. There were 135 received signal strength data points in total. During the test, the distance between the node and the gateway ranged from 300 m to 1300 m. Among the valid data points obtained, there were 18 at 300 m, 22 at 500 m, 19 at 650 m, 26 at 800 m, 19 at 1000 m, 5 at 1100 m, and 26 at 1300 m. The signal strength of these data ranged from -142.5 dBm to -119.8 dBm, including 1 data point greater than -120 dBm, 57 data points less than -120 dBm and greater than -130 dBm, 66 data points less than -130 dBm and greater than -140 dBm, and 11 data points less than or equal to -140 dBm. From the signal strength analysis, 92% of the test signals were greater than -140 dBm, while the received signal sensitivity of LoRa gateway was -142 dBm. Therefore, the RSSI of this test was within the acceptable range. However, when combined with the packet loss rate data analysis, the coverage radius of a single gateway should not exceed 1100 m.

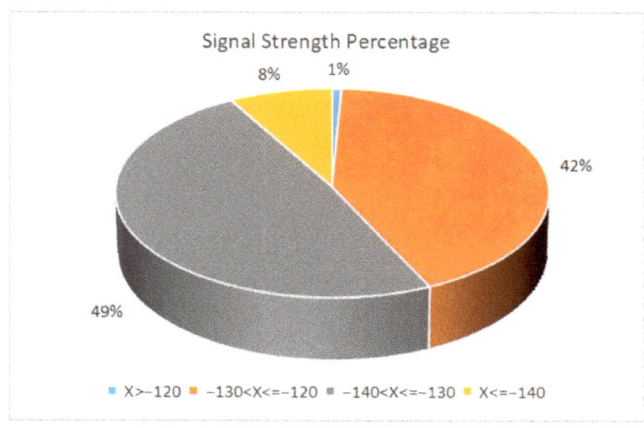

Figure 16. Signal strength distribution.

Compared with the theoretical parameters, the actual test data parameters, especially the communication distance, had large gaps. Many factors restrict wireless communication distance.

In an ideal environment, wireless communication satisfies the Friis transmission equation [41,42]. After considering the loss of the free space path, the Friis transmission equation can be transformed into the following Equation (7):

$$\text{Pt} - \text{Pr} + \text{Gt} + \text{Gr} = 20\lg\frac{4\pi f d}{c} + Lc + L0 \qquad (7)$$

where Pt is the transmission power of the transmitter, Pr is the sensitivity of the receiver, Gt is the transmitter antenna gain, Gr is the receiver antenna gain, f is the carrier frequency, d is the distance between the receiver and transmitter antennas, c is the speed of light, Lc is the feeder loss of the transmitter antenna at the base station, and L0 is the air propagation loss. Here, π and c are constants; therefore, Equation (7) can be easily converted into the following Equation (8):

$$\text{Pt} - \text{Pr} + \text{Gt} + \text{Gr} = 20\lg(f) + 20\lg(d) + Lc + L0 - 147.56(\text{dB}) \qquad (8)$$

Equation (8) can be converted to Equation (9) to calculate the distance:

$$d = 10^{\frac{\text{Pt} - \text{Pr} + \text{Gt} + \text{Gr} - 20\lg(f) - Lc - L0 + 147.56(\text{dB})}{20}} \qquad (9)$$

Therefore, according to the theoretical calculation formula, the factors affecting the wireless communication distance include the system's own factors, such as receiving sensitivity, transmission power, transmitter, and receiver antenna gain, as well as environmental conditions such as obstacles, transmitter and receiver antenna height, electromagnetic interference and weather influence. In the system parameter setting designed in this paper, considering the data transmission rate and battery life, we set the maximum SF and transmission power to maximize the sensitivity of the node. Therefore, the main factors affecting the communication distance of the system came from the antenna gain and the air propagation loss caused by the obstacles between the sensor node and the gateway. In our system, we chose an antenna with high gain as much as possible; however, due to the consideration of the overall waterproof and anti-corrosion properties of the sensor node, the transmitting antenna was encapsulated inside the node, which led to increased propagation loss. As mentioned in the previous communication test section, the LoRa gateway was deployed on a billboard about 2.5 m above the ground. The sensor node test location passed through the village, and there were tall and dense trees on both sides of the test route. These test conditions well simulated the low-cost deployment mode in a typical suburban farming environment, but objectively caused the propagation loss of wireless communication and greatly reduced the wireless communication distance. Due to the implementation environmental conditions, deployment difficulty, and cost, we did not deploy the LoRa gateway at a higher position for testing, but from the calculation formula, we could show that by deploying the gateway at a commanding height over the environment, we could reduce the obstacles between communications, reduce the air propagation loss, and improve the communication distance exponentially.

Overall, when a sensor device is designed as the node of an LPWAN, the transmission power, reception sensitivity, and carrier frequency are subject to the node power consumption and chip performance factors. In practical applications, considering the implementation environmental conditions, economic cost, and deployment and maintenance difficulty, we cannot blindly pursue the ideal communication transmission distance; thus, we need to find a balance and deploy the network reasonably [43].

6. Conclusions

With the idea of saving resources and costs, this paper applied LPWAN technologies to an inventoried proximal soil sensor device by designing an attachment hardware system

(AHS) and realized technical upgrades. Compared with conventional sensors based on wired communication technologies (such as RS-485 and SDI-12), mobile cellular technologies (such as 2G, 3G, 4G, etc.), and short-range wireless communication technologies (such as Bluetooth, ZigBee, etc.), it not only retained the original mature sensing technology of the sensor device but also exhibited ultralow power consumption and long-distance transmission, while having the advantages of an LPWAN. At experimental level, it can be seen from the actual energy consumption test and analysis that a single-battery power supply could work for more than 3 years without natural attenuation; thus, meeting the requirements of general applications. Additionally, flame-retardant epoxy resin can be used for integral molding and pouring, and this would make the system more compact as a whole, with high mechanical strength, strong heat resistance, and easy deployment, as well as being maintenance-free, waterproof, and anti-corrosion. However, traditional sensors need to be powered by mains power supply, which were troublesome to deploy in applications and required a lot of maintenance in the later stages. Even if some used a battery power supply, the sensor devices were still based on the traditional application design, and their power consumption could not use a long-term battery power supply, and were troublesome to maintain. Furthermore, through the communication distance test, signal coverage test, and gateway capacity analysis, it was shown that in a typical suburban farming environment, a single gateway could carry more than 5000 nodes within 1100 m, which could easily and quickly deploy a large-scale wireless sensor network; whereas, the traditional types would require a huge cost to achieve a large-scale sensor network. Finally, the sensor designed in this paper could obtain data remotely in real time, while the latter needed to obtain data manually on site.

The technical means to instantly obtain various soil physicochemical parameters in a field is not only an important research direction in soil science but also an important technical support tool for the development of conventional laboratory-based physicochemical soil testing and analysis procedures for field-based measurements [2]. The development and application of LPWAN technology has enabled low-cost, low-power, wide-coverage, and real-time soil field monitoring. In this paper, an AHS with LPWAN technology based on LoRa was designed and applied to an inventoried soil moisture sensor, to upgrade the technology so that it, not only retained the performance, accuracy, and reliability of the original sensor, but also had the ultralow power consumption and long-distance wireless transmission function of an LPWAN. After loading the AHS, the sensor device could be built and deployed as a node in a wireless sensor network in an economical, flexible, and convenient manner; this not only expanded the applicability of the LPWAN, but also prevented the elimination of inventoried soil moisture sensors, due to their outdated technology. It is further concluded that not only soil moisture sensors, but also other inventoried proximal soil sensor devices based on conventional communication methods (such as RS-485, SDI-12 and other data communication methods) or devices whose outputs are standard voltages or currents could be designed with, or adopt, AHSs with technical designs that require ultralow power consumption; in this way, they can not only possess the technical advantages and application capacities of an LPWAN, but also retain their original mature sensing technology and give full play to the application value and economic value of inventoried proximal soil sensor devices, to avoid a waste of resources.

Author Contributions: The contributions to this manuscript were as follows: conceptualization: Y.T.; methodology: W.H.; hardware and software development: Y.T. and H.T.; data analysis: Y.T.; field trials: Y.T.; writing: Y.T.; review and editing: H.T.; supervision: W.H. All authors have read and agreed to the published version of the manuscript.

Funding: This research was funded by the National Key Research and Development Project of China (Grant No. 2017YFF0108201), the Open Project of the State Key Laboratory of Soil and Sustainable Agriculture (Institute of Soil Science, Chinese Academy of Sciences (Grant number Y20160005), and the Nanjing Science and Technology project of China (Grant No. 2020011002), and the Youth Innovation Promotion Association of the Chinese Academy of Sciences (Grant No. 2019312).

Institutional Review Board Statement: Not applicable.

Informed Consent Statement: Not applicable.

Data Availability Statement: Not applicable.

Conflicts of Interest: The authors declare no conflict of interest.

Abbreviations

IoT	Internet of Things
LPWAN	Low-Power Wide-Area Network
AHS	Attachment Hardware System
ECa	apparent Electrical Conductivity
ISEs	Ion Selective Electrodes
ISFETs	Ion Sensitive Field Effect Transistors
MCU	Microcontroller Unit
PaaS	Platform as a Service
VWC	Volumetric Water Content
MQTT	Message Queuing Telemetry Transport
RAAEP	Red Azalea Agricultural Ecological Park
RSSI	Received Signal Strength Indication
SNR	Signal-to-Noise Ratio
DTU	Data Transfer Unit
ERP	Effective Radiated Power
BW	Band Width
SF	Spreading Factor
CR	Coding Rate
LBT	Listen Before Talk

References

1. Adamchuk, V.I.; Rossel, R.A.V. Development of On-the-Go Proximal Soil Sensor Systems. In Proceedings of the 1st Global Workshop on High Resolution Digital Soil Sensing and Mapping, University of Sydney, Faculty of Agriculture, Food & Nat Resources, Sydney, Australia, 5–8 February 2008.
2. Shi, Z.; Guo, Y.; Jin, X.; Wu, H. Advancement in study on proximal soil sensing. *Acta Pedol. Sin.* **2011**, *48*, 1274–1281.
3. Rossel, R.A.V.; McBratney, A.B. Laboratory evaluation of a proximal sensing technique for simultaneous measurement of soil clay and water content. *Geoderma* **1998**, *85*, 19–39. [CrossRef]
4. Rossel, R.A.V.; McBratney, A.B.; Minasny, B. *Proximal Soil Sensing*; Springer Science + Business Media B V: Berlin, Germany, 2010.
5. Myers, D.B.; Kitchen, N.R.; Sudduth, K.A.; Grunwald, S.; Miles, R.J.; Sadler, E.J.; Udawatta, R.P. Combining Proximal and Penetrating Soil Electrical Conductivity Sensors for High-Resolution Digital Soil Mapping. In Proceedings of the 1st Global Workshop on High Resolution Digital Soil Sensing and Mapping, University of Sydney, Faculty of Agriculture, Food & Nat Resources, Sydney, Australia, 5–8 February 2008.
6. Besson, A.; Cousin, I.; Richard, G.; Bourennane, H.; Pasquier, C.; Nicoullaud, B.; King, D. Changes in Field Soil Water Tracked by Electrical Resistivity. In Proceedings of the 1st Global Workshop on High Resolution Digital Soil Sensing and Mapping, University of Sydney, Faculty of Agriculture, Food & Nat Resources, Sydney, Australia, 5–8 February 2008; pp. 275–282.
7. Adsett, J.F.; Thottan, J.A.; Sibley, K.J. Development of an automated on-the-go soil nitrate monitoring system. *Appl. Eng. Agric.* **1999**, *15*, 351–356. [CrossRef]
8. Vu, V.A.; Trinh, D.C.; Truvant, T.C.; Bui, T.D. Design of automatic irrigation system for greenhouse based on LoRa technology. In Proceedings of the International Conference on Advanced Technologies for Communications (ATC), Ho Chi Minh City, Vietnam, 18–20 October 2018; pp. 72–77.
9. Bembe, M.; Abu-Mahfouz, A.; Masonta, M.; Ngqondi, T. A survey on low-power wide area networks for IoT applications. *Telecommun. Syst.* **2019**, *71*, 249–274. [CrossRef]
10. Pitu, F.; Gaitan, N.C. Surveillance of SigFox technology integrated with environmental monitoring. In Proceedings of the 15th International Conference on Development and Application Systems (DAS), Suceava, Romania, 21–23 May 2020; pp. 69–72.
11. Rachmani, A.F.; Zulkifli, F.Y. IEEE In Design of IoT Monitoring System Based on LoRa Technology for Starfruit Plantation. In Proceedings of the IEEE-Region-10 Conference (IEEE TENCON), IEEE Reg 10, Jeju, Korea, 28–31 October 2018; pp. 1241–1245.
12. Co, J.; Tiausas, F.J.; Domer, P.A.; Guico, M.L.; Monje, J.C.; Oppus, C. IEEE In Design of a Long-Short Range Soil Monitoring Wireless Sensor Network for Medium-Scale Deployment. In Proceedings of the IEEE-Region-10 Conference (IEEE TENCON), IEEE Reg 10, Jeju, Korea, 28–31 October 2018; pp. 1371–1376.

13. Hardie, M.; Hoyle, D. Underground Wireless Data Transmission Using 433-MHz LoRa for Agriculture. *Sensors* **2019**, *19*, 4232. [CrossRef] [PubMed]
14. Babaeian, E.; Sadeghi, M.; Jones, S.B.; Montzka, C.; Vereecken, H.; Tuller, M. Ground, Proximal, and Satellite Remote Sensing of Soil Moisture. *Rev. Geophys.* **2019**, *57*, 530–616. [CrossRef]
15. Liu, J.; Ge, H.; Zhou, G.; Wu, Y.; Bao, B. Research status of lithium/thionyl chloride battery. *Battery* **2005**, *35*, 408–410.
16. Gao, D. *Low Power Design for Key Circuits of Baseband Processor in Wireless*; Graduate School of Chinese Academy of Sciences (Shanghai Institute of Microsystem and Information Technology): Shanghai, China, 2007.
17. Xu, Z.; Yang, L. Design Methodology for Low-Power CMOS integrated CIrcuits. *Microelectronics* **2004**, *34*, 223–226.
18. Chandrakasan, A.P.; Sheng, S.; Brodersen, R.W. Low-Power Cmos Digital Design. *IEEE J. Solid-St Circ.* **1992**, *27*, 473–484. [CrossRef]
19. Najm, F.N. IEEE In Low-power design methodology: Power estimation and optimization. In Proceedings of the 40th Midwest Symposium on Circuits and Systems, Sacramento, CA, USA, 3–6 August 1997; pp. 1124–1129.
20. Wang, C.C.; Hsiao, Y.H.; Huang, M.C. Development of MSP430-based ultra-low power expandable underwater acoustic recorder. *Ocean Eng.* **2009**, *36*, 446–455. [CrossRef]
21. Raza, U.; Kulkarni, P.; Sooriyabandara, M. Low Power Wide Area Networks: An Overview. *IEEE Commun. Surv. Tut.* **2017**, *19*, 855–873. [CrossRef]
22. de Oliveira, F.C.; Rodrigues, J.; Rabelo, R.A.L.; Mumtaz, S. Performance Delay Comparison in Random Access Procedure for NB-IoT, LoRa, and SigFox IoT Protocols. In Proceedings of the IEEE 1st Sustainable Cities Latin America Conference (SCLA), Arequipa, Peru, 26–29 August 2019.
23. Mekki, K.; Bajic, E.; Chaxel, F.; Meyer, F. A comparative study of LPWAN technologies for large-scale IoT deployment. *ICT Express* **2019**, *5*, 1–7. [CrossRef]
24. Thoen, B.; Callebaut, G.; Leenders, G.; Wielandt, S. A Deployable LPWAN Platform for Low-Cost and Energy-Constrained IoT Applications. *Sensors* **2019**, *19*, 12. [CrossRef]
25. Haxhibeqiri, J.; De Poorter, E.; Moerman, I.; Hoebeke, J. A Survey of LoRaWAN for IoT: From Technology to Application. *Sensors* **2018**, *18*, 3995. [CrossRef]
26. Chen, J.J.; Liu, V.; Caelli, W. An Adaptive and Autonomous LoRa Gateway for Throughput Optimisation. In Proceedings of the Australasian Computer Science Week Multiconference (ACSW), Sydney, Australia, 29–31 January 2019.
27. Emeakaroha, V.C.; Cafferkey, N.; Healy, P.; Morrison, J.P. In A Cloud-based IoT Data Gathering and Processing Platform. In Proceedings of the 3rd International Conference on Future Internet of Things and Cloud (FiCloud)/International Conference on Open and Big Data (OBD), Rome, Italy, 24–26 August 2015; pp. 50–57.
28. Du, J.Y.; Guo, J.B.; Xu, D.D.; Huang, Q. A Remote Monitoring System of Temperature and Humidity Based on OneNet Cloud Service Platform. In Proceedings of the IEEE Electrical Design of Advanced Packaging and Systems Symposium (EDAPS), Haining, China, 14–16 December 2017.
29. Zhang, X.S.; Han, M.D.; Meng, B.; Zhang, H.X. High performance triboelectric nanogenerators based on large-scale mass-fabrication technologies. *Nano Energy* **2015**, *11*, 304–322. [CrossRef]
30. Farooq, M.O. Multi-hop communication protocol for LoRa with software-defined networking extension. *Internet Things* **2021**, *14*, 100379. [CrossRef]
31. Muthanna, M.S.A.; Muthanna, A.; Rafiq, A.; Hammoudeh, M.; Alkanhel, R.; Lynch, S.; Abd El-Latif, A.A. Deep reinforcement learning based transmission policy enforcement and multi-hop routing in QoS aware LoRa IoT networks. *Comput. Commun.* **2022**, *183*, 33–50. [CrossRef]
32. Mayer, P.; Magno, M.; Brunner, T.; Benini, L. LoRa vs. LoRa: In-Field Evaluation and Comparison For Long-Lifetime Sensor Nodes. In Proceedings of the 8th IEEE International Workshop on Advances in Sensors and Interfaces (IWASI), Otranto, Italy, 13–14 June 2019; pp. 307–311.
33. Zhang, C.; Wang, S.; Jiao, L.; Shi, J.; Yue, J. A Novel MuLoRa Modulation Based on Fractional Fourier Transform. *IEEE Commun. Lett.* **2021**, *25*, 2993–2997. [CrossRef]
34. Semtech Corporation. "137 MHz to 1020 MHz Low Power Long Range Transceiver" DS_SX1276-7-8-9_W_APP_V7 Datasheet, September 2013 [Revised May 2020]. Available online: https://www.semtech.com/products/wireless-rf/lora-core/sx1276#datasheets (accessed on 15 October 2020).
35. Lavric, A.; Popa, V. Performance Evaluation of LoRaWAN Communication Scalability in Large-Scale Wireless Sensor Networks. *Wirel. Commun. Mob. Comput.* **2018**, *2018*, 6730719. [CrossRef]
36. Elshabrawy, T.; Robert, J. Capacity Planning of LoRa Networks With Joint Noise-Limited and Interference-Limited Coverage Considerations. *IEEE Sens. J.* **2019**, *19*, 4340–4348. [CrossRef]
37. Available online: https://www.miit.gov.cn/jgsj/wgj/wjfb/art/2020/art_792ab84586c34b64bb391eb64 (accessed on 19 October 2020).
38. Van den Abeele, F.; Haxhibeqiri, J.; Moerman, I.; Hoebeke, J. Scalability Analysis of Large-Scale LoRaWAN Networks in ns-3. *IEEE Internet Things* **2017**, *4*, 2186–2198. [CrossRef]
39. Kam, O.M.; Noel, S.; Kasser, P.; Tanougast, C.; Ramenah, H.; Adjallah, K.H. Supervision and energy management system for smart telecom tower based on the LoRaWAN protocol. In Proceedings of the 6th International Conference on Control, Decision and Information Technologies (CoDIT), Conservatoire Nat Arts Metiers, Paris, France, 23–26 April 2019; pp. 539–544.

40. Zhang, J.-l.; Wang, M.; Tian, E.-m.; Li, X.; Wang, Z.-b.; Zhang, Y. Analysis and Experimental Verification of Sensitivity and SNR of Laser Warning Receiver. *Spectrosc. Spect. Anal.* **2009**, *29*, 20–23.
41. Shaw, J.A. Radiometry and the Friis transmission equation. *Am. J. Phys.* **2013**, *81*, 33–37. [CrossRef]
42. Bush, R.T. The antenna formula—An application of single-slit diffraction theory. *Am. J. Phys.* **1987**, *55*, 350–351. [CrossRef]
43. Zhang, Z.; Cao, S.; Zhu, J.; Chen, J. Long range low power sensor networks with LoRa sensor for large area fishery environment monitoring. *Trans. Chin. Soc. Agric. Eng.* **2019**, *35*, 164–171.

Article

AI Based Digital Twin Model for Cattle Caring

Xue Han [1,*,†], Zihuai Lin [1,†], Cameron Clark [2,†], Branka Vucetic [1] and Sabrina Lomax [2]

1 Centre of IoT and Telecommunication (CIoTT), School of Electrical and Information Engineering, Faculty of Engineering, University of Sydney, Camperdown, NSW 2006, Australia
2 Livestock Production and Welfare Group, School of Life and Environmental Sciences, Faculty of Science, University of Sydney, Camden, NSW 2570, Australia
* Correspondence: xhan3023@uni.sydney.edu.au; Tel.: +61-0498160158
† These authors contributed equally to this work.

Abstract: In this paper, we develop innovative digital twins of cattle status that are powered by artificial intelligence (AI). The work is built on a farm IoT system that remotely monitors and tracks the state of cattle. A digital twin model of cattle based on Deep Learning (DL) is generated using the sensor data acquired from the farm IoT system. The physiological cycle of cattle can be monitored in real time, and the state of the next physiological cycle of cattle can be anticipated using this model. The basis of this work is the vast amount of data that is required to validate the legitimacy of the digital twins model. In terms of behavioural state, this digital twin model has high accuracy, and the loss error of training reach about 0.580 and the loss error of predicting the next behaviour state of cattle is about 5.197 after optimization. The digital twins model developed in this work can be used to forecast the cattle's future time budget.

Keywords: digital twin; AI; deep learning; LSTM model

1. Introduction

Digital twins are virtual digital representations of physical objects, in which the physical object and its corresponding virtual digital representation interact remotely in real time [1]. A digital twin model incorporates multi-disciplinary, multi-physical quantity, multi-scale, and multi-probability simulation processes and fully utilises physical models, sensor updates, operation histories, and other data [2]. In addition, digital twins complete the mapping in virtual space so that the full life cycle process of associated entity equipment is reflected [3]. Digital twins are a transcendental idea that can be regarded as one or more crucial and interdependent digital mapping systems for the actual object [4,5].

Connectivity, modularity, and autonomy between virtual and actual items can all be realised with digital twins. It can be accomplished across the whole production process from product design through product system engineering to production planning, implementation and intelligence, resulting in a self-optimizing closed loop [6]. To put it another way, by connecting the actual object with the virtual number, the real object may offer real information to optimise the digital model, and the digital model can foresee potential situations to alter the real object. The two complement each other to create a self-closing optimisation mechanism [7]. Nowadays, digital twins have been increasingly employed in a variety of industries, including product design, product manufacturing, medical analysis, engineering construction and other areas [8]. As a result, digital twins can be seen as a major force behind the intelligent manufacturing paradigm [9]. Digital twins have recently been deployed in a variety of fields, including livestock farming [10,11].

Deep learning (DL) is a new direction in machine learning that is being introduced to bring it closer to the goal of AI and has made tremendous progress in solving issues that were previously unsolvable in AI. It has proven to be so effective in detecting complicated structures in high-dimensional data that it might be used in a wide range of scientific, business and government applications [12–14]. The long short-term memory network (LSTM) is a type of cyclic neural network and one of the deep learning algorithms that can analyse and forecast critical time with very long intervals and delays in time series [15,16]. In a long time series, the LSTM neural network algorithm can determine which information should be stored and which should be discarded [17]. The development of digital twins relies heavily on accurate time series prediction. Internal and external disruptions might result in time series that are exceedingly nonlinear and random. Complex object time series prediction may be employed at any stage of their life cycle, which is also a major component of the digital twin model [18,19]. Therefore, it is extremely dependable to use the LSTM model to build digital twins.

This research is primarily focused on the direction of intelligent livestock monitoring in the agricultural environment. For a long time, Australia has been a major producer of animal husbandry, and milk and beef production and export have also been at the forefront of the global [20]. Cattle statuses may be monitored in real time to enable breeders better determine their cattle's health and enhance meat and milk output correspondingly [21]. As a result, agriculture's evolution toward intelligence is a critical stage of growth [22]. This project aims to create a digital twin model for each individual bovine, which will allow for improved monitoring of cattle status at the digital and information levels, as well as the advancement of Australian animal husbandry. The main contribution of this study is that it developed an intelligent digital twins approach using an LSTM neural network to give a range of behavioural detection and prediction of cattle's state, such as impending physiological cycles, among other things. The digital twins model is significantly based on massive volumes of data reflecting cattle location, movement and free grazing time, etc, collected by the farm IoT monitoring system. The digital twin model has some limitations; for example, when the amount of sampled data is inadequate, the model's accuracy is unsatisfactory. As a result, this model needs a considerable amount of sample data.

The outline of the paper is given below. First in Section 2, the current related work of digital twin is summarized. After that, in Section 3, necessary data mining and data analysing for the IoT system are carried out. In this part, most of the data processing work is accomplished with the help of MATLAB. In Section 4, cattle's behaviour states are modelled by training the LSTM neural network in the digital twin model and cattle's states in the next cycle are predicted by using this deep learning technique. In Section 5, The accuracy of the trained LSTM model is discussed and verified. Finally, Section 6 deduces a proper conclusion.

2. Related Work

The concept of digital twins can be applied in many areas. For example, for wind power plants, cloud-based technology integrates technological and commercial data into a single digital twin through augmented reality (AR) and applies to multiple wind power plants to realize real-time monitoring of power plants [3]. In manufacturing, Schleich et al. [2] propose a conceptual integrated reference model for design and manufacturing that provides the first theoretical framework for digital twins in industrial applications [2]. In addition, digital twins are also used for product prediction and health management. This method effectively utilizes the interaction mechanism and fusion data of digital twins [8].

In the field of agriculture, more and more farmers are committed to the establishment of intelligent farms. The concept of intelligent agriculture mainly includes sensors, tracking systems, innovative digital technologies, data analysis and so on. The application of modern digital technology to farms can improve the efficiency of farm management [9,10]. More specifically, Yang et al. [7] come up with a digital farm management system that can effectively track production. In particular, this system uses smartphones to collect data, this is an efficient solution for precise vegetable farm management [7]

However, digital twins are rarely used on farms, there are only a few isolated cases. Digital twins are already being used in innovative internet-based applications, and digital twins can influence farm management [4]. A digital disease management system for dairy cows has been established, which realizes the digital management of dairy cows, systematic management of basic information of dairy cows, health assessment, electronic medical records and disease prevention. This system can effectively manage the disease of cows on the dairy farm [23]. In [13], Wagner et al. [13] use machine learning to detect the health of cows and predict when they would behave. They use different algorithms in machine learning to predict the activity duration of cows, including the K-neighborhood algorithm, the LSTM algorithm, the H-24 algorithm, and so on. The K-nearest neighbour algorithm performs the best after analysis and comparison. However, their study needs to be based on a much larger data set and needs to take into account the circadian nature of activity rhythms.

In addition, the application of an LSTM neural network to the establishment of the digital twin model used on a farm is also rare, but it's been used in many other ways and has been very successful. Hu et al. [18] propose a hybrid time series prediction model based on global empirical mode decomposition, LSTM neural network and Bayesian optimization, and apply it to the establishment of the digital twin model. They use their digital twin models to predict wind speeds in wind turbines and wave heights in ocean structures. The results show that the proposed model can obtain accurate time series prediction [18]

Although the application of digital twins in farm management is still in the early stage of development, it is not impossible to establish digital twins for each cow on this bovine disease digital management system. With the establishment of digital twins, the cattle farm can become an autonomous, adaptive system in which intelligent digital twins can operate, decide and even learn without human on-site or remote intervention [4].

3. Data Mining and Analysing

This section primarily discusses the processing method of the data sets, i.e., the original data measured by sensors of the farm's IoT system. This data set is systematically treated in preparation for future use of the modelling. Particularly, the data sets of the cattle's states are analysed, and a digital twin model of the cattle is produced using these data sets. A vast amount of data may be used to evaluate the model's correctness, and the state of the cattle can then be predicted.

3.1. Data Processing

The raw data set for the sensor contains 98 cattle of various breeds and genders. There are eight categories used to classify cattle's status: Resting, Rumination, High Activity, Medium Activity, Panting (Heavy Breathing), Grazing and Walking. Detailed descriptions for different cattle states are shown in Table 1. Each sensor takes a minute-by-minute reading of the cows' real-time status, with each cow having 74,455 data points collected between AU_time 8:06 a.m. on August 10 and 1:01 a.m. on 1 October 2019. Five cattle breeds are represented in the data sets: Angus, Brahman, Brangus, Charolais and Crossbred. This section focuses on the systematic processing of these data, including data segmentation, data cleaning, and data calculating.

Table 1. The explanation of the state.

State	Description
Rest	Standing still, lying, and transition between these 2 events, while lying, allowed to do any kind of movement with head/neck/legs (e.g., tongue rolling).
Rumination	Rhythmic circular/side-to-side movements of the jaw not associated with eating or medium activity, interrupted by brief (<5 s) pauses during the time that bolus is swallowed, followed by a continuation of rhythmic jaw movements.
Panting (Heavy Breathing)	Fast and shallow movement of thorax visible when looking animal from side, along with forward heaving movement of body while breathing. May or may not have open mouth, salivation, and/or extended tongue.
High Activity	Includes any combination of running, mounting, head-butting, repetitive head-weaving/tossing, leaping, buck-kicking, rearing and head tossing.
Eating	Muzzle/tongue physically contacts and manipulates feed, often but not always followed by visible chewing.
Grazing	Eating (see above definition) growing grass and pasture, while either standing in place or moving at slow, even or uneven pace between patches.
Walking	Slow movement, limb movement, except running.
Medium Activity	Any activity other than the above states.

3.1.1. Data Segmentation

The first step in data processing is the segmentation. The data are grouped by cattle of the same sex and breed. Because the original data are massive, we segment the data using RStudio and R programming language. Table 2 shows the number of cows segmented and integrated. A vast amount of data facilitates the analysis of overall data characteristics and avoid errors caused by individual and particular data. As a result, the resting state of Brahman's Female is used to demonstrate data processing and prediction.

Table 2. The number of cattle of different breeds and genders.

Category	Number
Angus Female	13
Angus Male	14
Brahman Female	14
Brahman Male	5
Brangus Female	10
Brangus Male	0
Charolais Female	3
Charolais Male	1
crossbred Female	38
crossbred Male	0
Total number	98

3.1.2. Data Cleaning

When the sensor detects and transmits the status of the cow, it also sends a lot of invalid data. The accuracy of the original data will be considerably influenced if using these data directly. Therefore, the initial step is to clear up the corrupted data.

Because the data returned by the sensor represents the cattle's states at a specific point in time, quantifying that state is critical for further design. In this work, the time of various states each hour in minutes is taken as the research object. Because corrupted or invalid data usually aggregate, identifying the point at which incorrect data arrives as 0 is not precise. For example, if there is a large amount of damage data in an hour, the rest state of the cattle for that hour will be marked as 0 min, which will affect the calculation of the average single period. Therefore, deleting corrupted data and the corresponding time serial number, to ensure that they are not included in the calculation of the average period.

The flow chart of data cleaning is shown in Figure 1. Data cleaning mainly focuses on the segmented data to clean and organize and finally obtains the cleaned data and its corresponding time series. This step primarily calculates the resting time of cattle in each hour. If it exists any corrupted data during the calculated hour, that hour's data will be destroyed.

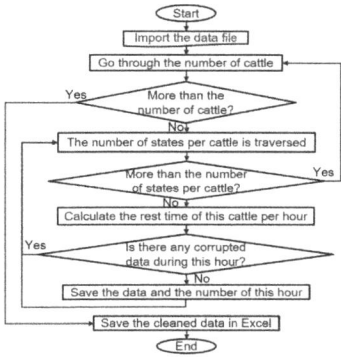

Figure 1. Process flow chart of data cleaning.

3.2. The State of Cattle throughout the Sampling Period

Acquiring the cattle's state changes across the sample period needs to average one group's data of cows due to large amount of discrete and lost data from a single cow. For example, averaging the resting time per hour of 14 Brahman females can determine variations in the resting state of Brahman treated during the sample period. The time series after data cleaning are different between each cattle's data set, since invalid data collected by sensors in the farm's IoT system is usually a random process. Therefore, the data processing in this step is to average the state data of the cattle with the same time serial number and obtain the state curve of the cattle in the whole cycle. The process flow chart of an average state time for several cattle can be found in Figure 2.

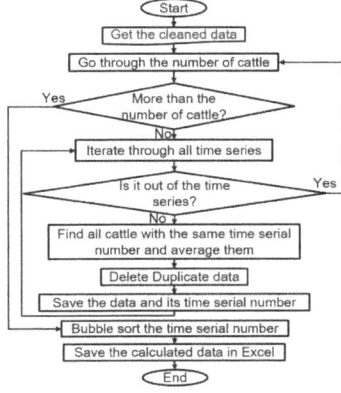

Figure 2. Process flow chart of average state times for multiple cattle.

The state diagram of cattle in the entire cycle can be obtained after the program has been executed. Figure 3 shows the calculated hourly rest time of the cattle in the whole cycle (Brahman Female). The number on the abscissa corresponds to the corresponding day, which includes all 24 h. The ordinate represents the rest time corresponding to this hour in minutes.

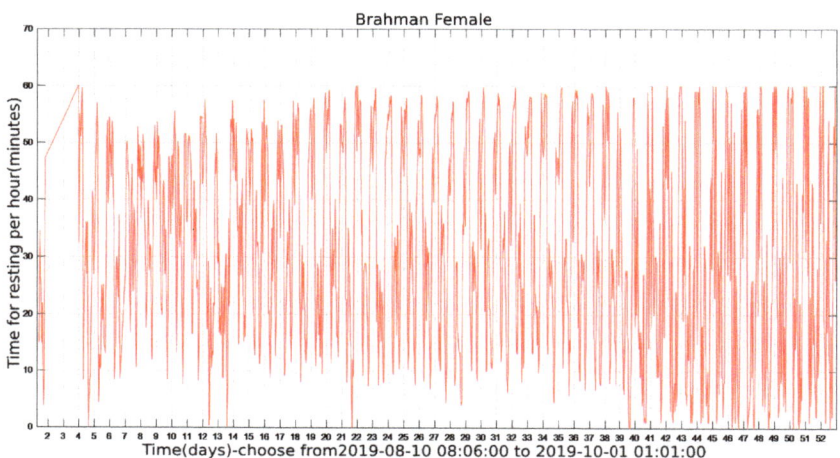

Figure 3. The resting time of Brahman Female during the whole sample period.

Figure 4 is a detailed zoomed-in part of Figure 3 and located between days 16 and 20. It is obvious that the rest time of cattle varies periodically with a cycle of one day. The peaks of the daily rest time can be found in both early morning and late-night while the valleys can usually be identified at forenoon and afternoon hours.

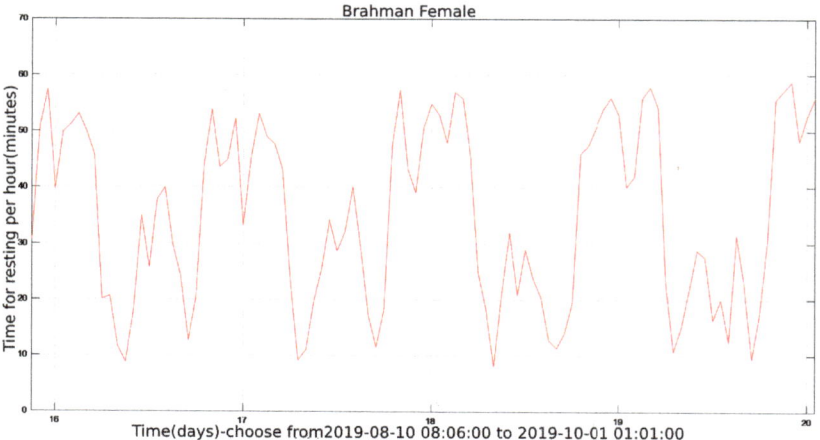

Figure 4. The enlarge vision shown 4 days.

3.3. The Average 24 h State of Cattle

The averaged single rest cycle data result (which is 24 h) of a single cattle is plotted in Figure 5. The entire sampling cycle is approximately 52 days as shown in Figure 3. The abscissa refers to the o'clock, i.e., from 0:00 to 23:59, and the ordinate relates to the rest period in minutes at this hour (Brahman Female). The average period's plot is flatter than a single period's plot. However, the trend and structure of these two are nearly identical, and a single cycle has more individual points and noises.

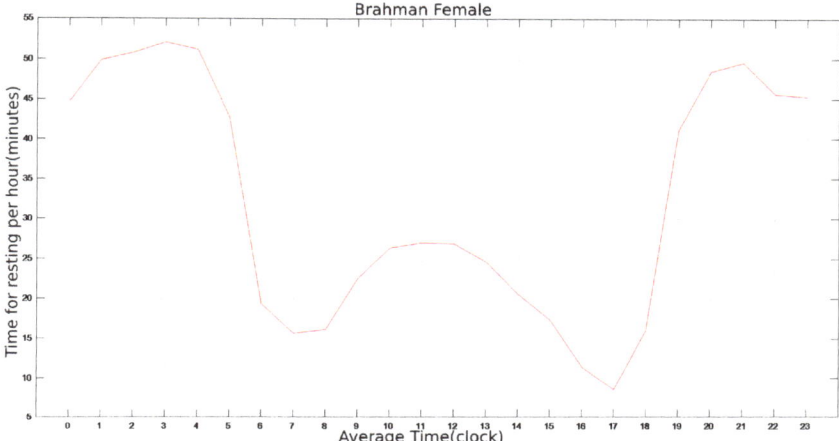

Figure 5. The average of one resting period for cattle(i.e., 24 h a day).

3.4. Fitting Curve for the Average State Period (24 h)

Curve fitting is commonly used to obtain the data relationship for such irregular curves. Typical fitting methods include minimum binomial fitting, exponential function fitting, power function fitting, and hyperbola fitting. Different fitting approaches are compared in this section to obtain the most ideal mathematical model [24,25].

Table 3. The results of different fitting methods.

Fitting Method	The Best Number of Items	Variance
Gaussian Fitting	8	3.0037
Sum of sine	8	20.1288
Polynomial	9	245.3264
Fourier	8	25.4590

Four fitting approaches are utilized to fit the 24-h average rest duration of cattle: Gaussian fitting, Sum of Sine fitting, Polynomial fitting, and Fourier fitting. The independent variable is the time, and the dependent variable is the rest period of cattle corresponding to that time while fitting the curve. The relationship between the time and the associated rest time can be established, and the curve of the cattle's rest period throughout the day can be obtained. As indicated in Table 3, Gaussian (item number 8) fitting is found to be the most accurate model among all candidates in terms of the fitting variance result. The error variance of Gaussian fitting is only 3.0037, which is much smaller than that of other fitting methods. The fitted curve shape is depicted in Figure 6, it is basically consistent with that of the average period in Figure 5.

The formula of the fitting curve (Gauss eight-term) formula is:

$$\begin{aligned} f(x) =& 51.29 e^{(-\frac{x-2.823}{2.957})^2} + 44.42 e^{(-\frac{x-24.19}{3.936})^2} + 1.378 \times 10^{14} e^{(-\frac{x+40.24}{7.546})^2} + 19.29 e^{(-\frac{x-13.55}{3.22})^2} \\ &+ 16.18 e^{(-\frac{x-19.06}{0.9367})^2} + 19.25 e^{(-\frac{x-4.588}{0.5802})^2} + 29.29 e^{(-\frac{x-20.39}{1.802})^2} + 20.45 e^{(-\frac{x-9.812}{2.834})^2} \end{aligned} \quad (1)$$

In Equation (1), x is the clock of a day, while $f(x)$ denotes the rest time within one hour of that clock. Regarding the low standard deviation and variance of this fitting result, this model is considered to be the proper candidate to describe the resting time of cattle in a day for Brahman Females. The models for other breeds, genders and states can be obtained in the same way.

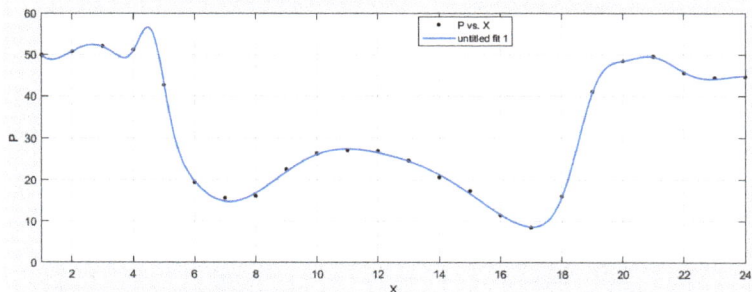

Figure 6. The Gaussian Fitting for one average period.

3.5. Noise Reduction Using Low-Pass Finite Impulse Response (FIR) Filter

Throughout the sample period, the cattle's condition varies on daily basis. The plot of the entire activity cycle contains noise and outliers in Figure 3. Therefore, denoising the sampled data is required.

FIR and Infinite Impulse Response (IIR) are two types of digital filters that are extensively employed. In theory, an IIR function's filtering effect is superior to that of an FIR function of the same order, but divergence can occur. The IIR digital filter has a high precision for amplitude-frequency characteristics, and with a non-linear phase, it is suited for audio signals that are insensitive to phase information. FIR digital filters have lesser amplitude-frequency precision than IIR digital filters. However, the phase is linear, meaning the time difference between signals of various frequency components remains unaltered after going through the FIR filter. In addition, the calculation time delay is relatively tiny, it is suited for real-time signal processing [26,26]. Because the state of the cattle is time-series data, it is critical to ensure that the filtered phase remains constant. Therefore, in this work, we use the FIR low-pass filter for denoising.

Cattle monitoring data are sampled once every 60 s in this study, resulting in a sampling frequency of around 0.0167 Hz. This is a low-frequency sampling signal, and the noise is present between each sampling. Noise frequency is more extensive than sampling frequency, so the signal between 0 and 0.0167 Hz is kept while the signal above 0.0167 Hz is eliminated. In Figure 7, the filter length is set to 5, and the filter's shape corresponds to its frequency. The filtered result is depicted in Figure 8, which uses the resting time of a Brahman Female's cow as an example.

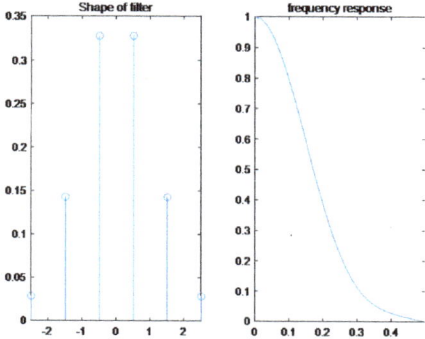

Figure 7. The shape of the FIR filter and the frequency response.

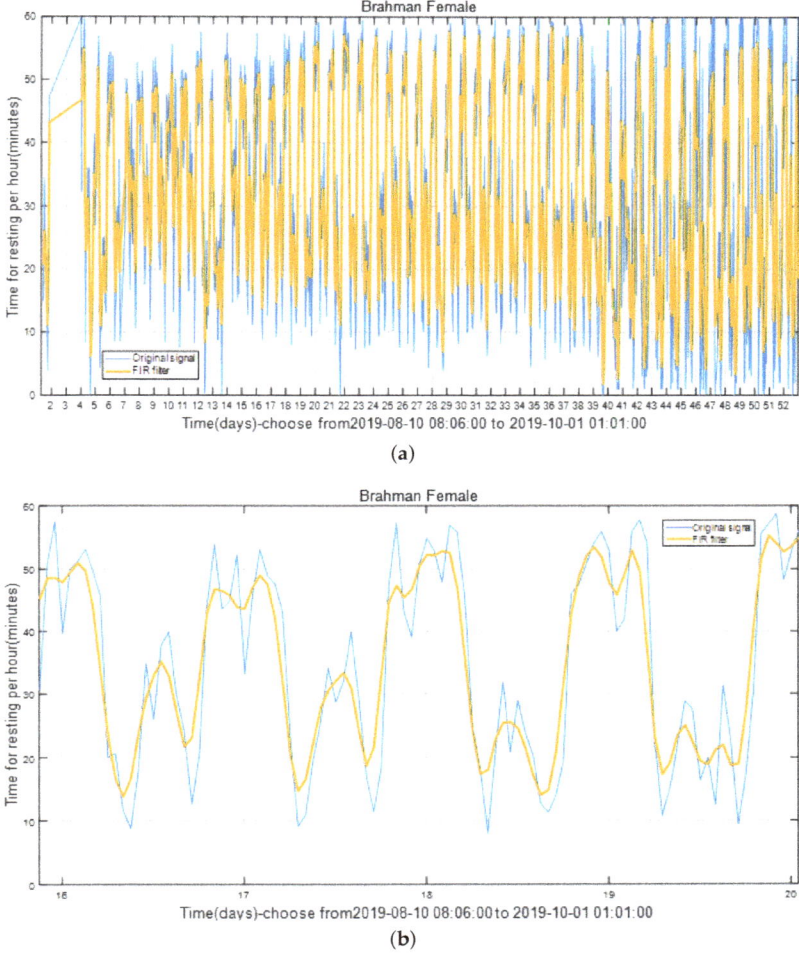

Figure 8. The resting time of cattle during the whole period after using the FIR filter. (**a**) The whole sample period. (**b**) The enlarge vision shown 4 days.

Figure 8b is a local detailed version of Figure 8a, focusing on the comparison of before using FIR filtering and after using FIR filtering from the 16th to the 20th day. Data performance is optimized after the introduction of the FIR filter for smooth signal processing, and the data trend can be clearly identified.

After going through the FIR filter, Figure 9 provides an image of a single rest period (one day, Day 17). In comparison to Figure 5, it exhibits the same trend, i.e., one day's rest time after filtering is nearly the same as one day's typical rest time. This feature demonstrates that the cattle's condition changes on a regular basis. It also indicates that the FIR filtered signal is effective and precise. The FIR filter effectively minimizes noise and eliminates outliers and gross inaccuracy. As a result, the signal filtered by the FIR filter can be used for subsequent modelling and prediction.

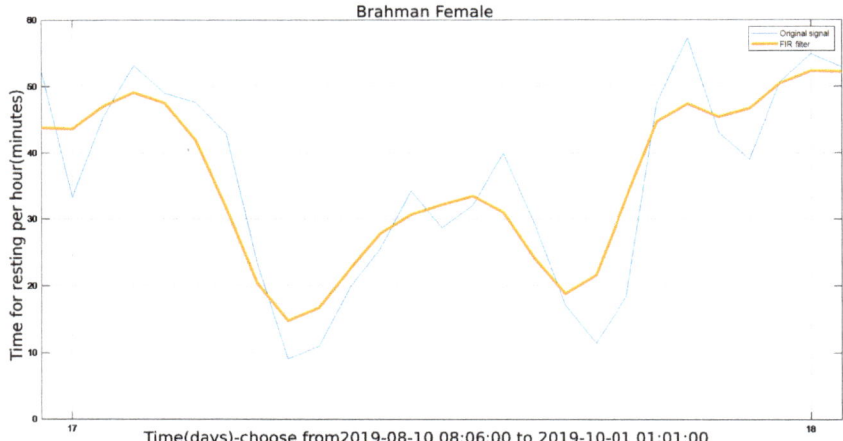

Figure 9. The single resting cycle after the FIR filter.

4. Prediction Based on LSTM Model

In DL, the LSTM network is a unique RNN model. Its unique structural design allows it to avoid long-term reliance. The default nature of LSTM is to remember information from a long time ago [12,17,27,28]. In this section, we employ the LSTM model to forecast the status of cattle based on the above research content. To be more explicit, the structure and properties of LSTM and how to construct an LSTM model are first discussed. Second, using the LSTM model, the cattle status is modelled and forecasted. Finally, the model is optimized in order to improve its accuracy.

4.1. Build the LSTM Model of the Cattle State

The program flow chart for establishing the LSTM model is shown in Figure 10. First, import the data previously filtered by the FIR filter, and divide it into a test set and a training set. Second, the LSTM model is created. Setting parameters: the number of input neurons, output neurons, hidden neurons, learning rate, batch size, epoch size (i.e., the number of training cycles) and the number of LSTM layers [29,30]. The loss error is chosen as the mean square error, and the LSTM neural network is trained using the Adam optimisation technique [31]. The cycle ends when the number of training times is reached, and the lowest loss error will be the output.

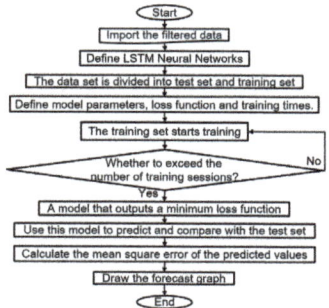

Figure 10. The code process of building the LSTM model.

4.2. Using the LSTM Model to Predict the State of Cattle

It is critical to determine the input, output, and time series before using the constructed LSTM model for cattle state prediction. The cattle's state must be presented as the output, and the number of the independent variable hours must be seen as a time series, according to the characteristics of the data sets. As a result, determining input variables is a challenging aspect of this approach. Because the output variable must be data with periodic changes, the input must be a known fixed periodic function. Time series as a fixed periodic function can be used as input. To be more specific, given that the state cycle of cattle is one day, it is appropriate to determine the input variable as the number of hours on the clock each day. The input and output variables, as well as the time series, for the resting time of Brahman Female's cattle are as follows:

Input: The number of hours on the clock each day (24 h).
Output: The resting time during this hour (e.g., The resting time at 7:00 means that the resting time during one hour from 7:00 to 7:59).
Time series t: The sequence number of this hour (e.g., 0:00 a.m. on the first day is the first hour, and t is 1. So on, 0:00am on the second day is the 25th h, and t is 25) [30,32].

- Training:
  Both the input and output data are periodicities. The distinction is that the input in this cycle has a set value and trend, whereas the output in each cycle has a varied value. For example, the input is 0 at 0:00 a.m. on Day 17th and 0:00 a.m. on Day 24th, as shown by the two red lines in Figure 11, but the output is different. In other words, the same input might result in multiple outcomes regardless of time. Although the input is the same, the input's matching time series is not. As a result, when a single input correlates to numerous outputs in a time series, the LSTM model can successfully handle the problem.

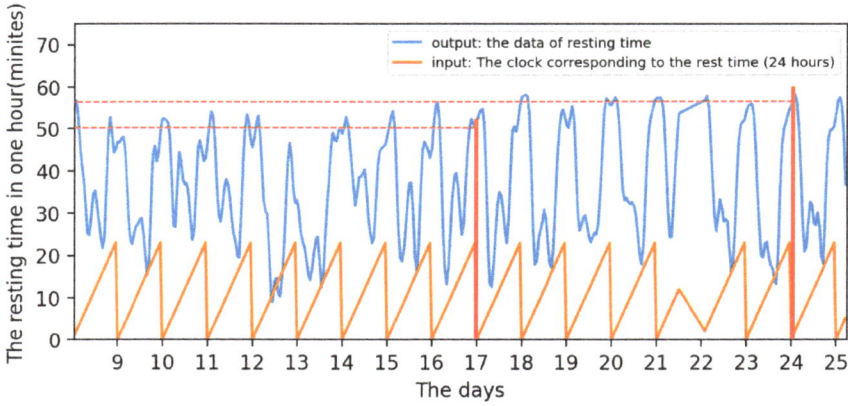

Figure 11. The input and output based on the LSTM model.

- Testing and prediction:
  In total, 90% of the data is used for training, and 10% for prediction and testing. For example, the input data sets for training are $input_{t_1}$ through $input_{t_{90}}$, while the data sets for testing are $input_{t_{91}}$ through $input_{t_{100}}$. The training outcomes are depicted in Figure 12.

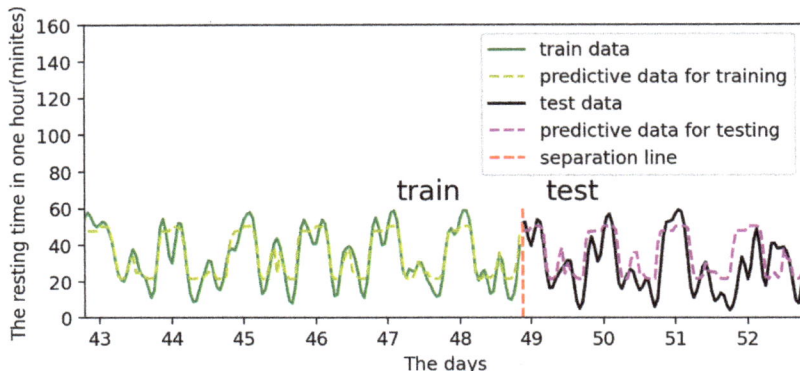

Figure 12. The predictive results after training and testing.

The predict and actual results are similarly shown in Figure 12. This means the digital twin model for individual cattle is basically established. The training loss reduces during the training process, showing that the model is converged and practical in Figure 13. However, the prediction results' error is relatively significant, which indicates further requirements of the parameter optimization in the model.

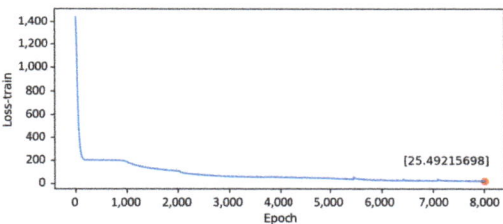

Figure 13. The training loss.

4.3. Parameter Optimization

For optimization and comparison purposes, the number of hidden units, LSTM layers, the batch size, and the epoch size were all modified [29,30].
Hidden units size: 4, 8, 16, 32, 64, 128, 256.
The number of LSTM layers: 1, 2, 3, 4, 5, 6, 7.
The batch size: 3, 6, 12, 24, 48, 96.
The epoch size: 100, 500, 1000, 2000, 5000, 10,000, 20,000.

- Selection of the number of LSTM layers

    The number of hidden units is 16, the batch size is 24, and the epoch size is 2000, all of which are randomly chosen. Only the number of layers in the LSTM is modified with the other parameters fixed: 1, 2, 3, 4, 5, 6, 7. The box diagram for the mean square deviation in the model learning process is shown in Figure 14.

    The top line and bottom line represent the edge's maximum and minimum values, respectively. The upper quartile is represented by the box's upper edge, while the box's lower edge represents the lower quartile. The orange line represents the median. Comparing the seven box charts, increasing the number of layers has a minor impact on the mean square error of model training [33]. When the number of layers is 5, 6 and 7, the error of the LSTM model will be stabilized to a fixed value immediately after a short training. As shown in Figure 14, the box plot has many outliers (that is, large outliers, black circles in the figure), and the median, upper quartile, and lower quartile overlap. However, in terms of model performance, using more LSTM layers, the running speed will be slower and it becomes more complex, and the result of the model operation is affected [34,35]. The loss error of the test set is positively correlated with that of the training set, and it is the smallest when the number of layers is 2. As a result, two layers of LSTM are best for this model.

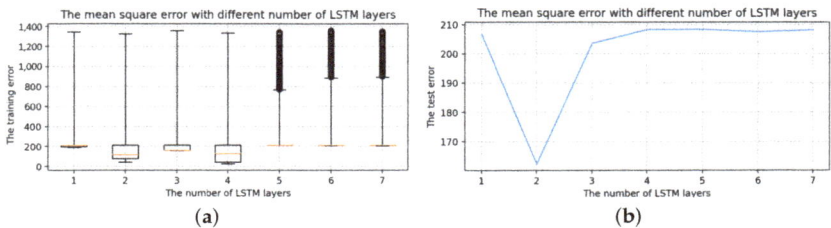

Figure 14. The mean square error with different numbers of LSTM layers. (**a**) The training error with different number of LSTM layers. (**b**) The test error with different number of LSTM layers.

- Selection of the hidden units size
  To determine the size of the hidden units, we keep the batch size and epoch size unchanged and run the LSTM model with different hidden units size, i.e., 4, 8, 16, 32, 64, 128, 256. The box diagram of the mean square is shown in Figure 15. In terms of error size and ultimate training effect, the choice of 128 hidden units is the best for training the data, with the majority of the mean square error values falling below 25, and the loss error of the test set is the smallest.

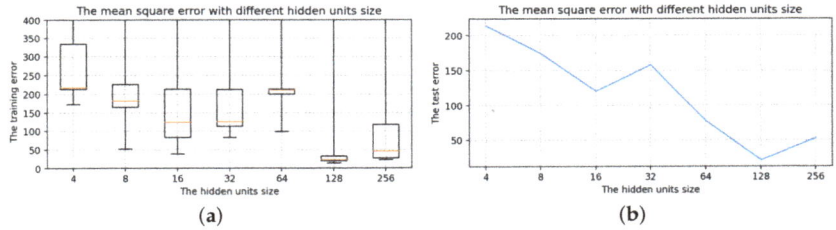

Figure 15. The mean square error with different hidden units size. (**a**) The training error with different hidden units size. (**b**) The test error with different hidden units size.

- Selection of the batch size
  The batch size, which can be 3, 6, 12, 24, 48, or 96, is altered when using two layers of LSTM with 128 hidden units. The box diagram is shown in Figure 16. The batch size refers to the number of samples fed into the model at once and divides the original data set into batch size data sets for independent training. This method helps to speed up training while also consuming less memory [36]. To some extent, batch size training can help to prevent the problem of overfitting [37]. As a result, when building the model, an acceptable batch size should be chosen. When the batch size is 24, the minimum value of the produced mean square deviation data set is the smallest in terms of minimum value and median, as well as the test error value.

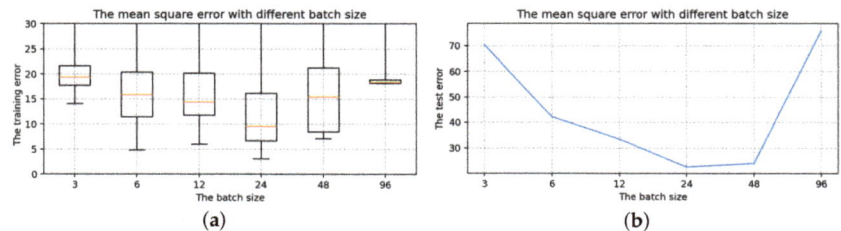

Figure 16. The mean square error with different batch sizes. (**a**) The training error with different batch sizes. (**b**) The test error with different batch sizes.

- Selection of the epoch size
  Select two layers of LSTM with 128 hidden units and the batch size is 24, but the epoch size can be any of 100, 500, 1000, 2000, 5000, 10,000, or 20,000. Figure 17 shows a box diagram for the mean square deviation in the model learning process.

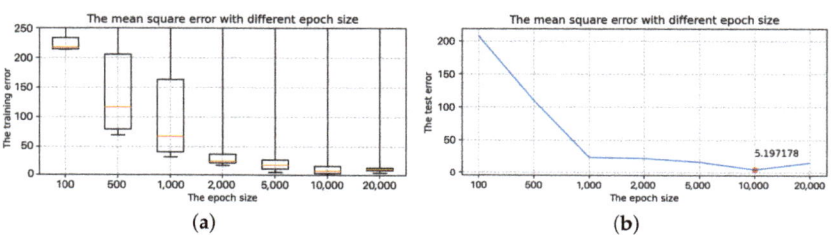

Figure 17. The mean square error with different epoch sizes. (**a**) The training error with different epoch sizes. (**b**) The test error with different epoch size.

The epoch size is the number of times the learning algorithm works in the entire training data set. An epoch means that each sample in the training data set has the opportunity to update internal model parameters [38]. In theory, the more training sessions there are, the better the fit and the lower the error. In practice, however, overfitting occurs when the epoch size exceeds a specific threshold, causing the training outcomes to deteriorate [39]. The epoch size of 100, 500, 1000, 2000, 5000, 10,000, and 20,000 is chosen in Figure 17. The inaccuracy rapidly decreases and approaches zero as the epoch size increases from 100 to 10,000. When the epoch size increases to 20,000, the error is still tiny, but it is greater than when the epoch size is 10,000, indicating an overfitting occurrence. Therefore, the model with a 10,000 epoch size has the best effect.

Figure 18 shows the training and prediction outcomes after optimizing model parameters, while Figure 19 shows the loss value after optimizing parameters. The best parameters for the LSTM model are shown in Table 4. The LSTM model has a good prediction of the resting state of cattle, which largely adheres to the periodic changes in cattle state and has a modest error. Therefore, the digital twin model for cattle has been established and optimized.

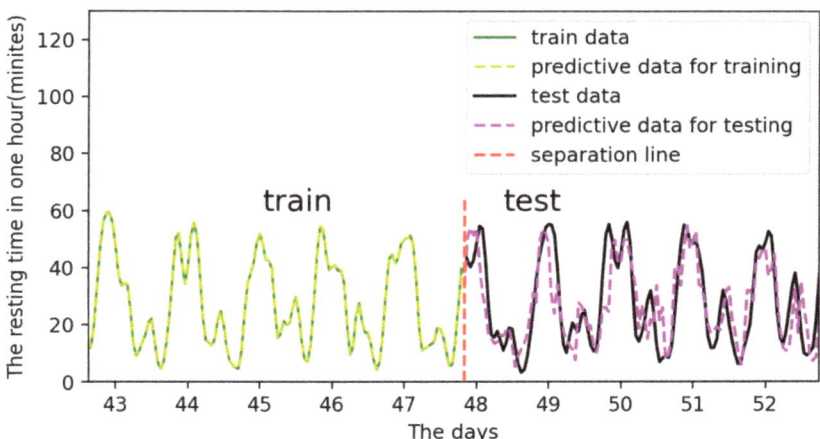

Figure 18. The training and prediction after optimizing model parameters.

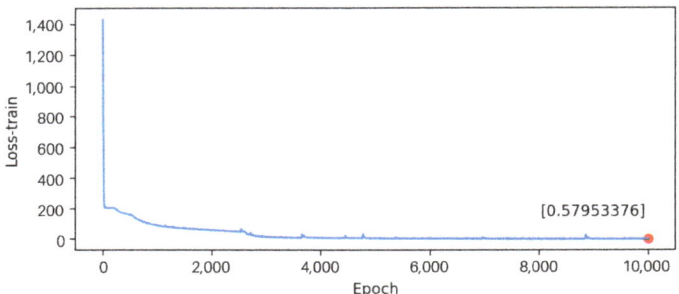

Figure 19. The training loss after optimizing model parameters.

Table 4. The best parameters for the LSTM model.

Hidden Neurons	128
Batch size	24
Epoch size	10,000
LSTM layers	2
Loss-train	0.57953376
Loss-test	5.197178

5. Results and Analysis

Figure 20 depict the LSTM model's training and prediction on different sexes, breeds, and states, respectively. This shows the applicability of this model, which can be used to predict various states of different cattle.

The trend of the results predicted by this LSTM model is nearly identical to the actual data. The model for Brahman males performs relatively poorly, which can be attributed to their relatively random rest state, poor cycle regularity, and other external environmental factors. It is possible that increasing the size of the data collection may result in improved predictions. Overall, the LSTM-based model for the cattle state cycle is accurate and effective, and it can accurately predict the dynamic trend of the next cattle state cycle.

In this way, the digital twin model can effectively predict the future time budget of cattle, which is conducive to efficient cattle breeding. Predict the future behaviour of cattle in advance so that appropriate preventive measures can be prepared.

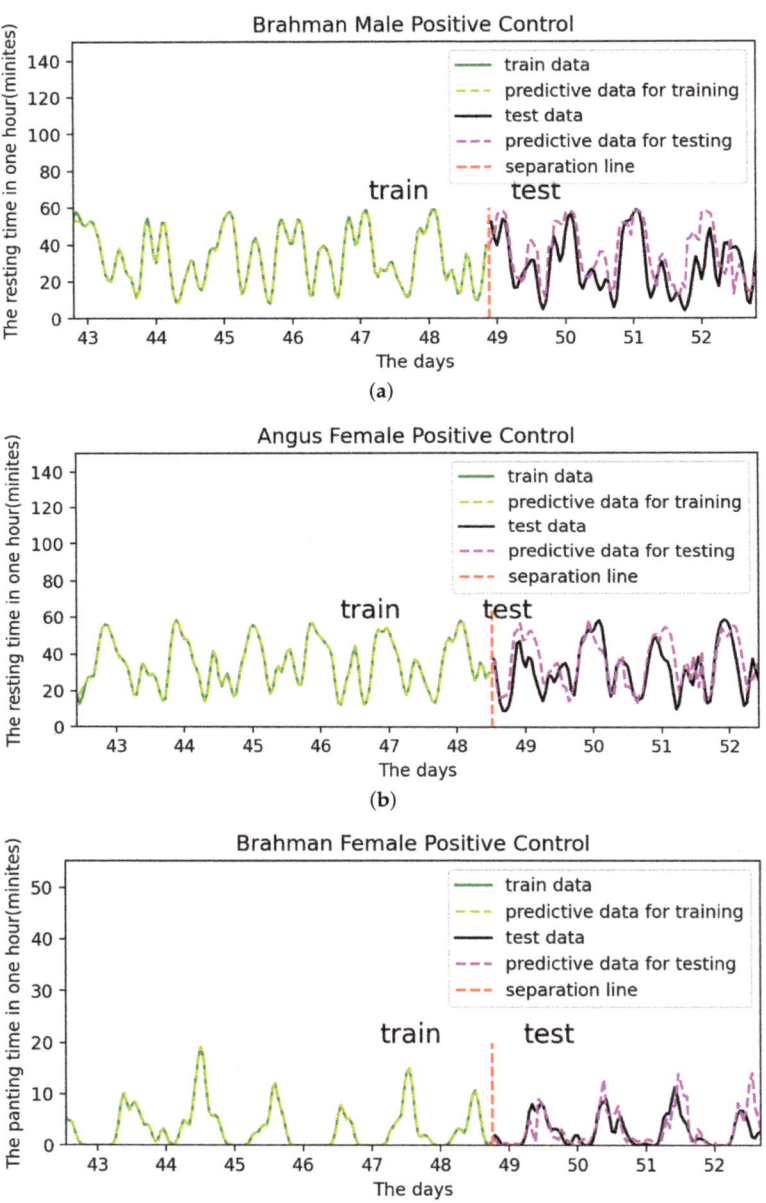

Figure 20. Applicability of the model. (**a**) Brahman Male cattle (Rest). (**b**) Angus Female cattle (Rest). (**c**) Brahman Female cattle (Pant).

6. Conclusions

The construction of a smart digital twin model of the state of cattle is primarily achieved in this work. It is primarily built on a farm IoT system to collect the state data of cattle under various combined treatments, with data cleaning and calculating. The average data of 24 h are fitted, and the data of the whole sampling period are de-noised. In addition, a deep learning-based LSTM model for cattle state dynamics is developed using the data after noise reduction, and the model can predict the state change of cattle in the next cycle. The model's accuracy and effectiveness are demonstrated when the prediction results are compared to the actual results. After optimization, the loss error of the training set is reduced to about 0.580, and the loss error of the prediction set is about 5.197. Using this digital twin model, the future time budget of cattle can be predicted quickly and accurately.

This model has certain limits as well, it requires a large quantity of data to learn, and a little amount of data will cause the model to be inaccurate. Furthermore, encapsulating the entire research into one system is a critical step toward commercializing digital twins in the future. In addition, estimating the time budget of cattle in advance necessitates human prediction of cow health conditions. Fully automated cow feeding and real-time monitoring of cattle condition and health are desirable in the future.

Author Contributions: Conceptualization, C.C., Z.L. and B.V.; methodology, X.H., Z.L. and C.C.; software, X.H.; validation, X.H., Z.L., C.C and B.V.; formal analysis, X.H., Z.L.; investigation, X.H.; resources, C.C. and S.L.; data curation, X.H.; writing—original draft preparation, X.H.; writing—review and editing, Z.L. and S.L.; visualization, X.H.; supervision, B.V.; project administration, Z.L. All authors have read and agreed to the published version of the manuscript.

Funding: This research received no external funding.

Institutional Review Board Statement: Not applicable.

Informed Consent Statement: Not applicable.

Data Availability Statement: Not applicable.

Conflicts of Interest: The authors declare no conflict of interest.

Abbreviations

The following abbreviations are used in this manuscript:

AI	Artificial Intelligence
DL	Deep Learning
LSTM	Long Short-Term Memory Network
RNN	Recurrent Neural Network
FIR	Finite impulse response
IIR	Infinite Impulse Response

References

1. Haag, S.; Anderl, R. Digital twin–proof of concept. *Manuf. Lett.* **2018**, *15*, 64–66. [CrossRef]
2. Schleich, B.; Anwer, N.; Mathieu, L.; Wartzack, S. Shaping the digital twin for design and production engineering. *CIRP Ann.* **2017**, *66*, 141–144. [CrossRef]
3. Pargmann, H.; Euhausen, D.; Faber, R. Intelligent big data processing for wind farm monitoring and analysis based on cloud-technologies and digital twins: A quantitative approach. In Proceedings of the 2018 IEEE 3rd International Conference on Cloud Computing and Big Data Analysis (ICCCBDA), Chengdu, China, 20–22 April 2018; pp. 233–237.
4. Verdouw, C.; Kruize, J.W. Digital twins in farm management: illustrations from the FIWARE accelerators SmartAgriFood and Fractals. In Proceedings of the 7th Asian-Australasian Conference on Precision Agriculture Digital, Hamilton, New Zealand, 16–19 October 2017; pp. 1–5.
5. Grieves, M.; Vickers, J. Digital twin: Mitigating unpredictable, undesirable emergent behavior in complex systems. In *Transdisciplinary Perspectives on Complex Systems*; Springer: Berlin/Heidelberg, Germany, 2017; pp. 85–113.
6. Boschert, S.; Rosen, R. Digital twin—The simulation aspect. In *Mechatronic Futures*; Springer: Berlin/Heidelberg, Germany, 2016; pp. 59–74.
7. Yang, F.; Wang, K.; Han, Y.; Qiao, Z. A cloud-based digital farm management system for vegetable production process management and quality traceability. *Sustainability* **2018**, *10*, 4007. [CrossRef]

8. Tao, F.; Zhang, M.; Liu, Y.; Nee, A.Y. Digital twin driven prognostics and health management for complex equipment. *Cirp Ann.* **2018**, *67*, 169–172. [CrossRef]
9. Grober, T.; Grober, O. Improving the efficiency of farm management using modern digital technologies. In *Proceedings of the E3S Web of Conferences*; EDP Sciences: Les Ulis, France, 2020; Volume 175, p. 13003.
10. Cojocaru, L.E.; Burlacu, G.; Popescu, D.; Stanescu, A.M. Farm Management Information System as ontological level in a digital business ecosystem. In *Service Orientation in Holonic and Multi-Agent Manufacturing and Robotics*; Springer: Berlin/Heidelberg, Germany, 2014; pp. 295–309.
11. Tekinerdogan, B.; Verdouw, C. Systems Architecture Design Pattern Catalogfor Developing Digital Twins. *Sensors* **2020**, *20*, 5103. [CrossRef] [PubMed]
12. LeCun, Y.; Bengio, Y.; Hinton, G. Deep learning. *Nature* **2015**, *521*, 436–444. [CrossRef] [PubMed]
13. Wagner, N.; Antoine, V.; Mialon, M.M.; Lardy, R.; Silberberg, M.; Koko, J.; Veissier, I. Machine learning to detect behavioural anomalies in dairy cows under subacute ruminal acidosis. *Comput. Electron. Agric.* **2020**, *170*, 105233. [CrossRef]
14. Schmidhuber, J. Deep learning in neural networks: An overview. *Neural Netw.* **2015**, *61*, 85–117. [CrossRef]
15. Gers, F.A.; Schmidhuber, J.; Cummins, F. Learning to forget: Continual prediction with LSTM **1999**.
16. Williams, R.J.; Zipser, D. A learning algorithm for continually running fully recurrent neural networks. *Neural Comput.* **1989**, *1*, 270–280. [CrossRef]
17. Sundermeyer, M.; Schlüter, R.; Ney, H. LSTM neural networks for language modeling. In Proceedings of the Thirteenth Annual Conference of the International Speech Communication Association, Portland, OR, USA, 9–13 September 2012.
18. Hu, W.; He, Y.; Liu, Z.; Tan, J.; Yang, M.; Chen, J. Toward a Digital Twin: Time Series Prediction Based on a Hybrid Ensemble Empirical Mode Decomposition and BO-LSTM Neural Networks. *J. Mech. Des.* **2021**, *143*, 051705. [CrossRef]
19. Schmidhuber, J.; Hochreiter, S. Long short-term memory. *Neural Comput.* **1997**, *9*, 1735–1780.
20. Greenwood, P.L.; Gardner, G.E.; Ferguson, D.M. Current situation and future prospects for the Australian beef industry—A review. *Asian-Australas. J. Anim. Sci.* **2018**, *31*, 992. [CrossRef]
21. Cabrera, V.E.; Barrientos-Blanco, J.A.; Delgado, H.; Fadul-Pacheco, L. Symposium review: Real-time continuous decision making using big data on dairy farms. *J. Dairy Sci.* **2020**, *103*, 3856–3866. [CrossRef] [PubMed]
22. Huang, Y.; Zhang, Q. *Agricultural Cybernetics*; Springer: Berlin/Heidelberg, Germany, 2021.
23. Li, L.; Wang, H.; Yang, Y.; He, J.; Dong, J.; Fan, H. A digital management system of cow diseases on dairy farm. In Proceedings of the International Conference on Computer and Computing Technologies in Agriculture, Nanchang, China, 22–25 October 2010; Springer: Berlin/Heidelberg, Germany, 2010; pp. 35–40.
24. Kolb, W.M. *Curve Fitting for Programmable Calculators*; Imtec: 1984.
25. Buttchereit, N.; Stamer, E.; Junge, W.; Thaller, G. Evaluation of five lactation curve models fitted for fat: protein ratio of milk and daily energy balance. *J. Dairy Sci.* **2010**, *93*, 1702–1712. [CrossRef] [PubMed]
26. Rabiner, L.; Kaiser, J.; Herrmann, O.; Dolan, M. Some comparisons between FIR and IIR digital filters. *Bell Syst. Tech. J.* **1974**, *53*, 305–331. [CrossRef]
27. Goodfellow, I.; Bengio, Y.; Courville, A.; Bengio, Y. *Deep Learning*; MIT Press: Cambridge, UK, 2016; Volume 1.
28. Xie, T.; Yu, H.; Wilamowski, B. Comparison between traditional neural networks and radial basis function networks. In Proceedings of the 2011 IEEE International Symposium on Industrial Electronics, Gdańsk, Poland, 27–30 June 2011; pp. 1194–1199.
29. Zhang, J.; Wang, P.; Yan, R.; Gao, R.X. Long short-term memory for machine remaining life prediction. *J. Manuf. Syst.* **2018**, *48*, 78–86. [CrossRef]
30. Tang, Y.; Yu, F.; Pedrycz, W.; Yang, X.; Wang, J.; Liu, S. Building trend fuzzy granulation based LSTM recurrent neural network for long-term time series forecasting. *IEEE Trans. Fuzzy Syst.* **2021**, *30*, 1599–1613. [CrossRef]
31. Yaqub, M.; Asif, H.; Kim, S.; Lee, W. Modeling of a full-scale sewage treatment plant to predict the nutrient removal efficiency using a long short-term memory (LSTM) neural network. *J. Water Process Eng.* **2020**, *37*, 101388. [CrossRef]
32. Domun, Y.; Pedersen, L.J.; White, D.; Adeyemi, O.; Norton, T. Learning patterns from time-series data to discriminate predictions of tail-biting, fouling and diarrhoea in pigs. *Comput. Electron. Agric.* **2019**, *163*, 104878. [CrossRef]
33. Spitzer, M.; Wildenhain, J.; Rappsilber, J.; Tyers, M. BoxPlotR: a web tool for generation of box plots. *Nat. Methods* **2014**, *11*, 121–122. [CrossRef]
34. Koutnik, J.; Greff, K.; Gomez, F.; Schmidhuber, J. A clockwork rnn. In Proceedings of the International Conference on Machine Learning, Bejing, China, 22–24 June 2014; pp. 1863–1871.
35. Wu, Y.; Schuster, M.; Chen, Z.; Le, Q.V.; Norouzi, M.; Macherey, W.; Krikun, M.; Cao, Y.; Gao, Q.; Macherey, K.; et al. Google's neural machine translation system: Bridging the gap between human and machine translation. *arXiv* **2016**, arXiv:1609.08144.
36. Radiuk, P.M. Impact of training set batch size on the performance of convolutional neural networks for diverse datasets. *Inf. Technol. Manag. Sci.* **2017**, *20*, 20–24. [CrossRef]
37. Mu, N.; Yao, Z.; Gholami, A.; Keutzer, K.; Mahoney, M. Parameter re-initialization through cyclical batch size schedules. *arXiv* **2018**, arXiv:1812.01216.
38. Brownlee, J. What is the Difference Between a Batch and an Epoch in a Neural Network. *Mach. Learn. Mastery* **2018**, *20*. Available online: https://machinelearningmastery.com/difference-between-a-batch-and-an-epoch/ (accessed on 26 April 2022).
39. Huang, G.; Sun, Y.; Liu, Z.; Sedra, D.; Weinberger, K.Q. Deep networks with stochastic depth. In Proceedings of the European Conference on Computer Vision, Amsterdam, The Netherlands, 11–14 October 2016; pp. 646–661.

MDPI
St. Alban-Anlage 66
4052 Basel
Switzerland
Tel. +41 61 683 77 34
Fax +41 61 302 89 18
www.mdpi.com

Sensors Editorial Office
E-mail: sensors@mdpi.com
www.mdpi.com/journal/sensors

www.ingramcontent.com/pod-product-compliance
Lightning Source LLC
LaVergne TN
LVHW070248100526
838202LV00015B/2192